Understanding Bacteria

Understanding Bacteria

by

Sheela Srivastava

Department of Genetics,
University of Delhi, New Delhi, India

and

P.S. Srivastava

Centre for Biotechnology,
Hamdard University, New Delhi, India

KLUWER ACADEMIC PUBLISHERS
DORDRECHT / BOSTON / LONDON

A C.I.P. Catalogue record for this book is available from the Library of Congress.

ISBN 1-4020-1633-6

Published by Kluwer Academic Publishers,
P.O. Box 17, 3300 AA Dordrecht, The Netherlands.

Sold and distributed in North, Central and South America
by Kluwer Academic Publishers,
101 Philip Drive, Norwell, MA 02061, U.S.A.

In all other countries, sold and distributed
by Kluwer Academic Publishers,
P.O. Box 322, 3300 AH Dordrecht, The Netherlands.

Printed on acid-free paper

Printed in the Netherlands.

Preface

The discipline of microbiology that deals with an amazingly diverse group of simple organisms, such as viruses, archaea, bacteria, algae, fungi, and protozoa, is an exciting field of Science. Starting as a purely descriptive field, it has transformed into a truly experimental and interdisciplinary science inspiring a number of investigators to generate a wealth of information on the entire gamut of microbiology. The later part of 20th century has been a golden era with molecular information coming in to unravel interesting insights of the microbial world. Ever since they were brought to light through a pair of ground glasses by the Dutchman, Antony van Leeuwenhoek, in later half of 17th century, they have been studied most extensively throughout the next three centuries, and are still revealing new facets of life and its functions. The interest in them, therefore, continues even in the 21st century.

Though they are simple, they provide a wealth of information on cell biology, physiology, biochemistry, ecology, and genetics and biotechnology. They, thus, constitute a model system to study a whole variety of subjects. All this provided the necessary impetus to write several valuable books on the subject of microbiology. While teaching a course of Microbial Genetics for the last 35 years at Delhi University, we strongly felt the need for authentic compiled data that could give exhaustive background information on each of the member groups that constitute the microbial world. For the present book, we chose bacteria (though archaea have also been dealt with but more for comparison), the organisms for which geneticists also have special affinity.

The book, "Understanding Bacteria" is intended to provide related information on basic as well as applied aspects and should cater to the needs of both undergraduate and postgraduate students. Besides Microbiology, bacteria form an integral component of various courses, such as Botany, Zoology, Biosciences, Life Sciences, Biochemistry and Biotechnology. This book thus should fulfil a long-standing demand of a whole variety of readers. The text of the book is organized in such a way that the flow of information appears spontaneous giving due weightage to classical, and molecular/applied aspects. Those who are seriously into microbiological aspects should find it a good resource book, and those who wish to use these bugs for biochemical, genetic, biotechnological, and applied research, they should get the required background information. Our endeavor has been to keep the curiosity of the reader on and make this book of multipurpose usage.

The contents of the book have been spread over nine chapters and information within has been regrouped as sections and subsections, as required. This allows the logical assemblage of contained information. The figures have been intentionally drawn free-hand so as to bring out the clarity of information. At the end of the text, few books and research/review articles have been listed which can serve as excellent reference material, though we would also like to leave it to the choice of the readers to select any other book that may become available and that fulfils their requirement. A comprehensive index has been appended at the end to help the readers.

Although, utmost care has been taken to present the facts as they are, responsibility for any unintentional mistakes that might have crept in is exclusively ours. We would feel obliged to our readers for suggestions and ways to improve the book.

While preparing this book we have freely indulged in acquiring academic inputs from the written materials and discussions with various experts, colleagues and students. Of equal importance have been the forthcoming help and assistance of a number of coworkers, associates and our doctoral and post-doctoral collaborators. While we take this opportunity to place on record the deep sense of gratitude to all of them, some deserve especial mention. Dr. T. Satyanarayana, Professor, Department of Microbiology at Delhi University has always been forthright in providing extra information. The idea that we could always fall back upon the support rendered by our post-doctoral and doctoral associates, Dr. Deepshikha, Dr. Sanjeev, Dr. Alka, Ms. Toolika, Ms. Sapna, Ms. Maryam and Ms. Nisha of Centre for Biotechnology, Hamdard University, New Delhi has been a constant source of strength throughout. Their ungrudging help at all levels of the preparation of manuscript made our job simpler.

Teaching a course on the subject brought us in close contact with a large number of graduate and post-graduate students. Their querries and inquisitiveness have guided us in deciding the direction of this book. We shall remain thankful to all of them.

Excellent computational support from Mr Satish, Ms Mansa, and Mr Rajesh has seen the book through to its present shape.

June 12, 2003
New Delhi

Sheela Srivastava
P S Srivastava

Contents

CHAPTER 3

CHAPTER 5

CHAPTER 1
THE EMERGENCE OF BACTERIOLOGY

Living organisms are conventionally classified into two large groups of plants and animals. These can be identified on the basis of mere observations by naked eye. The earth is, however, teeming with a vast multitude of life forms whose existence can not be established with an unaided eye. These minute, microscopic organisms are microorganisms that include viruses, rickettsiae, bacteria, protozoa, algae, and fungi. Unlike higher plants and animals, whose cells are unable to live independently, and thus exist as part of a multicellular organization, microbial cells exist as single cells or cell clusters. These single cells are able to carry out all life functions, such as, growth and metabolism, energy generation, and reproduction, and, therefore, are true representatives of life.

1.1. The Science of Microbiology (Bacteriology)

Microbiology is the Science that deals with the study of microorganisms or microbes, with each group forming a separate branch. For example, bacteria, which are unicellular or exist in small clusters, forming a predominant group of microbes, are dealt with by a branch called bacteriology. Although simple in organization, they are functionally at par with higher organisms. Their study has, therefore, not only provided important leads in understanding other microbes but also the biology of higher organisms, including humans. Microbiology is a combination of both basic and applied biological science. As a basic science, its members have contributed significantly towards the understanding of chemical and physical principles behind life processes. As an applied science, microbes have helped deal with important aspects in agriculture, industry and medicine. The vast information collected on these two aspects has culminated in the development of a whole new discipline called biotechnology. In this book, an integrated approach has been taken on these two aspects.

1.2. Bacteria- The First Inhabitants of Earth

Of the earth's 4.6 billion years of age, bacterial forms have shared 3.5 billion years. In other words, 80% time the life existed on this earth, it consisted of these single celled organisms. Arriving first on the face of earth, they spread rapidly in all the elements of earth, viz, air, water and soil. Inhabiting almost all the types of ecosystems, including highly harsh or unlivable, which other life forms fear to tread, they have evolved to generate an unbelievable range of diversity in structure, nutrition, metabolism and reproduction. Flexibility, adjustment, and adaptation to the existing conditions or changes there in are the basis of their success story of existence. The reducing conditions prevailing on the early earth was tackled by carrying out some form of anaerobic metabolism. The conditions suited chemolithotrophic, chemoorganotrophic and even phototrophic metabolism. The first phototrophic organisms were undoubtedly anoxygenic but with the evolution of second light reaction of photosynthesis, that provided O_2 by photolysis of water (oxygenic photosynthesis), O_2 gradually accumulated converting reducing atmosphere to oxidizing one. This was an extremely important point in the evolution of life as it led to the evolution of aerobic microorganisms and eventually to animals and plants. Reflections of all these metabolic versatilities and changeover exist even today in the bacterial kingdom.

1.3. Discovery

Although, bacteria were the first inhabitants of the earth, the science dealing with them is a relatively young branch. Being unseen with the eye, their discovery was linked to the invention of instruments like microscope. Although Robert Hooke described the fruiting structure of molds in 1664 using his microscope, the credit for observing and describing the microorganisms, in some detail goes to a dutchman called Antony van Leeuwenhoek. In 1674, he viewed through his crude microscope a number of materials, such as drops of saliva, water from various sources, blood, semen, and muscle tissues, and observed a whole range of life forms, like, bacteria, protozoa, yeasts, spermatozoa, red blood cells and muscle cells. So fascinated was he with these structures that he named them "dierkens" and wrote a series of letters to Royal Society of London. His observations along with the diagrams were published by Royal Society in 1684 where these organisms were translated in english as "wee animalcules" (Fig. 1.1).

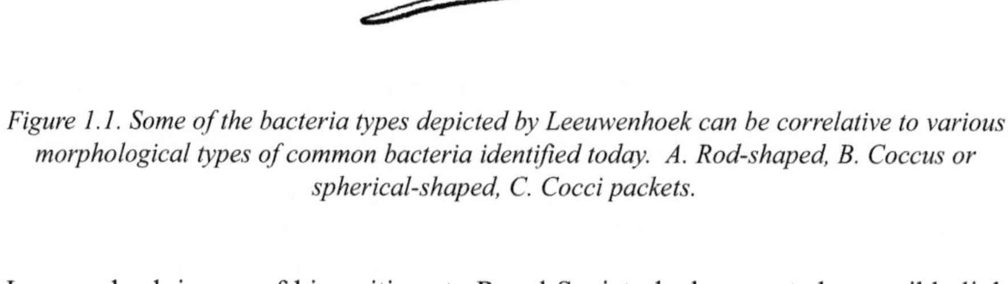

Figure 1.1. Some of the bacteria types depicted by Leeuwenhoek can be correlative to various morphological types of common bacteria identified today. A. Rod-shaped, B. Coccus or spherical-shaped, C. Cocci packets.

Leeuwenhoek in one of his writings to Royal Society had suggested a possible link between these small organisms and fermentation, decay and disease. The disease causing nature of some invisible organisms was infact postulated in 13th century itself by Roger Bacon. Several such ideas were put forward for almost two centuries after Leeuwenhoek but in the absence of any experimental support they remained mere postulations. It was only when, Louis Pasteur discovered the involvement of these creatures in fermentation, and Koch, Smith, Pasteur and many others traced the link between these microbes and diseases, that the science of Microbiology got a firm experimental footing.

1.4. Fall of Spontaneous Generation- Microbiology an Experimental Science

One of the most significant contributions of bacteriologists was to discredit the theory of "Spontaneous generation". As early as 384-322 B.C., Aristotle, the famous philosopher and scientist of his time, suggested that insects arise spontaneously from decaying earth, some food stuffs, and excretory materials. So deep was the impact of this idea that it not only remained alive for almost 2000 years but was extended to other animals as well. The first major blow to this notion came from Redi in 1668 when he showed that the spontaneous origin of insects from rotting meat is because the eggs are laid on them by flies. As late as 1745, Needham showed that living creatures can be detected in boiled infusion of meat. He reasoned that since boiling is supposed to kill all living things, the origin of these creatures must be spontaneous.

In 1776, Spallanzani repeated Needham's experiment but kept the boiled meat extract in hermetically sealed flask. He demonstrated that the extract remained free of any organism. His opinion that the air is the source of these creatures in a normal stoppered bottle, though did not receive much attention, was the first direct proof against the spontaneous origin. Schultze in 1836, Schwann in 1837, and Schroder and Dusch in 1853 refined Spallanzani's experiment by allowing the entry of air into boiled meat extract by passing it through sulphuric acid, fusible metal, and cotton plug, respectively. Schwann categorically stated that air contain germs of the fungi and infusoria which nourish and grow at the cost of the organic substances and bring about such a decomposition that causes putrefaction. It can be easily conceived today that all the three conditions will render the air free of microbes and thus the meat extract will remain uncontaminated. This is, what was demonstrated by these three independent experiments.

At about the same time, Nicholas Appert (1750-1841), a french confectioner and distiller, demonstrated that fruits and vegetables can be preserved better if they are kept sealed in jar and then immersed in boiling water. This turned out to be one of the major practices for canning industry and for which Appert was suitably awarded by the French Government.

It was finally left to Louis Pasteur to demonstrate in 1862 that all putrefactions are caused by microorganisms and the controversy over spontaneous generation was finally settled. Pasteur made several contributions to the field of microbiology but he is remembered to change microbiology from a speculative to experimental science. From his time onwards a spurt in the microbial experiments coupled with the availability of more refined instruments and techniques, led to a surge of information on various aspects of microbes. This eventually culminated into some well defined branches of microbiology, one of which, the bacteriology, would be dealt with in this book.

Pasteur's initial work was directed towards confirming the experiments of his predecessors on spontaneous generation. He demonstrated that the boiled infusion of meat remains clear (uncontaminated) if the air that is allowed to enter is first heated and then cooled. He got the same result when he attached a narrow S-tube to the mouth of the flask (Fig. 1.2). In such a tube, microbe entering along with the air settled with the dust particles at the curvature, causing no spoilage of the meat extract.

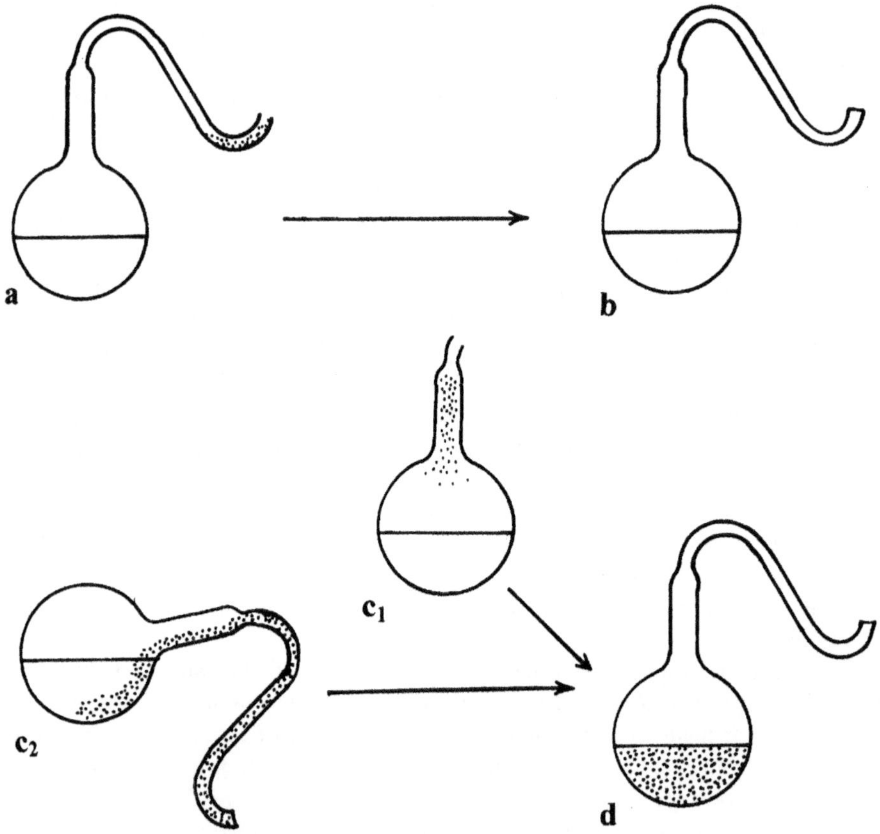

Figure 1.2. Pasteur's experiment with swan-neck flask that discredited the theory of spontaneous generation. a. A flask contained broth sterilized by heating and its neck was drawn out in a long, curved structure with an open end. The dust containing microorganisms when enters is settled in the bend (b). b. Such a broth remained sterile even when stored for a long time. However, when either the tip was broken so that broth was in direct contact with air (c_1) or the flask was tilted so that dust along with microorganisms came in contact with the broth (c_2), even in a short time it led to the growth of the microorganisms (d).

In 1876, Tyndall, a physicist (known for Tyndall effect or the scattering of light by air and water) proposed that tiny particles or motes in the air seen during scattering of light are nothing but microorganisms. The air can, therefore, be made optically empty by passing it slowly through a tight cotton plug. The root cause of spontaneous generation could thus be uncovered.

1.5. Metabolic Versatility Recognized- Good vs Bad

Another common microbial activity that engaged the microbiologists was the decomposition of a variety of materials. What was important was the observation that material lifted from the sites of decomposition was able to endow similar characteristics to fresh material. Bizio (1823) attributed the appearance of red spots on porridge and the ability to cause similar characters on fresh porridge, when transferred, to a fungus *Serratia marcescens*. This name was subsequently retained even though the causative agent was identified as a bacterium. Fuchs (1841) observed that milk often turns blue on standing. He found that this is because of the bacterium *Pseudomonas syncyanaea* (Fuchs named it *Vibrio cyanogenus*), which can be maintained in the infusion of mallow plants and had the capability to cause spoilage of fresh milk. Fuchs experiments were the beginning of what we call today as "Dairy Bacteriology".

Pasteur's another significant contribution was the discovery of an important microbial process, fermentation. He demonstrated that microbes can ferment fruits and grains and produce alcohol. He was also able to show that different microorganisms possess different fermentative capabilities leading to different end products. So while yeast produces alcohol, lactic-acid bacteria form lactic acid and *Vibrio butyrique* yields butyric acid. Pasteur strongly opined that fermentation is a life process.

With his interest in fermentation, Pasteur decided to tackle some problems of wine industry. He found that souring and bitterness of wine is caused by microorganisms and industry was incurring huge losses of money and material due to same. Taking clue from Appert's work, Pasteur subjected the bottled wine to a temperature of 50-60°C for a brief period. He found that this killed the undesirable microorganisms but did not affect the flavour. This process of preservation came to be known as pasteurization and is universally used for preserving milk and other food products. The most commonly used contemporary pasteurization process involves heating at 71.6°C (161°F) or little higher for 15-30 sec and is called high temperature-short-time (HTST) method.

1.6. Microbial Association with Diseases- Medical Microbiology

By the middle of nineteenth century, the focus shifted to the disease causing nature of bacteria. Such a role of microorganisms was hinted at by Leeuwenhoek and others. In 1850, Davaine and Rayer observed large numbers of tiny rods in the blood of animals suffering from anthrax disease. Some ten years later, Delafond (1861) was able to demonstrate the outbreak of anthrax in healthy animals when inoculated with the blood of diseased ones. Delafond considered anthrax to be a fermentation product. He faced the objection of blood itself being diseased but got support from the work of Klebs and Tiegel, though much later. They filtered the blood from diseased animal through a porous porcelain filter which could hold back even the finest particles and showed that two fractions fared differently in transferring the disease. As the filtered blood was incapable of causing disease, the cause was assigned to solid- the corpuscles or bacteria.

These experiments categorically rejected the prevailing medical notion of diseases being caused by foul airs or miasmas. Budd in 1862 strongly stated that diseases are caused by poisons that propagate by the law of continuous succession and equated them to microorganisms. Pasteur was also able to demonstrate the role of bacteria in disease. While studying the "pebrine" disease of silk worms in 1865, he showed that the diseased worms carry the corpuscles (bacteria). By carefully nurturing the eggs laid by moths free of bacteria, he raised a population of healthy silk worms. In 1869, he showed that healthy worms could contract the disease if fed on the debris of the diseased moths.

The real credit for unrolling the bacterial cause and transmission of some diseases goes to Koch. In 1876, Koch was able to demonstrate that anthrax disease is transmitted through bacterial spores. The spores liberated by the carcass of the diseased cattle find their way into healthy cattles through infected food. His simple but logical experiment was able to put at rest the notion that anthrax is caused by diseased blood. He diluted the blood from infected animals several folds with fresh blood serum and reasoned that such a blood should not cause the disease if blood is the root cause. Instead, he showed that diluted blood was infectious.

This work got further support when bacterial basis of several other plant and animal diseases was worked out in the next two decades. Also in many cases, the causative bacterial species could be cultured in pure form. Koch was the leading bacteriologist in this area and proposed what came to be called as Koch's postulates. These postulates laid down the conditions needed to be fulfilled if the microbial basis of a disease is to be established. These were: (i) organism must be present in the lesions of the disease, (ii) it must be possible to culture these organisms in pure form outside the host body, (iii) the pure cultures should be able to produce typical symptoms of the disease, and (iv) the same organism should be present in the disease caused by the inoculum.

1.7. Bacteria Land Up in Laboratory

Koch is also remembered for his contributions towards enunciating the concept of "pure culture". He designed and used a medium containing broth, serum or other suitable materials which was solidified with gelatin. The medium was heat sterilized, cooled and the bacterial inoculum was added to it. The isolated colonies appearing on this medium were perhaps derived from single spores and were thus pure. Koch was clear that if this process is repeated purity can be ensured. Lister's dilution method to eliminate the bacteria present in low numbers was a further refinement. Replacement of gelatin by agar-agar was another landmark contribution of Koch's group. Taking clue from some home-recipes, his student, Hesse in 1883, used agar-agar in the nutrient medium. Such a medium proved much better as many microbes were known to decompose gelatin. Subsequently, even silica gel was used but so far agar-agar is the most popular gelling agent used in different media. Through another modification, Koch's cumbersome plates were replaced by shallow glass dishes by another of Koch's students, R J Petri. These dishes were named after him as petri dishes.

Koch's pure culture concept really caught the fancy of a large number of microbiologists. In 1896, Hensen devised a dilution technique to get pure cultures of microorganisms. He mixed a drop of yeast culture with a drop of water or broth on a coverslip. A droplet from this was mixed with a second drop of water. Such repeated transfer allowed him to get a single yeast cell that grew up to produce a pure culture. Similarly, Joseph Lister used a dilution method to count the number of bacteria in a solution. This method, though slightly modified, is still the most popular way for the quantitative estimation of microbes. One of the significant contributions of these methods was to provide a pure population of a microbe for not only the morphological studies but also for subsequent genetic and biochemical work. Pure culture concept is still the most basic aspect of any study conducted on microorganisms.

1.8. Concept of Immunity

By the turn of the 18th century, another important concept emerged in relation to microbial diseases. This came in the form of phenomenon of immunity. It was a common observation that people, such as milkmaids, who may have contracted mild cowpox, become virtually resistant to (immune) virulent small pox. It is said that as early as 1718, in Turkey, people were immunized against small pox by infecting healthy individuals with the scabs from a diseased person. Edward Jenner was the first to foresee the scientific basis of this observation and in 1796 devised a safe process of immunization and named it vaccination (vacca=cow). So clear and confident he was of the success of this process that he inoculated 23 healthy persons with cowpox and demonstrated that they are protected from the outbreaks of small pox.

Pasteur was highly impressed by Jenner's work and devoted a number of years working in this area. In 1879, Pasteur selected fowl cholera, and found that prolonged incubation weakens the pathogenicity of the organism, which he termed as "attenuation". He was of the opinion that cowpox strain used by Jenner represented the mild form of small pox organism and that this was the basis of endowing the immunity. In the following years, he successfully demonstrated the immunization conferred by mild, weakened, or attenuated pathogenic strains. With his continued interest in anthrax disease, Pasteur along with Chamberland and Roux found that eight days incubation at 42-43°C is sufficient to inactivate anthrax bacterium. At this temperature, bacterium loses its capability to form endospores and thus can be safely inoculated to healthy cattles immunizing them against anthrax. His experiments leading to vaccination of over 85,000 cattles against anthrax within a year was hailed as one of the most outstanding work of the time. In 1883, Pasteur and Thuiller showed that attenuation of the causative organism for swine erysipelas can be brought about by a simple passage of the organism through rabbit. Pasteur then took up the case of rabies and his group was able to show that dried, two-week-old spinal cord of a dog that had died of rabies could be used as vaccine. This led to his famous experiment in 1885, when he vaccinated and saved a 9-year-old boy who was bitten by a rabied dog. He reported that even dogs can be protected by this vaccine who may have received a violent strain of rabies. This received a wide acceptance amongst the people and many availed this protection measure.

One striking aspect of Pasteur's work on vaccination was that he never used the pure cultures of concerned bacteria. The astounding capability of microorganisms to over grow others in a particular niche of environmental conditions led to the success of his experiments. Pasteur's technique of producing vaccine, however, was not hundred percent safe and was cumbersome. The discovery by Salmon and Smith during 1884-1886, that heat killing the hog cholera bacillus provides the desired attenuation paved the way for large scale production of vaccines.

Two theories were put forward to explain the basis of immunity. While humoral theory, proposed by Buchner, Pfeiffer and others, emphasized the role of blood and body fluids in combating the infectious agent, the cellular theory, propounded by Metchinkoff, suggested engulfing and destroying (phagocytosis) of the infectious agent by specialized cells.

Another area, fast developing during this time was associated with the problems relating to microbial infections during surgery. The art and science of surgery was still in its infancy. Many patients used to die of post-surgical infections. As early as 1847, Semmelweis used aseptic methods to reduce the chances of such infections. It was, however, Joseph Lister, a Glasgow surgeon, who taking clue from Pasteur's work suggested the use of phenol-soaked bandages to overcome the post-surgery infections. He also emphasized upon the use of sterilized instruments and conduct of surgery under a spray of disinfectants to prevent the access of microorganisms to the wounds. Such methods, though highly refined, have become an integral part of the present day surgical operations.

Thus, with these inputs, medical microbiology got a solid footing in experimental science. This field has taken rapid strides in 19th and 20th century and will continue to do so, so much so that it is difficult to bring them all under a historical purview.

1.9. Agricultural Microbiology

Another field in which microbes have contributed significantly is soil and agricultural microbiology. Soil has always been a preferred home for a host of microorganisms, but the role that they play started unfolding only in the late 19th century. Initially, they were considered to indulge in some chemical reactions for their own benefit. The process of nitrification, for example, in which organic nitrogenous compounds and ammonia are converted to nitrate was regarded as an oxidation process. Pasteur himself had suggested that microorganisms can bring about oxidation and that organic compounds can be preserved if they are protected from microbial growth. In 1862, he also recommended further research to reveal the role of microbes in nitrification. Based on this, Schloesing and Muntz (1877) demonstrated that nitrification is a microbial function and any inhibition of growth of microbes leads to inhibition of nitrate production from nitrogenous compounds. Russian microbiologist, Winogradsky isolated autotrophic bacteria capable of nitrification and later the sulphur bacteria that obtain energy from the oxidation of H_2S and elemental S, and iron-oxidizing bacteria. Following this, a number of studies revealed the unbelievable scavenging activities of microorganisms that not only maintains the all life-sustaining biogeochemical cycles but also keeps the earth clean of unwanted wastes. Thus, came into existence a field called environmental microbiology.

Winogradsky is also remembered for his isolation of an anaerobic bacterium, *Clostridium* in 1894, that could fix atmospheric nitrogen, a function restricted to bacterial kingdom. Soon thereafter, Beijerinck discovered the free-living, aerobic, nitrogen fixer, *Azotobacter*, in 1901. Beijerinck had earlier (1888) established the role of *Rhizobium* in fixing nitrogen in association with (nodules) legumes. The idea that many plants, such as legumes, can fix nitrogen came as early as 1838 by a French chemist, Boussingault. He faced severe criticism by many workers including Liebig till his work was confirmed by Atwater in 1885 followed by that of Beijerinck. Ever since its discovery, biological nitrogen fixation has created immense interest and has been a subject area of intense research with the following major goals: (1) to increase the food production without spending on the expensive production of chemical fertilizers, (2) to improve the dietary protein deficiencies, and (3) to make the important crop plants fix their own nitrogen. Inspite of various bottlenecks, the latter remains the dream project for agricultural scientists.

Winogradsky and Beijerinck were also instrumental in designing the technique of enrichment culture, a routine procedure in microbiology today. They showed that if conditions are provided that will preferentially support the growth of one type of microorganism, they will outgrow the others and thus be enriched. The enrichment culture method is one of the strongest microbial tools for the isolation of microorganisms with a desired set of physical and chemical attributes.

1.10. The Science of Genetics and Biotechnology- Bacteria Hog the Limelight

The other major contribution of microbes, in general, and bacteria in particular, is in the development of the science of Genetics. Though arriving late on the scene (only by the middle of the 20th century), their haploid nature, the short life cycle, their simple organization, and the ease with which they can be grown and manipulated under laboratory conditions have helped them strike a deal with the geneticists. Geneticists soon discovered their favourite organism, *Escherichia coli*, named after its discoverer Theodor Escherich, a German pediatrician who isolated it from faeces in 1885, and its habitat, the colon. In this association of over half a century this tiny Gram-negative bacterium provided such informations that not only revolutionized biology but also helped better understanding of life and its processes, the end of which is yet to be seen. Genetically, the best understood organism, bacteria also enabled the biologists to develop a field called biotechnology. Although, the microbes are not the only type of living organisms that come under the precincts of biotechnology, they have surely provided the most important inputs, ingredients and leads. Getting the basic clues from them, today the molecular biologists are at the brink of manipulating genes at will, cloning cells, tissues, and organisms with desired attributes, that will help human society at large.

1.11. Development of Physical and Biochemical Techniques

Microbes have also been instrumental towards the development of highly sophisticated physical instruments and techniques, starting from the famous Leeuwenhoekian microscope, to highly sophisticated microscopes, other instruments and several physico-chemical techniques. Their usage and Man's continued inquest for better understanding of bacterial structure and structure-function relationships have led to the continued refinement of these instruments and techniques. In return, man has arrived at better understanding of biology of living systems. Moreover, microbes have served as foundation stone for the development of modern branches of cell and molecular biology. Though small and simple, they continue to spring new surprises serving as model systems for both fundamental and applied research.

CHAPTER 2
KNOWING BACTERIA-METHODS OF STUDY

For proper characterization of a cell or a living organism, the first set of informations are derived from direct observations. The small size of bacterial cell is, therefore, an impediment for such studies. However, the same restrain provided the impetus for the development of several biophysical and biochemical techniques so as to make it possible to look at their cells for some structural details.

2.1. Microscopy

As we have seen in Chapter 1, the existence of bacterial kingdom was brought to light through microscopy. Starting from the crude Leeuwenhoekian microscope, microscopy is still one of the commonest type of such techniques of which a whole range is available now. These vary in range of magnification, resolution, and the use of visible light or a beam of electrons. Except for the electron microscope in which magnification of 10,000-1,000,000 and resolution of 1-10 nm can be achieved, others fall in the range of 1,500x magnification and 100-200 nm resolution (Fig. 2.1).

The most common type of microscope available in the microbiology laboratory is the light microscope (=Bright field microscope). A condenser lens is used to focus the light on the specimen. Based on the combined strength of ocular and objective lenses, magnification upto 1000 times can be achieved. Staining of the sample facilitates the viewing and characteristic size, shape, and arrangement of microbial cells can be observed. Fluorescence microscopy involves staining the microbes with a fluorescent dye and examining through a microscope that does not employ a condenser. A series of exciter and barrier filters are used to obtain light of a particular wavelength that illuminates the specimen. The viewing wavelength corresponds to the one emitted by the dye. This microscopy has become particularly useful as the fluorescent dye can be conjugated with antibodies directed towards a particular structure, providing greater target specificity to staining.

Better contrast between the specimen and the background without staining can be achieved by dark-field microscopy. This permits the visualization of living specimens. In a dark-field microscope, the light that reflects off the specimen only will be seen so that the background appears dark. This is specially useful for visualization of even small bacteria but no details can be seen.

Another type of microscope that can be used for visualizing unstained live samples is phase contrast microscope. This microscope works on the fact that when light passes through a microorganism, a slight alteration in the phase of the light wave occurs. The interference of light waves creates the visible light intensity differences that leads to the visualization of microorganisms. The cell structures are thus easily seen.

Interference microscopes also rely on light interference to produce a visible image. Unlike phase-contrast microscope, however, where only a single beam of light passes through the specimen, in interference microscope, two beams of light after passing through the specimen, combine to give a better contrast and can use the colours to increase the contrast. This is based on a stereoscopic effect giving a pseudo-three-dimensional image. Different structures may appear in different colours giving it a better clarity.

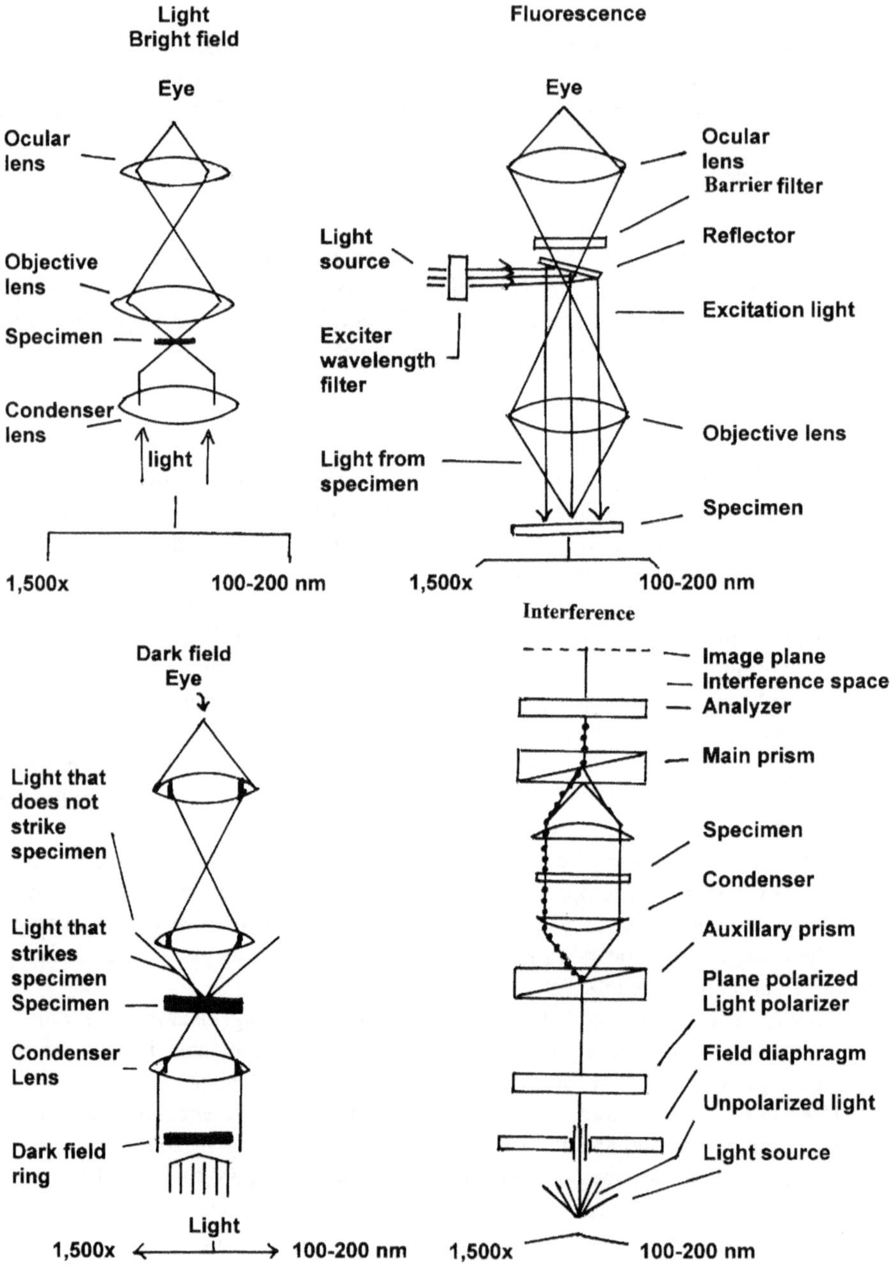

Figure 2.1. Working plan of different types of microscopes.

The most significant of the microscopic techniques is the electron microscopy which provides not only better resolution but also a very high magnification. This has allowed the elucidation of cell structures of microorganisms in extensive details. Two types of such microscopes are available: Transmission electron microscope (TEM) and Scanning electron microscope (SEM) (Fig.2.2).

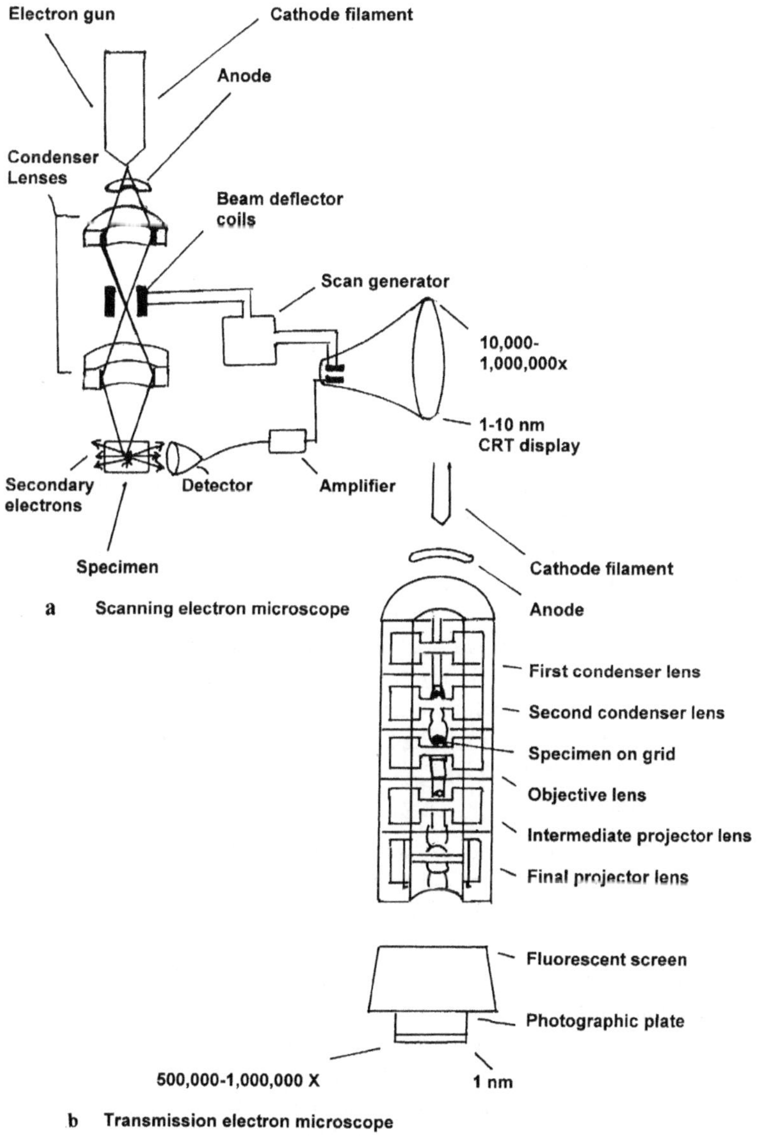

Figure 2.2. Two type of electron microscopes. a. Scanning electron microscope, b. Transmission electron microscope.

In these microscopes, light is replaced by an electron beam, generated at a high accelerating voltage. When an electron beam collides with the specimen, electron scattering proportional to the electronic density of the specimen occurs, producing the contrast. This can be further improved by staining. The staining used for electron microscopy does not consist of dyes, but electron-dense heavy metals such as tungsten. This improves the scattering and thus the contrast. TEM essentially requires thin sectioning of the fixed and dehydrated material. TEM is useful for ultrastructural details and even the visualization of molecules. In SEM, an electron beam is scanned across the surface of the specimen and is primarily used for viewing the surfaces. The contrast in SEM is determined by surface topography and can be improved by metal coating. The operation principles and design of SEM are quite different from TEM.

Some special preparative procedures have been devised to facilitate refined resolution by electron microscope. One such technique is freeze etching. The specimen frozen in liquid nitrogen are fractured by a knife blade and then etched by allowing some of the ice to be evaporated. The exposed surface is then treated with vapours of a heavy metal and subsequently with vaporized carbon. This produces a carbon replica of the structure that can be viewed with an electron microscope. This technique is very useful in studying surface structures.

2.2. Availability of Material

Bacteria are ubiquitous in nature, yet they generally occur in mixed forms. For any detailed study, these forms will have to be obtained in pure-form, which in turn is produced by growing them on synthetic substrates called media. The process of growing them on synthetic media is known as culturing. The type of the media to be used, though vary from organism to organism, is based on nutritional spectrum. Broadly speaking, it consists of a combination of macro-and micronutrients as required by any living cell. It should be borne in mind that all bacteria cannot be cultured as their growth requirement may not be fully known. In fact, the culturable bacteria comprise a very small fraction (~ < 1%) of all the bacteria existing in nature (**see Chapter 8**).

2.3. Culture Media

Bacteria are highly versatile as far as their nutritional adjustment is concerned. While higher plants are photoautotrophs and animals, chemoheterotrophs, bacteria span this whole range, some times growing even on weird substrates, like rocks, oil, some plastics and wood. An optimal medium must provide the following:

(i) **Carbon** - It is the principal component of all living cells forming the structural backbone and other organic constituents. While photoautotrophs or autotrophs can utilize CO_2 by fixing it, the others can be grown on different C-sources. Although some species can utilize higher hydrocarbons (petroleum, for example), hexoses like glucose or disaccharide such as sucrose may be preferred.

(ii) **Hydrogen and Oxygen** - The other two ubiquitously found elements are normally metabolism-derived. However, O_2 from air can be utilized and gaseous H_2 can also be used by some bacteria.

(iii) Nitrogen - As an important constituent of nucleic acids and proteins and several other macromolecules, it is generally obtained from NO_3^- or NH_3. While in nature, some bacteria can utilize number of organic compounds, others can fix N_2 into usable forms like NO_3^- and NH_3, a function limited to some members of prokaryotic kingdom.

(iv) Phosphorus - This is an important component of nucleic acids, phospholipids, coenzymes and energy-storing compounds. The requirement is generally met from phosphates.

(v) Sulphur - This is present in some essential amino acids and, therefore, in proteins and some coenzymes. Most common source of sulphur is sulphates but can also be derived from organic compounds such as proteins and appropriate amino acids. Some are uniquely capable of utilizing elemental sulphur.

(vi) Metal Ions - Several metal ions like potassium, sodium, magnesium and calcium comprise important inorganic cations. These required in large amounts (macronutrients), function as cofactors in several enzymatic reactions, as structural components, and in membrane transport. The others like Fe, Cu, Ni, Co, B, Mn, Zn, and Mo also participate in important metabolic reactions and are required in trace amounts (micronutrients). An important anion is chloride and must be provided in the medium.

Table 2.1. Composition of some commonly used complex media

S.No.	*Medium*	*Constituents*	*Amount (g/l)*
I	Nutrient	Beef extract	3.0
	Broth	Peptone	5.0
II	Luria-Bertani	Tryptone	10.0
	Medium	Yeast extract	5.0
		NaCl	10.0
III	Tryptic Soy Broth	Pancreatic digest of casein	17.0
	(Aerobic heterotrophs)	Papaic digest of soybean meal	3.0
		NaCl	5.0
		K_2HPO_4	2.5
		Glucose	2.5
IV	Reinforced	Yeast extract	3.0
	Clostridium	Beef extract	10.0
	Medium	Pancreatic digest of casein	10.0
	(Anaerobic heterotrophs)	Glucose	5.0
		NaCl	5.0
		Na-acetate	3.0
		Soluble starch	1.0
		L-cysteine-HCl	0.5

A specific mixture of macro-and micronutrients in appropriate amounts constitute a defined mineral salt or minimal medium. A number of complex media have also been designed which contain complex nutrients like beef extract, meat extract, peptone/tryptone, casein hydrolysate and yeast extract. These compounds essentially furnish all the chemicals needed for the growth of various types of microorganisms. Such a medium is generally used for culturing and maintaining the microbial cultures (stocks). The composition of some of these media is given in Table 2.1, and 2.2.
Many of these media are now available as readymade preparations from commercial sources, and can be prepared as per the instructions of the manufacturer.

Table 2.2. Some defined mineral salt media

S.No.	*Medium*	*Constituents*	*Amount (g/l)*
(I)	M 9	$Na_2HPO_4.2H_2O$	6.0
		NH_4Cl	1.0
		KH_2PO_4	3.0
		NaCl	0.5
		$MgSO_4.7H_2O$ (1M)	2.0 ml
		$CaCl_2.2H_2O$ (0.1M)	1.0 ml
		Glucose (20% stock)	10.0 ml
(II)	E medium	$MgSO_4.7H_2O$	0.2
		Citric acid.H_2O	2.0
		K_2HPO_4.anhydrous	10.0
		$NaHNH_4PO_4.4H_2O$	3.5
		Glucose or other sugars (20% stock)	10.0 ml
(III)	Cyanobacterial medium	$NaNO_3$	1.5
		K_2HPO_4	0.04
		$MgSO_4.7H_2O$	0.075
		$CaCl_2.H_2O$	0.036
		Citric acid	0.006
		Ferric ammonium citrate	0.006
		EDTA (disodium salt)	0.001
		Na_2CO_3	0.02
		Trace metals solution	1.0 ml
(IV)	Medium for N_2-fixing bacteria	KH_2PO_4	0.4
		K_2HPO_4	0.1
		$MgSO_4.7H_2O$	0.2
		NaCl	0.1
		$FeCl_3$	0.01
		$Na_2MoO_4.2H_2O$	0.002
		Sodium malate	5.0
		Yeast extract	0.05

These and other media can either be prepared in liquid form or jellified by adding 1.0-2.0% agar-agar to provide the solid support.

2.3.1. Specific Media

Various types of media have been designed based on the purpose that they serve. Some of these are:

(i) **Enriched Media**- Media like nutrient broth/agar can be supplemented with complex substances like extracts of plants, blood, serum. Such media are used for the cultivation of fastidious heterotrophs.

(ii) **Selective Media**- Supplementation of the medium with some specific substances can facilitate the growth of one type and prevent the growth of other types of bacteria. For example, dyes like basic fuchsin, crystal violet and brilliant green will preferentially inhibit the growth of Gram-positive bacteria. Salt and so also antibiotics can be used to selectively differentiate between tolerant and sensitive type of bacteria.

(iii) **Differential Media**- Supplementation of medium with certain compounds can change the growth characteristics and allow the identification of certain type of bacteria. For example, when a mixture of bacteria is plated on blood-agar medium only hemolytic bacteria will produce a clear zone by lysing the blood cells.

(iv) **Assay Media**- Based on characteristic growth behaviour of microorganisms, media can be designed to assay for antibiotics, amino acids, enzymes, vitamins, etc.

(v) **Enumeration Media**- These are employed for determining the bacterial load of milk, water, and other solutions.

(vi) **Characterization Media**- Different types of media have been formulated to characterize bacteria by their specific growth responses.

(vii) **Indicator Media**- A variety of colour indicator agars are available to determine the capability of a carbon source utilization. On MacConkey agar medium that contains a dye sensitive to pH and lactose, both Lac^+ and Lac^- bacteria can grow. However, the utilization of lactose by Lac^+ cells leads to acid production and thus a consequent local drop in pH causing the colonies to turn pink. Lac^- cells, on the other hand, can not utilize lactose, instead use amino acids as a carbon source with ammonia being produced. This increases the local pH that decolourizes the dye and the colony appears white (Fig. 2.3).

2.3.2. Preparation of Culture Media

Once the composition of the desired medium is available, preparation of the same is a simple process. The appropriate amounts of media components are accurately weighed, dissolved in double glass distilled water and the final volume of the medium is obtained. Such a medium can either be used as liquid or broth or solidified by adding 1-2% agar-agar powder. The pH of the medium is adjusted according to the requirement for optimal growth which is usually a pH of near neutrality. One important aspect of microbial media preparation is that it should be free of any unwanted microorganism (contaminant). For this, not only the media but also the accompanying implements such as pipettes, tubes, petridishes, and other materials must be sterilized. While glasswares can be heat

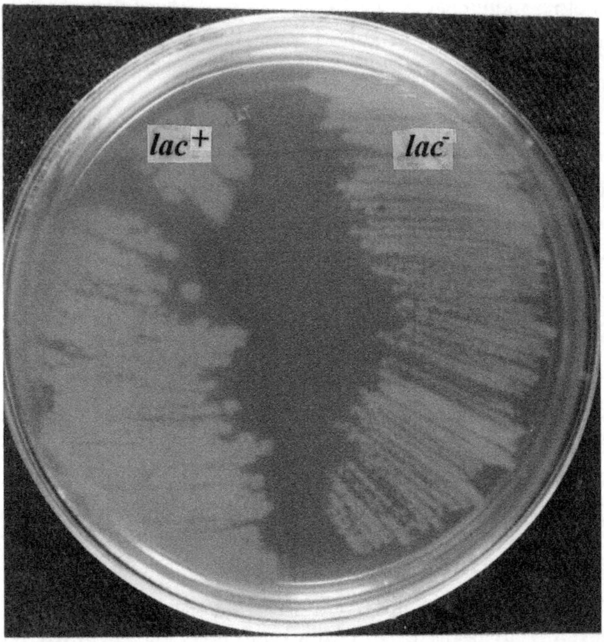

Figure 2.3. Growth of lactose-utilizing and lactose-non-utilizing E.coli strains on MacConkey agar plate. On this indicator agar lac⁺ produces purple and lac⁻ colourless colonies.

sterilized by placing them in an oven at 150°-200°C overnight, cloth, rubber, plastic and media which get destroyed by direct heating are sterilized by steam under pressure (autoclaving). The medium is dispensed into appropriate containers which are properly closed, and placed in an autoclave and sterilized at 15 lb/inch^2 pressure and 121°C temperature for 15-30 minutes (Fig.2.4).

This combination of temperature and pressure is effective in killing microorganisms present in the medium. The temperature and pressure may vary depending upon the nature of the material to be sterilized. Once the medium is sterilized it can be stored in the same containers. Solid medium is melted prior to use either by autoclaving or placing the container in boiling water.

Many components of the media, such as buffers, amino acids, other organic compounds and blood sera are destroyed at the temperature of autoclave. These materials in solution can be filter sterilized. Various types of filters are used but they all are composed of material that contain pores which are so small (usually 0.2 μm) that bacterial cells cannot pass through them. Thus a solution passed through such filters becomes microbe-free. The common membrane filters are made up of cellulose or plastic films but others may be derived from fused glass (sintered glass filter), compressed asbestos discs (seitz filter), unglazed porcelain (Chamberland and Selas filter) or diatomaceous

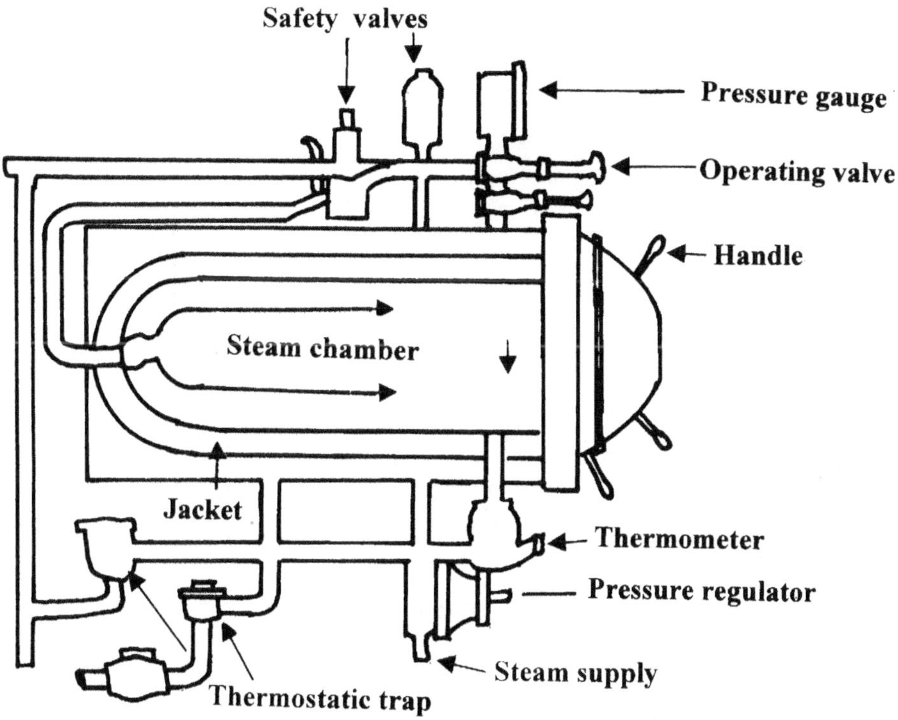

Figure 2.4. Autoclave (Horizontal) used for sterilization.

earth (Mandler filters). Many of these filters are fitted in a funnel that can be attached to a suction flask. Many of the membrane filters are commercially available with a simple filteration attachment for one time use (Fig.2.5). A filter sterilized component can be added to sterilized medium so that the desired composition can be obtained.

Many heat-labile materials such as surgical instruments, thermometers, lensed instruments, polyethylene and other rubber tubings and implements, inhalation and anaesthesia equipments can also be sterilized by "cold sterilization". This is performed in enclosed chamber resembling an autoclave but employs chemical agents such as ethylene oxide, formaldehyde or hydrogen peroxide. Ethylene oxide is a very popular agent as it is not only toxic to bacteria, viruses, and fungal cells but also effective against heat-resistant bacterial endospores. It is non-corrosive, non-deleterious, relatively inexpensive and easy to handle and residual amounts can be easily removed by standing or aeration. Though highly inflammable, its mixture with carbon dioxide (1:9) or with freon can prevent explosion and inflammability without reducing its effectivity as a sterilizing agent.

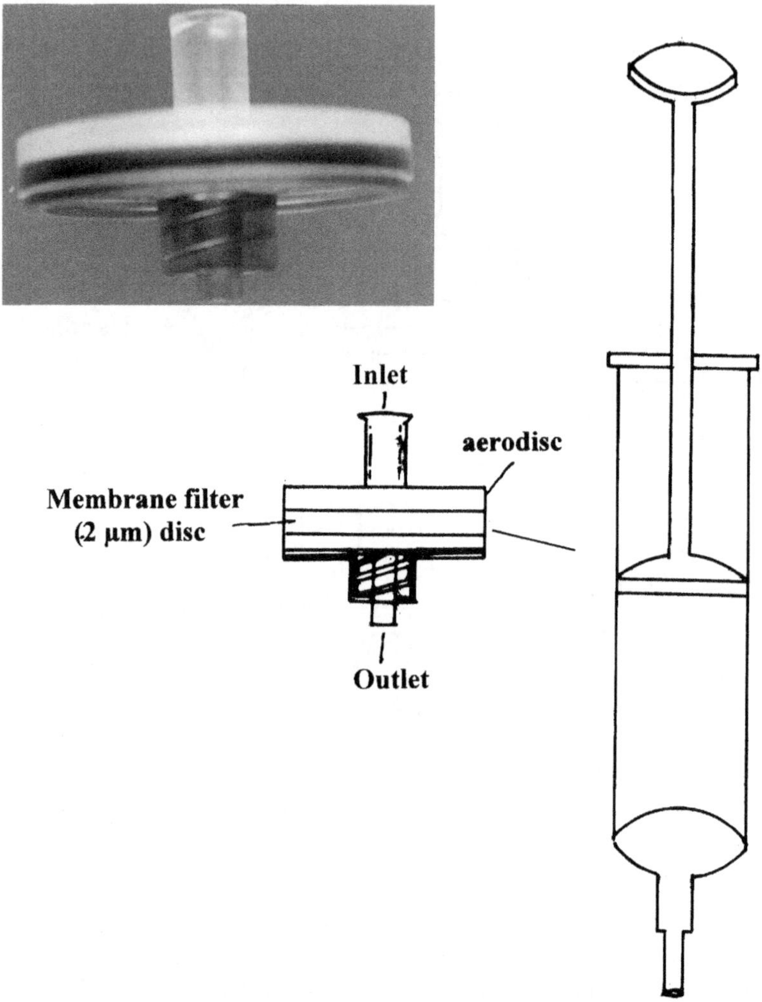

Figure 2.5. Minisart assembly for direct filtration to remove particulate matter or microbes from liquids. The syringe is connected at the inlet end of the minisart. If the need be, a sterile needle can be attached at the outlet end.

2.3.3. Pouring of Media

The hot sterilized medium is dispensed in flasks, tubes, or petridishes as per the requirement. This exercise, called pouring, is carried out with utmost precaution so that the medium does not get contaminated. The use of a Laminar Air Flow bench and a burning bunsen burner are part of the requirements to maintain aseptic conditions. The molten agar media in tubes is generally solidified in the form of slopes or slants by keeping them in a slanting position. The agar plates are prepared by pouring about

30-40 ml of molten medium in sterile petridishes. During any kind of pouring the glass vessel is opened for the minimum time so that the medium does not come in contact with the outside air for very long. It is also customary to cool the medium to about 45°C before pouring so as to avoid the loss of moisture by condensation.

2.4. Inoculation

The transfer of microbial material on to an appropriate medium is inoculation. With the basic requirement of maintaining the purity of the culture, aseptic conditions are to remain maintained throughout this process. Not only the medium has to be sterilized but also all the instruments and implements should be sterilized so as to prevent any kind of contamination. While aseptic conditions inside the Laminar Air Flow bench is created by switching on the UV Lamp at least 30 min. before use, the working surface can be wiped with some germicidal solution or rectified spirit. The needle, inoculation loop generally used are sterilized by heating them to red hot on a bunsen burner whereas glass spreader and other glass articles are dipped in rectified spirit and then flamed. During inoculation, a portion of the biomass (inoculum) from a stock culture is aseptically transferred to fresh medium (Fig.2.6).

2.5. Incubation

Like all other living organisms, bacterial biomass also requires a specific temperature or temperature for optimum growth. This temperature should be known before hand. All inoculated materials are then placed in an incubator, the temperature of which is specifically adjusted. The time required for the microorganism to grow will also vary but bacteria are amongst the fastest growing organisms.

2.6. Pure Culture

Microbes in nature are generally found in mixed forms. For any type of meaningful study, however, each microbial species should be obtained in pure form. Various types of techniques have been designed to achieve this goal.

2.6.1. Streak Plate Method

A small amount of bacterial culture is streaked over the surface of complex agar medium (Fig.2.7). The basis behind this technique is to achieve gradual dilution of the biomass so that at some places well-spread single colonies are obtained. Being unicellular, each colony is derived by the clonal multiplication of one cell, and being haploid each one of them will be genetically pure.

2.6.2. Spread Plate Method

In this method, microbial cells in a suspension are diluted serially and an appropriate dilution is plated on complex agar plates. The plating involves the transfer of a small inoculum (100-200 μl of cell suspension) on solid medium plates and its spreading by a sterile glass spreader all over the surface. The basis once again is to obtain well-spread single colonies (Fig. 2.8a) Material from each of these colonies can then be checked for

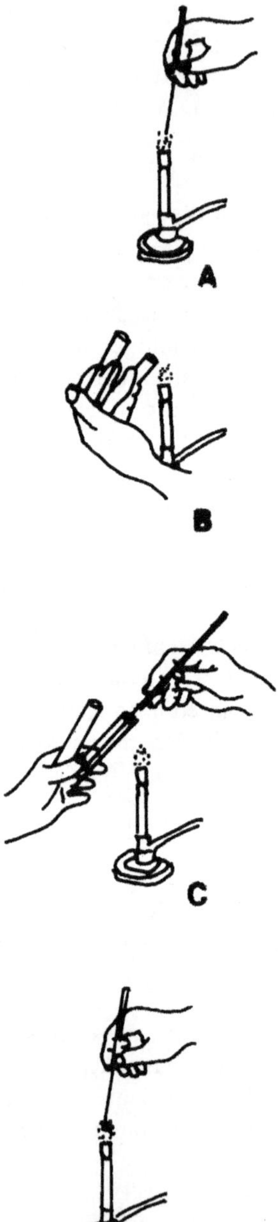

Figure 2.6. Aseptic culture of bacteria. A. Loop is heated red hot and allowed to cool, B.Culture tube is uncapped over the flame, C. Culture sample is transferred to the tubes and is recapped, D. The loop is resterilized over the flame.

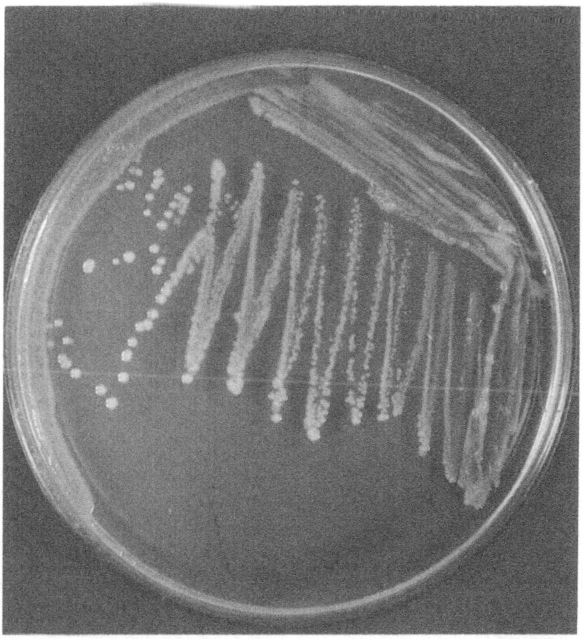

Figure 2.7. A pure bacterial culture can be produced by streak plating- a small inoculum is streaked over the surface of the agar medium in different directions so as to get the dilution of the biomass and thus the single colonies.

some specific characters of the strain and then maintained on agar slants or plates. This technique is also used in quantitative enumeration of bacteria. In this, the number of colonies are counted from a plate where the colonies are well dispersed. When the same is multiplied with the degree of dilution, total cell count of the original sample can be determined. This technique is also useful in estimating the viable cell counts, as colony forming units (CFUs/ml), as only the viable cells will be capable of forming colonies (Fig.2.8b).

2.6.3. Pour Plate Method

Originally developed by Koch, in which the cells were mixed with molten and cooled gelatin (now replaced by agar-agar) medium and poured in sterilized dishes. The solidifying agent (gelatin or agar-agar) then immobilizes cells so that each one of them could produce an isolated colony. The technique also known as seeding, has been slightly modified in which a diluted sample is used to ensure the development of isolated colonies.

All these techniques are aimed at the formation of isolated colonies, derived from a single viable unit that will specify a particular genotype. Repeated application of these techniques coupled with characterization ensures genetic purity. Pure cultures are then stored as stock cultures.

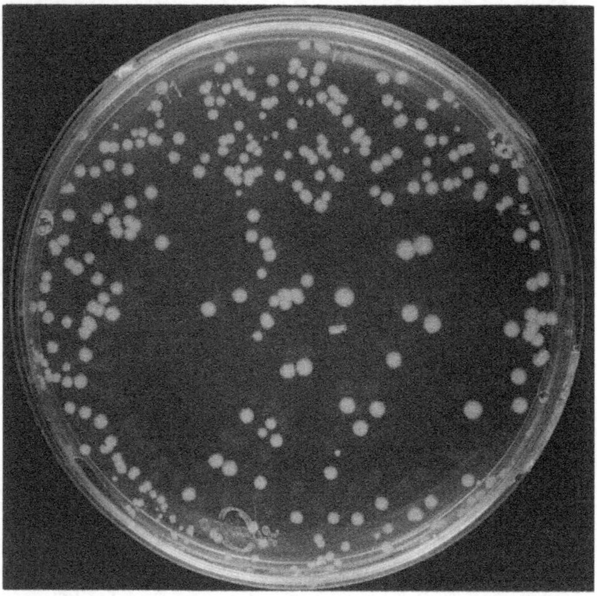

Figure 2.8a. Spread plating. A small amount of diluted cell suspension is spread over the agar surface to obtain well spread colonies. Each cell will develop into an isolated colony.

2.7. Culturing Aerobic/Anaerobic Organisms

Availability of oxygen greatly influences the growth. Oxygen diffuses only slowly into liquid and, therefore, growth of aerobic and facultative anaerobes is greatly facilitated by putting them on a shaking platform or by an impeller within the culture vessel or even by bubbling oxygen through the culture medium. On solid media, these organisms have to be placed on the surface.

For obligate and strict anaerobes, oxygen is highly inhibitory. Removal of O_2 is accomplished by adding either certain chemicals such as sodium thioglycollate that reacts with oxygen or amino acid cysteine and other compounds containing sulphahydryl group that can scavenge molecular oxygen. For liquid culture, N_2 is bubbled to replace oxygen and then the container is sealed airtight.

Some special types of chambers, such as gas pack system are available for anaerobic culture. In these H_2 and CO_2 are generated, while the former reacts with O_2 to produce H_2O, CO_2 replaces the volume of O_2 depleted (Fig.2.9).

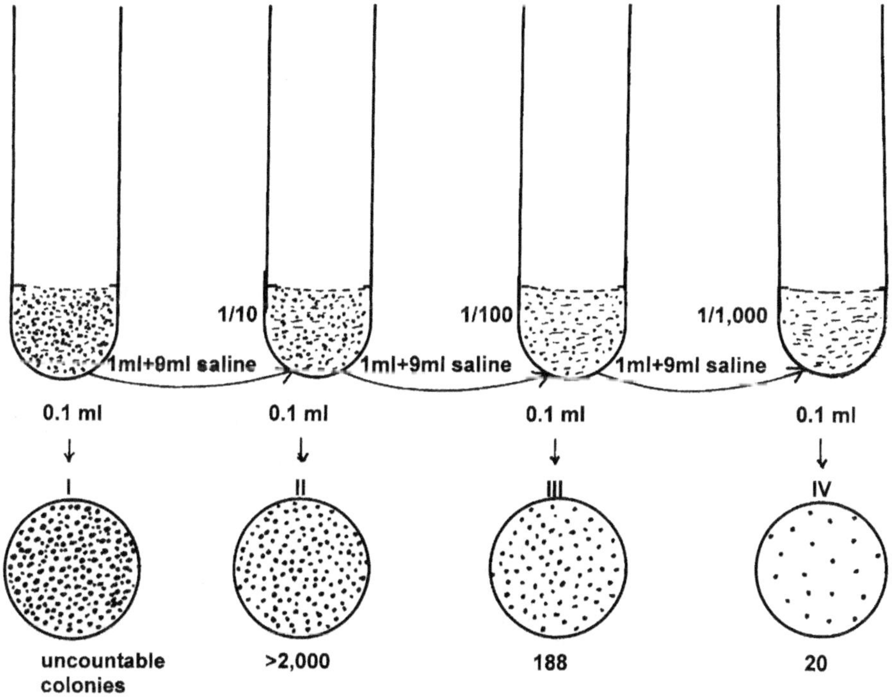

Figure 2.8b. Serial dilution technique. A thick bacterial suspension is diluted in a series by taking 1 ml of suspension and mixing it with 9 ml of saline. This leads to 10 times dilution in cell number at each step. Of each 0.1 ml is plated on nutrient agar plate, on which the colony number will show a gradual reduction. Thick suspension will produce innumerable colonies that would often form a lawn. From the plate showing countable number of colonies one can determine the viable cell count in the suspension (for example in the 4th plate 20 colonies were present in 0.1 ml suspension that is equivalent to 200 colonies/ml and with 10^3 times dilution the total cell count in this suspension will be $2x10^5$. Since only viable cells form colonies, it is also represented as CFUs/ml). The cell count in the original suspension can be derived by multiplying with dilution factor. In this example it will be $2x10^8$.

2.8. Enrichment Culture Techniques

Once it is decided to work with a specific microorganism, it is possible to design the medium and/or create specific conditions that will allow the preferential growth of that microorganism. Generally known as Enrichment culture technique, it is based on the metabolic capabilities of a specific microorganism. For example, if one wishes to isolate microorganisms capable of utilizing higher hydrocarbons, a medium can be designed in which that hydrocarbon is added as a sole carbon source. Such a medium will not support the growth of most other microorganisms incapable of utilizing such a hydrocarbon and will specifically enrich the desired type of microorganisms.

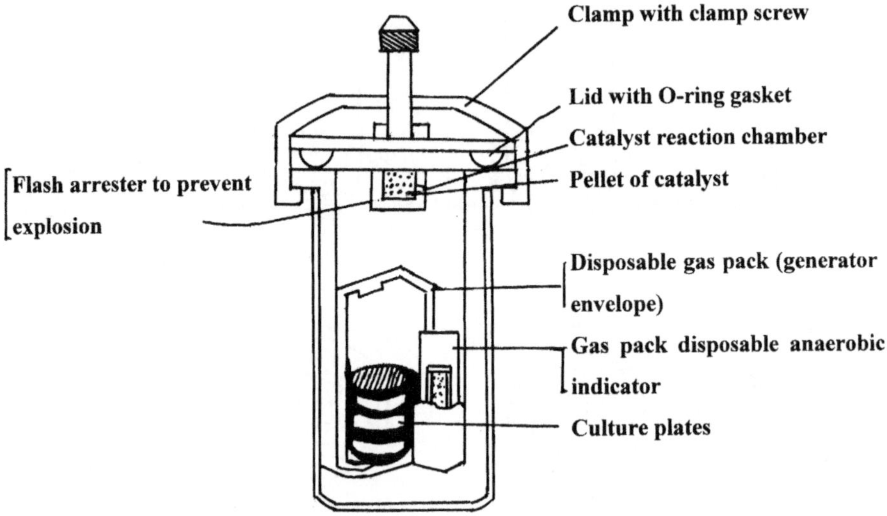

Figure 2.9. Anaerobic gas pack system for the culture of anaerobic bacteria.

Besides the medium, environmental factors such as temperature, pH, aeration etc. can be manipulated to obtain the growth of a specific type of microorganism. Manipulation of temperature can lead to the growth of psychrophilic (living in cold temperature), mesophilic (moderate temperature like 37°C), or thermophilic (above 50°C) microbes. Aeration by shaking or sparging with air will favour the aerobic and exclusion of oxygen will favour the growth of anaerobic microorganisms.

The conditions used for enrichment essentially mimic those of natural environment which tends to favour the occurrence of different types of microorganisms.

2.9. Culture Storage and Preservation

Once a pure culture of a desired microbe is obtained it is very essential to maintain them under appropriate conditions from which these microbes can be made available for future use.

One of the commonest method of maintaining a bacterial culture is the periodic subculturing on a fresh medium. Though highly suitable for short term storage, it is time consuming and labour intensive if a large number of pure cultures are to be handled. Moreover, at every subculturing, problems of contamination will have to be taken care of.

Therefore, a variety of methods have been developed for proper preservation of pure cultures. These include refrigeration at 0 to 5°C for short term storage, at -20°C to increase the time of storage, and freezing in sterile glycerol at -70°C or in liquid nitrogen at -196°C for prolonged storage. Lyophilization or freeze drying is another popular method for long term preservation. In this, the culture is frozen at a very low temperature and placed under high vacuum. Under such conditions, water in the culture and microbial cells is directly converted from the frozen solid to the gaseous state. This dehydration process precludes microbial growth but viability in dormant state is maintained. Glycerol is often used as an "antifreeze" agent to prevent damage due to ice crystal formation and ensure the viability of the cells when frozen cultures are thawed.

The cultures are usually deposited at Culture Collection Centres, where they are preserved, maintained and catalogued. Such Type Culture Collection Centres are the sources for obtaining the authentic cultures of a microbe by any scientist around the world.

2.10. Visual Observation

The direct visual observation of bacteria is possible with different types of microscopes as described in the beginning of this chapter. The extent of information derived depends upon the type of microscope used.

Several morphological and physiological characteristics of bacteria can also be studied by culturing them on a specific type of medium. The phenotypic traits such as colony shape and size and in some cases the colour can be ascertained by direct visual observations. Similarly, physiological characters such as requirement for specific nutrient, capability of utilizing a particular compound, resistance to certain antimicrobial compounds can also be determined on an appropriate medium.

2.11. Study of Live Bacteria

A bacterial cell in living state can be examined by the "hanging drop technique" through a microscope. For this, a drop of the bacterial culture or suspension is placed on a clean coverglass and the same is inverted on a cavity slide (Fig. 2.10). The cover slip is fixed in place by applying grease on the region of slide where cover slip is rested. This slide can now be placed on the stage of the microscope and examined.

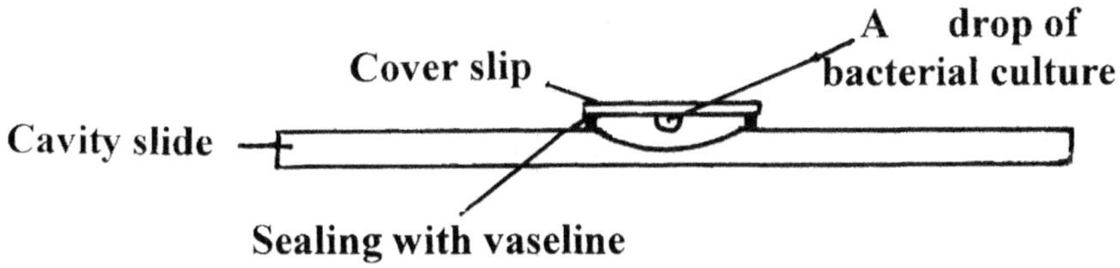

Figure 2.10. Hanging drop method for studying live bacteria.

2.12. Staining

Even with a refined microscope, the unstained specimens do not reveal any substantial details. The staining improves the contrast between the cells and the background and thus, produces a clearer picture of the specimen.

Various stains and dyes have been recommended for staining bacterial cells or any subcellular structure/component. Three such stains have been classified as acidic, basic, and neutral. One primary requirement for staining is the fixation of the material and the most commonly used method for the same is heat fixation. A small amount of bacterial culture is spread on a clean glass slide to form a thin film. The film is first air dried and then subsequently passed rapidly through the flame of a burner, a few times. The heating may kill the bacterial cell and coagulate the protoplasm but does not distort the structure. Often, other fixatives like alcohol, acetone etc. are also used.

As the bacterial cell carries a slight negative charge, it reacts best with dyes carrying a positive charge. The common stains used are crystal violet, methylene blue, carbol fuchsin, and safranine. Of these, the best results are obtained with crystal violet which penetrates the cells very rapidly. Carbol fuchsin reacts most rapidly but tends to overstain the cell.

The acidic or negative stains like eosin Y and nigrosine do not react with the cell but get deposited around it. In such a staining, also called negative staining, the cell remains unstained but background is coloured. This staining is useful in determining the size and the shape of the cells. India ink-mediated negative staining is often used to visualize the capsules. A neutral stain like Sudan III is used to specifically stain the fat inclusions and toluidine blue for staining polyphosphate granules.

One of the most remarkable staining procedures devised for bacteria is a differential stain called Gram stain, as it was introduced by Hans Christian Gram in 1884, for the identification of bacteria in animal tissues. In this, the fixed bacterial specimen is first treated with a basic dye like crystal violet. The cells are then treated with iodine which acts like a mordant, and subsequently with 95% alcohol. This last step is used to differentiate bacteria into two broad groups: those which retain the basic dye are known as Gram-positive ($Gram^+$) and those which get decolourised by alcohol are the Gram-negative ($Gram^-$). The $Gram^-$ bacteria are then counterstained with other dyes like safranine (Fig. 2.11).

Although initially designed for the identification of bacteria, the utility of this staining procedure increased several fold when it was found that the differential response of bacteria towards gram-staining reflects a number of differences in some basic properties of the members of these two groups (Table 2.3).

The differential staining per se is based on the differences in the nature of the cell wall of $Gram^+$ and $Gram^-$ bacteria. The cell wall of Gram-negative bacteria is comparatively thinner and has a higher content of lipids. The decolourization step with 15% alcohol leads to distortion and disorganization of the cell walls of Gram-negative bacteria. This enhances the permeability leading to the release of crystal violet-iodine complex. The same alcohol dehydrates and tightens the cell wall of Gram-positive bacteria, decreasing its permeability and thus, retention of the stain.

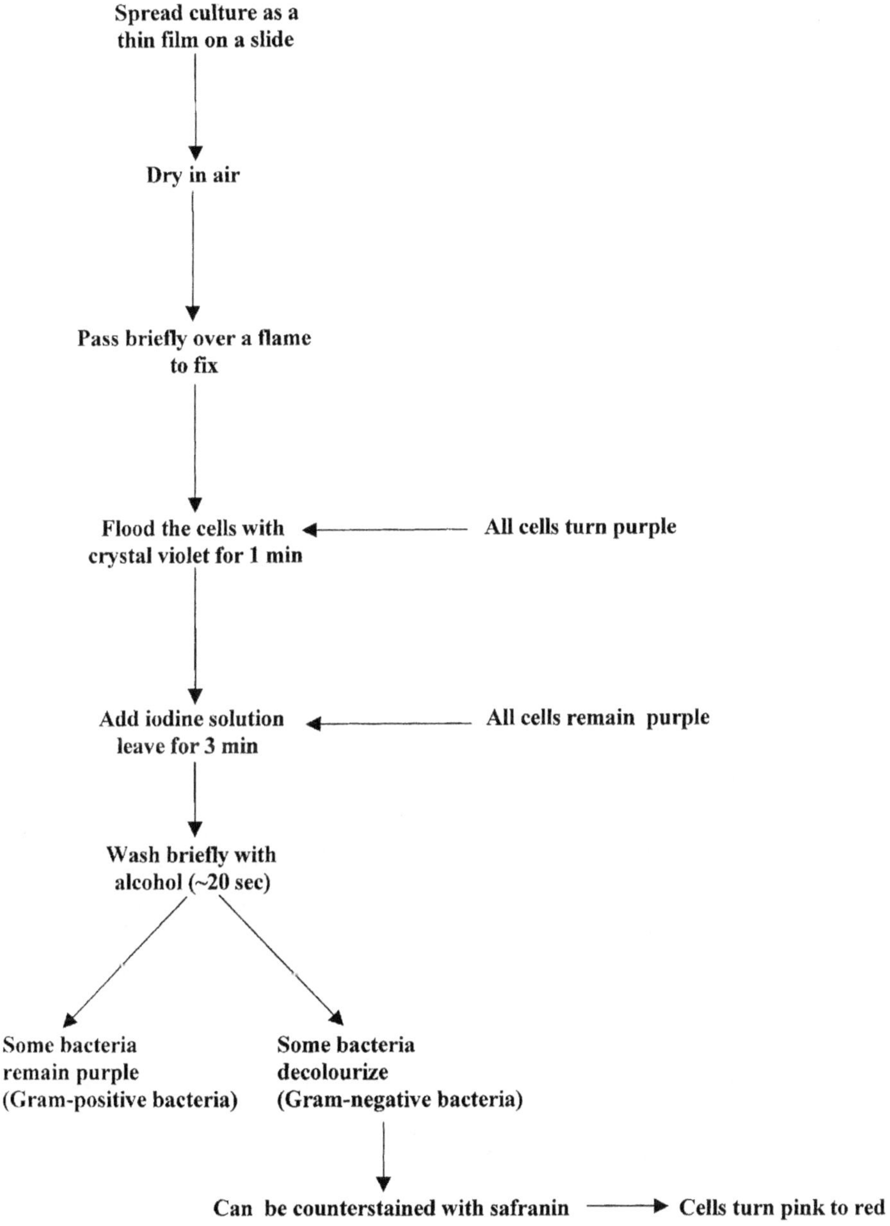

Figure 2.11. Flow chart for Gram staining of bacteria.

Table 2.3. Some differences between Gram$^+$ and Gram$^-$ Bacteria

Gram-positive	*Gram-negative*
1. The cell wall or peptidoglycan or murein is relatively thick.	Peptidoglycan constitutes a thin layer.
2. No cell envelope.	An envelope or outer membrane consisting of LPS, phospholipids and proteins outside the peptidoglycan is present.
3. No distinct periplasmic space.	A distinct periplasmic space separating outer and inner membrane is discernible.
4. Lysine is the third amino acid in tetrapeptide chain.	Lysine is replaced by diaminopimelic acid.
5. The cross linking between tetrapeptide chain involves a pentaglycine bridge.	Cross linkage occurs by direct bonding between diaminopimelic acid of one and terminal alanine of another tetrapeptide chain.
6. Teichoic acids are covalently bonded to the peptidoglycan.	Teichoic acids are absent.
7. Penicillin sensitive.	Generally penicillin insensitive.
8. Cell wall can be easily digested by lysozyme.	Lysozyme reacts only after pretreatment.
9. Thick murein is able to trap the dyes used for staining.	Thin murein when treated with decolourizing agent becomes porous and the stain washes out.
10. After division new cell wall is laid first at the equator and the old wall is subsequently pushed apart.	New cell wall is laid randomly intercalated between the old wall.
11. The basal body of the flagellum has a single set of rings for attachment to cell membrane.	Two sets of rings are employed for attachment to outer membrane and cell membrane.
12. LPS content is virtually none.	LPS content is high (in outer membrane).
13. Lipid & lipoprotein contents are low.	Lipid & lipoprotein contents are high (in outer membrane).
14. Resistance to physical disruption is high.	Resistance to physical disruption is low.
15. Toxins produced primarily are exotoxin.	Toxins produced primarily are endotoxin.
16. Susceptibility to Streptomycin, Chloramphenicol, and Tetracycline is low.	Susceptibility to Streptomycin, Chloramphenicol, and Tetracycline is high.
17. Susceptibility to anionic detergents is high.	Susceptibility to anionic detergents is low.
18. Resistance to sodium azide is high.	Resistance to sodium azide is low.
19. Resistance to drying is high.	Resistance to drying is low.

Staining procedures are also available to reveal the structure of different cellular components. For example, carbol fuchsin and nigrosine can be used for endospore staining, leifson stain for flagella, crystal violet for capsule and feulgen stain for chromatin. Treatment with cationic surface agents like acetyl pyridinium chloride can introduce positive charge on the cell wall. Such cells when stained with a basic dye greatly improves the contrast between cytoplasm and the cell wall. Various other dyes like fluorescent dyes can also be employed for specific purposes.

CHAPTER 3
DOMAIN PROKARYOTA-TAXONOMIC DELINEATIONS

The evolution of prokaryotes is expected to show a relationship with the geochemical development of the earth. Contemporary theories of the origin of life has led to the inference that the first organisms were anaerobic. On the basis of well-preserved fossil microorganisms obtained from shallow marine cherts from the Archaean period, it has been suggested that the ancestor of the bacteria was an anaerobic thermophilic autotroph or phototroph. From about 2.5 billion years ago as oxygen accumulated in the earth's atmosphere, aerobic phenotypes appeared among bacteria.

Our earth is teeming with microorganisms. Arriving first on the face of earth, some 3.5 billion years ago, they monopolized the planet for the next two billion years. Spreading to every of the earth's elements, soil, water, and air, they inhabited areas not normally treaded by others and got into all kinds of associations with other living organisms that evolved much later. All this is reflected in the tremendous diversity in structural and functional organization. Based on this diversity, the task of grouping them into organizational systems becomes a daunting one. Moreover, the lack of substantial morphological information, normally used in such systematics, adds on to the difficulties. Systematics is the comparative study of the diversity of organisms, with the aim of establishing an orderly system within which the organisms can be described and classified.

Taxonomy is the process based on established procedures and rules of describing groups of organisms, their interrelationships, and defining the boundaries between groups of organisms. The taxonomic exercise comprises classification (ordering or placing of organisms into groups based on their relationship), nomenclature (assigning names to the units described in a classification system), and identification (applying the system of classification and nomenclature to assign the proper name to an unknown organism and to place it in its proper position within the classification system). Instead of relying on scanty direct morphological data, bacterial taxonomists have often used characters such as biochemical, physiological, ecological, and genetical and supplemented them with morphological attributes derived from the application of techniques, devised and refined from time to time (Fig 3.1).

3.1. Species Concept

The taxonomic organization of bacteria follow the same hierarchy, as in higher organisms, of species, genus, family, order, division or class, phylum and kingdom. Large variations taking place in a species have even led to the division of species into subspecies or types.

While in higher organisms, members of a species can be readily distinguished as a result of their reproductive barrier, the same is not readily applicable to microbes. The most appropriate description of a bacterial species thus would be: an assemblage of strains showing a high order of phenotypic similarity that can be distinguished from other assemblages by a large number of independent characters. One of the better characterized strain is known as the type strain.

3.2. Identification

Once a microorganism is isolated, certain parameters are employed to decide about its

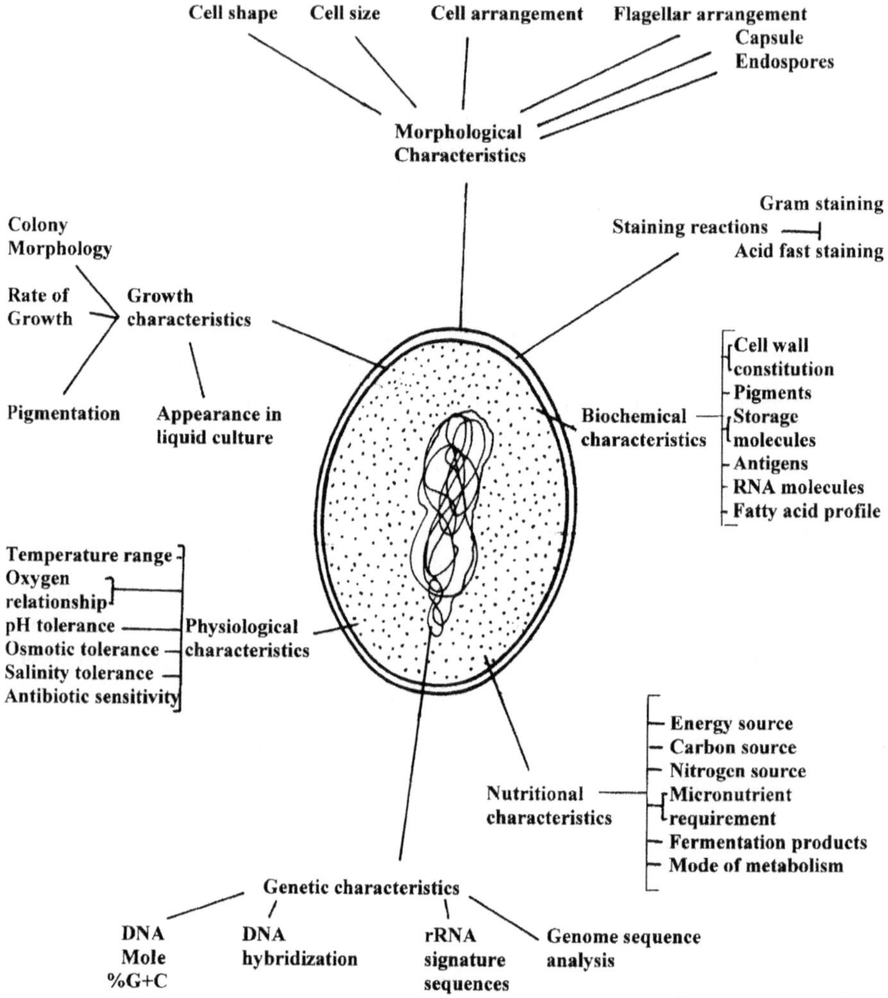

Figure 3.1. Some parameters used for bacterial classification/identification.

placement in a particular taxonomic group. This exercise is known as identification. The first requirement is the availability of pure cultures. The characteristics to be used for identification are such that clearly distinguish members of one group from another. These features are obviously different from those employed for classification.

3.2.1. Identification Key and Diagnostic Tables

This is the most classical approach and consists of a series of questions to design an identification key. This leads through a classification system to arrive at the identity of an organism. In a dichotomous key, a series of yes-no question is asked and then going through the branches of a flow chart the microorganism is placed in a specified group (see Table 3.1). The dichotomous key is still the most popularly used method of identification.

Table 3.1. Identification of species of the genus Pseudomonas by a dichotomous key

S.No.	*Test*	*Response*	*Key*	*Identification*
1	Oxidase	- +	Move to 2 Move to 4	
2	Lysine	+ -	 Move to 3	 *P. maltophila*
3	Motility	+ -		*P. paucimobilis* *P. malleii*
4.	Fluorescence	+ -	Move to 5 Move to 8	
5.	Pyocyanin & Pyorubin	+ -	 Move to 6	 *P. aeruginosa*
6	Growth at 42°C	+ -	 Move to 7	 *P. aeruginosa*
7	Gelatinase	+ -		*P. fluorescens* *P. putida*
8	Motility	+ -	Move to 9	 Not a pseudomonad
9	Peritrichous Polar flagella	+ +	 Move to 10	Not a pseudomonad
10	Glucose oxidation	+ -	Move to 13 Move to 11	
11	Two or more flagella Single flagellum`	+ +	 Move to 12	 *P. diminuta*
12	PHB	+ -		*P. testosteroni* *P. alcaligenes*
13	Mannose	- +	Move to 14 Move to 16	
14	Ornithine	+ -	 Move to 15	*P. putrefaciens*
15	Mannitol	+ -		*P. acidovorans* *P. pseudoalcaligenes*
16	Arginine	+ -	Move to 17 Move to 19	

S.No.	*Test*	*Response*	*Key*	*Identification*
17	6.5% NaCl	+ -	 Move to 18	*P. mendocina*
18	Gelatinase	+ -		*P. fluorescens* *P. putida*
19	Galactose	- +	Move to 20 Move to 21	
20	Mannose	+ -		*P. maltophila* *P. vesicularis*
21	Lactose	- +	Move to 22 Move to 23	
22	6.5% NaCl	+ -		*P. stutzeri* *P. pickettii*
23	Nitrogen production	+ -	 Move to 24	*P. pseudomallei*
24	Citrate	+ -		*P. cepacia* *P. paucimobilis*

Similarly, diagnostic tables can be designed for identification. In such a table, the characteristics of the taxonomic groups are summarized (Table 3.2).

Computer assisted identification has become very common as large number of characters can be handled and compared rapidly. Also the information generated on an unknown organism can be easily compared to a data bank consisting of information on defined taxa. Such identification system is based on the use of probabilistic identification matrices developed by characterizing large number of strains belonging to each taxonomic group. Sometimes a numerical profile is created from probability matrix and the unknown organism is simply compared with the known ones.

Several commercial systems, such as Enterotube, API 20-E, Minitek, Micro-ID, Enteric Tek, and r/b enteric systems have been developed for rapid identification of a microbial system, specially a clinical isolate. The pattern of test results obtained in these systems is converted to a numerical code that is used to arrive at the identity of the isolate. These tests are not only rapid but also reliable. These may consist of different media, miniature reaction chambers or reagent impregnated paper discs. Many of them have been automated to process and assess the results.

3.2.2. Whole Cell Fatty Acid Profile : FAME Analysis

Another very common method of bacterial identification is based on whole cell fatty acid profile. Most bacteria have a large number of fatty acids, and the combination of these fatty acids is characteristic of each strain. For actual analysis, a bacterial culture is grown under standardized conditions, and fatty acids are extracted from bulk lipids. The fatty acids are then chemically modified to their corresponding methyl esters

(FAME). These volatile derivatives are identified by gas chromatography and a very distinctive and stable fatty acid profile can be established. The fatty acid profile of the unknown organism is then compared with the library of a large number of bacterial strains and a 'match' is selected. The identity (genus, species, and often subspecies) can be obtained based on a similarity index through a comparison chart to the library entry (Fig. 3.2).

Table 3.2. Identification scheme for some Gram-negative bacteria [adapted from Palleroni, N. J. (1984)]

Characteristic	*Response*							
	1	**2**	**3**	**4**	**5**	**6**	**7**	**8**
a. Rods	+	+	+	D	D	+	+	D
b. Muramic acid in peptidoglycan	+	+	+	+	—	+	+	+
c. Motility	+	D	D	D	D	D	+	—
d. Catalase	+	+	—	—	+	+	+	D
e. Oxidase	D	D	—	—	+	—	D	D
f. Some sugars utilized	+	+	+	—	D	+	—	D
g. Pathogenic to man	D	—	—	—	—	—	+	D
h. Pathogenic to plants	D	—	D	—	—	D	—	—
i. Marine organisms	D	—	—	—	—	—	—	—
j. Yellow colonies	D	D	—	—	—	—	—	D
k. Water soluble fluorescent pigment	D	D	—	—	—	—	—	—

N_2 fixing

Use methane as C-source

Require 12-15% NaCl for growth

Ethanol oxidized to acetic acid

Cysteine + iron required for growth

(1) Pseudomonadaceae
(2) Azotobacteraceae
(3) Rhizobiaceae
(4) Methylococcaceae
(5) Halobacteriaceae
(6) Acetobacteriaceae
(7) Legionellaceae
(8) Neisseriaceae
D= may deviate

3.3. Nomenclature

If a previously undescribed organism is identified, certain principles of nomenclature

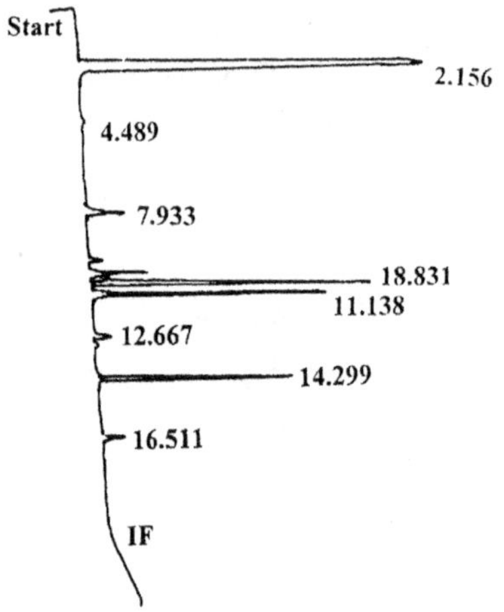

ISBR [Rev 3.30]	Pseudomonas................................0.901
	P. pickettii....................................0.901
	P. solanacearum 0.525

Comparison with ISBR [Rev 3.30]: Pseudononas pickettii Distance: 1.474

```
                       0  5  10  15  20  25  30  35  40  45  50  55  60  65  70  80  85  95  100
                       .  .  .   .   .   .   .   .   .   .   .   .   .   .   .   .   .   .   .
14:0                   .  x-  .  .   .   .   .   .   .   .   .   .   .   .   .   .   .   .   .
14:0 2OH              -0  .  .   .   .   .   .   .   .   .   .   .   .   .   .   .   .   .   .
16:1 CIS 9             .  .  .   .   .   .   .-x+---     .   .   .   .   .   .   .   .   .   .
16:0                   .  .  .   .   .   .x-- --  .   .   .   .   .   .   .   .   .   .   .
17:0 CYCLO            -+x- .  .   .   .   .   .   .   .   .   .   .   .   .   .   .   .   .
16:1 2OH              -x- .  .   .   .   .   .   .   .   .   .   .   .   .   .   .   .   .
18:0                   x . .  .   .   .   .   .   .   .   .   .   .   .   .   .   .   .   .
18:1 2OH               .  +x .   .   .   .   .   .   .   .   .   .   .   .   .   .   .   .
SUMMED FEATURE  3      .  . -x-  .   .   .   .   .   .   .   .   .   .   .   .   .   .   .
SUMMED FEATURE 7       .  .  .  .------+-x---  .   .   .   .   .   .   .   .   .   .   .   .
```

Figure 3.2. Gas chromatographic profile of fatty acids and the identification of a bacterial strain as Pseudomonas pickettii.

are applied to name the isolate. The nomenclature is based on a binomial system, as with other living organisms, in which the organism is identified by a genus and species name. The rules of nomenclature established by international committees ensure that the name of the microorganism specifically and unambiguously designates the said organism. A variety of criteria are used in assigning a name. For example, a generic name may represent an unique feature or function or the name of the scientist who isolated and described the organism. The species epithet then reveals either an additional important feature of the genus or the one which may modify the generic name meaningfully (Table 3.3).

Table 3.3. Characteristic features from which certain bacterial and archaeal names have been derived

Binomial name	*Characteristic features*
Aeromonas	Gas producing
Aquifex	Water maker
A. pyrophilus	Extreme thermophile
Bacillus	Rod-shaped
B. subtilis	Slender
B. cereus	Wax coloured
B. albus	White
Beijerinckia	Named in the honour of M. Beijerinck
Clostridium	Spindle-shaped
C. botulinum	Sausage-shaped
C. perfringens	Breaking through
Escherichia	Named in the honour of T. Escherich
E. coli	Lives in colon
Lactobacillus	Grows in milk, rod-shaped cells
Methanobacterium	Produces methane
Micrococcus	Small spheres
Photobacterium	Emitting light
Pseudomonas	False (pseudo), an unit (monod)
P. fluorescens	Producing a fluorescent pigment
P. marina	Grows in oceans
Serratia	Named in honour of Serrati
S. marcescens	Produces decaying
Staphylococcus	Irregularly clustered spheres
S. aureus	Produces golden pigment
Streptococcus	Flexible chains of spheres
S. pyogenes	Produces pus
S. equi	Causes disease in horses
Thermus aquaticus	Grows in hot aquatic environment

3.4. Systems of Classification

Inspite of the difficulties faced in describing bacteria phenotypically, taxonomy has always attracted bacteriologists. A number of different approaches have been employed to develop a classification system. Starting with the classical morphological approach, bacteriologists subsequently emphasized on physiological properties. Both are also considered in modern systems but the emphasis has shifted to molecular similarities or genetic relatedness. Thus, these systems could be classified as phenetic (assessing similarity) or phylogenetic (assessing evolutionary relationships).

As early as 1866, Haeckel coined the term protista to place bacteria differently from plants and animals. Stanier *et al.* (1957) further divided the protista into lower protists containing blue-green algae and bacteria and higher protists represented by protozoa, fungi and other algae.

Dougherty (1957) classified living organisms into two broad groups, procaryota, and eucaryota, on the basis of absence or presence of a well-defined nucleus (nuclear membrane) and placed bacteria along with blue-green algae and viruses in the former group. These terms are still being used to define the living organisms on the same basis.

In 1969, Whittaker proposed a five kingdom classification and raised a new kingdom Monera to accommodate blue-green algae and bacteria.

3.5. Bergey's Manuals

(i) Of Systematic Bacteriology

(ii) Of Determinative Bacteriology

Many systems of classification were proposed till Bergey in his Manual of Determinative Bacteriology (1923-1957) divided the plant kingdom into: Prosophyta, Thallophyta, Bryophyta, Pteridophyta, and Spermatophyta. Prosophyta contained primitive plants like bacteria and viruses. This manual has been updated a number of times to include more and more, new meaningful information as it becomes available. The eighth edition published in 1974 consisted of 19 major groups which were further divided into systematic units of orders, families, tribes, genera, and species. The edition that came out in 1994, described 35 groups including 5 groups of Archaeobacteria (archaea). In 1984, the Bergey's Trust published a four volume treatise, the Bergey's Manual of Systematic Bacteriology. These manuals are considered a recognized authority on bacterial taxonomy. This compilation comprising both classical and molecular information on all recognized species of prokaryotes is useful for identification purposes and for taxonomic descriptions.

Volume I- Deals with Gram-negative bacteria of medical and industrial importance.This volume consists of 11 sections which describe spirochetes, spiral and curved bacteria, Gram-negative aerobic and facultatively aerobic rods, Gram-negative obligate anaerobic cocci, sulphate and sulphur reducing bacteria, rickettsias and chlamydias, and mycoplasmas.

Volume II- Includes Gram-positive bacteria of medical and industrial importance. This volume contains Gram-positive cocci, Gram-positive endospore-forming and non-sporing rods, mycobacteria and non-filamentous actinomycetes. These are grouped in six sections.

Volume III- Deals with the diverse group of remaining Gram-negative bacteria and the archaea. This volume describes, phototrophic, gliding, sheathed, budding and appendaged bacteria, cyanobacteria, chemolithotrophic bacteria, methanogens, extremophiles and other archaea. These are divided into eight sections.

Volume IV- This principally deals with filamentous actinomycetes and related bacteria. This volume describes 8 sections showing the diversity of actinomycetes.

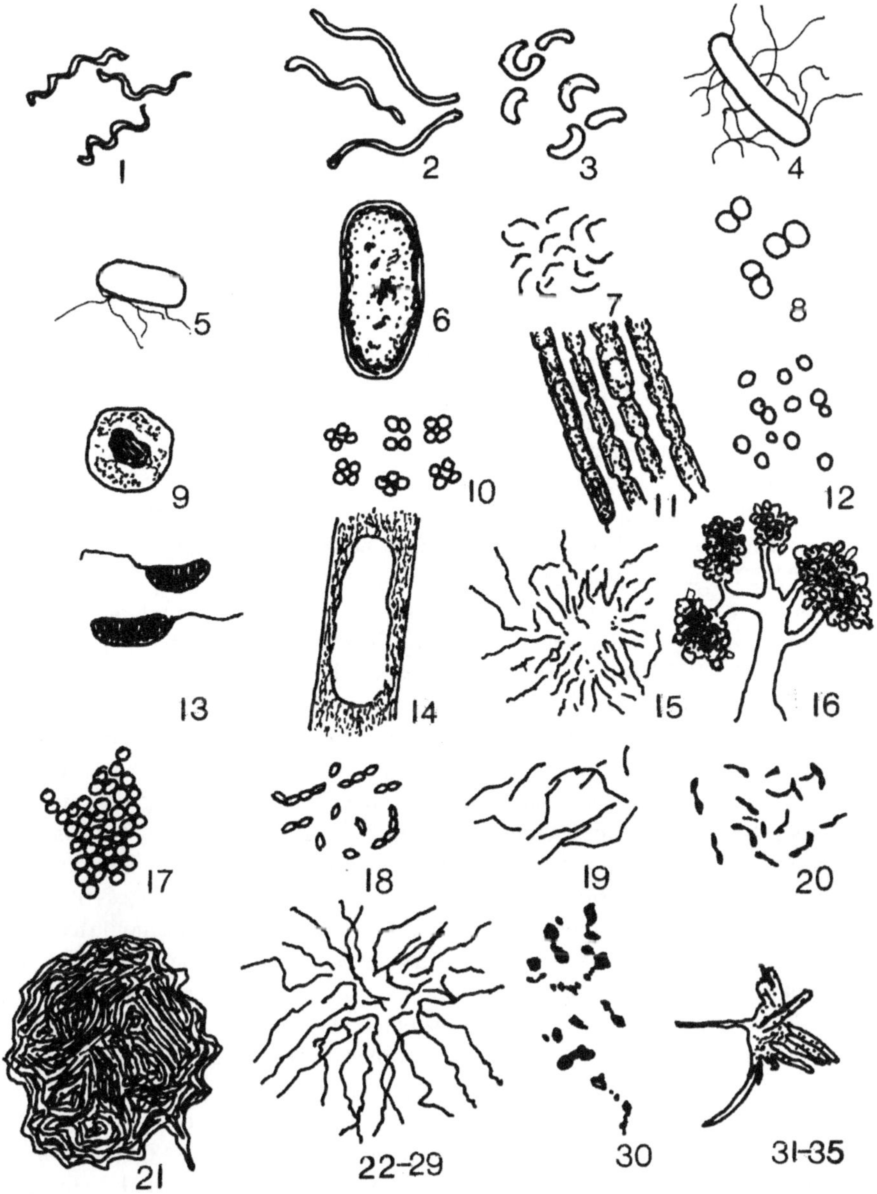

Figure 3.3. Representative bacterial genera (not drawn to scale). The figure depicts some special features. The number represents the group to which they belong.

Some of the characteristic features of the members belonging to 35 groups (Fig. 3.3) along with few examples are described below:

Group 1- Spirochetes- *Spirochaeta, Borrelia, Leptospira-* These bacteria possess unique morphology, as the cells are typically slender, flexuous and helical (coiled) often very long. They are widely spread in aquatic environments and body of animals and humans. Some are strongly pathogenic. The cell consists of an axial fibril which is attached to the cell poles and is wrapped around the protoplasmic cylinder. The protoplasmic cylinder consists of regions enclosed by the cytoplasmic membrane and the cell wall. The two are surrounded by a triple layered membrane also known as external sheath. The fibrils are used for special spinning or whirling movement. The members may be obligate anaerobic, to facultatively aerobic, microaerophilic or even strongly aerobic. *Treponema pallidum* causes syphilis in humans, species of *Leptospira,* leptospirosis in domestic animals and humans and *Borrelia* is the causative agent of relapsing fever and lyme disease transmitted through insect vectors. They may be found in fresh, marine or polluted aquatic environments, and some are associated with molluscs.

Group 2- Aerobic/Microaerophilic, Motile, Helical/Vibroid Gram-Negative Bacteria or Spirilla- *Azospirillum, Aquaspirillum, Bdellovibrio-* These are Gram-negative, motile, spiral-shaped bacteria with a high level of physiological versatility. They possess single or multiple polar flagella, and may be obligately aerobic or microaerophilic to facultatively aerobic. Both symbiotic as well as plant and animal pathogenic members are known. Some members have nitrogen-fixing ability, ability to utilize methane as electron donor, halophilic, or thermophilic and some may be luminiscent. *Azospirillum* inhabiting the rhizosphere of cereals is an important nitrogen fixer. *Aquaspirillum* demonstrating a dramatic directed movement in a magnetic field (magnetotactic) also has a curved-rod cell shape. The genus *Bdellovibrio* consists of bacterial predators.

Group 3- Non-motile (or rarely motile), Gram-Negative Curved Bacteria- *Spirosoma-*These are closely related in cell shape to the members of group 2 but form rings of curved rods such as in *Spirosoma.* The members, however, are non-motile.

Group 4- Gram-Negative Aerobic Rods and Cocci- *Pseudomonas, Xanthomonas, Azotobacter, Rhizobium, Agrobacterium, Methylococcus, Halobacterium, Acetobacter, Legionella, Neisseria, Thermus, Alcaligenes, Bordetella-* This group consists of diverse taxonomic members. Pseudomonads are straight or curved rods that are motile by means of polar flagella and are non-fermentative in nature. Some pseudomonads can, however, carry out anaerobic respiration such as forming molecular nitrogen through a process called denitrification. They are widely distributed in soil and aquatic habitats and are known by their nutritional versatility in being capable of degrading several natural and synthetic organic compounds. They may occur as free-living or in association or even as pathogens of plants and animals. *Xanthomonas* is another plant pathogenic member of this group. This group also consists of nitrogen-fixing members both free-living such as *Azotobacter,* as well as obligately symbiotic such as root nodule-forming, *Rhizobium.* Another important genus is *Agrobacterium*, which produces tumorous growth on infected plants called galls. This bacterium plays a significant role in plant genetic engineering research.

Utilization of 1-carbon-containing (C1) compounds such as CO, methane and methanol, is the special character of some members. This reaction can be industrially exploited as microorganisms can be grown on methane or methanol as source of proteins.

Legionella and *Neisseria* are proven human pathogens. *N. gonorrhoeae* causes gonorrhoea and *N. meningitidis* is the causative agent of meningitis. *Bordetella pertussis* is responsible for whooping cough. Several other genera of uncertain affiliation such as *Alcaligenes, Acetobacter,* and *Thermus* are included in this group. While *Alcaligenes* is the hydrogen-oxidizing, carbon-dioxide-fixing bacterium, *Thermus* is known because of its growth at temperatures over 70°C and for thermostable enzymes such as *Taq* polymerase.

Group 5- Gram-Negative Facultatively Anaerobic Rods- *Escherichia, Salmonella, Shigella, Klebsiella, Enterobacter, Erwinia, Vibrio, Pasteurella, Yersinia, Zymomonas, Proteus-* Two major families of this group are the Enterobacteriaceae and Vibrionaceae. In the former, the motile members possess peritrichous flagella whereas in the latter the arrangement is polar type. The family Enterobacteriaceae consists of members which are important for several reasons. Members of *Escherichia* inhabit human intestinal tract and one, *E. coli,* occupies a prized position as being the organism of choice for various bacterial physiology, genetics, and molecular genetics studies.

Many species of *Salmonella* and *Shigella* are important human pathogens causing typhoid and other gastrointestinal upsets and bacterial dysentery, respectively. All members of the genus *Erwinia* are plant pathogens.

Vibrio species are generally aquatic consisting of curved rod-shaped cells. One species, *V. cholerae* causes cholera. *Photomicrobium* possesses luciferase that through an ATP-driven reaction creates bioluminescence. The association of some of these bacteria with certain fishes provides them the nomenclature of flash light fish.

Group 6- Gram-Negative Aerobic Rods- *Bacteroides, Selenomonas-* The genus *Bacteroides* is an important member of the normal microbiota of humans and some of them are pathogenic. Many genera occur in the rumen of cattles and other animals where they play an important role in the digestion of cellulosic materials and nutrition of the animal.

Group 7- Dissimilatory Sulphate or Sulphur-Reducing Bacteria- *Desulfobacter, Desulfovibrio-* Members are characterized by their ability to reduce sulphate, or other oxidized sulphur-containing compounds to H_2S. *Desulphomonas* can even use elemental sulphur as the electron acceptor. They constitute important members of biogeochemical cycling of sulphur. The cells are straight to curved rods to even ellipsoidal to spherical.

Group 8- Gram-Negative Anaerobic Cocci- *Veillionella-* Members contain very few species, comprising a part of the normal human microbiota specially in oral cavity. Some are pathogens causing infections in the oral cavity, intestinal and respiratory tracts. Spherical cells typically occur in pairs (diplococci).

Group 9- Rickettsias and Chlamydias- *Rickettsia, Anaplasma, Chlamydia-* Members of both the groups are intracellular obligate parasites. Having limited metabolic capability to generate ATP, they are a drain on the ATP pool of hosts. Both are very tiny in size and chlamydias are known to exist into small, rigid-walled, infectious as well as large thin-walled non-infectious forms. Many members cause disease in humans and other animals.

Phototrophic Bacteria- These bacteria are able to use light energy in order to produce ATP. Most members are autotrophs fixing CO_2. They have been divided on the basis of whether they liberate oxygen (oxygenic) or do not liberate O_2 (anoxygenic) during photosynthesis.

Group 10- Anoxygenic Phototrophic Bacteria- Purple bacteria- *Chromatium, Thiocapsa;* Green bacteria-Green sulphur bacteria- *Chlorobium*, and multicellular filamentous green non-sulphur bacteria- *Chloroflexus, Heliobacterium-* As described above, they do not use water as an electron donor but sources of reduced sulphur or organic compounds act as electron donors for the generation of reduced coenzymes. Their photosynthesis is thus anaerobic. These include purple sulphur and purple non-sulphur bacteria and green sulphur and green flexibacteria. While the former two utilize reduced sulphur compounds such as H_2S as electron donor, in the latter two, organic compounds serve the same purpose. The members are characterized by the possession of bacteriochlorophyll a and b and carotenoids that impart characteristic colour.

Group 11- Oxygenic Phototrophic Bacteria- The members have been further divided into: Cyanobacteria- *Gloeothece, Spirulina, Lyngbya, Anabaena, Nostoc, Calothrix;* Prochlorales- *Prochloron-* Like higher plants, the members of this group, split water to form oxygen. They occupy a position intermediate to phototrophic bacteria and green algae providing a probable evolutionary link. Some cyanobacteria do show anoxygenic photosynthesis. They contain chlorophyll a and Prochlorales members also have chlorophyll b, with auxiliary photosynthetic pigments in the form of phycobilliproteins and ß carotenes. Cyanobacteria consist of Gram-negative unicellular rods or cocci or filamentous structures, composed of vegetative cells known as trichomes. The latter forms may possess or lack heterocysts. Many members possess nitrogen fixing capability. *Prochloron* occurs as single-celled extracellular symbiont of marine invertebrates.

Group 12- Chemolithotrophic Bacteria- This is a diverse group consisting of: A. Nitrifying bacteria- *Nitrobacter, Nitrosomonas*; B. Colourless sulphur bacteria- *Thiobacillus, Acidiphilium*; C. Obligate hydrogen oxidizers- *Hydrogenobacter*; D. Iron and manganese oxidizing and/or depositing bacteria- *Siderocapsa*; and E. Magnetotactic bacteria, *Aquaspirillum-* These bacteria possess diverse metabolic activities. The common feature between them is the oxidation of inorganic compounds to generate ATP. The nitrifying bacteria consist of two classes. One class oxidizes nitrite to nitrate (names with the prefix Nitro-) and another ammonia to nitrite (names with the prefix Nitroso-). The combined action of these bacteria has a major influence on soil fertility. Several genera can metabolize sulphur and sulphur-containing inorganic compounds. Some members of genus *Thiobacillus* oxidize only sulphur compounds whereas others also oxidize ferrous to ferric iron. This metabolic activity leads to mobilization of various metals so that they can be extracted from even low-grade ores. Other members oxidize iron or manganese, depositing iron and/or manganese oxides in capsules or extracellular material. Members of this group have important functions in several biogeochemical cycles.

Group 13- Budding and/or Appendaged Bacteria- The members have been further distinguished into A. Prosthecate bacteria - Budding type - *Hyphomicrobium*; non-budding- *Caulobacter*; B. Nonprosthecate bacteria- *Planctomyces*, non-budding- *Gallionella*; C. Morphologically unusual budding bacteria- involved in iron and manganese deposition- *Metallogenium*; and D. Spinate bacteria- This is a heterogenous group with members sharing the common feature of possessing extensions or protrusions from the cell. The appendages called prostheacae may have in some members a reproductive function and in others a physiological purpose. Some members reproduce by budding and others by binary fission. Some members use these appendages to efficiently concentrate the available nutrients. The appendages in *Caulobacter* referred to as stalks provide a holdfast by which the cell can attach to a substratum. The arrangement of appendages provides a characteristic shape such as star-shaped structure.

Group 14- Sheathed Bacteria-*Leptothrix*- In these bacteria, the cells are contained within filamentous structures called sheaths which enable them to attach to a soiled surface. Sometimes the sheath is encrusted with metal oxides. Sheath may provide protection against predators and parasites.

Group 15-Nonphotosynthetic, Non-fruiting, Gliding Bacteria- *Cytophaga, Lysobacter, Beggiatoa, Leucothrix, Achronema*- These bacteria are highly diverse but all exhibit gliding motion. Some forms like *Beggiatoa* may form filaments, oxidize H_2S and deposit sulphur intracellularly. *Cytophaga* sp. on the other hand, do not form filaments but can hydrolyze agar, cellulose and chitin. Their ecological importance has been realized in the decomposition of the organic matter.

Group 16- Fruiting, Gliding Bacteria- *Myxococcus, Polyangium*- These are another group of bacteria exhibiting gliding motion. They along with the non-fruiting group cover the entire range of 30-70 mole % G+C content. The small rod-like cells are embedded in slime aggregate to form fruiting bodies. These can be seen with naked eyes as brightly coloured slimy growth on decaying plant material, bark of living trees, or on animal dung. These bacteria produce various hydrolytic enzymes such as cellulases.

Group 17- Gram-Positive Cocci- *Micrococcus, Staphylococcus, Streptococcus, Enterococcus*- The coccoid cells may occur singly, in pairs, chains, or regular or irregular clusters. Members of *Staphylococcus* are found on skin surfaces and some are capable of causing wound infection and food poisoning. Species of *Streptococcus* may cause dental caries and some may cause pneumonia.

Group 18- Gram-Positive Endospore-forming Rods and Cocci- *Bacillus, Clostridium*- Members of this group produce heat-resistant endospores. Most endospore forming bacteria belong to Gram-positive rods (*Desulphotomaculum* is the only Gram-negative endospore former reported so far). The members of *Bacillus* species are strict aerobes or facultative anaerobes but those belonging to *Clostridium* are obligate anaerobes. Several members of the two genera are capable of food-spoilage. *C. botulinum* causes botulism, *C. tetani*, tetanus and *C. perfringens*, gas gangrene.

Group 19- Regular, Non-sporulating Gram-Positive Rods- *Lactobacillus, Listeria*-One important member of this group is *Lactobacillus* which produces lactic acid as the major fermentation product and is extremely important in dairy industry. Some members are also found inhabiting human oral cavity, vaginal tract, and intestine.

Several species of *Listeria* are animal pathogens and are also known to be responsible for some human food-borne infections.

Group 20- Irregular Non-spore forming Gram-Positive Rods- *Corynebacterium, Arthrobacter, Actinomyces*- This group also known as coryneform bacteria exhibit characteristic pleomorphic morphology. This is brought about by the continued association of the cells after divisions. Many species are plant or animal pathogens. For example, *Corynebacterium diphtheriae* causes diphtheria (only when it itself is infected by a specific temperate phage). *Arthrobacter* changes from rod-shaped to coccoid cells during growth.

Group 21- Mycobacteria- *Mycobacterium*- These are slow-growing, aerobic bacteria which are highly pleomorphic. They contain waxy mycolic acid in their cell walls and some members produce a pigment. Several mycobacteria are important human pathogens. For example, *Mycobacterium tuberculosis* causes tuberculosis and *M. leprae*, leprosy.

Group 22-29- Actinomycetes- *Nocardia, Rhodococcus, Frankia, Actinoplanes, Streptomyces, Actinomadura, Thermomonospora, Glycomyces*- The assemblage of members belonging to **group 22 to 29** are all placed in actinomycetes. These are Gram-positive bacteria having high mole % G+C. They produce branched mycelia that superficially resemble the mycelia of filamentous fungi but the cells are distinctly prokaryotic. **Nocardioform- (Group 22)**- This is a group of actinomycetes that typically produce brightly coloured fungal-like filament that often fragment into non-motile cells. Most members are saprophytes or facultative parasites with *Frankia* species forming symbiotic N_2 fixing root nodules of non-leguminous plants (**Group 23**). The spore may (**Group 24**) or may not be borne in sporangia. Under latter conditions spore may be formed singly or in short or long chains (**Group 25**). The members of **group 26** contain mesodiaminopimelic acid and madurose and some are brightly pigmented. They may bear short chains of aerial spores or spores inside sporangia. In some, the mycelium divides transversely to form motile cocci (**Group 27-29**). The main ecological role of actinomycetes is in the decomposition of organic matter in soil, although some may be found associated with humans and other animals (**Group 26**). Some saprophytic members are thermophilic (**Group 27-28**) with members of latter group forming endospores. Gram-positive bacteria forming branching filaments and bearing chains of aerial spores are placed in actinomycetes (**Group 29**). Many species of *Streptomyces* are important sources of antibiotics. Nearly 75% of the antibiotics is contributed by the members of this genus.

Group 30- Mycoplasma- *Mycoplasma, Mycoplasma*-like-organisms of plants and invertebrates- Endosymbionts of Protozoa, insects, invertebrates other than Arthropods and Fungi. *Hotospora, Caedibacter,* and number of others, which have not been placed in known functional groups- Mycoplasmas are cell-wall less bacteria bounded by a single triple layered membrane. This endows them with a soft skin phenotype called as mollicutes. These are small organisms capable of self reproduction. Most members require sterol for growth and produce characteristic fried-egg colonies on artificial media. Several of them are human pathogens causing some forms of pneumonia.

Several bacteria live as obligate endosymbionts of invertebrates without adversely affecting the host. They have been described from protozoa, insects and various other invertebrates. Mycoplasma-like organisms have also been found in plants. Although readily seen with the host, they have not been cultured in artificial media and thus are poorly understood. This has also been the reason why they have not been placed in defined functional groups with other bacteria.

Group 31-35- Archaebacteria (= formerly known as Archaeobacteria)- Members of archaebacteria are phylogenetically related to each other on the basis of the analysis of their 16S rRNA. On the same basis they are so different from eubacteria and eukaryotes that they have been placed in a separate kingdom, archaea. Unlike other prokaryotes, they lack murein in their cell wall and possess an unusual ether linkage in their phospholipid molecule. They are also known to inhabit extreme environments and show characteristic adaptation their of. Members have been grouped in five groups. One group (**Group 31**) consists of methane producing bacteria or methanogens. Generally H_2 serves as an electron donor and CO_2 as the electron acceptor but some can utilize formate or acetate as substrate for generating ATP. The methanogens often form consortia in association with other microorganisms which create conditions conducive for methanogens, such as the rumen of cattles.

These constitute important members of biogas production. The archael sulphate reducers (**Group 32**) have a requirement for salt and high temperatures and have been isolated from anaerobic submarine hydrothermal regions. SO_4^{2-}, SO_3^{2-} or $S_2O_3^{2-}$ (thiosulphate) is reduced to H_2S by them. Extremely salt loving obligate halophiles (**Group 33**) grow on a medium containing at least 15% NaCl.They may be divided into alkaliphilic (growing at pH around 10) and neutrophilic (nonalkaliphilic) members. The majority of members are purple due to the presence of intracellular bacteriorhodopsin or red, pink, orange due to carotenoid pigments. They are normally aerobic but can grow anaerobically in light. The latter property is due to bacteriorhodopsin driven ATP synthesis under anaerobic conditions. Some acidophilic, thermophilic members (**Group 34**) lack a cell wall. Thus, they may vary in size and shape. They grow both aerobically and anaerobically and under latter conditions sulphur is reduced to H_2S.

Another branch consists of members that are extremely thermophilic (**Group 35**) and convert elemental sulphur to H_2S using organic compounds or H_2 as electron donors. Some members are extreme acidophiles and prefer to grow between pH 2-3. Members could be aerobes to facultative aerobes or even anaerobes. *Sulpholobus*, a strict aerobe is found in sulphur-rich hot acidic springs in volcanic areas. It oxidizes S° to H_2SO_4. Some anaerobic genera found in marine and terrestrial sulphur-rich regions, are extreme thermophiles and convert S° to H_2S. One member grows at 105°C, clearly the upper temperature limit for life.

Even though this classification system is primarily based on morphological and physiological characteristics, Bergey's Manual is a recognized authority on bacterial taxonomy. The conventional or classical approach to taxonomy has undergone several changes over the years to make it more comprehensive and phylogenetic. One such approach suggested by Michael Adanson is known as Numerical or Adansonian taxonomy.

The latest edition of Bergey's Manual spread over 5 volumes came out in 2000 and will appear over next several years. This edition incorporates rRNA sequencing data (see later section) along with scores of conventional taxonomic information. The

Prokaryotes have been divided into two kingdoms, instead of domains: I- Archaeota, and II- Bacteria. Rest of the classification follows the same hierarchy of phyla, section, class, order (sometimes also suborder), family, genus, and species.

Volume I- The Archaea, Cyanobacteria, Phototrophs, and Deeply Branching Genera: Phyla I-VI; Sections I-XIV.

Volume II- The Proteobacteria: Phylum VII; Sections XV-XIX.

VolumeIII- The low G+C Gram-positives: Phylum VIII; Sections XX-XXII.

Volume IV- The high G+C Gram-positives: Phylum VIII Contd; Section XXIII.

Volume V- The Planctomyces, Spirochaetes, Fibrobacters, Bacteroides and Fusobacteria: Phyla IX-XVI; Sections XXIV-XXXI.

The multivolume treatise, The Prokaryotes is currently being updated and will soon become available on line.

3.6. Numerical Taxonomy

This approach relies on overall similarity using a large number of characteristics. Thus equal weightage is given to each characteristic and similarities and differences among organisms are assessed. These measures express indices of similarity or similarity coefficients. The most common of such indices of similarity are the simple matching (S_{SM}) and Jaccard (S_J) coefficients. The simple matching coefficient is based on all measured characteristics, positive as well as negative. The Jaccard coefficient, on the other hand, does not take into consideration the features for which the organisms in question are negative. They are represented as follows:

$$S_{SM} = \frac{(++)+(--)}{(++)+(--)+(+-)+(-+)}$$

S_{SM} = Simple matching coefficient
++ = Positive matches
- - = Negative matches
(+ -) / (- +) = Mismatches

$$S_J = \frac{(++)}{(++)+(+-)+(-+)}$$

S_J = Jaccard coefficient
++ = Positive matches
(+ -)/ (- +) = Mismatches

The classification based on numerical taxonomy is graphically represented by a similarity matrix or dendrogram (Fig. 3.4). In this, organisms with high similarity are grouped in close geometric proximity, whereas those with low similarity are distanced. Until recently, numerical taxonomy was considered impractical and time consuming as a wide-range of characters need to be studied. With the advent of computer this problem has been largely solved. A computer can be programmed to compare a large number of

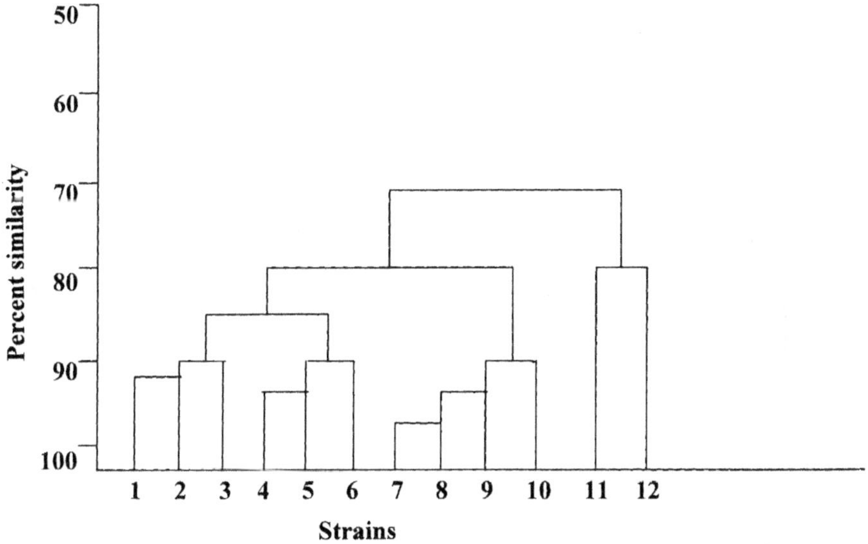

Figure 3.4. A dendrogram showing hierarchial relationships between twelve strains. The horizontal line expresses the level of similarity and vertical lines depict the individual strains. Strains showing highest similarity are grouped together.

characters and arrive at the degree of similarity and relationship.

3.7. Genetic Analysis

In order to derive phylogenetic relationships, genetic analysis can provide significant information. Various such analyses are known and are employed in bacterial taxonomy.

3.7.1. Base composition of DNA

The genetic material of bacteria like other living organisms is composed of DNA. Determination of the relative proportion of guanine (G) and cytosine (C) compared to the total number of nucleotide base pairs in the DNA is one parameter that can give some idea on the microbial genome. One should remember that specifying G + C also tells about adenine + thymine (A+T) content. The composition of DNA is normally described as mole % G + C.

When DNA is slowly heated it melts, that is the double stranded DNA is converted into single strands or denatured. The single strands absorb 40% more UV than when the same strands form a double helix. This change in UV absorption is known as

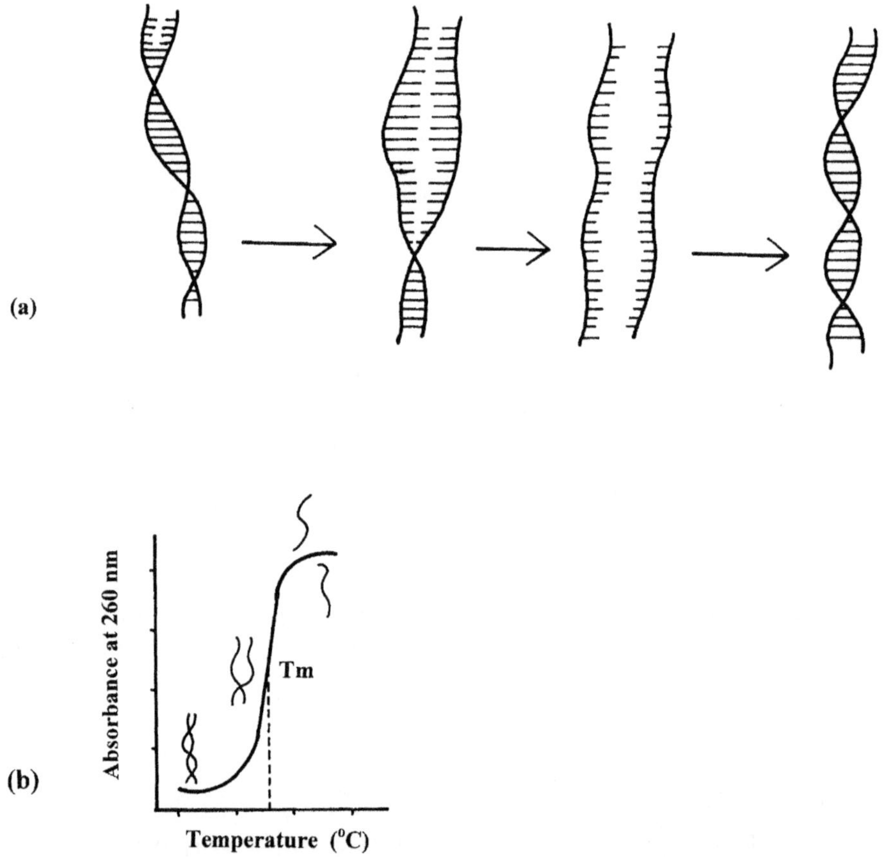

DNA Melting and its Relationship with the GC Content of the DNA Sample

Figure 3.5a,b. DNA melting and its relationship with the GC content of a DNA sample. a. When a double stranded DNA is heated, the two strands start separating (melting) due to break-down of the hydrogen bonds, holding the bases in two strands, producing single strands. If such a sample is slowly cooled the hydrogen bonds are re-formed generating a double helical structure. This process is generally referred to as denaturation and renaturation, respectively, b. The double stranded DNA absorbs less UV as compared to single stranded DNA. If the UV absorbance is monitored over a temperature range, the mid-point temperature Tm can be estimated. This is the temperature at which strand separation occurs and is a function of the GC content of the DNA.

hyperchromatic shift and the mid-point temperature of the denaturation curve is known as the melting temperature (Tm) of the DNA (Fig. 3.5).

The Tm of a DNA depends heavily on base composition thus, the DNA with a high G + C have a higher Tm than those rich in A + T. The empirical relationship between Tm and mole % G + C is given by the equation: Tm=0.41% (mole % G+C)+16.6 log M + 81.5, where M is the concentration of monovalent cations in the range of 0.0001-0.2 molar.

The mole % G + C can also be determined by measuring the buoyant density (s) of the DNA. The density is determined by using CsCl density gradient ultracentrifugation. The higher the mole % G + C, the greater is the density (molecular weight of G + C is more than A+ T). The relationship is: s =1.66 + 0.00098 (mole % G + C).

The mean DNA base composition of the nuclear DNA in plants and animals is very similar (35-40 mole % G + C). Among the protists this range is wider and the widest range is seen amongst bacteria (Table 3.4). A close similarity in the mean base composition does not always suggest genetic similarity (e.g., plants and animals). A high divergence, however, does get reflected in a number of other differences. Microorganisms showing phenotypic similarities but differing in their G + C content by 2-3% represent different species. For example, gliding bacteria, cytophagas, and myxobacteria share the motility mechanisms and many other characters but their G + C content is highly diverse.

3.7.2. Nucleic Acid Hybridization

Another way of looking at genetic similarities is to carry out nucleic acid hybridization. When the heat denatured DNA is slowly cooled the two strands reassociate (annealing or renaturation) to form original double stranded DNA. This is because the complementary bases in the two strands rejoin by employing the H-bonds. As soon as this technique became available, it became clear that this could be applied to the DNA preparations from two related organisms. Thus, such DNAs should be able to form hybrid molecules under these conditions. The extent of hybridization will be determined by the degree of relatedness. In order to differentiate the two DNAs, one of them could either be labelled with heavy D_2O or by a radioisotope. A hybrid molecule can then be ascertained either by CsCl density gradient centrifugation or amount of associated radioactivity. This technique is generally used in conjunction with a reference homologous pairing. The similarities shown by heterologous molecule can then be quantified in comparison to the reference. Many a times, binding of mRNA to DNA can also provide similar information as it is complementary to one of the DNA strands.

DNA-DNA or DNA-mRNA binding of *Salmonella typhimurium* and *E. coli* to *Aerobacter aerogenes* is about 50% of that of homologous reference of *A. aerogenes*. However, that of *Pseudomonas aeruginosa* is only 1-2%. It can thus be inferred that *S. typhimurium* and *E. coli* are relatively closer to *A. aerogenes* and *P. aeruginosa* is unrelated to all of them.

The DNA-DNA homology can now be directly compared by doing genome sequencing. The genome sequence of a number of microorganisms is now available and several others are nearing completion (**refer Chapter 4**). With the advent of automated nucleotide sequencers, this technique has become further simplified.

Table 3.4. DNA base composition in terms of mole % G+C ($=\frac{G+C}{A+T+G+C}$ *x100) of some bacteria and archaea*

Name	*mole % G+C*	*Name*	*mole % G+C*
Actinomyces	59-73	*Anabena*	38-44
Bacillus	32-62	*Bdellovibrio*	42-51
Chlamydia	41-44	*Chromatium*	48-70
Deinococcus	62-70	*Derxia*	69-73
Escherichia	48-52	*Enterobacter*	52-60
Halobacterium	66-71	*Hyphomicrobium*	59-67
Lactobacillus	32-53	*Leptospira*	33-43
Mycobacterium	62-70	*Methanobacterium*	30-61
Mycoplasma	23-40	*Myxococcus*	68-71
Nitrobacter	59-62	*Nitrosomonas*	45-53
Proteus	38-41	*Psuedomonas*	58-70
Rickettsia	29-33	*Rhodopseudomonas*	64-72
Salmonella	50-53	*Spirillum*	38
Spirochaeta	51-65	*Staphylococcus*	30-38
Streptomyces	69-73	*Sulfolobus*	31-37
Thermoplasma	46	*Thiobacillus*	52-68
Treponema	25-54	*Thiospirillum*	45

Note: For comparison: Algae- 35-68; Fungi- 20-60; Protozoa- 20-65; Plants- 32-48; Animals- 35-45

3.8. RNA Fingerprinting (Molecular Phylogeny)

In order to determine the phylogenetic relationships, certain cellular macromolecules can be employed as evolutionary chronometers. Of the different macromolecules, ribosomal RNA (rRNA) has turned out to be excellent candidate for discerning evolutionary relationships. These are not only ancient molecules, they are functionally constant, universally distributed, and moderately well conserved across broad phylogenetic distances (Fig. 3.6). The degree of similarity in rRNA sequence of two organisms indicates their relative evolutionary relatedness. The prokaryotic rRNA consists of three species, 5S, 16S and 23S. The 16S and 23S rRNA consists of approximately 1500 and 3000 nucleotides, respectively, whereas 5S rRNA is a small molecule ~120 nucleotides in size. Because 16S rRNA is more manageable than 23S molecule, the former has been extensively developed as a tool in phylogenetic systematics.

This concept was pioneered by Carl Woese and his group in 1970s at the University of Illinois. They produced "fingerprints" of 16S rRNA of one organism and compared it with similar fingerprint from another organism. For this, they grew the bacterial cells in a medium containing radiolabelled ^{32}P so that all nucleic acids, including 16S rRNA

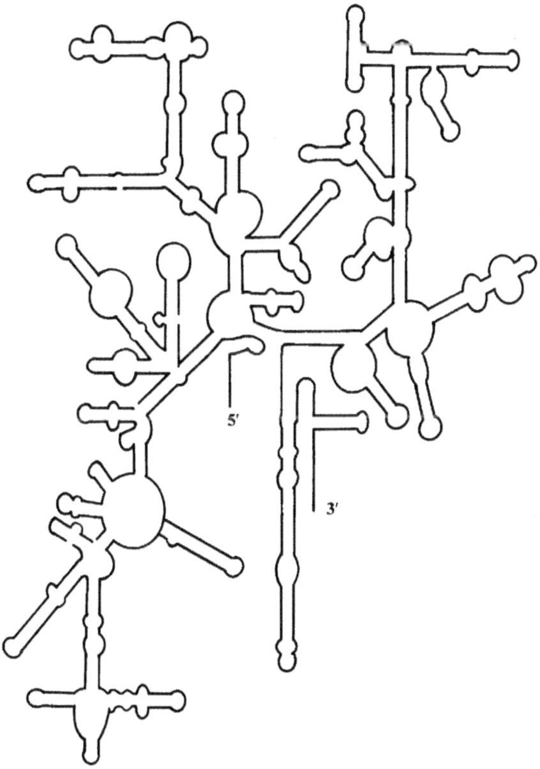

Figure 3.6. A typical structure of 16S rRNA of E. coli.

acquire the label. The rRNA was isolated and enzymatically cleaved into oligonucleotides. The fragments were separated by gel electrophoresis and oligonucleotides, 6 or more nucleotides in length were considered as they are likely to occur only once in any 16S rRNA molecule. The RNA sequences derived from two sources thus can be aligned and compared.

The computer analysis of rRNA sequences has greatly facilitated these comparisons. Through this analysis, short oligonucleotide sequences unique to a certain group or groups of organisms have been identified (Table 3.5).

Such sequences called as signature sequences have in fact been identified not only for the three primary domains (Archaea, Bacteria, and Eukarya) but also for major taxa in each domain. These sequences are located in the specific region of 16S rRNA and can also be used for constructing genus- and species-specific nucleic acid probes, so commonly employed now for identification purposes.

Such nucleic acid probes called phylogenetic probes can either be universal, binding to the conserved sequences in the ribosomes of all organisms, or some can be specific to, for example, bacteria, archaea and eukarya. The binding of probes to ribosomes in a cell can be visualized under the microscope if a fluorescent dye is attached to the probe. In this technique called FISH or Fluorescence-in-situ hybridization, cells in culture or from an environmental sample are permeabilized and treated with a mixture of oligonucleotide probes/dye. After these labeled probes are taken up and hybridize, cells can be visualized under fluorescence microscope. This technique has a wide spread usage in microbial ecology and clinical diagnostics. While, on one hand, it allows the microscopic identification and tracking of microbes in different environments, on the other, it can also be used for rapid diagnosis of specific pathogens from patient's specimens.

A phylogenetic universal tree can be generated from 16S rRNA sequence analysis. The evolutionary distance (ED) is calculated as the percent of non-homologous sequences between the two organisms. The corrected ED is a statistical correlation necessary to account for either back mutations or additional forward mutations. The tree is ultimately generated by computer analysis of the data to give the best fit. The length of the branches separating any two organisms is proportional to the evolutionary distance between them.

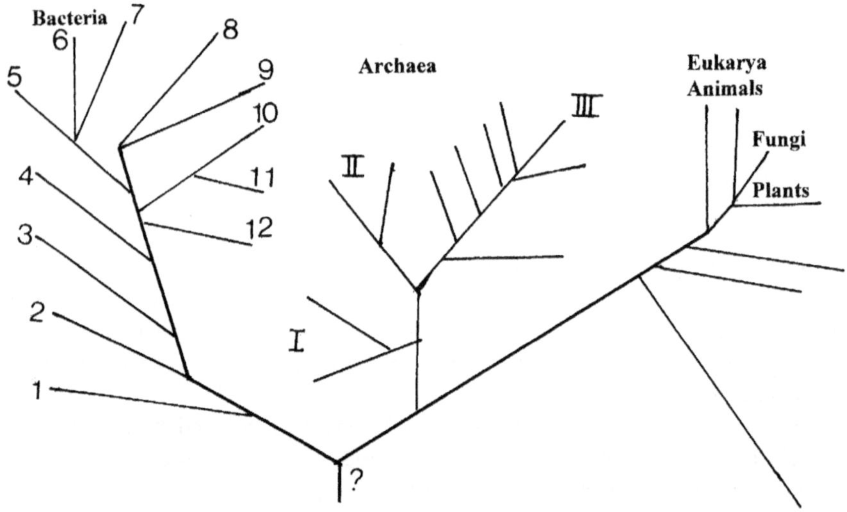

Figure 3.7. The universal phylogenetic tree derived from ribosomal RNA sequencing. Starting with the root of the tree that represents the common ancestor of all cells there are three distinct domains, Bacteria, Archaea, and Eukarya. Twelve major groups of bacteria have been defined on the basis of rRNA analyses, (1) Aquificales, (2) Thermotogales, (3) Green non-sulphur bacteria, (4) Deinococci, (5) Proteobacteria, (6) Gram-positive bacteria, (7) Cyanobacteria, (8) Chlamydiae, (9) Planctomyces, (10) Bacteriodes and relatives, (11) Green sulphur bacteria, (12) Spirochetes. Three major groups have been identified in Archaea: I-Korarchaeota, II-Crenarchaeota, and III- Euryarchaeota.

This method revealed an entirely new pattern of phylogenetic relationships. Biologists have often grouped the living organisms in five kingdoms, only one of which is assigned to prokaryotes. On the basis of rRNA analyses, these five kingdoms do not represent five major evolutionary lines. Instead, three major lineages can be identified, two of which are exclusively microbial and composed only of prokaryotic representatives (Fig. 3.7). These three are Bacteria (=Eubacteria), Archaea (=Archaebacteria) and Eukarya (=all eukaryotic organisms). They define domains of life and are placed at the top of the hierarchial organization.

Table 3.5. Characteristic signature sequence(s) from 16S or 18S rRNA demarcating the three domains*

Signature sequence	*Approximate position in rRNA*	*Percentage occurrence among members examined*		
		Archaea	*Bacteria*	*Eukarya*
CACYYG	315	0	>95	0
AAACUCAAA	910	3	100	0
AAACUUAAAG	910	100	0	100
UCCCUG	1380	>95	0	100
CUCCUUG	1390	>95	0	0
UACACACCG	1400	0	>99	100
CACACACCG	1400	100	0	0
U	546	98	0	0
A	675	0	100	2
U	880	0	2	100

* *only a few are represented here, Y= any pyrimidine*

The root of this phylogenetic tree, though unknown, based on the common 16S rRNA sequences, represents a point in evolutionary time, when all extant life on earth shared a common ancestor. This is presumed to have evolved sometimes in the first billion years of the earth's existence and gave rise to the photosynthetic bacteria found in 3.5 billion years old sedimentary deposits. The three domains, therefore, appear to have diverged quite early. Woese has suggested that the common progenitor was a progenote, a rapidly evolving rudimentary organism that had primitive and inaccurate gene replication and translation systems. This explains the drastic changes that were required to make these three domains so distinct from each other from the very beginning. The initial evolution took place in two directions, exemplified by Bacteria and the Archaea/Eukarya. The latter group subsequently diverged from each other to give two lineages. It can also be noticed that Archaea branched off from a point closer to the root, suggesting that they are most primitive. It can also be inferred that Archaea are actually more closely related to Eukarya than they are to Bacteria.

Ribosomal RNA sequencing technology can be put to a more direct use in phylogenetic analysis of microbial communities. As we shall see in **Chapter 8**, microorganisms in nature are found in a mixture called a community. In a community, its constituent members bring about different physical and chemical transformations by interacting with each other and the environment. Readers may also recall that for determining the phylogenetic status, we must be able to culture them in pure form (**see Chapter 2**). However, only a very small fraction of microbial types have been successfully cultured, leaving a large microbial world still unidentified and unknown. Some molecular techniques can be applied to identify these microbial types. This involves DNA extraction from the entire population in a microbial community and PCR amplification using a group-specific (bacteria, archaea, and eukarya) oligonucleotide probes. The PCR product consists of a mixture of different types of rRNA which can be sorted out by cloning, PCR amplification, and sequencing. The rRNA sequence is then matched with those available in the database and identification can be arrived at.

3.9. Phylogenetic Classification

On the basis of sequence-distance analysis, characteristic signature sequences, and structural features of 16S rRNA, a modern phylogenetic classification system of prokaryotes has emerged. Accordingly, the domain Bacteria has been divided in atleast twelve divisions or phyla.

3.9.1. Proteobacteria

It contains the purple non-sulphur and purple sulphur bacteria as well as most of the familiar Gram-negative bacteria. Very large genera, such as *Pseudomonas* has been split with many species getting distributed in some subdivisions of the class. Based on DNA-rRNA hybridization, several 'RNA superfamilies' corresponding to 16S rRNA based subclasses have been identified, each one represented by different subgroups.

(i) The Alpha subclass (Superfamily IV)- This contains most of the purple non-sulphur bacteria intermixed with a number of metabolically diverse non-photosynthetic organisms which are believed to have evolved from photosynthetic ancestors. Some common members are *Azospirillum, Nitrobacter, Agrobacterium, Brucella, Rhizobium* and some rickettsia. Prosthecate organisms such as *Caulobacter* also belong to this group.

(ii) The Beta subclass (Superfamily III)- Its members comprise the remaining purple non-sulphur bacteria and many non-photosynthetic organisms. One subgroup contains *Alcaligenes, Bordetella*, some "*Pseudomonas*" sps (now named differently, *Burkholderia*) and ammonia oxidizers such as *Nitrosomonas*. The others include members such as *Neisseria* and some gliding organisms, sulphur- and iron-oxidizers such as *Thiobacillus*.

(iii) The Gamma subclass- This is closely associated with the ß subclass and contains several subgroups. One contains purple sulphur bacteria such as *Chromatium* and another constituting the **superfamilies I and II** is represented by enterobacteria, vibrios, and fluorescent (true) pseudomonads, *Acinetobacter, Xanthomonas,* some gliding organisms like *Beggiatoa, Legionella* and some rickettsias.

(iv) The Epsilon subclass- The diverse members consist of intracellular bacterial parasites, *Bdellovibrios,* gliding bacteria of myxobacterales and sulphur- and sulphate-reducing bacteria.

(v) Campylobacteraceae and related bacteria (Superfamily VI)- This subclass consists of *Campylobacter*, some wrongly named *Bacteroides* species and some sulphur oxidizers.

3.9.2. Gram-positive bacteria.

These are aerobic as well as anaerobic organisms; the two of this division's four recognized subdivisions contain poorly characterized members that do not have Gram-positive cell walls. The two other branches are specified by their G + C content. The one with less than 50 mole % G + C is represented by *Bacillus, Clostridium, Lactobacillus, Staphylococcus, Streptococcus*, and mollicutes (Mycoplasma). Those with 55 or more mole % G + C contain actinomycetes, the anaerobic genera such as *Propionibacterium,* and others like *Corynebacterium, Mycobacterium* and sulphate-reducing endospore-formers. *Bacillus* species span this range of G + C content.

3.9.3. Cyanobacteria

The widely distributed and highly diverse members contain bacteriochlorophyll a and carry out oxygenic photosynthesis. Green chloroplasts have been found within the division, confirming their endosymbiotic origin. The division appeared to have evolved late and related more to the Gram-positive bacteria.

3.9.4. The Spirochaetes

This division comprises the helically coiled bacteria with periplasmic flagella. They are very ancient and some members closely resemble enterobacteria in their 16S rRNA sequences. *Leptonema, Treponema* and *Borrelia* are some of the representative members.

3.9.5. Gram-negative anaerobic rods, Cytophagas and Flavobacteria (Superfamily V)

This odd group consists of two subdivisions. One contains Gram-negative anaerobic rods such as the members of *Bacteroides* and the other of aerobes with members like *Flavobacterium* and some gliding bacteria like *Cytophagas*. The gliding motility appears to be an ancestral property of this group. Members having ring like morphology also belong to this phylum. They contain sphingolipids and characteristic fatty acids, and menaquinone system. Members have a low mole % G + C (30-45).

3.9.6. Green sulphur bacteria

The members are anoxygenic phototrophic organisms like *Chlorobium.* A distant relationship with the *Bacteroides* group has also been inferred. They carry out autotrophic carbon dioxide fixation via the reductive tricarboxylic acid cycle.

3.9.7. Chlamydiae

This small group consists of members which are all parasites of homotherms and lack peptidoglycan in the cell wall. A remote relationship with other peptidoglycan-less organisms like *Planctomyces* has been suggested. They are capable of biosynthetic activities but depend entirely on the host cell for energy.

3.9.8. Planctomyces and relatives

The members are budding, non-prosthecate organisms and possess some highly conserved rRNA sequences. Their 5S rRNA is significantly shorter than most of the bacteria, and their cell walls lack peptidoglycan. In fatty acid composition they show resemblance to thermophilic archaeans.

3.9.9. Deinococcus and relatives

These Gram-positive, radiation resistant organisms are similar to Gram-negative bacteria in their lipid profiles and outer membrane structure. The thermophilic genus, *Thermus* also shares several features of this group.

3.9.10. The green non-sulphur bacteria and relatives

The members consist of flexible, gliding bacteria, such as, *Chloroflexus*. Some are oxygenic phototrophs and others consist of non-motile, pleomorphic thermophiles. The members share a distinctive rRNA signature and structure.

3.9.11. Thermotoga and relatives, and other divisions

Members of the genus *Thermotoga* are extremely thermophilic anaerobes, with unique lipids and sheath-like structures and are found in marine waters and springs. They along with the related genera represent some of the early forms of the bacteria. This group probably represents a much larger group with many members yet to be discovered. Some prosthecate organisms and a genus of thermophilic, sulphur reducing rods which appear to have unclear relationship with any of the other subdivision may finally find a place in this group.

3.9.12. Aquificales

The *Aquifex-Hydrogenobacter* lineage comprises the oldest branching of the bacterial domain. *A. pyrophilus* is an extreme thermophile, with a temperature optimum of 85°C. The members are Gram-negative, rod-shaped, microaerophiles which can use H_2, $S_2O_3^{2-}$ and S° as electron donors to reduce oxygen to water (hence named *Aquifex* = water maker). They contain low-density protein-rich ribosomes and have glycerol diethers in their cytoplasmic membranes. These differ from archeal diethers and their glycerolipids are like other bacteria. *Hydrogenobacter* consists of at least five different species which are obligate autotrophs. They possess a relatively small genome size (ca. 1.0×10^6 bp) and an outer protein coat. They are characterized by an unusual lipid profile, an unique quinone and unique carotenoid pigment. The placement of this group at the earliest branch point suggests that the ancestral bacterial progenitor was probably thermophilic and may have fixed carbon chemo-autotrophically. Their dependence on free O_2 is considered to represent the later stages of evolutionary adaptations of early bacteria to an aerobic environment.

3.10. Archaea

Before the recognition of the Archaea as the second prokaryotic kingdom or domain, its members were thought to represent the adaptations of eubacteria to special, often extreme, environmental niches. The domain is broad and contains organisms with diverse metabolic capabilities, a wide range of cell wall types, shapes, sizes, and pigmentation and multiplication by a variety of mechanisms.

There are two main archaeal divisions based on distinctive DNA-dependent RNA polymerases and different degrees of tRNA modifications. Some have suggested that the divisions be treated as kingdoms, Crenarchaeota and Euryarchaeota while some others recognizing the Thermococcales as a third major division. The members of both the divisions consist of deeply branching groups of slowly evolving extremely thermophilic, sulphur-reducing organisms. This has led to the inference that the archaeal ancestor was also a sulphur-reducing thermophile.

CHAPTER 4
CELL STRUCTURE AND ORGANIZATION

All living organisms are made up of basic structural and functional unit called cell. Barring the acellular organisms such as viruses, viroids, and prions, the simplest form of organization known as unicellular is exemplified by bacteria, protozoa and some lower algae. All other organisms consist of a large number of cells and are thus known as multicellular. A cell is a self-contained unit surrounded by a membrane, that separates it from its surrounding, and has the ability to obtain and process the energy and the material required for sustaining life. The organization and the metabolic capability is determined by the genetic information contained in the cell, that also allows it to reproduce its own organizational and functional pattern.

A cell is broadly classified into two types: (i) the prokaryotic cell lacks the membrane bound organelles particularly the genetic material is not contained within nucleus (the nuclear membrane is absent), and (ii) the eukaryotic cell is more complex containing well-defined organelles including nucleus. Other differences also exist between them. Both archaebacteria (=archaea) and eubacteria have prokaryotic cell organization. In contrast, protozoa, algae, fungi, plants and animals, all consist of eukaryotic cells. Thus, the differences between prokaryotic and eukaryotic cells represent a fundamental division of living cells with distinct evolutionary implications.

4.1. Shape and Arrangement

The majority of bacteria exist in unicellular forms though assemblages of different kinds are also reported. Three general forms of cell shape, viz., spherical or ellipsoidal, rod-like or cylindrical and spiral or helicoidal are found.

The circular, spherical or ellipsoidal bacterial cell is commonly known as coccus (pl. cocci). These may exist singly (monococcus), or form various types of associations. An association of two such cells is known as diplococcus, a short chain as streptococcus and an irregular arrangement to give an appearance of a bunch is called staphylococcus. Sometimes, eight or more cells may form a cuboidal or packet arrangement called as sarcina (Fig.4.1).

The rod-shaped cylindrical cells are called bacilli (singular bacillus). These are normally found singly though small associations like a pair (diplobacilli) or streptobacilli with short chains can also be seen. Associations in bacillus type of bacteria may depend upon the culture conditions and stage of growth. Diphtheria bacterium may organize into a plate like or palisade arrangement and tubercle bacilli may occur in short chains of three giving an impression of short branched structures. Many rod-shaped bacteria may acquire very irregular shapes called pleomorphic and some very small rods look spherical and have been termed as coccobacillary type (Fig.4.1).

The helical bacteria (spirillum, pl. spirilla) consists of spirally twisted cells. The length, number and amplitude of turns in a spiral, however, vary from species to species. The short half-spiral is known as comma or vibrio. In some cases, the curvature may produce a ring-like structure (Fig.4.1).

Cell shape therefore has been a character of special consideration and many bacteria derive their generic names from such cellular arrangements.

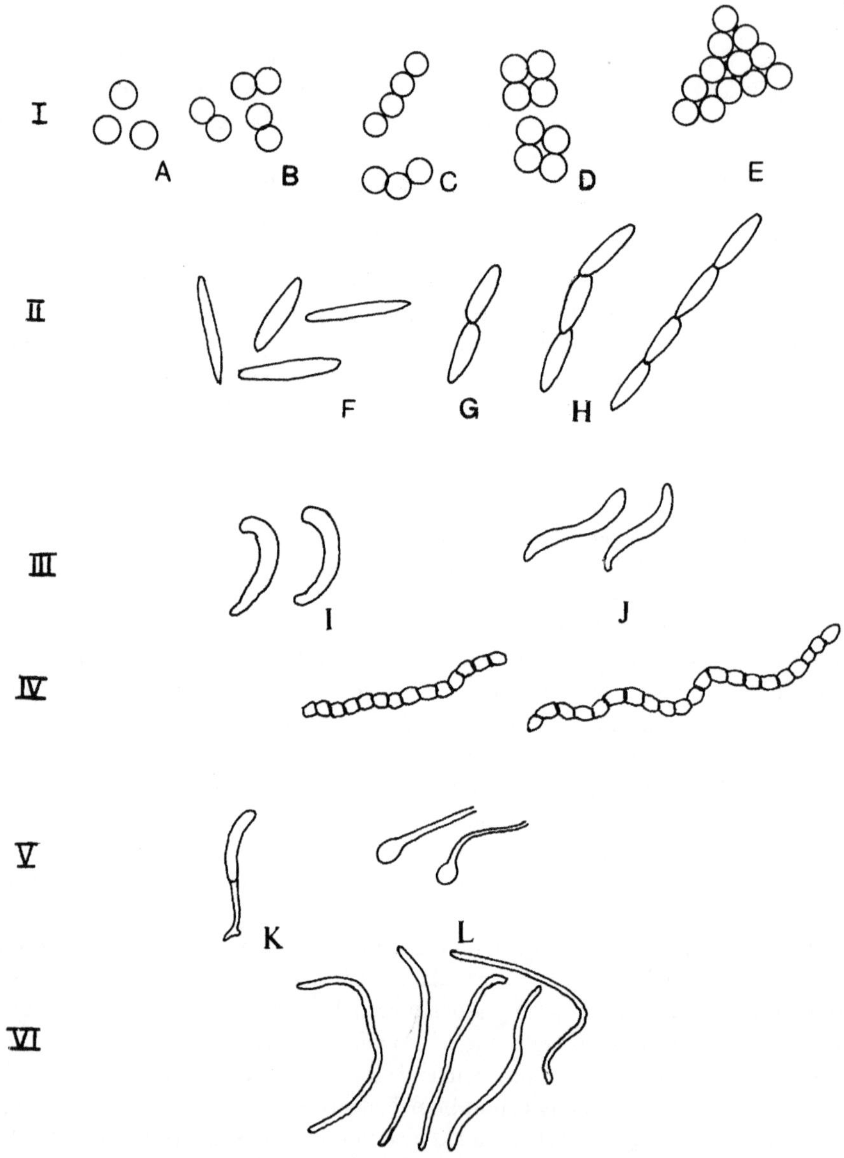

Figure 4.1A-L. Different types of cell shape and their associations in bacteria: I- Coccus-spherical/ellipsoidal. A. Monococcus, B. Diplococcus, C. Streptococcus, D. Sarcina, E. Staphylococcus, II- Bacillus-rectangular cells. F., G., H. Single, paired or in chains. III- Curved cells. I. Comma-shaped or Vibrio, J. Spirilla-spirally twisted cells, IV- Spirochaete- highly twisted cells, V- K., L. Two types of budding or appendaged cells, VI- Long filamentous cells.

The shapes described above are commonly found in eubacteria. Members of other groups may vary in shapes. Some form straight or spiral long filaments that may be divided by closely-spaced cross walls, and others may have a curved rod-shaped body resting on a small stalk. The actinomycetes produce mycelium-like filaments and myxobacteria have flexuous cells because they lack a cell wall (see also Fig. 3.3).

4.2. Size

Bacteria are microscopic. Their range of size may lie between minute coccobacillary types to long filamentous forms (Table 4.1). The average size of a bacterial cell measures 0.5-1.0 μm in diameter and 2.0 to 5.0 μm in length. Bacteria of staphylococcus and streptococcus type show a slightly larger diameter and some filamentous forms may be as long as about 100 μm. The concept about the minute size has changed due to discovery of the bacterium, *Epulopiscium fishelsoni* found in the guts of surgeonfish. This bacterium has cells which are 10^6 times larger in volume (50 ~100 μm in diameter and 0.5 mm in length) than the average bacterial cell. Another bacterium, *Thiomargarita namibiensis,* recently discovered, has cells with average diameter ranging from 100-300 μm. Hence, this appears to be the largest cell size encountered amongst bacteria. The age of the cell may also be important, as an actively growing cell is often found to be larger than the aged cells. In comparison, an eukaryotic cell may be 2 to 200 μm in diameter.

Their small size can be gauged from the fact that about a trillion (10^{12}) bacteria can be packed into a one millilitre pipette and about 5 times more can be accommodated in a teaspoon (~5 ml). Thus, a teaspoonful of bacteria will constitute approximately 2,000 times as many individuals as there are human beings on earth.

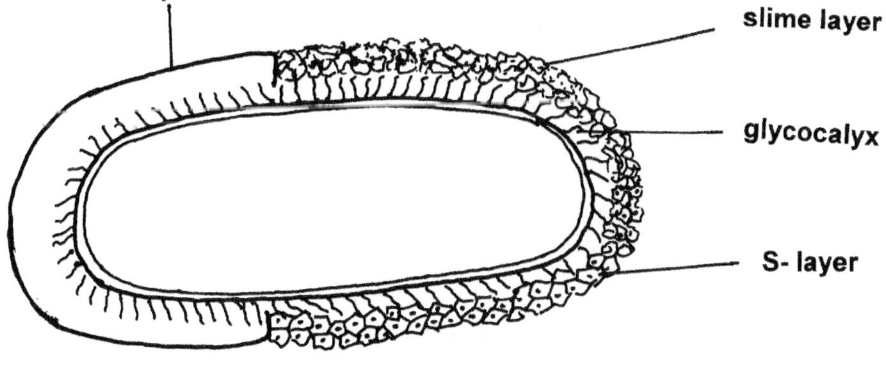

Figure 4.2. Cell surface layers in different bacteria.

4.3. Cell Structure

The prokaryotic cell is comparatively simple in organization than an eukaryotic cell with membrane-bound organelles so typically found in the latter are completely missing. A typical prokaryotic cell of either bacteria or archaea consists of the following major structures: cell wall, cytoplasmic membrane, ribosomes, some inclusions, and nucleoid. In some cases, an additional extraneous surface layer may also be seen, and similarly the motile bacteria are endowed with appendages called flagella, and many others with appendages like pili that can facilitate their attachment to a surface.

Table 4.1. Cell size of some representative genera

Name	*Diameter (μm)*	*Length (μm)*
Anabena	5.0	Long trichomes
Aquaspirillum	0.3	2.0
Bacillus anthracis	1.0-1.5	3.0-10.0
Beggiatoa	2.0-5.0	Several cms (trichomes)
Bordetella	0.2-0.3	0.5-1.0
Caryophanon	3.0-4.0	6.0-40.0
Caulobacter	0.5	-
Chlamydia	0.3	2.0
E. coli	0.4-0.7	1.0-3.0
Haemophilus	0.25	1.2
Lactobacillus	0.8	2.0
Leptospira	0.1	6.0-20.0
Pseudomonas	0.5-1.0	1.5-4.0
Rickettsia	0.3	-
Sarcina ventriculi	3.5-4.0	-
Spirochaeta	0.2-0.75	5-250
Staphylococcus	0.8-1.0	-
Streptococcus	0.6-1.0	-
Streptomyces	0.5-1.0	Long filaments
**Epulopiscium fishelsoni*	50	600
***Thiomargarita namibiensis*	100-300 (With some reaching 700)	

* The largest single celled rod-shaped bacterium reported so far

** On the basis of cell diameter clearly the largest cell size

4.3.1. Surface Layers (capsules and slime layers)

Many types of surface layers have been described in different bacteria (Fig. 4.2). Capsule may be found in many bacteria (e.g. *Acinetobacter*) but it is not an universal structure. Capsule is usually made up of polysaccharides and in some cases it may consist of polypeptide. This layer is typically thin and less dense and so it can be best visualized by negative staining. This is a protective layer and often acts as a major factor in pathogenicity (e.g. *Streptococcus pneumoniae,* formerly *Diplococcus*). The encapsulated cells can be protected against phagocytosis by eukaryotic host cells that leads to virulence. In many such bacterial species, the cells may lose capsule and with it the pathogenicity but viability is not affected. Some Gram-negative bacteria possess a relatively thinner protective layer called microcapsule.

Another external layer which is similar to capsule in chemical composition but is of less structural integrity and only loosely bound to the cell is known as slime layer (e.g. *Rhizobium trifolii*). This layer may be used by bacteria to attach to a substratum available in favourable conditions. These external layers may protect the cell against adverse environmental conditions, such as dehydration and loss of nutrients. In some cases, they even act as a trap by controlling the flow of substrates away from the cell. In addition to these layers, some eubacteria (e.g. *Aquaspirillum serpens*) have a crystalline protein layer called the S-layer. In archaebacteria, S-layer is found among species of all groups (*Methanospirillum*) but in some, S-layer may be seen as the only layer surrounding the cytoplasmic membrane.

4.3.2. Appendages

Many bacterial cells are surrounded by a specialized structure called the glycocalyx that plays an important role in attachment processes. This consists of a mass of tangled fibers of polysaccharides or branching sugar molecules. The glycocalyx may be present around individual cells or colony of cells forming multicellular aggregates. Some pathogenic bacteria adhere to the animal tissue with the help of glycocalyx (e.g., enteropathogenic *E. coli*). Similarly in aquatic habitats, they attach to the rocky substratum through the glycocalyx.

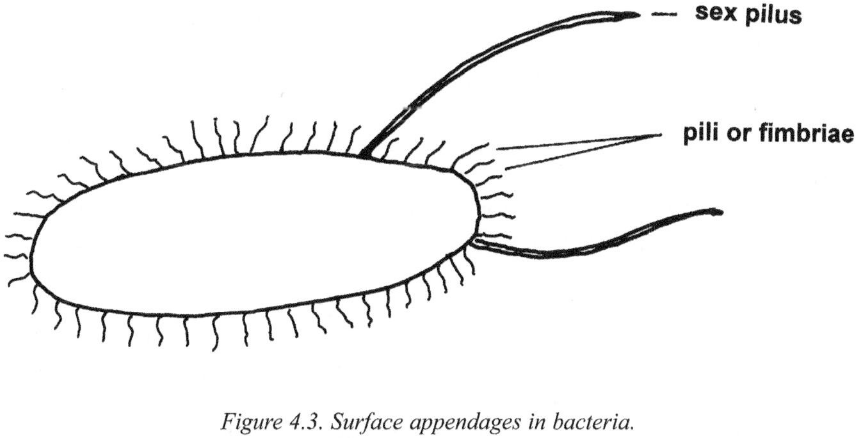

Figure 4.3. Surface appendages in bacteria.

In addition to slime layer and glycocalyx, some hair like appendages present all along the cell (e.g. *E. coli, Salmonella typhi*) are called pili or fimbriae (Sing. pilus) (Fig. 4.3). There are several different types of pili each serving a different function, but attachment or adhesion is the primary function. A special type of pilus, known as F or sex pilus is involved in bacterial matings. These are found exclusively on the cells that act as donors of the genetic material. Pili also provide the site of attachment to some bacteriophages. They have also been implicated in the recognition of the specific receptor sites available on the cytoplasmic membrane of host cells so as to facilitate colonization and subsequently the disease. *Neisseria gonorrhoeae*, the bacterium causing gonorrhea in humans, utilizes pili for this purpose.

The pili are more common in Gram-negative bacteria. An *E. coli* cell possesses 100 to 300 of these pili each with a width of 7nm and a length of 0.2 to 2.0 μm arranged peritrichously around the cell. They are thinner than flagella and do not serve the purpose of motility. Different types of pili have been distinguished on the basis of diameter which ranges from 30 Å (type I) to 250 Å (type V). The nature of piliation is genetically determined but one species can possess different types of pili. Pili are made up of one protein called pilin with a molecular weight of ~17kDa. A pilus consists of ~1000 repeating subunits of pilin, helically coiled around a central "hole" or core about 20 Å wide. Each turn of the helix is formed by 3.14 subunits.

4.3.3. Cell Wall

The universal external most layer in bacteria is cell wall. It is a rigid structure that imparts the characteristic shape to a cell. Unlike the cellulosic cell walls of plants, bacterial cell wall is more complex and unique. It also protects the cell and is generally so porous that it does not hinder the passage of small molecules in and out of the cell. Under electron microscope the average thickness of the cell wall is between 15-20 microns, representing ~1% of the total thickness of the cell and constituting 20-30% of the cell's dry weight.

Cell wall is composed of two polymers, one consisting of saccharide subunits and the other of amino acid subunits. Thus, bacterial cell wall is a glycopeptide also known as peptidoglycan, murein (murus in latin = wall) or mucopeptide. Peptidoglycan is unique to bacteria as it is not found in archaebacteria or eukaryotes. The saccharide component comprises alternating repeating units of two amino sugars related to glucose. One of them is N-acetylglucosamine (NG) and another N-acetylmuramic acid (NM), the two together forming the backbone. Glucosamine is a natural polymer present in chitin that principally forms the external skeleton of insects.

While making the backbone, the alternating units of NM and NG are linked together by a glycosidic bond (ß-1,4 linkage), as shown in Fig. 4.4 a, b. This contributes to the stability and the strength of the cell wall. The glycosidic bond is the target of the enzyme lysozyme, which can destroy the cell wall completely or sometimes partially. This type of polysaccharide organization resemble both, cellulose of plants and chitin of insects. The additional rigidity to the cell wall is provided by small peptides attached to the NM. These peptides in Gram-positive bacteria and actinomycetes consists of 4 amino acids: L-alanine, D-glutamic acid with an additional-OH group in some cases, L-lysine, and D-alanine. In Gram-negative bacteria, myxobacteria, some Gram-positive bacteria and cyanobacteria, on the other hand, L-lysine is replaced by diaminopimelic acid (DAP). The D-configuration of glutamic acid and alanine, makes the peptide bond resistant to digestive enzymes like pepsin and trypsin. The peptide chains are cross-linked with each other either through a direct bond (Gram-negative bacteria) or through a pentaglycine bridge that is formed between terminal D-alanine of one peptide chain and the third amino acid (lysine or DAP) of the other (Fig. 4.4a, b). There is a great variation in the nature of amino acids that participate in the cross bond but some like aromatics, branched-chain, and S-containing, histidine, proline, arginine are never found. Variations also exist in another form as cross linking can occur between tetrapeptides of different chains or even between two adjacent tetrapeptides. All this contribute to the strength and rigidity of the murein layer.

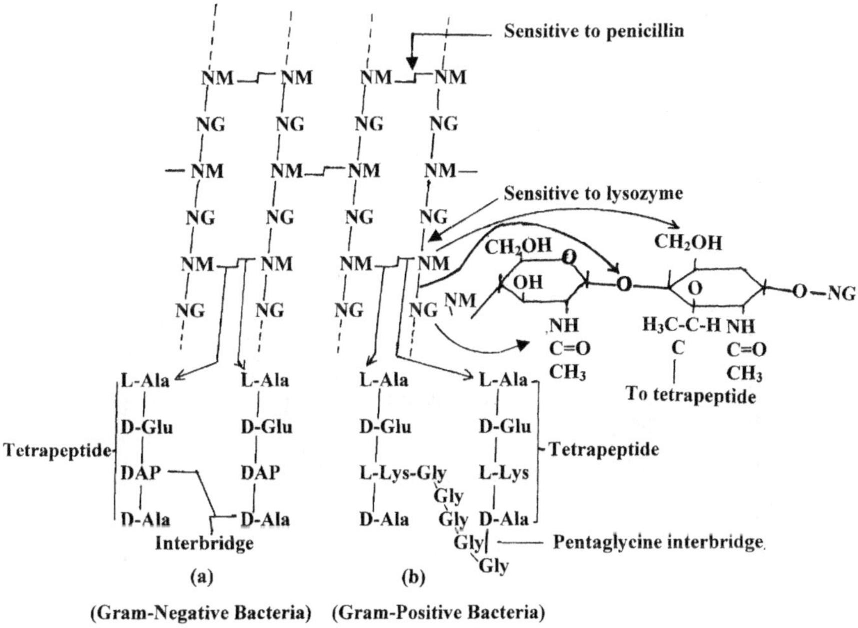

Figure 4.4a,b. The ground plan of cell wall peptidoglycan in bacteria. Figure shows some components in detail and depicts the differences in Gram-negative and Gram-positive bacteria. NM=N-acetylmuramic acid; NG=N-acetylglucosamine; Ala=alanine both D- and L-forms; D-Glu=glutamate; DAP=diaminopimelic acid; Lys=lysine; Gly=glycine. Sites of two important cell wall inhibitors are: Lysozyme sensitive β-1,4 linkage between NM and NG, and penicillin- sensitive interbridge between tetrapeptides.

The Gram-positive cell wall is completely destroyed by lysozyme. The cellular contents enclosed within plasma membrane is then termed as protoplast. On the Gram-negative cell walls, lysozyme action is not so drastic leaving some remnants of outer membrane. Such a structure is referred to as spheroplast. Both are typically spherical and require specific conditions to prevent osmotic lysis.

The antibiotic penicillin is a potent inhibitor of the cross linkages and, therefore, is capable of disrupting the cell wall. Unlike lysozyme, however, penicillin does not affect the pre-existing cross linkages but destroys only the newly forming peptidoglycan. *Mycoplasma*, that lacks the cell wall entirely, is not inhibited by penicillin for this reason only and so also the non-growing bacterial populations that possess a cell wall. *Mycoplasma* is unique as it fortifies its cytoplasmic membrane with sterols that it acquires from the eukaryotic host cell.

The bacterial cell wall is not only a protective layer but serves an important purpose of distinguishing bacteria into two major groups, i.e., Gram-negative and Gram-positive (see also **Chapter 2**). The major differences, therefore, are expected in the structure and composition of this layer (Figs. 4.4 – 4.6).

Gram-positive bacteria possess a massive peptidoglycan layer comprising approximately 90% of the cell wall (Fig. 4.5). The other characteristic is the presence of an acidic polysaccharide, the teichoic acid that is covalently bonded to the peptidoglycan. Teichoic acid is made up of carbohydrate, such as glucose, an alcohol, such as glycerol or ribitol and phosphate (Fig. 4.6a). This component serves an important purpose of binding the protons and thus maintaining the cell wall at relatively low pH. This protects the cells from self-produced autolysins. It also endows the capability of binding metal ions and acting as receptor sites for some viruses.

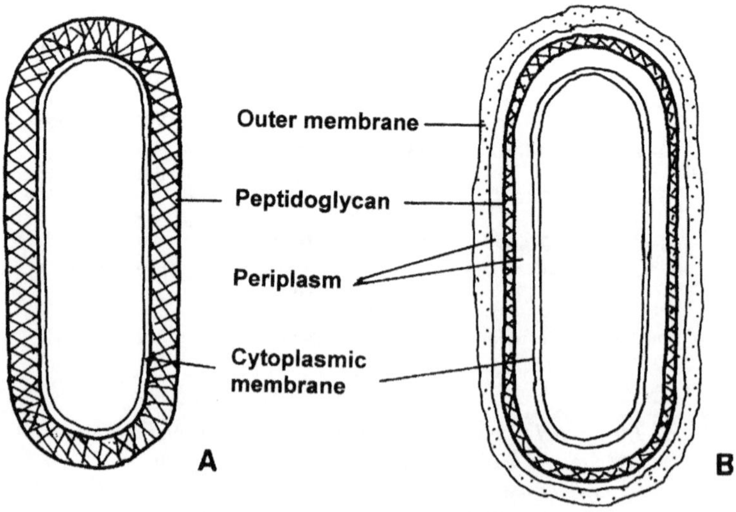

Figure 4.5. Relative contribution of peptidoglycan layer in Gram-positive (A) and Gram-negative (B) bacteria.

The Gram-negative cell wall is more complex although the peptidoglycan is very thin contributing only 10% or less of the cell wall (Fig. 4.5). In place of teichoic acid, peptidoglycan is bonded with lipoproteins. Additional layers of lipopolysaccharides (LPS), phospholipids and proteins are present outside the peptidoglycan. This multilayered complex structure is generally referred to as cell envelope of the Gram-negative cell wall. The cell envelope can be distinguished into an outer membrane that like the inner or cytoplasmic membrane is made up of bilipid layer containing phospholipids and proteins along with LPS. LPSs are made up of three distinct regions: lipid A, R core region, and the O side chain (Fig.4.6b), the last two being polysaccharide in nature.

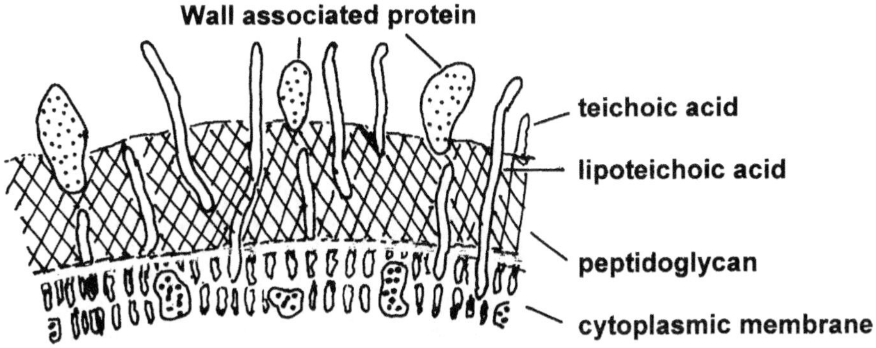

Figure 4.6a. Structure of cell wall in Gram-positive bacteria.

The lipid A is not a typically glycerol lipid but is composed of fatty acids linked to a disaccharide, N-acetylglucosamine phosphate. In constituting the outer half of the membrane, LPS gets associated with a number of proteins. On the inner side, a lipoprotein complex consisting of small proteins anchors the outer membrane to the peptidoglycan layer. In the inner leaf, the predominant component is phospholipid which in the outer leaf is replaced by LPS. The core polysaccharide is characterized by ketodeoxyoctonate, some heptoses (7-carbon sugar) and hexoses like, glucose, galactose, and N-acetyl glucosamine. Attached to this layer is the O-polysaccharide that is usually made up of hexoses, galactose, glucose, rhamnose, mannose and one or more unusual dideoxysugars, such as abequose, colitose, paratose, or tyvelose. These sugars are present in the form of 4 or 5 membered sequences which are often branched. Long O-polysaccharide chains can be formed by repeated sugar sequences. The outer membrane of many Gram-negative bacteria, like *Escherichia, Salmonella,* and *Shigella* elicit toxic reaction in animals. The toxicity is assigned to LPS and is often referred to as endotoxin.

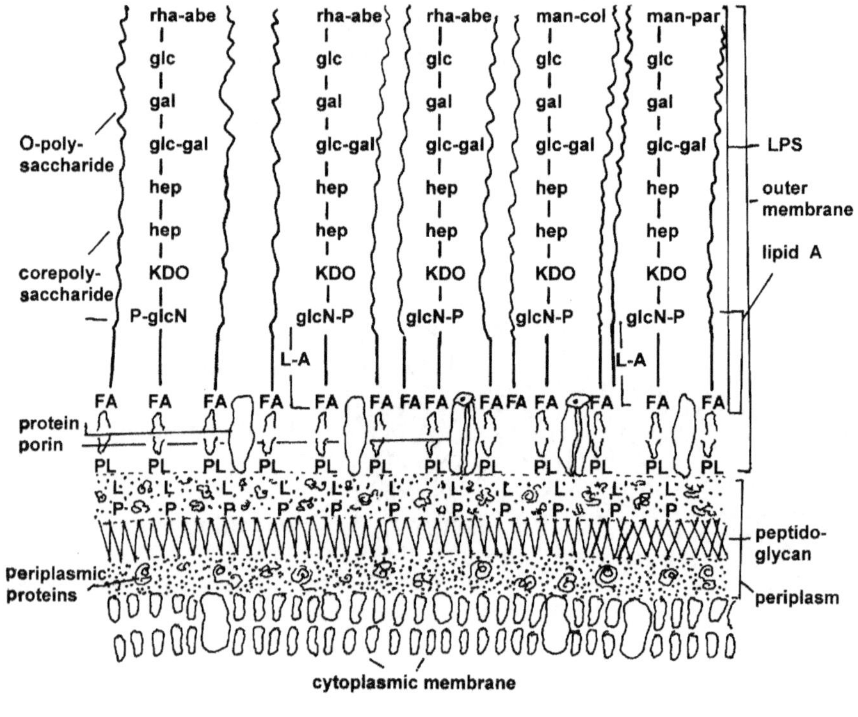

Figure 4.6b. Organization of cell wall in Gram-negative bacteria. A thin peptidoglycan layer lodged into the periplasmic space is overlaid with a multilayered complex, the outer membrane. Small lipoproteins (L) anchor the outer membrane to peptidoglycan. PL-phospholipids producing a bilayer and that also contains proteins and porins. External to this is the LPS layer consisting of N-acetylglucosamine phosphate (glc N-P) linked to fatty acids (FA) (lipid- A LA). To glc N-P is attached a variable length chain of sugars consisting of ketodeoxy octonate (kDo), heptoses (hep), glucose (glc) and galactose (gal) and constituting the core polysaccharide. This is followed by another chain of sugars that is made up of glc, gal, rhamnose (rha), mannose (man) and some unusual deoxy sugars such as abequose (abe), colitose (col), and paratose (par). This layer is called O-polysaccharide, the length of which varies in different species. The periplasmic space is filled with gel-like substance with a number of important proteins. The detailed structure is only partly represented: fatty acid +glC N-P=L-A; core polysaccharide; and O-polysaccharide.

Unlike cytoplasmic membrane, outer membrane is relatively permeable to small hydrophilic and hydrophobic molecules. This permeability is attributed to the presence of proteins called porins that form membrane channels for the passage of many molecules (Fig. 4.7). Both specific and non-specific porins are known. While the latter are the water-filled channels for many general purpose transports, specific porins act by containing specific binding sites for one or more substances. Large porins can allow the entry of molecules upto 5000 molecular weight. Porins are transmembrane proteins consisting of three identical subunits that align to form pores about 1 nm in diameter. A mechanism exists to open and close these pores as required. The outer membrane, however, is impermeable to protons and certain large molecules like enzymes.

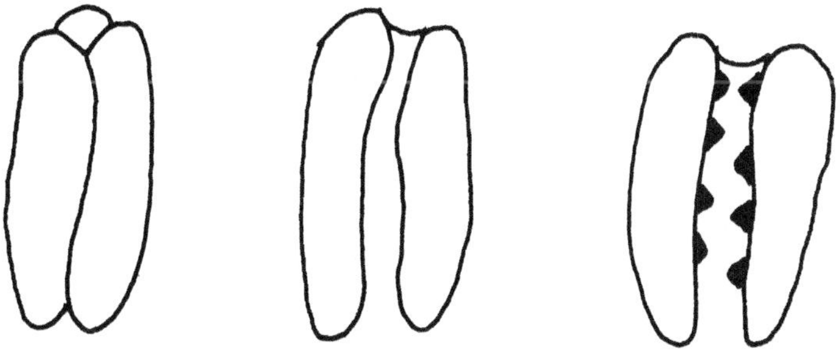

Figure 4.7. Porins. Specific transport system in Gram-negative bacteria may involve a complex of outer membrane proteins. These complexes or porins function as a pore.

The space between the outer and cytoplasmic membrane is known as periplasm, though because of its gel-like consistency it is often referred to as periplasmic gel. The peptidoglycan layer of Gram-negative bacteria is located within periplasm which measures to approximately 12-15 nm as in *E. coli.* Three types of proteins are found in periplasm: the hydrolyzing enzymes with a role in initial degradation of food molecules, binding proteins that help in transporting various substances, and chemoreceptors involved in chemotactic response. A defined periplasmic space and LPS layer are absent in Gram-positive bacteria.

The archaebacteria lack murein in their cell walls and exhibit extreme variation in wall composition. Some are composed of pseudomurein which resembles the peptidoglycan of eubacteria but contains N-acetyltalosaminuronic acid in place of N-acetylmuramic acid, and lacks D-amino acids. Some have walls composed of protein subunits and still others have variable composition. The cell wall is able to protect the cytoplasmic membrane under very harsh conditions inhabited by archaebacteria, such as hot, acidic and saline environments.

4.3.4. Cytoplasmic Membrane

One of the unifying features of all living cells is the presence of a cytoplasmic membrane around the cytoplasmic contents of the cell. This membrane regulates the flow of material in and out of the cell based on differential permeability barrier. The cytoplasmic or inner membrane is essentially organized as a lipid bilayer largely consisting of phospholipids (Fig. 4.8). The nonpolar hydrophobic lipids consist of fatty acids containing 16-18 carbon atoms. They constitute an internal matrix of a lipid bilayer oriented towards each other and slightly overlapping, with polar phosphate component on both the sides. In addition, some proteins that are integrated into or associated with the basic membrane structures are also found. Many of these proteins are involved in constituting transport channel with some degree of selectivity.

Though various models to explain the membrane structure and function have been proposed, the fluid mosaic model is currently the accepted one. According to this model, membrane is a dynamic structure, with both proteins and lipids having the ability to move within the phospholipid matrix. Lipids held together by weak bonds can move laterally more easily in comparison to proteins. Some proteins are found on the surface (extrinsic proteins) whereas some are partially or totally buried within the membrane matrix (intrinsic proteins).

The prokaryotic cell membranes generally lack sterols so characteristic of eukaryotes. In contrast to bacteria and eukaryotes that contain straight chain fatty acids linked to glycerol by ester linkage, archaebacterial membrane is composed of branched lipids linked to glycerol by ether bonds. The structure of the membranes makes them resistant to the extreme conditions inhabited by archaebacteria.

In all living cells, the cytoplasmic membrane plays the central role of regulating the transport. This involves passive diffusion or facilitated diffusion down a concentration gradient or energy requiring active transport. When thin sections of bacterial cells are viewed through electron microscope extensive invaginations of the cytoplasmic membranes, called mesosomes, have often been observed. Although earlier many functions, such as a role in cell division, various metabolic processes, enzyme secretion, functions of membranes of organelles in eukaryotic cells, have been assigned to mesosomes, they appear to be more of an artifact than the real cell structure.

In the absence of defined organelles, the bacterial cytoplasmic membranes are multipurpose structures. They are involved in energy transformations, namely the generation of ATP. Some groups of bacteria contain extensive internal membranes that are also involved in chemiosmotic generation of ATP. In many cases, these specialized internal membranes may completely fill the bacterial cell. The specialized membranes found in photosynthetic bacteria exhibit great variation in organization. They may occur as simple extensions of the cytoplasmic membrane as in the purple sulphur bacteria, or form cylindrical vesicles called chlorosomes as in the green photosynthetic bacteria or may form an extensive multilayered membrane structures, known as thylakoids as in the cyanobacteria (see Fig. 4.12). The organization of photosynthetic membranes in these systems suggest the possible evolutionary developmental sequence culminating in the present day eukaryotic organelle, the chloroplast.

4.3.5. Locomotory Organs

Many bacteria are motile, and this ability is provided by a structure called flagellum (pl. flagella). This movement also allows the bacterium to search for an appropriate microenvironment and respond to a large number of environmental stimuli. The position of the flagellar attachment provides a marker to classify bacteria into different types.

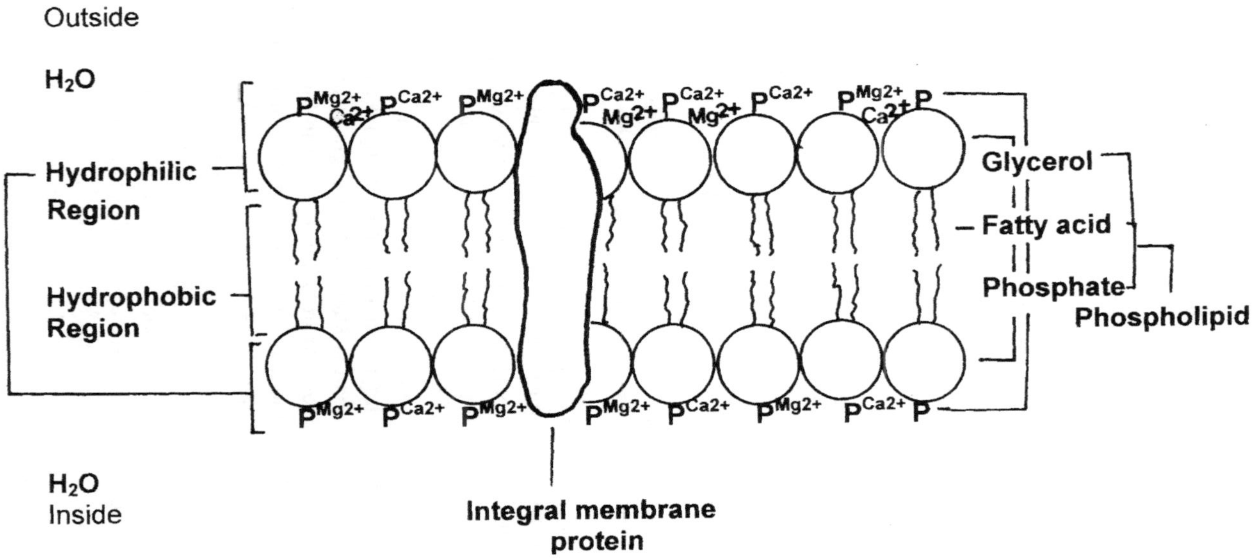

Figure 4.8. *Typical structure of a cytoplasmic membrane showing a phospholipid bilayer, the integral membrane proteins and the differentiation of hydrophilic and hydrophobic regions.*

There are two main types of flagellar pattern: polar, in which the flagella are present at one or both ends of the cell, and peritrichous, when they are distributed all over the cell surface. The former may be subdivided into monotrichous (single flagellum at one end), lophotrichous (two or more flagella at one or both ends), and amphitrichous (one flagellum at each end) (Fig. 4.9).

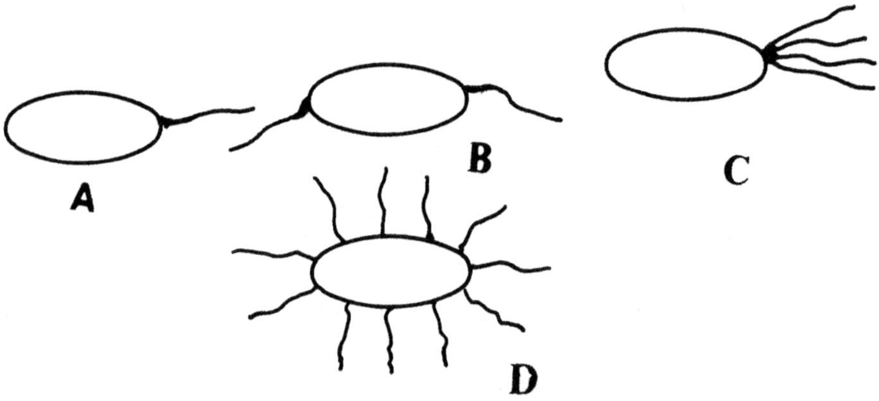

Figure 4.9. Flagellar pattern in bacteria. A. Monotrichous, B. Amphitrichous, C. Lophotrichous, D. Peritrichous.

The length of the flagella ranges up to 10 times the cell length and the diameter is between 120 and 185 Å. In vibrio group, the flagella are distinctly stouter (300 Å) and possess a basal sheath which is formed by the extension of the cell wall. A flagellum can be divided into three parts (Fig. 4.10). The long thread-like structure, seen outside the cell in the appropriately stained preparations, is known as filament. The filament is made up of a twisted fibre of a protein, the flagellin. The number of flagellin subunits in the fibre is variable. The attachment region is more complex than the rest of the length. At or near the base, the filament is attached to a slightly thickened basal hook that penetrates the cell wall. One end of the hook is fixed to the basal body that finally attaches the hook and the filament to the cell membrane. The basal body is a motor like structure formed by spindle shaped rings surrounding a cylindrical rod. In Gram-negative bacteria, there are two tiers of such rings, the outer (L&P) is anchored to the outer membrane and the inner, S and M, to the cytoplasmic membrane. In Gram-positive bacteria, on the other hand, there is only one ring that attaches the structure to cell membrane alone. Some spirilla have a tuft of polar flagella and the basal body is connected to a specialized polar membrane system lying just below the cell membrane. The whole system, therefore, ensures that flagellum remains in direct contact with the cytoplasmic membrane.

The "flagellin" that constitutes the filament, is a protein of molecular weight ranging from 30-60 kDa, particularly rich in glutamic and aspartic acid, has low levels of aromatic amino acids and completely lacks cysteine. Three to six flagellin fibres are helically

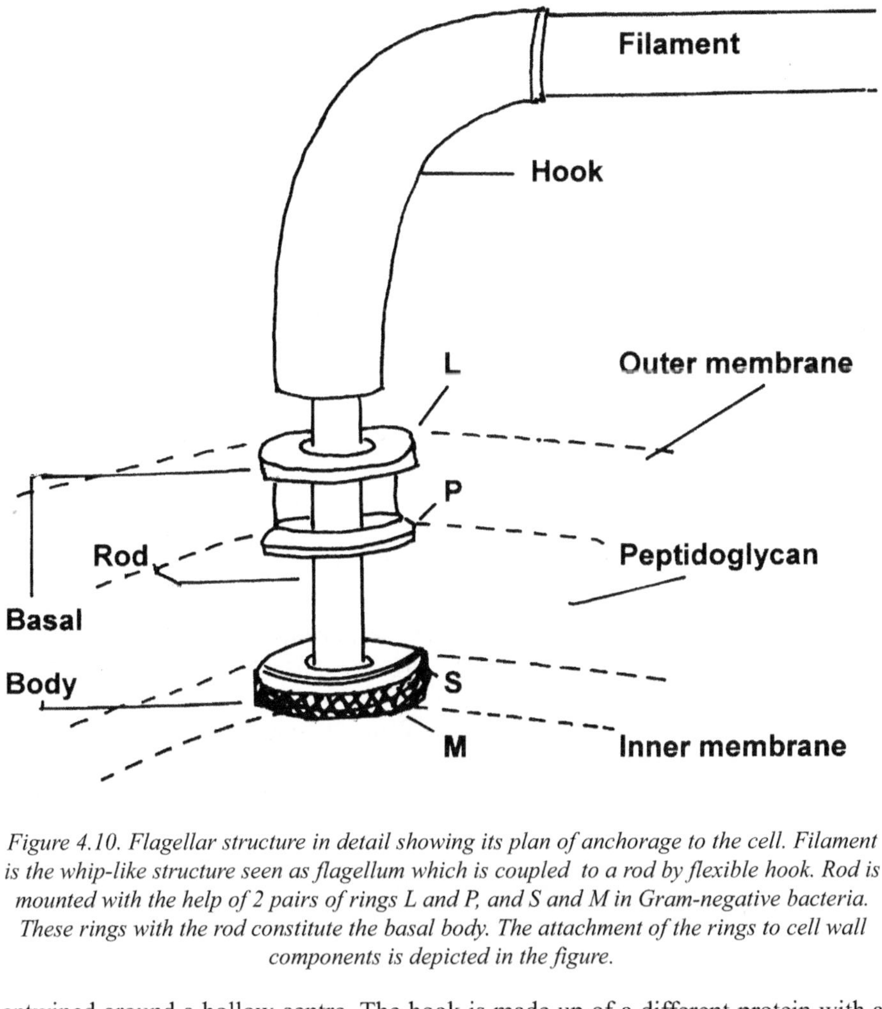

Figure 4.10. Flagellar structure in detail showing its plan of anchorage to the cell. Filament is the whip-like structure seen as flagellum which is coupled to a rod by flexible hook. Rod is mounted with the help of 2 pairs of rings L and P, and S and M in Gram-negative bacteria. These rings with the rod constitute the basal body. The attachment of the rings to cell wall components is depicted in the figure.

entwined around a hollow centre. The hook is made up of a different protein with a single polypeptide of ~42 kDa. The basal body consists of atleast 9 polypeptides with molecular weight ranging from 9 to 11 kDa.

The flagellar synthesis and function require over 40 genes and are known as *fla* genes. These genes code for structure, assembly, and regulation. Each individual flagellum is actually a rigid structure that moves by rotation in a manner similar to that of propeller. This action of movement is imparted by basal body, and the required energy is provided by the proton motive force. The maximum rate of flagellar rotation is about 200 revolutions per second (12,000 rpm). The motion of organisms having polar and peritrichous flagella differ. While the latter move and rotate slowly in a straight line, the former show rapid movement.

4.3.6. Cytoplasm

In all living cells, the intracellular content enclosed within the cytoplasmic membrane consist of a matrix of gel-like consistency called the cytoplasm. In bacteria and other prokaryotes, the cytoplasm is relatively uniform and does not contain the majority of organelles found in an eukaryotic cell. The only other structure seen in the cytoplasm is the nucleoid or chromatin body and that too is not separated by a nuclear membrane. The cytoplasm is almost uniformly occupied by ribosomes and few may contain vesicles or vacuoles as well as some reserve materials.

4.3.7. Ribosomes

Ribosomes are ribonucleoprotein particles of ~70Å in diameter and a mass of $2.5x10^6$ Da. In all living cells, they constitute the sites of protein synthesis. A metabolically active cell may contain as many as 10,000 ribosomes per cell, but the number may go down several hundred folds in a starved cell. During protein synthesis, many ribosomes may be held together to form polyribosomes or polysomes. Each functional prokaryotic ribosome is 70S (Svedberg unit) with two subunits, the smaller squash-shaped, 30S, and the larger, more spherical 50S subunit (Fig. 4.11). In comparison the eukaryotic ribosome is 80S and the two subunits 40S and 60S, respectively. The Svedberg units refer to their sedimentation coefficients when subjected to the centrifugal force in an ultracentrifuge. The subunits of ribosomes require high levels of Mg^{2+} ions to remain associated. In *E. coli*, the ribosomes are made up of 66% RNA, known as ribosomal RNA (rRNA) and 34% proteins. The 30S subunit contains one copy of 16S rRNA of ~1540 nucleotides, and 21 different proteins. The 50S subunit, on the other hand, consists of two different rRNAs, the 23S (~2900 nucleotides) and the small 5S rRNAs of ~120 nucleotides along with 31 different proteins. With the exception of one protein that is present in four copies all other proteins are represented by one copy per ribosome. Genes for rRNA in bacteria are repeated so that multiple copies, forming *rrn* operon, are present per haploid genome.

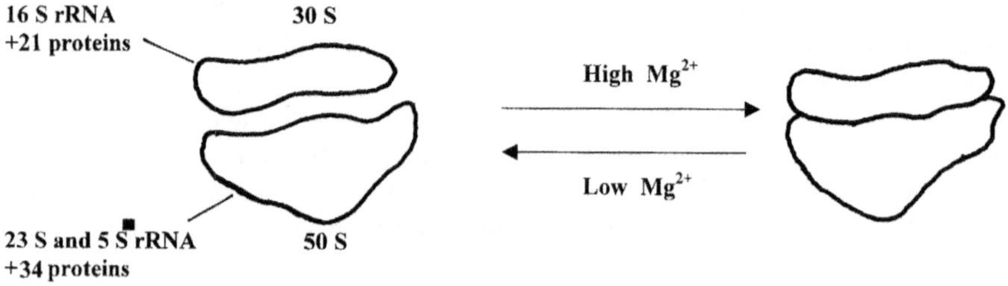

Figure 4.11. 70S ribosome of prokaryotes composed of 30S and 50S subunits. The two subunits constitute the functional ribosome in the presence of high Mg^{2+} ions.

4.3.8. Photosynthetic Membrane Systems

The photosynthetic pigment in phototrophic bacteria is lodged into membrane systems the nature of which differ in different groups (Fig. 4.12). In the purple bacteria, the internal photosynthetic membrane systems are elaborate and are connected to or produced as extensions of cytoplasmic membrane, often occupying much of the cell interior. In some cases, the systems consist of an array of flat sheets called lamellae (e.g. *Ectothiorhodospira*), whereas in others they may be like round tubes or vesicles (e.g. *Chromatium*). In green bacteria, such as *Pelodictyon*, a series of cylindrical structures lying just beneath and attached to the cytoplasmic membrane constitute the photosynthetic apparatus and is referred to as chlorosomes. These vesicles are bound by a thin membrane that does not have the usual bilayered organization (non-unit membrane). Bacteriochlorophylls c, cs, d or e (depending upon the species) are present inside the chlorosome but bacteriochlorophyll a (BChl a) and the constituent photosynthetic electron transport chain are located in the cytoplasmic membrane. In unrelated heliobacteria, bacteriochlorophyll g is present in the cytoplasmic membrane itself. In oxygenic cyanobacteria, the lamellar membrane systems are often complex and form multilayered structures known as thylakoids. In some simple cyanobacteria, lamellae located around the periphery are regularly arranged in concentric circles. The purple and green bacteria are anoxygenic phototrophs, and so also the heliobacteria. On the basis of their ability to obtain reducing power for CO_2 fixation, the former have been further divided into respective sulphur and non-sulphur groups.

4.3.9. Inclusions and Storage Products

The bacterial cells often possess granules or other inclusions. Although their nature may vary, they serve the function of a reservoir of energy and other structural building blocks. They can be easily spotted in suitably stained preparations. Most cellular inclusions are bounded by non-unit, thin, lipid-based membrane and are thus referred to as inclusion bodies.

One of the commonest of such inclusion bodies are the ones consisting of poly-ß-hydroxybutyric acid (PHB, e.g., *Alcaligenes, Rhodospirillum*). The monomer may be as small as C-4 to as long as C-18 in some organisms. This class of polymer that may aggregate as granules is often referred to as poly-ß-hydroxyalkanoate (PHA) (Fig. 4.13). Lot of interest has been generated in PHAs in recent years as many of them have plastic-like consistencies and have thus been promoted as biodegradable plastic. These granules show an affinity for fat-soluble dyes such as sudan black.

Another storage product, glycogen, is a starch like polymer of glucose subunits. Although smaller in size than PHB granules they can be easily stained with dilute iodine. Both PHB and glycogen are storage granules for carbon and energy.

Many bacteria accumulate large reserves of inorganic phosphates in the form of polyphosphate granules. These granules are metachromatic as they turn reddish violet when stained with toluidine blue.

Similarly, a variety of prokaryotes (phototrophic green and purple bacteria) are capable of oxidizing reduced sulphur compounds, such as hydrogen sulphide, thiosulphate, and others. Such bacteria accumulate large amounts of elemental sulphur

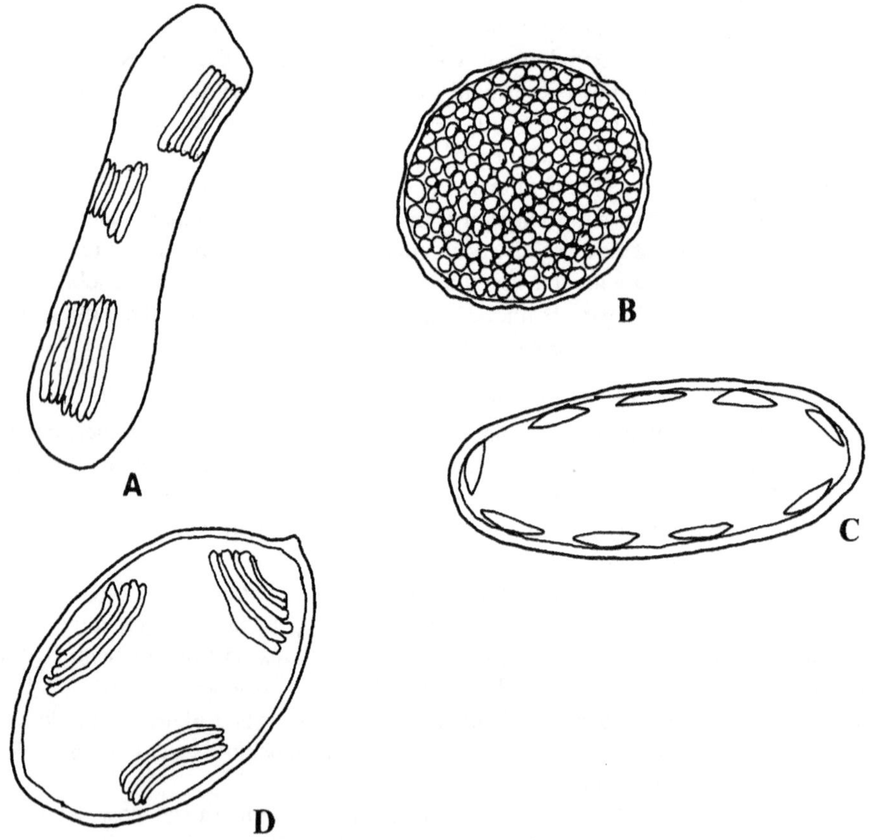

Figure 4.12. Arrangement of lamellar membranes in phototrophic bacteria. A. Purple bacterium, Ectothiorhodospira shows membranes in flat sheets or lamellae, B. In purple bacterium, Chromatium, the membranes are arranged as isolated vesicles (chromatophores), C. The green bacterium, Pelodictyon has chlorosomes in which the photosynthetic pigments are lodged, D. In cyanobacterium, Synechococcus, the membranes are arranged as lamellae or thylakoids. Note: Heliobacterium lacks an internal membrane system.

in the form of conspicuous readily visible granules deposited internally (e.g. *Chromatium*) or externally (e.g. *Chlorobium*). The stored sulphur is oxidized under limiting reduced sulphur conditions (Fig. 4.13).

Some aquatic bacteria, such as *Aquaspirillum magnetotacticum,* possess intracellular crystals of iron oxide, magnetite and Fe_3O_4. These crystals are organized into membrane bound particles called magnetosomes (Fig. 4.13). They exhibit very different morphology ranging from square to rectangular to even spike-shaped. The membrane consists of phospholipids, proteins, and glycoproteins. Such bacteria referred as magnetotactic, possess an interesting capability of orienting and migrating along geomagnetic field, or magnetotaxis.

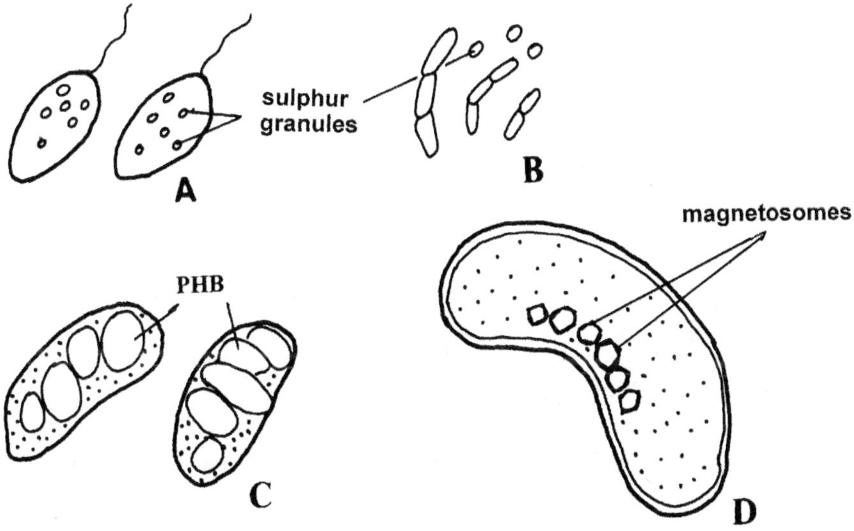

Figure 4.13. Material deposition (storage granules) in different bacteria. A. Intracellular sulphur granules (Chromatium), B. Extracellular sulphur granules (Chlorobium), C. Polyhydroxybutyrate deposition (Alcaligenes), and D. Magnetosomes containing crystals of magnetite (Fe_3O_4) *(Aquaspirillum magnetotacticum).*

4.3.10. Gas Vesicles

Many prokaryotic organisms (e.g. *Anabena, Microcystis*) living in different water bodies possess gas vesicles. With the help of these vesicles the cells are able to float up and down in a water column in response to different environmental factors. The large blooms produced by certain cyanobacteria are endowed with these gas vesicles. Gas vesicles are proteinaceous, hollow but rigid, spindle-shaped structures (Fig. 4.14). They may vary in length from 300-700 nm and in width from 60-110 nm, but their size in a species is constant. Their number may also vary from a few to several hundreds per cell. The vesicle membrane is about 2 nm thick and is composed of only protein which is organized into a rigid ribbed structure. This membrane is permeable to gases but does not allow the entry of water and solutes, thus conferring the required buoyancy to the cells. Gas vesicles contain only two different types of proteins, GVPa, the major hydrophobic protein and GVPc the minor protein that strengthens the shell of the gas vesicles. The genes coding for these proteins from taxonomically diverse gas vesiculate bacteria show a remarkable degree of DNA sequence homology suggesting great evolutionary conservation of these unique structures.

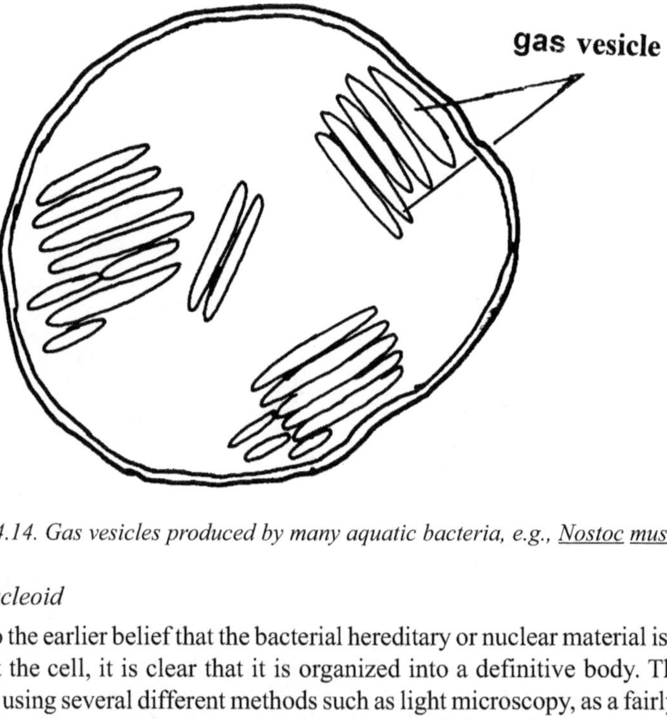

Figure 4.14. Gas vesicles produced by many aquatic bacteria, e.g., Nostoc muscorum.

4.3.11. Nucleoid

Contrary to the earlier belief that the bacterial hereditary or nuclear material is dispersed throughout the cell, it is clear that it is organized into a definitive body. This can be visualized, using several different methods such as light microscopy, as a fairly compact clump, occupying roughly one third of the volume of the cell, and is called nucleoid or nuclear body (Fig. 4.15a). Fluorescence microscopy provides one useful approach to delineate the gross boundaries of the nucleoid. A fluorescent probe, like 4',6-diamidino-2-phenylindole dihydrochloride (DAPI) which readily enters the cell and binds to DNA at a concentration that is nontoxic, greatly helps in visualization. Electron microscopy has revealed that nucleoid of *E. coli* is confined to a ribosome-free zone in the cell and has a highly clefted appearance (Fig. 4.15b). Since transcription and translation in bacteria are coupled, it may indicate that little or no transcription could occur from DNA sequences located deep in the nucleoid. Alternatively, transcriptionally active

DNA is located on nucleoid surface or on DNA loops extending from the core. Through immunogold staining by antibodies directed towards RNA polymerase and topoisomerase I, location of both the enzymes primarily on the surface was suggested. If transcription is indeed localized in this manner, it will require the nucleoid to be a highly dynamic structure so that potentially all the genes must cycle from their interior location on to the surface or on to the extended loops. In Gram-negative bacteria, the nucleoid is attached to the cell membrane by the region that corresponds to the origin of replication. A peculiar feature of bacterial nucleoid is that it is naked i.e., no nuclear membrane surrounds it. This is the basis for classifying them as prokaryotes, unlike the higher organisms or eukaryotes which have a well-defined nuclear membrane enclosing the nuclear material. Condensed state of nucleoid material is maintained by counter ions, and thus, the DNA unfolds during lysis.

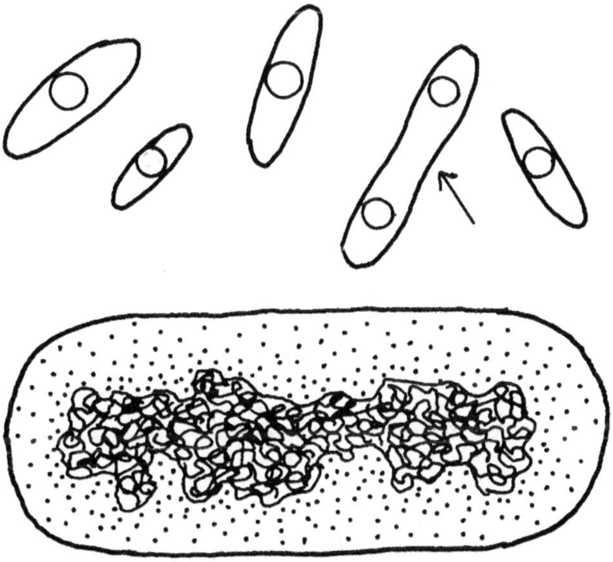

(a)

(b)

Figure 4.15a,b. Bacterial nucleoid. a. Light photomicrograph of E. coli cells stained to locate the nucleoid (nuclear body). One of the cells is under the process of cell division (→), b. Representation of a transmission electron micrograph of E. coli showing a compact but highly clefted nucleoid.

4.3.12. Chromosome

The nucleoid carries the bacterial genome that represents the total genetic complement of the cell. The genome principally refers to the bacterial chromosome as the extrachromosomal element, such as plasmids, may or may not be present.

The bacterial chromosome represents a covalently closed circular DNA molecule which in *E. coli* consists of ~4.6x10^6 bp (=4.64 mb in length) and is approximately 1mm long. With the bacterium measuring to only about 1-2 μm, the bacterial chromosome must be condensed about 1000-fold to fit into the cell. This is brought about by organizing the DNA into a series of loops which are supercoiled in a plectonemic (underwound) fashion (Figs. 4.15, 4.16). The folding pattern, however, ensures that transcription, recombination and other functions are carried out unhindered. Most DNA in bacteria is negatively supercoiled with an average of one negative supercoil per every 300 bp. The supercoiling, however, can be unevenly distributed. We know that two DNA strands are wound around each other about once every 10.5 bp to produce the helix. The supercoiling is introduced when DNA duplex is twisted in space around its own axis. In negative supercoiling, the DNA is twisted in opposite direction to the clock-wise turns of the right handed double helix, whereas in positive supercoiling it is twisted in the same direction. While DNA in the former state is supposed to be in underwound, in the latter it is in overwound state. Negative supercoiling can lead to

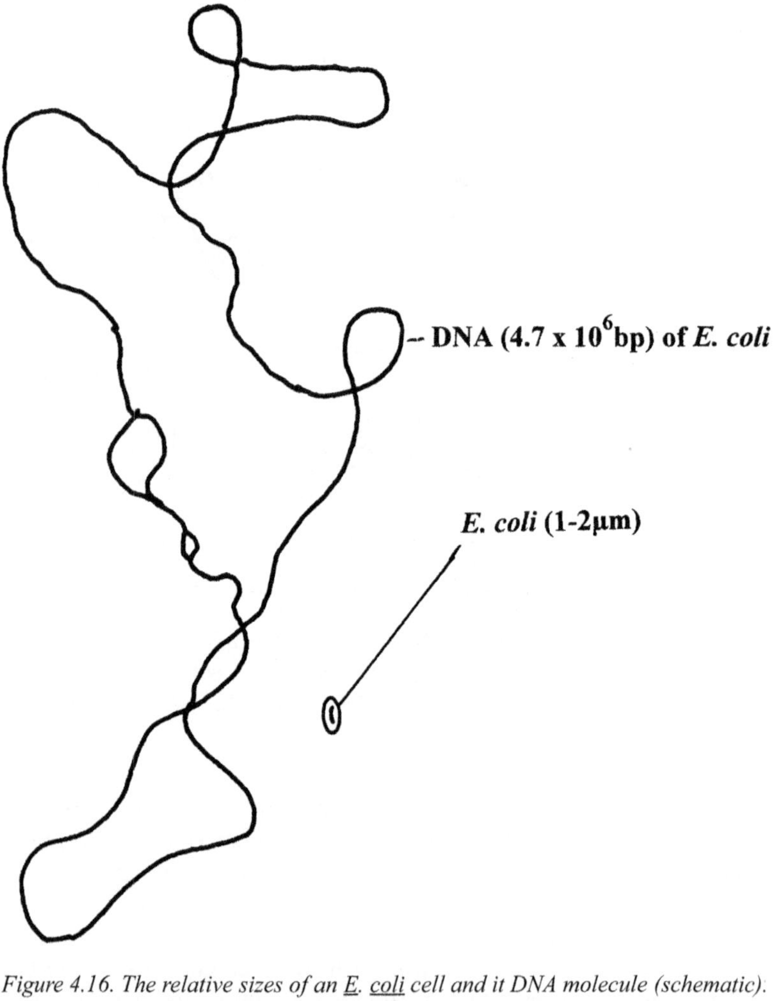

Figure 4.16. The relative sizes of an E. coli cell and it DNA molecule (schematic). The DNA is folded to about 1000 times.

strand separation, so important for many functions of DNA. DNA in the chromosome is overwhelmingly in the form of right handed B-DNA, although left-handed Z DNA, cruciforms, and triple helix may also exist in a localized manner. Other localized structures that may be important to organization and function of the bacterial chromosomes are bends and loops.

The nucleotide composition of prokaryotic genomes varies greatly ranging from as low as 25% G+C in *Mycoplasma capricolum* to as high as 75% G+C in *Micrococcus luteus* with *E. coli* falling in the intermediate range of 51% G+C. The size of the genome among bacteria may range from 585 kb in *Mycoplasma genitalium* to 9554 kb in *Myxococcus xanthus*. In majority of the cases, the chromosome is a covalently closed circle though in a few cases linear chromosome, as described later, have also been reported. Usually a single chromosome is encountered in a cell but *Rhodobacter sphaeroides* contains 2 circular chromosomes (3046 kb and 914 kb).

Given the large size of the chromosome and the relatively small volume of the cell, it is obvious that DNA in the nucleoid is highly compacted. Condensation appears to occur at two levels: 1) Organization of DNA into independently supercoiled domains, and 2) Condensation due to binding of histone-like proteins.

The isolated nucleoid shows 30-50 loops of DNA emerging from a more condensed region, or core (Fig. 4.17). Most of this DNA is in highly negatively supercoiled form. Negative supercoiling is introduced by DNA gyrase imparting torsional tension to the DNA molecule that favours the loop formation. Introduction of a single nick by DNAase is unable to relax all supercoiling, suggesting that DNA is organized into independent supercoiled domains. Evidence for such an organization has come from a number of sources. The average size of a domain in *E. coli* is about 100 kb. It is of interest to note that similar-sized domains have also been measured in the chromosomes from diverse biological systems, including eukaryotes. Such a domain organization prevents relaxation from one chromosomal domain from being propagated into the others and provides elements of structural flexibility to the chromosome. Thus, replication, repair, recombination in one domain, all of which affect DNA topology, will not affect DNA organization in other domains. Such a structure also helps in achieving long range compaction of DNA. The fully extended circular chromosome of *E coli* would have a diameter of 430 μm, but when organized into 50 fully extended equal sized loops, it measures about 17 μm in dimension.

Supercoiling and domain formation do not account for the total compaction of bacterial DNA. Association of DNA with small abundant proteins and estimation of superhelical density similar to that found in eukaryotic DNA, provided the compelling evidence that DNA in prokaryotic chromosomes is coiled in nucleosome-like structures. These evolutionarily conserved proteins with amino acid composition resembling eukaryotic histones have been called histone-like proteins. Those identified so far are HU, IHF, H-NS, and FIS. All these are able to bend or coil DNA and have higher affinity for bent DNA. They thus provide mechanisms for distorting or introducing flexibility in the short range structure of DNA in the nucleoid. Deletions affecting HU, IHF, and H-NS individually are not lethal suggesting that they may have partly overlapping functions.

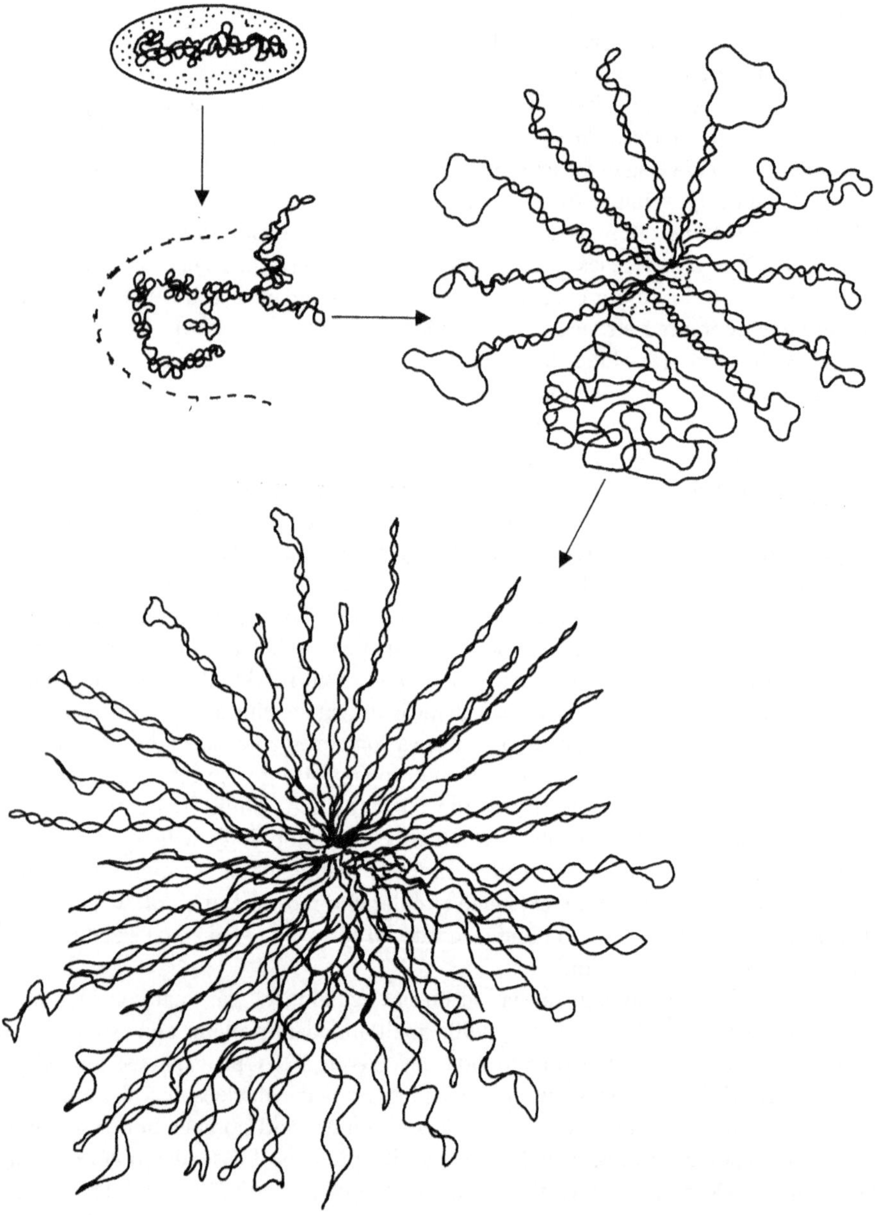

Figure 4.17. The highly folded nature of the bacterial chromosome and the relative compaction of the same. The folded chromosome is showing independent domains.

HU protein distributed throughout the nucleoid is the major DNA-binding protein known. There is enough HU protein to bind about 1 HU dimer per ~300-400 bp. Like histones, its binding reduces or restrains the effective superhelical density that leads to the appearance of plectonemic supercoils. Like histones, HU also binds with little sequence specificity and the DNAase protection assay of DNA-HU complex suggests the possible wrapping of DNA on the HU. Certain features of HU-DNA interactions make them similar to eukaryotic HMG-1 and HMG-2 proteins. HU bends DNA and prefers to bind to bent or kinked DNA. HU also binds preferably to regulatory sites cooperatively with other regulatory proteins. HU thus facilitates short range structure transitions, and other protein-DNA interactions including site-specific recombination and the initiation of DNA replication at *oriC.*

IHF has 30% amino acid sequence identity with HU and can substitute for each other *in vitro*. DNA binding arm of IHF, however, has homologies with eukaryotic TATA-binding factor, TFIID. IHF also facilitates DNA bending but recognizes a specific DNA sequence. IHF is also involved in regulation of different genes (positive regulation). H-NS localized in the nucleoid, can compact DNA and is apparently involved in organizing short range DNA structures. H-NS shows little sequence specificity but like HU binds preferentially to bent DNA. H-NS is shown to change DNA topology and thus can affect transcription by modulating required levels of DNA supercoiling. It can also function as a transcriptional repressor directly and interferes with the formation of open promoters.

4.3.12.1. Functional Organization of Chromosome. The chromosome of *E. coli* like those of yeast, but unlike those of higher eukaryotes, is densely packed with genes. This fact was verified when the contiguous sequences analyzed by 1994 showed that about 80% of this DNA is likely to code for proteins and another 4% is transcribed to yield non-messenger RNAs. Much of the remaining intergenic sequence is in very short stretches and is probably comprised mainly of sequences concerned with the control of transcription and translation. Only about 2/3rds of the genes detected are organized into operons (for further details see later section on genome sequencing). Many of these operons may further be present in multiple copies (e.g. rrn operon).

Analysis undertaken as part of the sequencing projects, has assigned definite functions to between 40 and 45% of the ORFs identified in *E. coli*. The majority of these correspond to previously defined or sequenced genes. Some functions could be assigned because of the near identity of new ORFs with those encoding better studied homologues in related organisms. A further 20-25% of translated ORFs have significant amino acid sequence similarity to existing data base entries which could be used to predict function. The remaining 30% of ORFs remain functionally uncharacterized.

(i) Cryptic Genes- These are phenotypically silent genes that are not normally expressed but upon mutation confer a selective advantage to the organism under certain environmental conditions. Instead of being lost they are thus actively maintained. For instance, *bgl* gene (ß-glucoside utilization gene) in *E. coli* is cryptic. The number of such genes per bacterial genome and their distribution in different bacteria is difficult to estimate.

(ii) REP Sequences (Repetitive Extragenic Palindromic Sequences or PU)- The

REP sequences or palindromic units (PU) constitute a family of highly conserved inverted repeat sequences measuring 30-40 nucleotides. The known REP sequences are always extragenic and occur in the middle or at the end of operons in transcribed but untranslated sequences. They may occur in clusters of 2-4 copies or more and as frequently as once per 13 kb. The tandem copies always alternate in orientation, i.e., in inverted orientation with respect to each other. Each cluster of REP sequences is referred to as REP element. By sequencing estimate, 581 REP sequences organized in 314 REP elements containing 1-12 tandem copies are distributed throughout the *E. coli* genome, constituting up to 0.54% of the genome.

Each REP sequence consists of a highly conserved dyad symmetry which bestows the potential to form a stem-loop structure in the DNA or in the transcribed RNA. Adjacent copies of the REP sequences are separated by less than 25 bp and their inverted orientation provides a potential for the formation of RNA hairpin structure. No example of REP sequence in the coding sequence for a protein has yet been found. These are highly conserved across the bacterial species barrier and implicated in the regulation of intraoperonic differential expression by affecting the efficiency of translation of downstream genes. As a consequence of palindromic structure formation, they may be involved in protection of mRNA from 3′ degradation. REP sequences could also play a role in chromosomal rearrangements and thus in the generation of genetic diversity. They have also been involved in chromosomal structure and organization, especially in maintaining the domain structure and appear to provide the sites at which DNA might attach to the core.

(iii) BIME (Bacterial Interspersed Moșaic Element)- REP sequences in association with other non-palindromic repetitive DNA sequences form complex repeats called BIME. These have been reported from *E. coli* and other Gram-negative bacteria.

(iv) ERICS (=IRU)- Enterobacterial Repetitive Intergenic Consensus Sequences: ERIC sequences are found in enteric bacteria such as *E.coli*, *S. typhimurium*, *Klebsiella*, *Vibrio cholerae* etc. These 126 bp sequences, like REP, are associated with transcribed regions of the genome. No function has been assigned to them as of now. A total of 19 copies have been detected from the genome sequence analysis of *E. coli*.

(v) Other Repeat Elements- *Mycoplasma pneumoniae* contains reiterated sequences of 300 bp called REPmp1 and of 400 bp called SDC1. The Gram-negative bacterium, *Haemophilus influenzae* harbours approximately 600 copies of an 11bp sequence in its genome. Their respective functions are not very clear.

In Gram-negative *E. coli*, 33 copies of Box C, 6 copies of RSA, 3 tandem and a lone fourth copy of LDR and a 29-bp long *jap* repeats found in three clusters of 14,2, and 7 copies have been shown. The largest repeated sequence in *E. coli* K-12 are the five Rhs elements which are 5.7 to 9.6 kb in length and together comprise 0.8% of the genome. Although no function has been assigned to them, strain comparisons suggest that they may be mobile. They show uncanny similarities to the mammalian transposable sequence, LINE-1.

(vi) Insertion Sequences- *E. coli* chromosome is dotted with as many as 10 different types of IS elements. They are a rich source of spontaneous mutations both by insertional inactivation and by generating chromosomal aberrations. For further details on these elements readers may refer to **Chapter 7**.

(vii) Cryptic Prophage and Phage Remnants- Number of cryptic prophages, such as defective lambdoid prophages DLP12, Rac, and Qin, the element e14, and "cryptic P4" phage CP4-57 and related CP4-6 and CP4-44 and a new cryptic prophage are present in *E. coli.* These have lost some functions essential for lytic growth, but retain other functional genes. They can thus rescue bacteriophages carrying mutation in these genes. Besides these, numerous instances of isolated genes that are similar to bacteriophage genes have also come to light. These have been called phage remnants but their origin may lie either in the bacteriophage or its host.

4.3.12.2. Genome Sequencing. Because of the extraordinary contribution of *E.coli* K-12 as a model system in biochemical genetics, molecular biology and biotechnology, this organism was the first candidate suggested for whole genome sequencing. Before *E. coli* genome was fully sequenced, however, five prokaryotic and one eukaryotic genome sequences became available. These include: *Haemophilus influenzae, Mycoplasma genitalium, M. pneumoniae,cyanobacterium, Synechocystis* sp. strain PCC 6803, the archaebacterium, *Methanococcus jannaschii* and the yeast, *Saccharomyces cerevisiae.* Since then a large number of other bacterial and archaeal genomes have been sequenced (Table 4.2) and several others are nearing completion. Being a well-studied experimental organism, *E. coli* genome sequence generated tremendous excitement as it will help the reevaluation of the vast information collected from this bug over the years and initiate a fresh look at genome organization and gene structure-function relationships.

The genome of *E.coli* consists of 4,639,221 bp of circular duplex DNA. The 4288 annotated protein-coding genes account for 87.8% of the genome, the share of stable RNAs is 0.8%, of non-coding repeats 0.7%, leaving ~11% for regulatory and other functions. The base pair 1 has been assigned to a region between genes *lasT* and *thrL.* The origin and terminus of replication divide the genome into oppositely replicating halves, which have been termed as replichores. One interesting feature revealed by genome sequencing is the frequent presence of oligomers in the form of a family containing the trimer CTG (often in the pentamer GCTGG) in leading stand. The DnaG primase binding sites is proposed to be CTG with T being the template for the first base of the RNA primer of Okazaki fragments. Even though a direct proof is lacking, their spacing is consistent with Okazaki fragment size. The high abundance of these sites on both strands supports a model that both strands are replicated discontinuously. Many features of *E. coli* genome are correlative to replication orientation. For example, all seven ribosomal RNA (rRNA) operons, and 53 of the 86 tRNA genes are expressed in the direction of replication. Similarly, 55% of the protein coding genes are also aligned in the direction of replication. This has not only confirmed an earlier observation but also given credence to the hypothesis that genes whose products may be required in abundance may be aligned this way.

Other interesting features include, a greater abundance of G in the leading strand of both the replichores in comparison to C, A or T, detection of new tRNA genes (six), identification of full two operons for degradation of aromatic compounds, nearly identical flagellar operons as in *Salmonella,* and assignment of gene functions to several open reading frames (ORFs). Moreover, it has become possible to arrive at homology between

E.coli proteins and those from other sequenced genomes. Maximum number of matches to *E. coli* is found in the *Haemophilus influenzae* genome both belonging to gamma subdivision of proteobacteria. Amongst the four eubacteria, *E.coli*, *H. influenzae*, *Synechocystis* sp. and *M. genitalium*, 111 proteins are in common. The number of matches across the kingdom, i.e., eubacteria, archaebacteria and eukaryotes, goes down considerably and that too is restricted to translational proteins. Numerous differences found amongst the proteins match represent the subset of proteins specific to a particular species.

Table 4.2. Completed and published genome sequences of some bacteria and archaea

S.No.	*Name & Strain*	*Size(~Mb)*	*Year*
1.	*Haemophilus influenzae* Rd KW 20	1.83	1995
2.	*Mycoplasma genitalium* G 37	0.58	1995
3.	*Methanococcus jannaschii* DSM 2661	1.66	1996
4.	*Mycoplasma pneumoniae* M 129	0.81	1996
5.	*Synechocystis sp* PCC 6803	3.57	1996
6.	*Archaeoglobus fulgidus* DSM 4304	2.18	1997
7.	*Bacillus subtilis* 168	4.20	1997
8.	*Borrelia burgdorferi* B 31	1.44	1997
9.	*Escherichia coli* K -12	4.60	1997
10.	*Helicobacter pylori* 26695	1.66	1997
11.	*Methanobacterium thermoautotrophicum* delta H	1.75	1997
12.	*Aquifex aeolicus* VF 5	1.50	1998
13.	*Chlamydia trachomatis* Serovar D	1.05	1998
14.	*Mycobacterium tuberculosis* H 37 Rv	4.40	1998
15.	*Pyrococcus horikoshi* OT 3	1.80	1998
16.	*Rickettsia prowazekii* Madrid E	1.10	1998
17.	*Treponema pallidum* Nichols	1.14	1998
18.	*Aeropyrum pernix* K1	1.67	1999
19.	*Chlamydophila pneumoniae* CWL 029	1.23	1999
20.	*Deinococcus radiodurans* R 1	3.28	1999
21.	*Helicobacter pylori* J 99	1.64	1999
22.	*Thermotoga maritima* MSB 8	1.80	1999
23.	*Campylobacter jejuni* NCTC 11168	1.64	2000
24.	*Chlamydia trachomatis* MoPn	1.07	2000
25.	*Neisseria meningitidis* Serogroup A, Z 2491	2.18	2000
26.	*Neisseria meningitidis* MC 58	2.27	2000
27.	*Chlamydophila pneumoniae* AR 39	1.22	2000
28.	*Chlamydophila pneumoniae* J 138	1.22	2000
29.	*Agrobacterium tumefaciens*	2.84	2001
30.	*Bacillus halodurans*	2.40	2001
31.	*Caulobacter crescentus*	4.01	2001
32.	*Chlamydia muridarum*	1.06	2001

S.No.	Name & Strain	Size(~Mb)	Year
33.	*Clostridium acetobutylicum*	3.94	2001
34.	*Escherichia coli* O 157:H 7	5.49	2001
35.	*Escherichia coli* O 157:H 7 EDL 393	5.52	2001
36.	*Lactococcus lactis* subsp. *lactis*	2.36	2001
37.	*Listeria innocus*	3.01	2001
38.	*Listeria monocytogenes EGD-e*	2.94	2001
39.	*Mesorhizobium loti*	7.03	2001
40.	*Mycobacterium leprae*	3.26	2001
41.	*Mycobacterium tuberculosis* CDSC 1551	4.40	2001
42.	*Mycoplasma pulmonis*	0.96	2001
43.	*Pseudomonas aeruginosa*	6.26	2001
44.	*Rickettsia conorii*	1.26	2001
45.	*Salmonella typhimurium* LT 2	4.85	2001
46.	*Sinorhizobium meliloti*	3.65	2001
47.	*Staphylococcus aureus* subsp. *aureus* Mu 50	2.87	2001
48.	*Staphylococcus aureus* subsp. *aureus* N 315	2.81	2001
49.	*Streptococcus pneumoniae* R 6	2.03	2001
50.	*Streptococcus pneumoniae* TIGR 4	2.16	2001
51.	*Streptococcus pyogenes*	1.85	2001
52.	*Vibrio cholerae*	1.07	2001
53.	*Xylella fastidiosa* 9 a 5 c	2.67	2001
54.	*Yersinia pestis*	4.65	2001

The other significant contribution is towards the identification of operons, promoters, and regulatory proteins. Of 2192 predicted operons, a surprisingly high (73%) have only one gene, 16.6% have two, 4.6% have three and 6% have four or more genes. Sixty-eight percent contain one promoter, 20% two promoters, and 13% contain three or more promoters.

Genome sequences can also be used to obtain significant evolutionary insights. For example, the proteins that appear to have evolved recently in comparison to those with much longer histories can be discerned by comparing Archaea, Eubacteria, and Eucarya. This information can be used for phylogenetic analysis and relationships between these three superkingdoms.

Similarly, the size of the 'minimal genome' can be ascertained. *M. genitalium* possesses the smallest known genome of any free-living organism. This cell wall-deficient fastidious prokaryote has a genetic complement of only 470 genes (~10-15% of *E.coli*). Although it obtains most of its nutrition directly from its host, it can grow independently. Thus, one can ask whether its small genome defines a basic set of genes that is shared by all free-living organisms. Sequence analysis can answer this question. One can also ascertain whether there are unique genes which define a bacterial species or a bacterial species is simply a unique combination of genes, each of which is shared with some other species.

Genome sequencing of typhus-causing *Rickettsia prowazekii* (1,111,523 bp) suggests that one of its ancestors might have contributed towards the formation of mitochondrion. Amino acid sequences of the proteins coded by mitochondrial and bacterial genes, involved in energy metabolism and genetic processes appear to be closely related. Both lack anaerobic glycolysis as well as most of the genes involved in amino acid and nucleotide biosynthesis and share the pathway of ATP production.

4.3.12.3. Linear Chromosome. It was initially believed that another important difference between prokaryotic and eukaryotic organisms is the presence of circular chromosome in the former and linear ones in the latter. However, linear chromosomes have now been reported in a number of bacteria, such as *Borrelia burgdorferi*, the causative agent of Lyme disease, *Streptomyces lividans*, and *Rhodococcus fasciens*. Ends of some of the linear chromosomes possess inverted repeat sequences that are sometimes covalently joined to each other. The replication of bacterial linear chromosomes appears to be different from that of eukaryotes. Interestingly, bacteria having linear chromosomes also contain linear plasmids.

4.3.13. Plasmids

Besides the chromosome, most bacteria contain an additional genome component. Although more modest in genetic complexity, this extrachromosomal, autonomously replicating DNA molecule, known as plasmid, play a significant role in bacterial adaptation and evolution. Ever since they have been used as important tools in the field of molecular biology they have acquired a house-hold name.

Plasmids, which vary widely in size, from a few kilobases to hundreds of kilobases, and in type, exist most often as covalently closed circular double stranded DNA molecules carrying genes that may have tremendous importance under particular environments (Fig. 4.18). Some bacteria possess linear plasmids, such as *Borrelia burgdorferi* and some members of genus *Streptomyces*. These have telomeric sequences at their ends, analogous to eukaryotic chromosomes. Like chromosomes, plasmids code for functions responsible for their replication, distribution when the cell divides, and transfer but some may depend upon many chromosomally encoded functions for these processes. The plasmid transfer can be to the cells of the same species or of the other species and sometimes also across the kingdom. The different degrees of host specificity indicate variations in the evolution of autonomy among plasmids. Unlike chromosome, plasmids, however, do not generally encode functions essential to bacterial growth, though exceptions to this rule are also becoming available. Some plasmids may give up their autonomous life-style by occasionally recombining with the host chromosome (e.g. F plasmid), a feature reminiscent of temperate phages like lambda. At the same time, some phages may also exist in a form of plasmid, like coliphage P1 (on rare occasions it can also integrate into the chromosome). This suggests evolutionary interrelationship between these two structures.

4.3.13.1. Detection and Isolation of Plasmids. Plasmids can be detected by both genetic and physical experiments. The genetic method of detection is based on the availability of phenotypic marker/s, whose transfer to a cell can be monitored. The initial identification of F plasmid was carried out by its ability to promote genetic recombina-

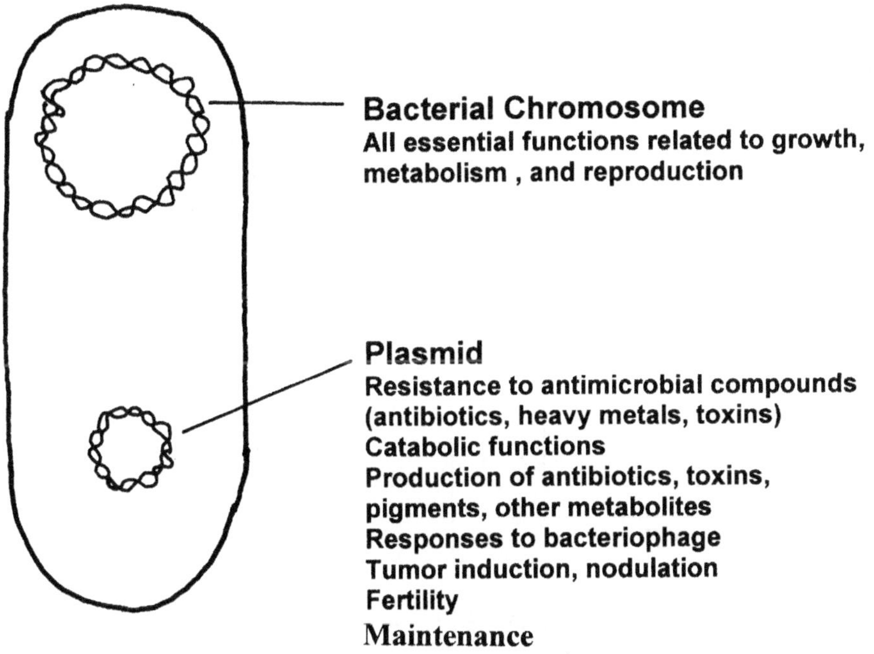

Figure 4.18. Comparison of chromosomal and plasmid-encoded functions. A cell may carry one or multiple copies of small- or large-sized plasmids. Moreover, plasmids may remain autonomous or may get integrated into the chromosome sometimes.

tion with the cell not carrying the plasmid. The plasmid-containing cell is thus known as donor and the plasmid-free the recipient. The plasmids carrying antibiotic resistance markers are easy to be studied as their transfer can be detected by a simple plating test.

For physical detection (Fig. 4.19), the bacterial cells are gently opened (lysed) by detergent treatment, and the resulting translucent solution (called a cell lysate) can be loaded for agarose gel electrophoresis. Because of large size differences between the chromosomal and plasmid DNA, the two can be easily separated. For isolating the plasmid DNA, the cell lysate is treated with alkali to denature DNA strands and subsequently the salt concentration is increased to precipitate out chromosomal DNA along with proteins. The supernatant obtained after removing the precipitate also called clear lysate can be resolved by gel electrophoresis for the presence of plasmids.

Other methods used for plasmid purification is based on the fact that CCC DNA binds less ethidium bromide (EtBr) than do linear or nicked circular DNAs. Thus, the strands of chromosomal DNA, which usually get broken during the initial steps of isolation, will bind much higher concentration of EtBr. EtBr binding to DNA makes it less dense in salt solutions of Cesium chloride (CsCl). Thus if the DNA mixed with a solution of CsCl and EtBr is centrifuged to establish a gradient, CCC plasmid DNAs will band lower than the chromosomal DNA. The plasmid DNA from the band can be recovered and purified (Fig. 4.19).

The methods described above work well with plasmids which have many copies per cell and are not too large. The large plasmids will pose the same problem as the chromosome necessitating modified methods for their detection. In one, the cells are lysed directly on agarose gels and the subsequent application of electric field leads to the separation of chromosomal DNA from the plasmid. The DNA bands are visualized by staining the gel with EtBr and then viewing the gel under UV light.

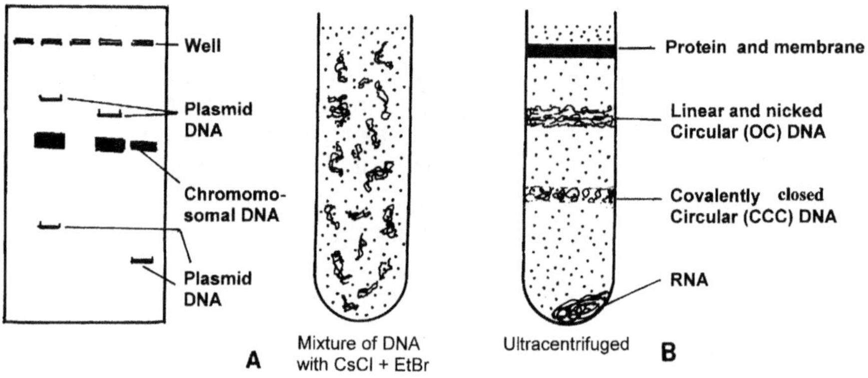

Figure 4.19. Two methods of detecting plamid DNA A. The crude lysate loaded in the wells can be run on agarose gel electrophoresis. Along with distinctive chromosomal DNA, plasmid DNA of different sizes are separated on the basis of their differential mobility, B. A mixture of DNA with CsCl and EtBr is resolved by ultracentrifugation. Covalently closed circles (CCC) of DNA consisting of plasmids is separated from nicked circular and linear chromosomal DNA. The plasmid DNA can then be retrieved and purified.

4.3.13.2. Plasmid Nomenclature. Plasmids were initially named on the basis of phenotypic function they expressed. For example, R-factor carries genes for resistance to several antibiotics, F-plasmid as it mediates fertility, ColE1 plasmid carries a gene for colicin E1 protein, Tol plasmid contains genes for toluene degradation, and Ti plasmid of *Agrobacterium tumefaciens* induces tumors in some plants. A more standardized method of plasmid nomenclature has now been evolved. A small letter "p" for plasmid precedes capital letters that describe either the plasmid or gives the initials of person or persons who isolated or constructed it. These letters are often followed by a number to identify the particular construct. Any subsequent modification in a plasmid is denoted by a number different from the original. For example, pBR322 was constructed by Bolivar and Rodriguez and the one used is the derivative number 322. When a chloramphenicol resistance gene was inserted in this plasmid, the plasmid was rechristened as pBR325.

4.3.13.3. Structure. Most plasmids are covalently closed circular structures consisting of double stranded DNA. Like the chromosome, plasmid DNA is also negatively supercoiled. To exist in free autonomous form, the plasmids must have the ability to replicate independently. This will demand an *ori* site, and proteins required for replication. Most plasmids contain an *ori* site and a few proteins to initiate replication. Most other proteins such as DNA polymerases, ligase, helicase, primase are borrowed from the host. Additional genes are carried for maintaining the copy number, partitioning, incompatibility, and maintenance. Another set of genes may be devoted towards the organization of transfer apparatus and execution and regulation of transfer. Genes conferring a particular phenotype are also located on plasmid. Examples of the last category would be genes for resistance to antimicrobial agents and/or virulence etc. (see Fig. 4.18).

4.3.13.4. Properties. The plasmid replication is initiated at an unique site called *ori*. The two modes of replication are theta replication and rolling circle. While the former is the most common mode (ColE1, F, RK2 and several others) the latter is commonly seen in some Gram-positive species.

The host range of a plasmid consists of all the species of bacteria in which a plasmid can replicate. On this basis, plasmids can be classified as narrow host range (pBR322, pUC) and broad host range (RK2, RSF1010) types.

The plasmids may also vary widely in their copy number per cell. While plasmid like F or P1 prophage exists in single copy, pBR322 occurs in 16, pUC series in 30-50 and *Streptomyces coelicolor* plasmid pIJ101 in 40-300 copies. Obviously, plasmids having low copy number have a tighter regulation of their replication (stringent control) as compared to those with high copy numbers (relaxed control). The plasmid copy number control can be mediated by RNA (ColE1), RNA and protein (ColE1, R1) or by proteins (iteron plasmids such as pSC101, F, RK2 and RP4-related plasmids).

Plasmids also express function to maintain themselves in a cell and resist spontaneous loss during cell division (called plasmid curing). This is brought about by a site-specific recombinase that destroys multimers (ColE1 and phage P1) or by involving *par* functions that ensures their proper partitioning at the time of cell division. The *par* system consists of a *cis*-acting *par* sequence and few proteins that act at this site, though the actual mechanism is still not very clear. Many plasmids, like F, P1, ColE, R1, have evolved mechanisms that resist plasmid curing. This has often been referred to as plasmid addiction system.

Many natural isolates may contain more than one type of plasmids. These plasmid types stably coexist in the same cell and maintained through many cell generations. However, there are plasmids that can not stably coexist, as they interfere with each other's replication or partitioning. When such plasmids are introduced in the same cell one or the other will be lost at a rate higher than normal. Such plasmids belong to same incompatibility or Inc group and the phenomenon is known as plasmid incompatibility. Plasmids have often been classified by their Inc group.

One of the earliest properties of the plasmid discovered is their capability to get transferred from a donor to recipient cell through cell-to-cell contact. Bacterial plasmids with this property are able to spread in a population easily.

Another important feature of plasmids like F is their rapid transfer to plasmid-free recipient cells. So much so that every recipient that receives a copy of F becomes potential donor which can transfer the plasmid to other F^- cells. Such rapid spread, however, is not seen for all plasmids. For example, only about 0.02% of a population of cells containing R-plasmids are competent donors. If after the transfer of R-plasmid, an excess of R^- recipients is added, R-plasmid quickly spreads through the recipient population. The kinetics of transfer suggests that an R^- cell which has just received an R-plasmid is able to transfer the same to a fresh recipient efficiently. This process known as fertility inhibition is caused by a multisubunit repressor, coded by two *fin* genes, that represses the genes required for transfer. While in a fresh recipient, where the repressor is absent, transfer process is derepressed. F also carries *fin* genes but due to an insertion of IS3, one of them (*fin O*) becomes inactive and fertility inhibition is not expressed.

A cell containing a plasmid limits its capacity to act as a recipient for the same or related plasmids. This property called as surface exclusion (*Sfx*) appears to be the fundamental requisite for a cell to act as donor. Five surface exclusion specificity classes, Sfx I to Sfx V, representing F, R, Col and others have been identified. Two genes implicated in this process *traS* and *traT* control two independent aspects of the phenomenon. The *traS* gene encoding a 16.9-kDa inner membrane protein appears to block the transmission of a mating signal and thus prevents triggering of donor conjugal DNA metabolism. The 26-kDa lipoprotein product of *traT* is the major outer membrane component of F-containing cells. Exposed on the cell surface it appears to function by inhibiting the cell aggregate formation. Capacity to transmit themselves typically requires three genetic functions, a transfer locus, *tra*, a mobilization locus, *mob*, and a sequence required in *cis* for transfer to initiate, called *oriT*. Including the *oriT*, 33.3 kb region of F constitutes the *tra* region. This region consists of over two dozens ORFs involved in pilus biogenesis, surface exclusion, mating aggregate stabilization, regulation, and conjugal DNA metabolism. Functions of some are still not known. F and several other plasmids possess this property, but many plasmids are not able to mediate transfer process. Accordingly, plasmids have been classified into four categories:

Non-transmissible plasmids lack genes necessary for effective contact and DNA transfer. A conjugative plasmid has genes to effect cell to cell contact. A mobilizable plasmid, on the other hand, can not make effective contact but can prepare its DNA for transfer. Such plasmids, like ColE1, however, are equipped to exploit the conjugative apparatus of self-transmissible plasmid and in turn, can be helped to get transferred. So if a cell contains both ColE1 and F, Col E1 plasmid can get transferred by utilizing the conjugative functions of F. This process is also known as donation. A self-transmissible plasmid normally cannot help a non-mobilizable plasmid. Transfer in this situation would require recombination between these two types of plasmids, a process known as conduction. A self-transmissible plasmid like, F, RK2, and R100 have both, the capability of effecting cell-to-cell contact and transfer its DNA.

This property of plasmid has contributed enormously to bacterial evolution, and at the same time has also been exploited as carrier for horizontal gene transfer not only between strains of the same species, but also to cross species barrier to move DNA between different species and even organisms from different kingdoms.

However, for one reason that plasmids have become extremely popular is their use as vectors in genetic engineering experiments. Plasmids as such or after appropriate "modifications" have been extensively used as cloning vectors in molecular biology. This is based on following properties:

i) They are small so that they can be easily isolated and introduced. Some of them can be tailored to an appropriate size. They are self-replicating.

ii) They usually have high copy number so that they can be purified and obtained in sufficient quantities.

iii) They often carry an easily selectable marker which can be used to select cells that contain the plasmid.

iv) They can have a whole spectrum of one or few sites for restriction endonucleases. This allows plasmids to be opened up and a foreign DNA can be inserted. These sites may occur in a selectable marker so that chimeric molecule (plasmid+foreign DNA) can either be identified by loss of trait (insertional inactivation) or by use of chromogenic substances.

v) They can be used in a form that will bring them under the purview of biological safety considerations.

Many plasmids can be modified to suitably express the properties stated above. Some additional modification can be brought about to significantly enhance their suitability. For example, addition of phage packaging systems (*pac* or *cos*) can lead to their packaging in phage heads (cosmid). Expression vectors can be used to express foreign proteins in a bacterial cell. They carry strong bacterial promoter and a translation initiation region (e.g. pUC and pET vectors). They can be derived from broad-host range plasmids, such as RP4 and RSF1010 which will extend the host range in which they can be applied. They can also be used in constructing a shuttle vector which is a composite carrying *ori* region from two organisms. They are so named as they can shuttle genes between two organisms. For example, shuttle vectors can replicate both in Gram-positive and Gram-negative, or say in *E. coli* and yeast. Some shuttle vectors having ColE1 origin and simian virus 40 origin of replication can replicate both in *E. coli* and mammalian cells. We shall discuss these vectors in detail in **Chapter 7**.

CHAPTER 5
BACTERIA AND LIFE PROCESSES- I
GROWTH AND MULTIPLICATION

Like any other living cell, a bacterial cell exhibits all the characteristics of life. These can be broadly classified as: (I) growth and multiplication, and (II) metabolism. These two form the subject matter of **Chapters 5 and 6**.

I - Growth and Multiplication

For an unicellular microbial cell, growth is synonymous to cell division. Thus, while in an eukaryotic organism the growth is estimated by increase in size, in prokaryotic organisms it deals with an increase in the cell numbers. Growth therefore has an important implication as both species maintenance and control of microbial proliferation/action, require an indepth knowledge of microbial growth.

In this Chapter we shall first discuss various aspects of biology of growth, measurement and factors affecting growth, the process of cell multiplication/division, and finally the control of microbial growth.

5.1. Population Growth

As described above, growth leads to an increase in the number of microbial cells in a population which can be equated with increase in the microbial mass. One of the parameters to assess growth is the growth rate that defines the change in cell number or mass per unit time. Another parameter is the estimation of generation time, which is the time required for doubling the cell number. The generation time, therefore, is also called as the doubling time. The generation time is highly variable ranging from few minutes to few hours or several hours to even days (Table 5.1). This is because growth is dependent upon number of factors, both nutritional and genetic. The microbial population exhibits a characteristic growth pattern known as exponential growth, in which the number of cells doubles during each unit time period. The increase in cell numbers can be estimated as a simple geometric progression of the number 2. Thus 2 cells doubling to form four can be expressed as 2^1 and 2^2, and so on. The number of cells (b) present at any given time can be expressed as $b = 1 \times 2^n$. However, b is also dependent upon the initial number of the cells (a) present in the inoculum so the relationship can be rewritten as $b = a \times 2^n$. When cell numbers are plotted against increasing time on arithmetic coordinates, a curve with constantly increasing slope can be seen. However, for a more convenient examination, the growth data are converted logarithmically. The $\log_{10}$ values plotted against time on arithmetic scale (semi-logarithmic graph) gives a straight line. The doubling time can be estimated directly from such straight lines (Fig.5.1).

Since there is a direct relationship between the number of cells present initially and the number that will be present after a period of growth, generation time can be calculated by the relationship: $N = N_0 2^n$.

Where N_0 is the initial number, N is the final number and n is the number of generations that have elapsed during exponential growth. The generation time (g) can be calculated by the relationship t/n, where t denotes the time of exponential growth in hours or minutes, n can be calculated in terms of readily measurable quantities by:

$N = N_o 2^n$

$\log N = \log N_o + n\log 2$

$\log N - \log N_o = n\log 2$

$$n = \frac{\log(N) - \log(N_o)}{\log 2} = \frac{\log(N) - \log(N_o)}{0.301}$$

Once the value of n is estimated, the generation time can be calculated by t/n.

The measurement of generation time is very useful in microbiological research (Table5.1). It reflects the physiological state of a cell population and therefore can be used to study the effect of different environmental factors. Moreover, the cells of a particular phase could be prepared for different physiological and genetic studies.

Table 5.1. Growth rate of some bacteria under optimal conditions

Organism	*Temperature (^oC)*	*Generation time (min.)*
Bacillus stearothermophilus	60	11
B. subtilis	37	27
Escherichia coli	37	20
Staphylococcus aureus	37	28
Streptococcus lactis	37	30
Pseudomonas putida	30	45
Lactobacillus acidophilus	37	75
Vibrio marinus	15	80
Mycobacterium tuberculosis	37	360
Bradyrhizobium japonicum	25	400
Nostoc japonicum	25	570
Anabena cylindrica	25	840
Treponema pallidum	37	1980

5.2. Measurement of Growth

The growth of a bacterial population can be measured either in terms of cell numbers or the cell mass.

5.2.1. Total Cell Count

One of the most direct measures to determine the cell number involves microscopic count. Special counting chambers are available for this purpose. In this a grid consisting of squares of known small area is marked on a glass slide. A known volume of liquid can be accommodated on the grid area. Thus when a small volume of suspension is introduced, number of cells per unit area of grid can be counted under the microscope (Fig. 5.2). From this the number of cells per milliliter of suspension can be calculated.

The space between cover slip and slide is 0.02 mm (1/50 mm). The grid consists of

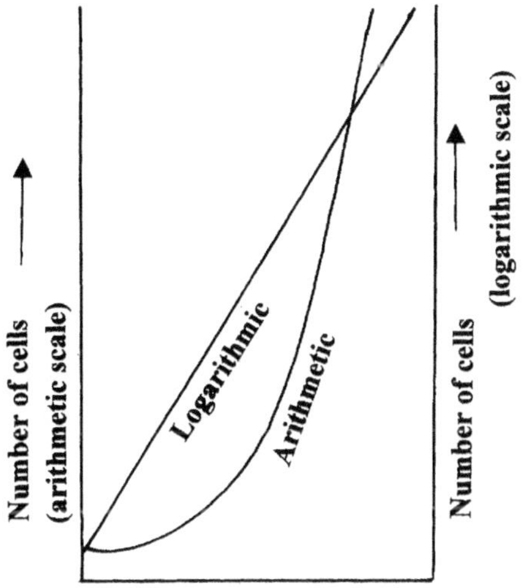

a

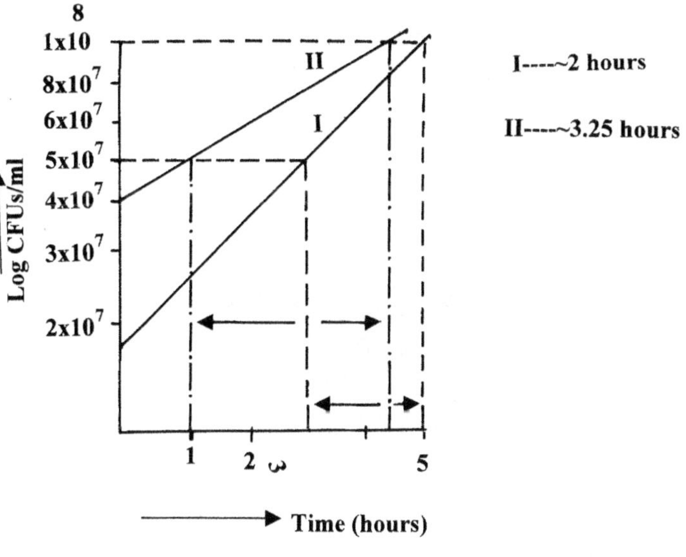

b

Figure 5.1a. Growth rate assessment of a bacterial culture when plotted on an arithmetic and a logarithmic scale. b. Growth profile of two bacterial strains used for calculating mean generation time. The difference reflects the rate of growth.

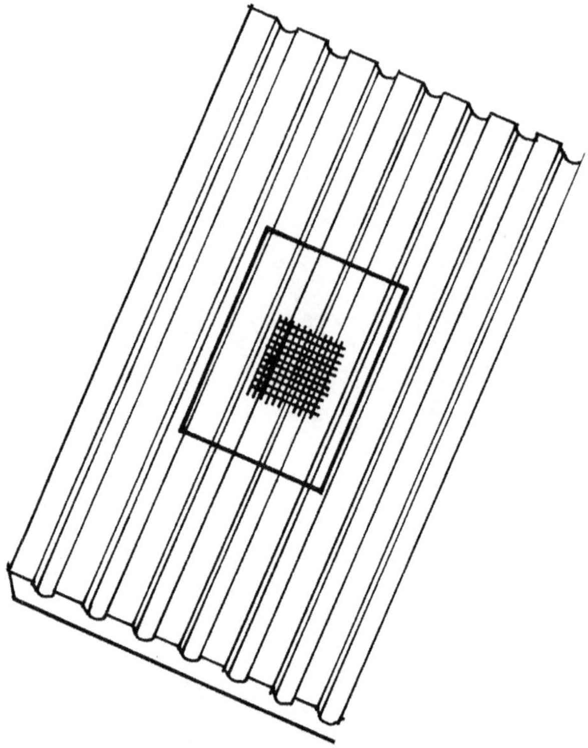

I

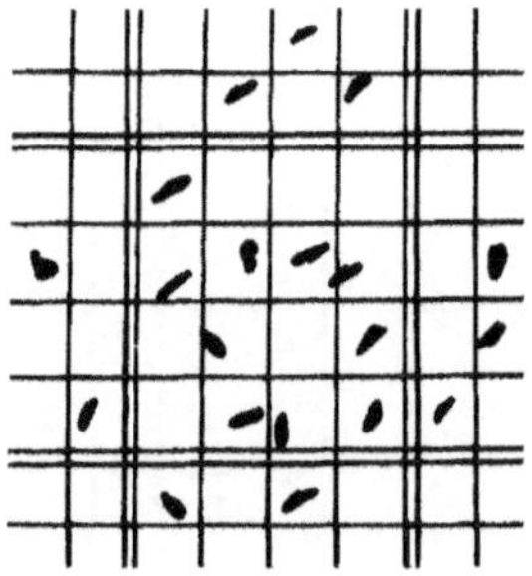

II

Figure 5.2. I. A haemocytometer with grooved slide, in the centre of which a grid containing 25 large squares is mounted. A small volume of bacterial cell suspension is placed in the grid area and covered with a cover slip. II. The slide is viewed under microscope and the number of cells is counted in all the squares and the average number is taken. The space between cover slip and slide is 0.02 mm (1/50 mm). The whole grid (25 squares) has a total area of 1 mm² and a total volume of 0.02 mm³. Thus the number of cells per millilitre of sample can be calculated: $10 \times 25 \times 50 \times 10^3 = 1.25 \times 10^7$.

25 large squares with a total area of 1 sq mm and a total volume of 0.02 mm^3. For example if, ten cells are counted in one large square (infact, several squares are counted and an average is taken), number of cells per ml can be calculated by:

Number (N_o) of cells per ml of sample = 10 cells x 25 squares x 50 x 10^3 = 1.25×10^7

10 cells	25 squares	50 x 10^3	1.25×10^7
Average number of cells counted per sq mm	Total number of squares	Total vol. of liquid	Number per cubic cm

Although it provides a quick measure of the cell number, the minute cell size poses a number of problems. Not only the viewing but also the counting becomes difficult and is less precise. Moreover, this method does not differentiate between live and dead cells.

5.2.2. Viable Cell Counts

In most studies, only the viable cells are important and since the microscopic count does not differentiate between a living and a dead cell, a viable cell counting method has been developed. This is based on the fact that all viable cells shall be able to form colonies on a suitable agar medium. This method is thus often called as the plate count or colony count, and the number of cells is obtained by counting the colony forming units (CFUs) per milliliter. There are two ways of performing a plate count both involving an appropriately diluted cell suspension. While in spread plate method, 0.1 ml of the diluted cell suspension is spread evenly on surface of an agar plate by a glass spreader, in pour plate the suspension is mixed with cooled molten agar and poured. In both cases the plates are incubated until the colonies appear (Fig.5.3).

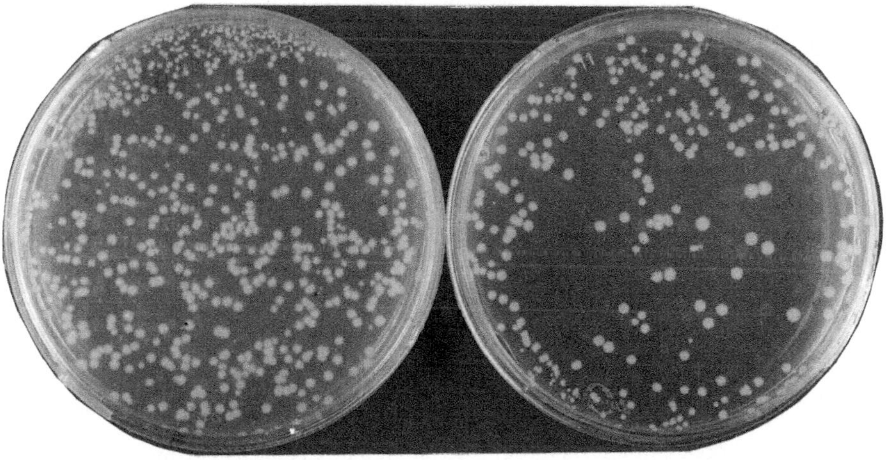

Figure 5.3. Plate counting- Bacterial growth assessment can also be made by counting the number of colonies developing from the plating of an appropriately diluted suspension. The dilution factor is then used to estimate the number of viable cells in the original suspension as colony forming units (CFUs/ml).

5.2.3. Cell Mass

In order to obtain the net cell mass, cells from a culture can be centrifuged and pellet of cells can be weighed to give the wet mass. Such cells can be dried in an oven at ~100°C and dry weight determined by weighing. Although it may vary from organism to organism and condition to condition, dry weight is approximately 10-20% of the wet mass.

The most common method of determining the cell mass is based on a turbidometric measurement. This is based on the fact that a cell suspension looks turbid because each cell placed in the path of light can scatter it. The more number of cells present will scatter more light and therefore more turbid the suspension will be. The turbidity is measured in terms of absorbance at a specific wavelength with a device called colorimeter or spectrophotometer (Fig. 5.4). Many times a standard curve between turbidity (optical density) and viable cell count is made beforehand over the growth period and the number of cells is ascertained by respective optical density. Although very quick and thus popular, this method is less sensitive as it may also not differentiate between living and dead cells.

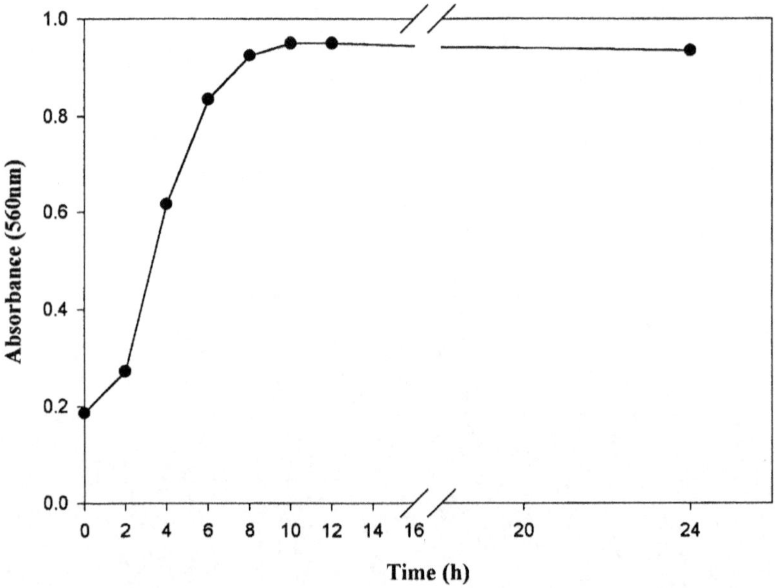

Figure 5.4. Assessment of bacterial growth by optical density measurement.

5.3. Growth Cycle

Each bacterial population goes through different phases during a growth cycle. Under the optimum physical and chemical conditions, a characteristic growth curve is produced as illustrated in Fig. 5.5. Based on the pattern of growth curve, it can be divided into four phases: lag phase, exponential or log phase, stationary phase, and death phase. In order to arrive at the growth curve, a population of cells is placed in a fresh medium and growth in terms of optical density or viable cell count is monitored over a period of time. Plotting the two gives the typical growth curve (Fig.5.5).

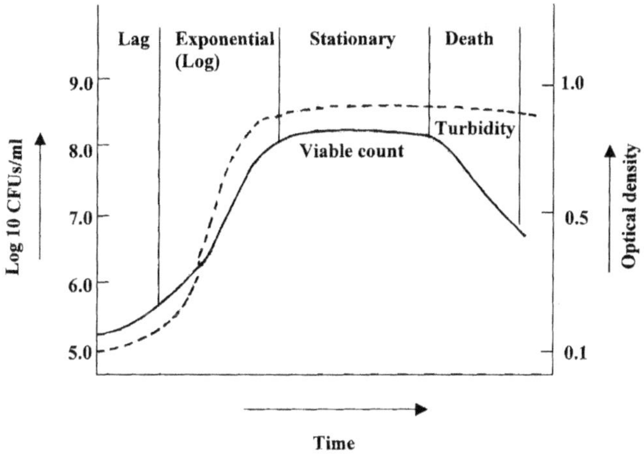

Figure 5.5. Typical growth curve of a bacterial population. In terms of viable cell count (CFUs/ml), four phase can be observed which are lag, exponential (log), stationary and death phase. When turbidity is measured the death phase may not be seen. The duration of each phase varies depending upon the bacterium and the environmental conditions.

When a cell population is inoculated into a fresh medium, the growth does not ensue immediately but only after a time has elapsed. This is reflected as the lag phase which depending on the condition may be brief or extended. For example, if the cells are taken from an exponential phase, the lag may not be seen at all but if the same are derived from old or stationary phase, a lag would usually appear. This is based on the fact that old cells may get depleted of some essential enzymes and other constituents and thus may require some time to re-establish their synthesis. Similarly, the cells which may have been exposed to deleterious conditions will require time to repair the damage. Also when cells are transferred from good growing conditions to adverse ones a lag will characteristically appear. A lag phase may also be seen when the exponentially growing cells are transferred to a medium with different nutrients, for the metabolism of which, the cells may have to make certain adjustments in terms of synthesizing some enzymes.

The exponential or the log phase has already been discussed earlier. This is the phase in which the cells multiply in geometric progression. The differences between the slow and fast growing organisms are most clearly seen during this phase. For example, *E. coli* with a generation time of 20-30 min grows very rapidly but *Mycobacterium tuberculosis* grows very slowly with one or two doublings per day. However, in comparison to eukaryotes, prokaryotes grow much faster. The rate of exponential growth is not only influenced by environmental conditions but also depends upon the genetic make-up of the organism. If an inoculum from exponential phase is transferred to fresh medium lag phase is bypassed and exponential growth continues.

The exponential growth, however, can not proceed for very long. The exponential growth is checked either by the exhaustion of a nutrient or synthesis of a toxic waste. This has important implications on population control. A bacterial cell growing with a generation time of 20 minutes will produce 10^3 cells in 3.3 h and in 48 h a population size, which with the average weight of a bacterial cell to be 10^{-12} gm, will come to 4000 times the weight of the earth.

Thus, the exponentially growing population eventually reaches a phase where there is no further net increase in the cell numbers. This is known as stationary phase. During this phase, in some cases the growth rate is equal to death rate. As the overall metabolic rate decreases, the cells may exhibit resistance to several environmental stresses, such as, elevated temperature and osmotic pressure. This physiological adaptation is partly due to expression of *katF* gene whose product regulates transcription of several genes. In some bacteria, like *Arthrobacter*, a morphological transition from rod-shaped to coccoid cells has been noticed during stationary phase.

Finally the growth is terminated by the onset of death phase during which the number of viable cells begin to decline rapidly. The rate of death is proportional to the number of survivors.

5.4. Diauxic Growth

When the bacterial cells are grown in the presence of a mixture of substrates, for example, two sugars, diauxic growth takes place. Diauxic growth is characterized by two exponential phases separated by a transient lag. For example, if *E.coli* cells are grown in presence of glucose and lactose, glucose will be utilized first. When glucose gets exhausted the growth will temporarily cease resulting in a lag. The system will get relieved of catabolite repression and will allow the synthesis of necessary enzymes for the utilization of lactose and a fresh round of growth will take place. Thus, two exponential phases reflect the active utilization of two sugars (Fig.5.6)

5.5. Arithmetic or Linear Growth

Under adverse physiological or metabolic conditions, the bacterial growth becomes linear, i.e. the cell division becomes proportional with time. Such a growth can be observed, for example, with an auxotroph grown in the absence of its nutritional requirement. Under such a situation, the cells will continuously get depleted of an essential metabolite so that the growth will reduce proportionately with time and will appear as linear or arithmetic. A similar situation can be observed when the cells are grown in the presence of an antimetabolite. For example, if *E. coli* cells are grown in the presence of ethionine, an analogue of methionine, the analogue is incorporated into proteins. Such proteins, however, may be faulty and devoid of the usual functions. For a cell to grow, the cell will depend upon the normal proteins synthesized before the analogue was added. The growth under such a situation will appear linear (Fig.5.7).

5.6. Growth of Bacterial Culture

The conditions used for growing bacteria in laboratory are very different from those

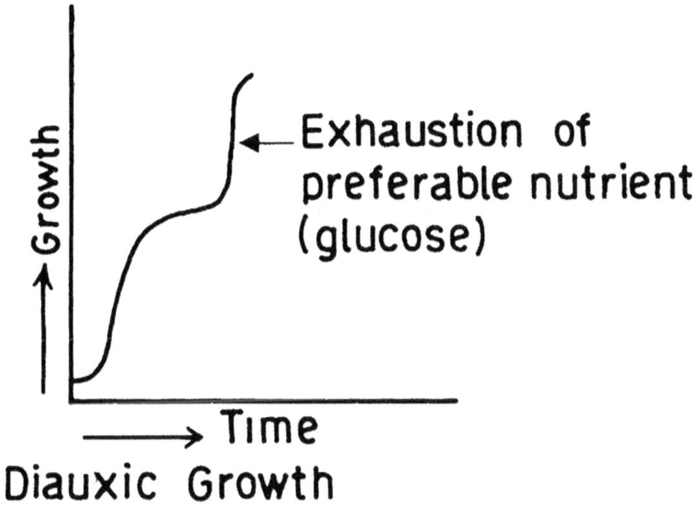

Figure 5.6. The diauxic or biphasic growth occurs when two carbon sources are present. An organism preferentially metabolizes one completely before utilizing the other.

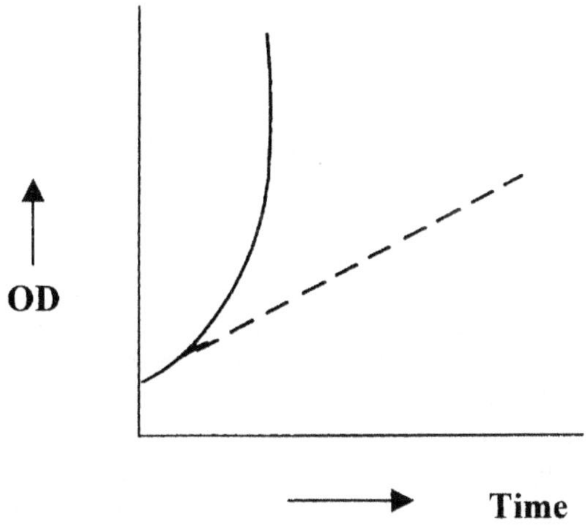

Figure 5.7. Linear growth- An exponential growth pattern changes to more linear growth when the cells face adverse conditions.

prevailing in nature. While in former the most optimal level of conditions are controlled and provided, in nature, they fluctuate and the growth rate varies as many species coexist. However, large scale applications of microbial biomass or the desired metabolic products demand the optimal growth of the particular bacterial species. Two such culture conditions have been designed for this purpose.

5.6.1. *Batch Culture*

Growing bacteria in a desired medium contained in a culture vessel is referred to as batch culture. The bacteria under such conditions exhibit the characteristic growth curve as described above. In such a culturing method, the nutrients are utilized and metabolic products accumulate. During exponential growth, the instantaneous, growth rate constant (μ) which reflects the growth behaviour of individual cells can be calculated by its relationship to generation time (g):

$$\mu = \frac{0.693}{g}$$

This relationship is based on the fact that during exponential growth, an initial given cell number (N) changes:

$$\frac{dN}{dt} = \mu N$$

The batch culture, though routinely utilized, is not suitable for mass production of biomass. In this process, each time a crop of a microorganism is needed, a new batch will have to be started.

5.6.2. *Continuous Culture*

The exhaustion of nutrients and accumulation of toxic products restricts the growth of a bacterial species within the limits of growth cycle. Such a method, therefore, cannot be used for continuous generation of biomass/products. However, if the spent medium is replenished by fresh medium continuous growth of a culture, thus referred as continuous culture, can be achieved. Two different set-ups are available to meet this objective (Fig. 5.8).

In a turbidostat, an optical sensing device measures the turbidity (growth) of a culture in the vessel. This generates an electric signal, that is employed to regulate the flow of fresh medium into the vessels and flow of the spent medium with the cells out of it. Thus, in a turbidostat, the rate of growth of the culture adjusts to this flow rate.

In a chemostat, the flow rate of the growth medium from a reservoir is fixed and the rate of growth of the culture adjusts to this flow rate. In a chemostat culture, nutrients are supplied and end products removed continuously. Thus, a stationary phase never reaches and bacteria remain in continuous growth. While bacteria may be continuously growing, a number of them are also being washed out and removed from the culture vessel. Thus, a constant number of bacterial cells are maintained in the chemostat. The instant growth rate of the bacterial population is measured by the relationship:

$$\frac{dN}{dt} = \mu N$$

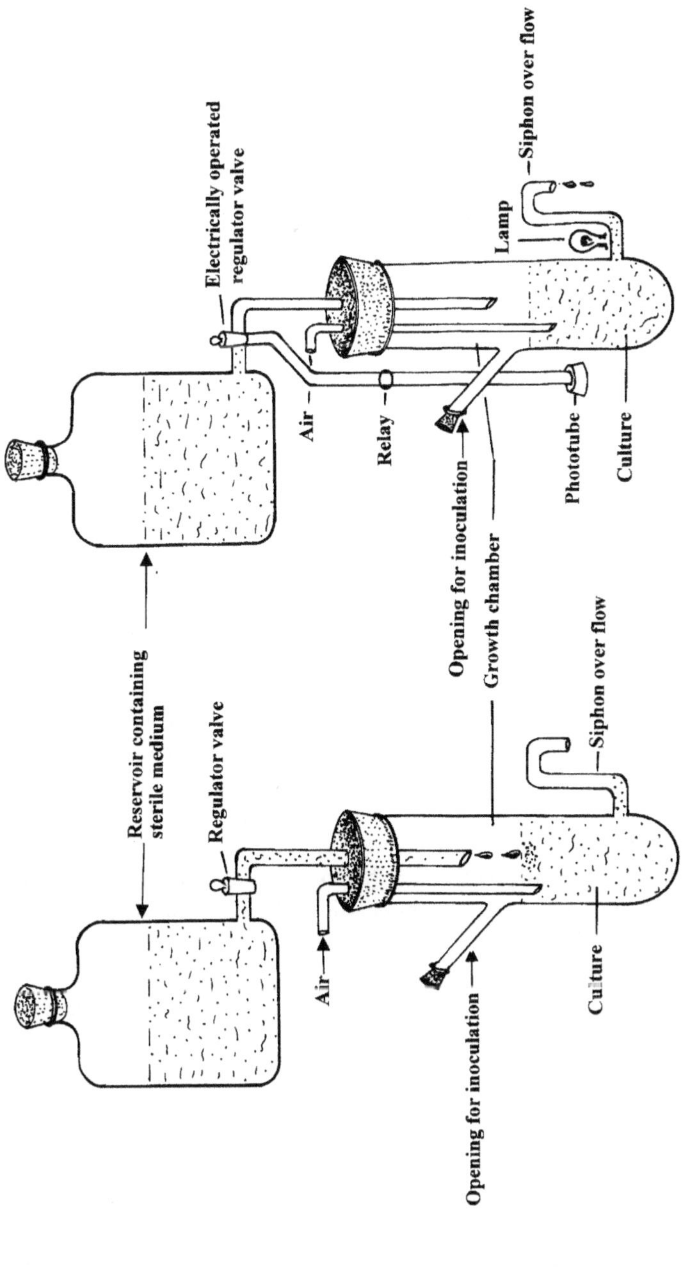

Figure 5.8. Two types of continuous culture systems. A chemostat and a turbidostat. A flow-through culture chamber receives continuous supply of nutrients allowing the uninterrupted growth.

This can be modified to make adjustments for cells being lost due to washing out. Where N is the size of the steady population and D is the rate of dilution:

$$\frac{dN}{dt} = \mu N - DN = (\mu - D)\,N$$

5.6.3. *Synchronous Culture*

When an inoculum is added to fresh medium, the growth begins almost synchronously but as the growth progresses, the population becomes highly heterogenous. In a synchronous culture, all the cells are at the same stage of growth and metabolic state. The synchronous state of a culture is highly desired as it makes the cells amenable to a wide range of analytical techniques, especially biochemical and genetic.

Synchrony can be brought about through two different techniques: induction synchrony and selection synchrony. In induction synchrony, the cells in a culture are brought to same stage of growth by physical or chemical manipulations. In selection synchrony, the cells at different stages of growth are physically separated and grown independently. Although the yields in selection synchrony are comparatively low, it is preferred as it is not fraught with any distortion of the growth cycle expected of induction synchrony.

One of the commonest method of inducing synchrony is to starve a culture or let it run into the stationary phase. Once such cells are placed in fresh medium, there is a burst of synchronous growth that may continue for one or two cycles. In *E. coli* and *Proteus vulgaris*, when early stationary phase cells are transferred to the fresh medium, synchrony is induced. Whereas in *E. coli* B, growing a culture at high cell density and a subsequent dilution is sufficient to bring out synchrony. Another way of inducing synchrony is to remove and re-add an essential nutrient of the culture medium. Thymine starvation in *E.coli* and *Lactobacillus acidophilus* results in cessation of DNA synthesis. A subsequent addition of this nutrient results in synchronous divisions. A temporary removal of a growth factor from the growth medium of a requiring strain, depletion and addition of glucose also leads to synchrony. Chemostat cultures of *E. coli* raised in phosphate-limiting medium become synchronous when phosphate pulses are provided at an interval coinciding with the doubling time of the culture.

Synchronization by cyclic changes in temperature or even a single heat shock has been successfully used in many bacteria. The synchronous cycles induced through these methods are much shorter than the normal cell cycle.

Different protocols of selection synchrony are also available. One is based on the filtration of a concentrated growing culture through layers of filter paper. The cells of different sizes are separated in different layers wherein the smallest ones pass through the filter and the largest are held on the top layer. A more refined method is to separate the cells on a sucrose gradient. A gradient of cells based on their sizes will be formed when a small volume of thick growing culture is layered on the top of a sucrose gradient and centrifuged. Any part of the gradient can be removed to start a fresh synchronous culture as the gradient linearly represents the cell cycle itself. These methods although applicable to large sections of microorganisms may run into practical difficulties as the size may not always reflect a specific stage of cell cycle and that the filtration may lead to metabolic disturbances.

An ingenious method for synchronizing *E.coli* B cells is based on membrane elution technique and popularly known as 'baby factory' method. A growing culture is passed through a membrane filter or an ion exchange paper. The filter is then inverted and fresh warm medium is run through slowly and continuously. During this flooding, the cells which are not properly adsorbed are washed off, leaving those which are strongly adhered to the membrane. When these cells divide, a crop of young, newly formed cells will be released in the circulating medium. Such cells can be collected periodically and used for raising fresh cultures (Fig.5.9).

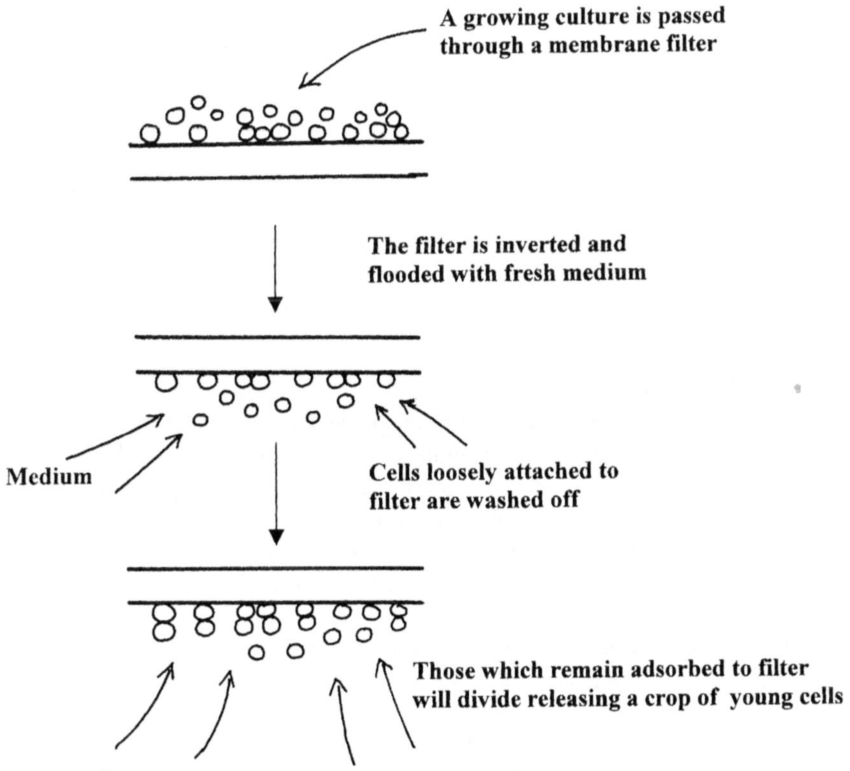

Figure 5.9. A technique for synchronization of bacterial growth.

Electronic Sorting: An electronic sorter which separates the cells by their volumes can also be used. Individual cells are dispersed in small droplets which are electrically charged and placed between deflector plates. The deflection which is proportional to the volume of a cell is measured and each droplet is collected based on the extent of deflection. Although this instrument can sort out about 1000 cells per second, it is cumbersome as the time required to collect a sufficient amount of material for any assay will be immensely long.

Some amount of natural synchrony has been observed in all living organisms. In many microorganisms, the stages during spore germination are naturally synchronized.

5.7. Growth on Solid Media

When a suspension of bacterial cells is spread on the surface of solid nutrient medium each cell grows to form a colony. In the colonies, basic growth curve is followed but since the cells do no disperse they are densely packed to give a characteristic shape and size. In the centre, the nutrients become limiting and the cells rapidly reach the stationary phase, but the peripheral cells continue to grow exponentially. The process of growth and division continues for a limited time leading to well defined colony. Each isolated colony is derived from a single cell and thus barring mutation, all bacterial cells in a colony will be genetically identical (see Fig. 5.3).

5.8. Effect of Environmental Factors on Growth

Growth is the direct outcome of an interaction between the cell's inherent capacity to grow and the environment in which it is placed. Thus, any fluctuation in these environmental factors will have a significant effect on growth. While under laboratory conditions these factors are available at their optimal levels, the natural conditions are expected to be entirely different, thus providing limitations of various types and at various stages.

5.8.1. Nutrient Concentration

The growth in biochemical terms is a composite process through which the cells produce energy and the essential cellular macromolecules. This cellular activity depends upon the utilization of various substances called nutrients which they obtain from the environment. Like all other living cells water forms the major component of a bacterial cell (~80-90% by weight). The role of several macro- and micro- elements essential for growth of a bacterial cell is already discussed in **Chapter 2**. Each nutrient plays an important role in overall growth.

Microbial growth remains unaffected at low nutrient concentration, until it becomes rate limiting. With an increase in the concentration the growth rate goes up until it becomes too high to be inhibitory (Fig.5.10). Such hyperbolic response can be obtained for all essential nutrients. The microorganisms thus have an affinity for the nutrients of the same magnitude as enzymes have for their substrates.

In nature, bacteria most often have to face situations of low availability of usable organic matters and other nutrients. Under such starvation conditions, bacteria have to

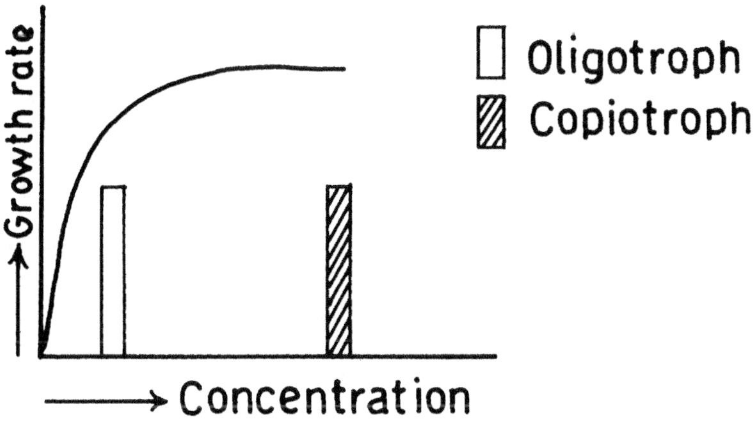

Figure 5.10. Relationship between the bacterial growth rate and nutrient concentration. The bars indicate two types of nutrient relationship amongst bacteria. Those growing at low nutrient concentrations are oligotroph and those preferring high nutrient concentrations are called copiotroph.

make several metabolic adjustments. Many bacteria like *Vibrio* species produce small minicells whereas *Bacillus* sps and others may form endospores. Both have low rates of metabolism and therefore are resilient to environmental stresses. Starvation survival is a critical strategy and several global regulatory mechanisms are invoked to cope up with such situations. Some of these mechanisms shall be discussed in **Chapter 6**.

The presence of a specific organic substrate may elicit the growth of particular organisms having specific catabolic activities. The amount of oxygen needed to oxidize organic matter completely to carbon dioxide is identified as COD (Chemical Oxygen Demand). COD is determined by measuring the amount of oxidizing reagent consumed during oxidation of organic matter with dichromate or permanganate. The BOD (Biological or Biochemical Oxygen Demand) determination reflects the readily usable concentration of organic matter. It is determined by the consumption of oxygen by microorganisms during oxidation of organic matter.

Bacteria that can preferentially grow at low nutrient conditions are called oligotrophs. In contrast, those growing with plenty of nutrients around, such as in a culture medium, are known as copiotrophs. The former have slow growth rates and may produce appendages to accumulate nutrients from low external concentrations. Prosthecae-forming bacteria are classical examples of this group. Mycobacteria produce a fruiting body as a response to low nutrient availability.

5.8.2. Temperature

Temperature is an extremely important factor that influences the growth significantly. Each bacterial species has a characteristic growth temperature range that comprises a minimum temperature below which the cells are metabolically inactive and a maximum temperature beyond which the cell does not grow or even becomes nonviable. Between the two extremes lies a temperature, at which a bacterial cell exhibits the highest growth rate, the optimal growth temperature. Both bacteria and archaea show extreme versatility as to the optimum temperature range that can support their growth (Table 5.2). This has allowed them to inhabit habitats ranging from frozen snow covers to boiling streams. Based on temperature preferences, they can be classified as psychrophiles (growing best at low temperatures, i.e, <20°C) or mesophiles (growing at moderate temperatures i.e., 20°-40°C) or thermophiles (could grow at high temperatures, >90°C); some preferring very high temperatures (>80°C) are known as extreme thermophiles or hyperthermophiles (Fig.5.11). Similarly, psychrotrophic bacteria can grow between 20-50°C, but in addition can also grow at low temperatures. Some bacteria, known as stenothermals prefer temperature at or near the optimal growth temperature, whereas eurythermal bacteria can grow over a wider range of temperature.

Table 5.2. Growth temperature range of some bacteria and archaea

Microorganism	*Temperature (°C)*
Micrococcus cryophilus	(-) 8 – 25
Vibrio marinus	(-)14 – 25
Pseudomonas avanae	0 – 40
Escherichia coli	7 – 42
Vibrio cholerae	12 – 42
Staphylococcus aureus	15 – 45
Streptococcus pyogenes	20 – 45
Lactobacillus lactis	20 – 47
Bacillus subtilis	20 – 50
Bacillus stearothermophilus	30 – 70
Thermus aquaticus	70 – 102
**Halobacterium salinarium*	20 – 55
**Thermoplasma acidophilum*	40 – 65
**Methanococcus jannaschii*	50 – 95
**Pyrococcus furiosus*	70 – 100
**Pyrococcus woesei*	70 – 105

* Belong to Archaea

Both psychrophiles as well as thermophiles exhibit characteristic physiological adaptations so as to thrive at extreme temperature. This involves the modification of enzymes and ribosomes and membrane structures so as to maintain the required metabolic activities. Many thermophiles have high G + C content making their DNA stable at such high temperatures. Several archaea are highly tolerant to high temperatures, giving credence to the suggestion that they evolved when Earth was still very hot. The critical determining factor in the ability of several archaea and some bacteria to grow at very high temperatures is the availability of liquid water rather than the actual temperatures. Many thermophilic archaeans, therefore, are capable of growing under high pressure, such as, thermal vents in deep oceans.

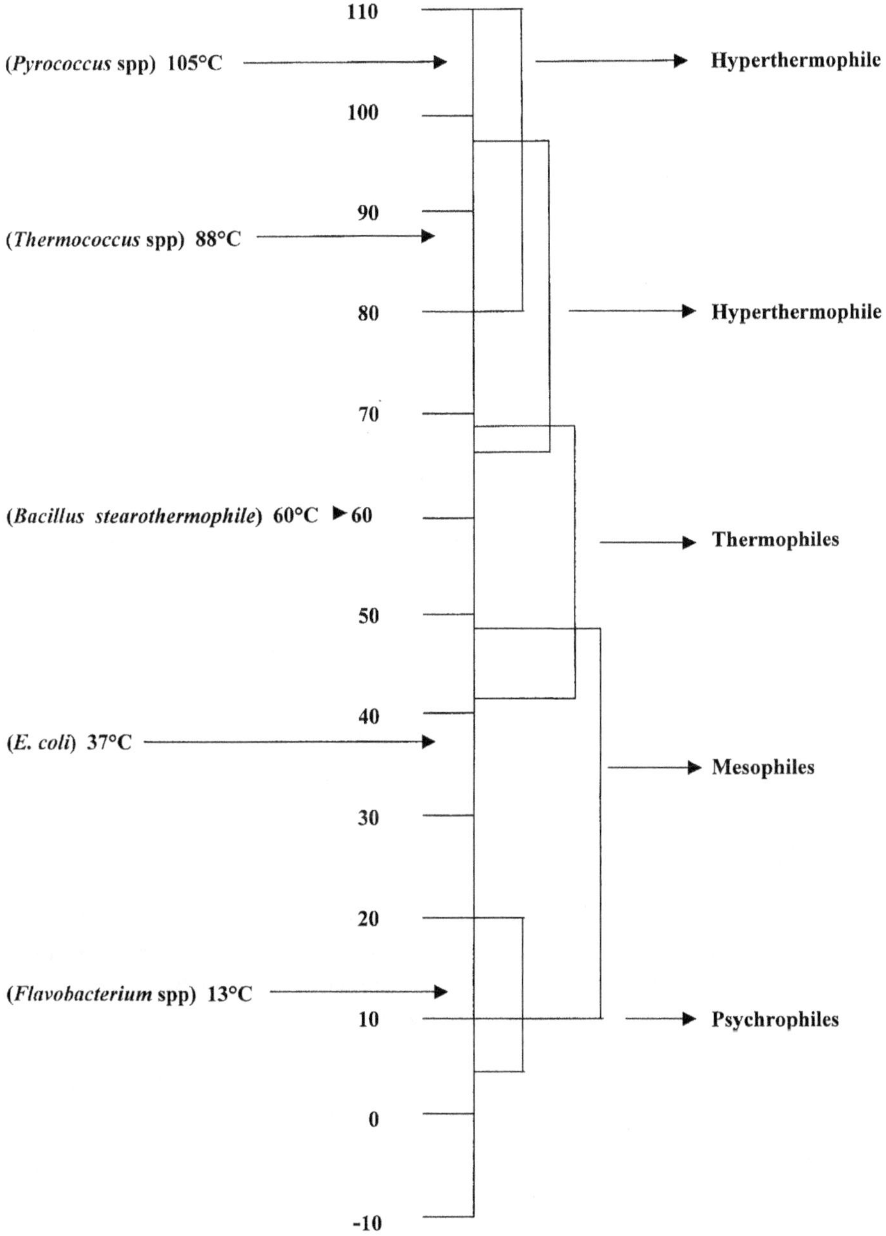

Figure 5.11. Growth temperature range of bacteria and archaea. Optimum growth temperatures of some representative members are given on the left.

The bacteria, archaea and also eukaryotes exhibit a heat shock response as a result of rapid shift to an elevated temperature. This heat shock response is largely conserved and is mediated through the production of a set of heat shock proteins (HSPs). In bacteria, a sudden increase in temperature results in increased cellular levels of a specific transcription factor σ^{32} that leads to transcription of genes coding for HSPs. In *E.coli*, 24 heat shock proteins are induced as a response to a change from 30°C to 40°C and are necessary for survival under temperature stress. One family of HSPs, HSP70, exhibits 50% homology between *E.coli* and the yeast, *Saccharomyces cerevisiae*. Under non-stress conditions HSPs perform diverse functions related to growth as well as molecular chaperons.

Psychrophiles produce enzymes that can optimally operate metabolic activities at low temperatures. They also possess suitably altered cytoplasmic membrane. Freezing can effectively prevent growth but does not always cause cell death.

5.8.3. Oxygen Concentration (Reduction Potential)

The presence or absence of oxygen in the environment greatly influences the growth of microorganisms. Microorganisms have evolved different types of metabolism so as to adjust to the availability of oxygen. Oxygen has limited solubility in water and in many aquatic environment it may become growth limiting factor.

So significant is the impact of oxygen that microorganisms have been grouped on the basis of their requirement or intolerance to O_2 (Fig.5.12). While aerobes grow in the presence of O_2, the obligate aerobes require O_2 for growth and carry out aerobic respiration. Some others called microaerophiles, grow only at reduced concentrations (~5%) of O_2 but not at atmospheric levels (20%). Some belonging to this group are able to grow at elevated CO_2 concentration (5-10%) and are called capnophiles. Facultative anaerobes can grow in the presence or absence of air. This is because they have the capability to switch from aerobic respiration to fermentation depending upon the presence or absence of O_2. The bacterium *E.coli* belongs to this group. Some strictly fermentative bacteria, like streptococci, are insensitive to oxygen (oxyduric) and hence can grow in the presence of O_2. Other bacteria that can grow only in the absence of air are called anaerobes. Of these, obligate anaerobes such as archaea, many bacteria and protozoa are strictly fermentative, and strict anaerobes are sensitive to O_2 so that even a brief exposure can be lethal to the cells. *Clostridium* sps are obligate, strict anaerobes.

The strong effect of O_2 on microbial metabolism is due to several factors that include the formation of toxic O_2 products and the presence and absence of enzymes that can detoxify these products. In atmosphere, oxygen exists in a triplet form which is highly electronegative molecule that can readily accept additional energy or electrons. The reduction of oxygen leads to the formation of singlet O_2 which has higher energy than triplet form and is more toxic to microorganisms. The reduction also results in superoxide radicals (O_2^-) and peroxides (O_2^{2-}), the latter may form hydroxyl free radicals, OH, as shown below:

$$O_2 + e^- \longrightarrow O_2^-$$

$$O_2^- + e^- + 2H^+ \longrightarrow H_2O_2$$

$$H_2O_2 + e^- + H^+ \longrightarrow H_2O + OH$$

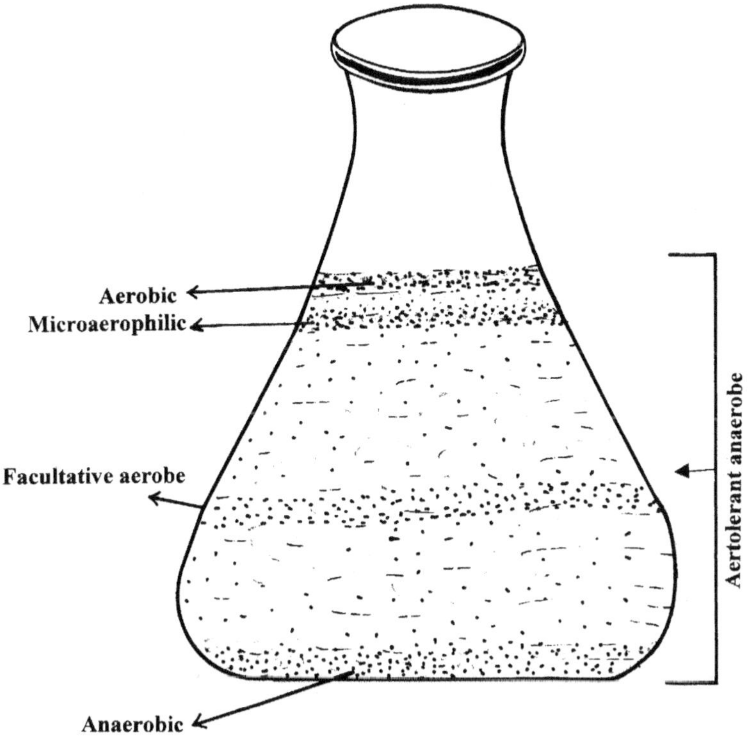

Figure 5.12. Bacterial growth responses to the presence or absence of oxygen.

All these are highly toxic as they are strong oxidizing agents. Many microbes synthesize enzymes, such as catalase and peroxidase, for detoxifying peroxides and superoxide dismutase to degrade superoxides.

Aerobic, facultative anaerobic, and microaerophilic bacteria possess superoxide dismutase and the first two groups also contain catalase. In contrast, strict anaerobes either do not synthesize these enzymes or synthesize them in low levels, thus explaining their sensitivity to O_2.

5.8.4. *Redox Potential*

Another effect of O_2 is to influence the redox potential. A positive redox potential (*E*h) value favour oxidation and a negative *E*h indicates reduced environment. The redox potential, therefore, determines the type of metabolic activities. The environment that exists in equilibrium with atmospheric O_2 have an *E*h of ~1800 mV, a value well below this will be found with reduced O_2 tension. Many heterotrophic bacteria in highly polluted ecosystems utilize and scavenge all available O_2 for decomposition and thus lower the reduction potential. Many sediments rich in organic matter may have an *E*h as low as 450 mV. This provides a congenial condition to obligate anaerobes to reduce SO_4^{2-} to H_2S and archaeal methanogens to reduce CO_2 to CH_4.

Many facultative anaerobes such as *E. coli* and *Bacillus* species switch from aerobic to anaerobic respiration when O_2 gets depleted. Adaptation to anaerobiosis is based on two regulatory mechanisms: the Arc (aerobic respiration control) and Fnr (formate-nitrate regulation) systems.

The Arc is an example of 2-component regulatory systems in bacteria (discussed in **Chapter 6**). It consists of Arc B, a histidine kinase, and ArcA, the response regulator. At low O_2 concentration, ArcB gets phosphorylated with a coupled conversion of ATP to ADP. Phosphorylated Arc B can transfer the phosphate group to Arc A which then represses the transcription of genes involved in aerobic growth.

Fnr system, on the other hand, functions by activating the genes coding for enzymes involved in anaerobic respiration and repressing those coding for electron transport carriers needed for aerobic respiration. This allows the cell to utilize alternative electron acceptors such as fumarate, nitrate and others. Additional systems (NarL/NarX/NarQ) determine the choice of electron acceptors under anaerobic conditions.

5.8.5. Water Activity

Water is an essential component of all living cells. It is an universal solvent and thus is required for all biochemical reactions. The amount of free water available to react is described as water activity (A_w) and is equivalent to the atmospheric measure of water availability known as relative humidity (RH; RH = 100 A_w). Binding of water to any solute will reduce its A_w. A_w of pure distilled water is 1.0, and that of saturated solution of NaCl and of sea water (~3% NaCl) is 0.8 and 0.98, respectively. The relatively low availability of water in air does not permit the bacteria to grow in air. By the same reason, availability of water has allowed the bacteria to grow on numerous surfaces sometimes the unconventional ones also. The non-availability of water has been one major factor for non-existence of life forms on other planets of our solar system. Drying has been a traditional method of preserving many food products from microbial deterioration.

Most bacteria require an A_w of 0.9 for optimum metabolism but xerotolerant organisms can survive and grow at much lower water activities. Fungi, in general, are more tolerant to low water activity as compared to bacteria. Many pathogenic bacteria like *Mycobacterium tuberculosis* have developed tolerance to drying to ensure their dissemination, whereas *Treponema pallidum,* syphilis causing bacterium, gets killed almost instantly in the air or dry surfaces. Many bacteria produce spores to survive high desiccation, low temperatures, and high irradiation levels. Formation of thick walls, production of extracellular polysaccharide layers, especially those containing trehalose, and slime layers are other adaptations that serve the protective functions.

5.8.6. Pressure

The external as well as internal pressures can affect the growth of all cells. The two major such pressures are osmotic and hydrostatic pressure.

5.8.6.1. Osmotic Pressure. The osmotic pressure in any habitat including the medium is controlled by the concentration of dissolved substances. The structure of the bacterial cell wall makes them relatively resistant to minor fluctuation in osmotic pressure. However, extreme osmotic pressures can result in the death of the cell. For example, in hypertonic solutions, bacteria may shrink and become desiccated and in hypotonic solutions they may burst. Organisms that can grow in solutions with high solute concentrations and thus low water activities are called osmotolerant. Some called as osmophiles may even require high solute concentration to grow. High sugar concentrations are generally used in laboratory to protect protoplast and spheroplast from bursting.

Salinity is also important in maintaining osmotic pressure. Some bacteria and archaea called halophiles require NaCl for growth. Moderate halophiles, like marine bacteria, may show optimum growth at 3% NaCl. They require at least 1.5% NaCl to maintain their integrity. Extreme halophiles can grow even in saturated brine solutions. Halophilic bacteria and archaea have been found growing in the presence of 15-25% NaCl, though most bacteria will not grow even at 3% salt concentration. Preservation of food articles by salting, in addition to reducing the water availability, is based on this response only. High salt concentrations disrupt membrane and denature many proteins. Many halophiles such as *Halobacterium*, therefore, have evolved suitable adaptations to overcome these effects. This includes unusual plasma membrane composition such as purple membranes and unusual enzymes which are not only salt tolerant but may also require salt for their activity, high internal concentration of KCl, and unique light-driven proton and chloride pumps, called bacteriorhodopsin and halorhodopsin, respectively.

5.8.6.2. Hydrostatic Pressure. The hydrostatic pressure exerted by a column of water due to its weight is an important factor for bacteria and other organisms found in deep oceans. In most natural systems, bacteria can withstand the available hydrostatic pressure. However, extreme pressures (>200 atm) may lead to disruption of membrane transport systems and inactivation of enzymes. Each 10 m of water depth is equivalent to ~1 atm pressure. Some barotolerant bacteria can grow at high hydrostatic pressure and some which are called barophiles even prefer such pressures. Very few bacteria are able to grow at deep depths of ocean. They have low metabolic rates to withstand both the pressure as well as low temperatures. Preservation of food articles often observed in ship wreckage submerged in deep sea is based on this microbial response. On the other hand, in hot thermal vents, high hydrostatic pressure which maintains water well below 100°C have high bacterial load.

5.8.7. Acidity and pH

pH is another important factor influencing growth (Fig.5.13). Most bacteria prefer near neutral pH for their metabolic activities. The effect of pH is exerted through protein alterations. The secondary and tertiary structure of proteins is largely maintained due to charge interactions within the constituent amino acids that is affected by pH. This results in changes in shape and thus the activity of the protein. Very high and very low pH are inhibitory. Moreover, bacteria are less tolerant to high temperature at low pH.

While describing the optimal pH for growth, it should be borne in mind that this represents the pH of the external environment only; the intracellular pH must be maintained around neutrality to protect the acid- or alkali-labile macromolecules inside the cell.

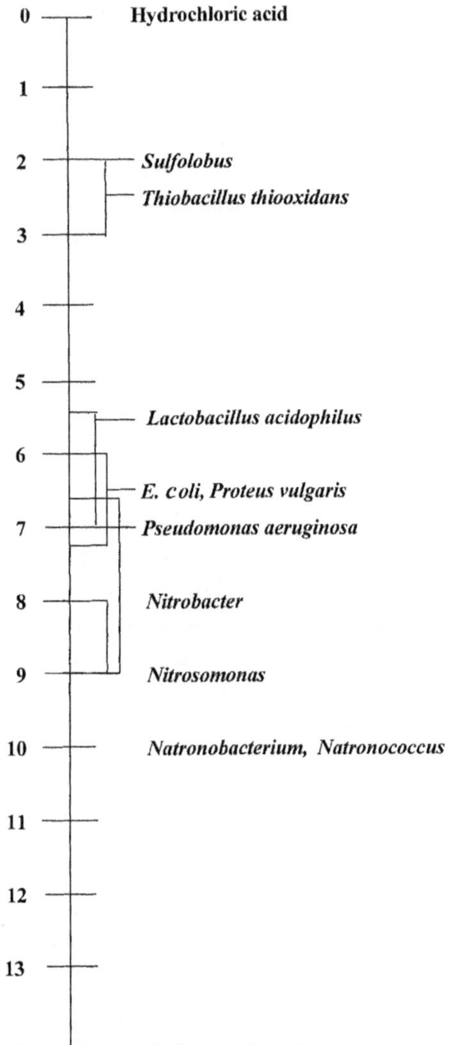

Figure 5.13. Optimum pH for growth of some bacteria and archaea.

The pH of the culture media as well as the industrial fermentors is controlled by buffering the solution. The pH tolerance range is highly variable but normally falls between 6.0-9.0. By this criterion, most bacteria are referred to as neutrophiles wherein the cytoplasm maintains near neutral pH. Others, however, can grow at low pH (acidophiles) and some at high pH (alkaliphiles). In some acidophilic bacteria, the cytoplasmic membrane remains intact and active only at very low pH. *Thiobacillus* and *Sulfolobus* are examples of acidophiles which not only maintain their structural integrity but also metabolic acidity at low pH. In extreme acidophiles or alkaliphiles, the intracellular pH may vary by 1.0-1.5 pH units from neutrality. Lot of interest has been generated in alkaliphiles like *Bacillus* species and other halophiles because several enzymes produced by them can work at alkaline pH. Such enzymes can find useful industrial applications.

The proton motive force, the essential energy source of a cell is composed of two interchangeable components: a Δ pH due to H^+ ion gradient and a ΔΨ (membrane potential) due to the charge separation across the membrane. All cells attempt to maintain their internal pH near neutrality. Acidophiles therefore have a very large Δ pH and a reversed ΔΨ to support cellular functions. Another group of bacteria the alkaliphiles, prefer to live under highly alkaline conditions. Such bacteria are found in salt lakes or in soils rich in sodium carbonate where pH ranges from 9.0-11.0. They adapt to grow in such environments by keeping Na^+ ions outside the cell. They are characterized by not having naturally-occurring Δ pH (the internal pH is 7.0-9.0, and external pH between 9.0-11.0). In order to drive the proton motive force they employ a Na^+/H^+ or K^+/H^+ antiporter to maintain a sufficiently high ΔΨ.

5.8.8. Light

Light is required by all photosynthetic bacteria for photoautotrophic generation of ATP. Many photosynthetic bacteria exhibit phototactic response to light. Both visible and ultraviolet light can be harmful to bacteria. They protect themselves by synthesizing carotenoids or other pigments that absorb light of certain wavelengths. The phototactic bacteria living in aquatic environment adjust their vertical position by synthesizing gas vesicles. In low light, the gas vesicles are formed and due to presence of certain gas, they float close to the surface of lakes and ponds and hence receive light. When light intensities are high these gas vesicles collapse and bacteria sink to a lower level.

5.9. Molecular Mechanism of Growth

As we have seen earlier in this chapter, growth in a bacterial population is a reflection of an increase in the population size. Moreover, in unicellular organisms like bacteria, this is brought about by cell multiplication. The process of cell multiplication is an outcome of cell division or binary fission, and at the biochemical and molecular level consists of two processes: nuclear division and cell division. Both these involve an orderly increase in all the cellular contents and their equidistribution in the resulting daughter cells.

5.9.1. Nuclear Division

Since bacteria do not possess a well-defined nucleus, nuclear division comprises the replication of the single circular DNA double helix or 'chromosome'.

In 1953, when Watson and Crick proposed the model of DNA structure, they suggested a semi-conservative mode of DNA replication. In this, the original strands of the double helix serve as the template allowing a blueprint of theirs to be synthesized (based on base complementarity). Thus, each daughter DNA contains one "old" and one "new" strand and both are exact replica of the original duplex (Fig. 5.14). Watson and Crick emphasized that this will allow the perpetuation of the genetic material and explain the genetic continuity.

Our understanding of DNA replication has been mainly derived from the experiments conducted initially on the bacterium, *E.coli*. In 1958, Meselson and Stahl based their experiments on a technique devised a year earlier by Meselson, Vinogradsky and

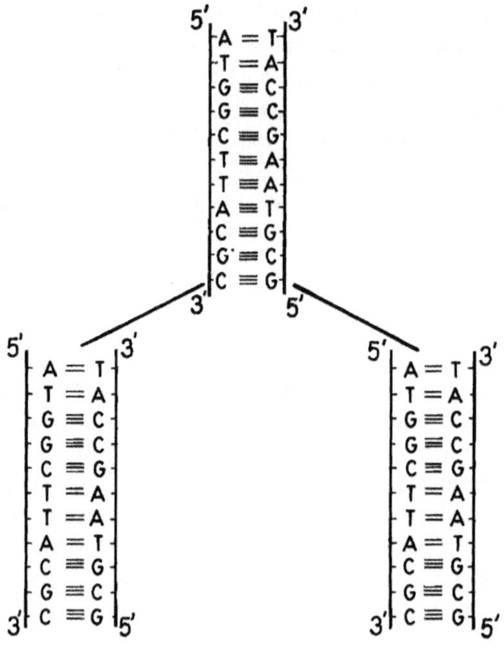

Figure 5.14. Semiconservative DNA replication- In this, both the strands of the duplex serve as templates allowing the complementary bases to be used to synthesize the new two strands. Thus, the original base sequence in the daughter DNAs is maintained, each consisting of one old and one new strand.

colleagues that allowed the separation of the DNA strands differing in their densities. The density differences can be created by growing, for example, *E. coli* cells in a medium containing heavy isotope of N (^{15}N) and compared with those grown in normal N (^{14}N). The heavier (^{15}N) DNA can be differentiated from normal (^{14}N) lighter DNA by centrifugation in a CsCl (Cesium chloride) equilibrium density gradient. On such a gradient, heavy DNA will settle at a denser region of the tube as compared to ^{14}N DNA. Meselson and Stahl grew *E.coli* cells in a medium that contained ^{15}N-compounds for several generations. Such cells will have their DNA now containing heavy isotope. These cells were then transferred to ^{14}N -medium and grown for one or two generations. Samples were removed at regular intervals (O=prior to the cultures being shifted to ^{14}N-medium and after 1 and 2 generations, respectively), their DNA was extracted and run on CsCl-equilibrium density gradient centrifugation.

Meselson and Stahl could predict the DNA banding pattern on the basis of semi-conservative replication. At time, 'O' both the strands will contain ^{15}N DNA and, therefore, settle at the denser region of the gradient. After one round of replication, both the daughter duplexes will have one ^{15}N and one ^{14}N strands which should show an intermediate density to $^{15}N/^{15}N$ and $^{14}N/^{14}N$ DNA. After second round of replication while one band of intermediate density will be retained, a fresh band of density equivalent to $^{14}N/^{14}N$ DNA will appear (Fig. 5.15). Interestingly, these predictions were experimentally proved, suggesting that DNA replication indeed follow the semi-conservative mode.

This work was soon followed by autoradiographic studies on replicating bacterial chromosome by Cairns in 1963 (Fig. 5.16). In his experiment, thymidine-requiring cells were grown for two division cycles in a medium containing tritiated (^3H) thymidine. This allowed the incorporation of radiolabelled thymidine in freshly synthesized DNA. The cells were then gently lysed and DNA was processed for autoradiography. The autoradiograms were analyzed on the basis of grain (radioactive label) distribution (Fig. 5.16). The autoradiography showed two types of label distribution. The joint regions represented one old non-radioactive and one newly synthesized labelled strand. In other parts of the closed circle there was heavy distribution of label, suggesting that both the strands carry ^3H-thymidine (showing second round of replication). The semi-conservative mode was once again confirmed.

In the light of the available evidences, the process of replication of a circular DNA in *E. coli* can be summarized as follows:

5.9.1.1. DNA Replication. The unit of DNA in which an individual act of replication occurs is called the replicon. The bacterial genome is thus a single circular replicon which must have an origin at which replication is initiated and a terminus at which the same must stop. The unique chromosomal replication initiation site is called *oriC* (~245 bp in *E.coli*) where the two growing points or replication forks are initiated that move around the genome till they meet at the termination region. Such a replication mode is called bidirectional in contrast to unidirectional where only one replication fork is involved (Fig. 5.17). The termination regions, *terD, A* and *terC, B* are located ~100 kb on either side of the meeting point. Each terminus is specific for one direction of fork movement and are arranged in such a way that each fork will have to pass the other in order to reach the terminus. Such a structure acts as a replication fork trap, so that if for some reason one fork is delayed, the more rapid fork will be trapped at the *ter* region to wait for the arrival of the slow fork.

An enzyme that can synthesize a new DNA strand on a template strand is DNA polymerase. Three DNA polymerases have been characterized in *E.coli.* PolymeraseI (Pol I) is involved in the repair of damaged DNA and in a subsidiary role in semi-conservative replication. Polymerase II (Pol II) is also implicated in repair. Polymerase III (Pol III) is a multisubunit replicase responsible for DNA synthesis *de novo.* All the bacterial enzymes (Polymerase I, II, III) possess a 3'-5' exonuclease activity that provides proof-reading function and thus a high degree of fidelity to the replication process. The polymerases are supported by other proteins during the entire process of replication.

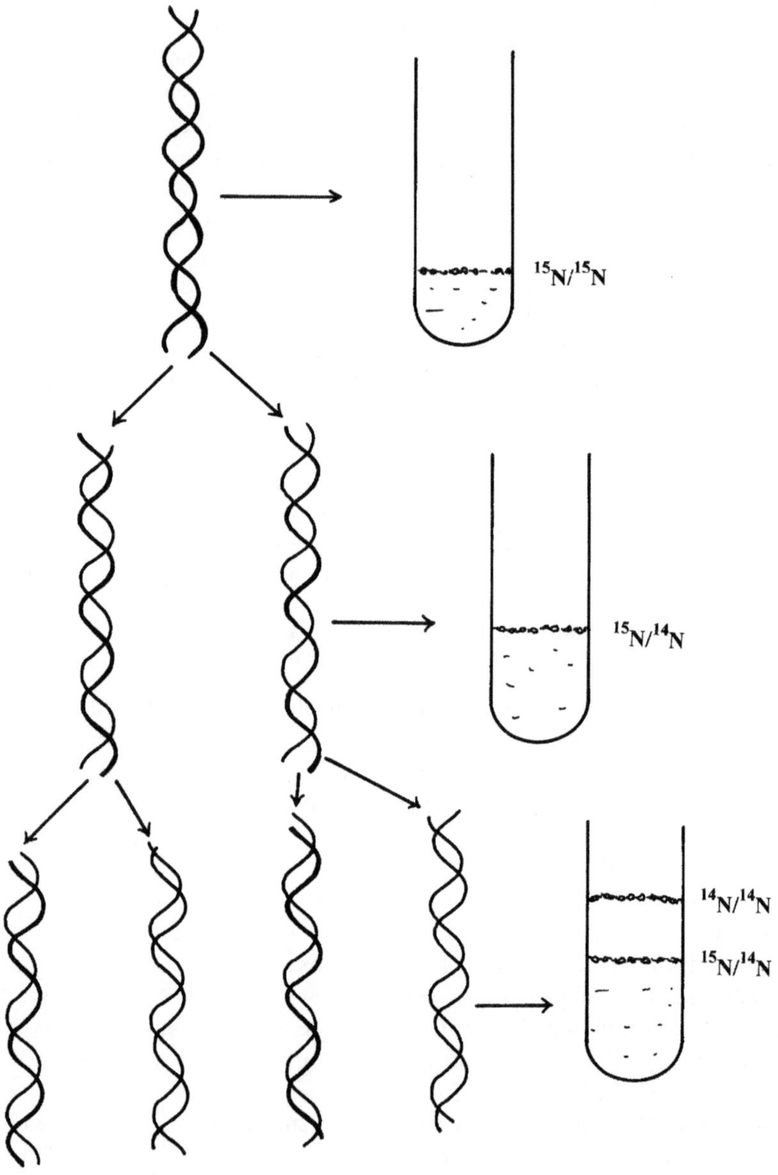

Figure 5.15. The Meselson-Stahl experiment to demonstrate semiconservative replication of DNA in E. coli. E. coli cells grown in heavy (^{15}N)-N medium has heavy DNA. Growth for one generation in ^{14}N-medium will produce DNAs of hybrid density. Growth for another generation on a similar medium will give DNA of both the hybrid and light (normal) density. They were able to demonstrate these different types of DNA by equilibrium density gradient centrifugation.

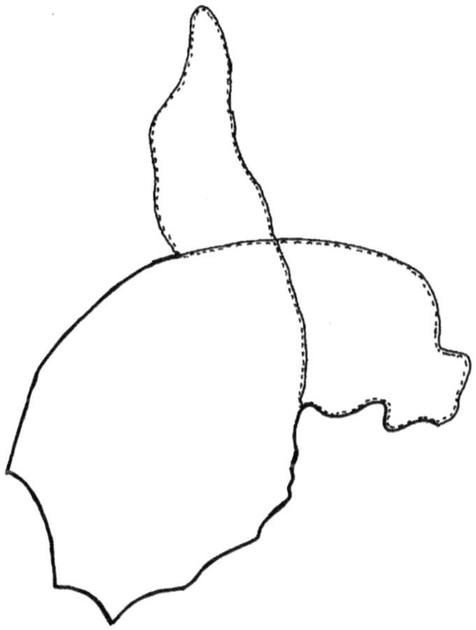

Figure 5.16. Diagrammatic representation of an autoradiogram of replicating E. coli chromosome labeled with ^{3}H-thymidine. Note: 1) E. coli chromosome is a closed circle, and 2) the thick lines represent the unreplicated and the thin ones the region under replication.

Polymerase I coded by *polA* is a single polypeptide of 103 kDa which can be cleaved by proteolytic treatment. The larger cleavage product (63 kDa) contains polymerase and the 3'-5' exonuclease activity. This fragment called Klenow fragment is extensively used in synthetic reactions *in vitro*. The small fragment of 35 kDa possesses a 5'-3' exonucleolytic activity. This activity coordinated with synthetic/proof-reading activity allows Pol I to carry out nick translation. This is principally used to incorporate radiolabelled nucleotides into DNA *in vitro*. Pol III α subunit encoded by *polC* (=*dnaE*) possesses DNA synthesizing activity and epsilon (ℑ) subunit with 3'-5' proof-reading activity is coded by the gene *dnaQ*.

None of the three polymerases can initiate DNA synthesis. They require a primer to provide a free 3'- OH end that can be extended by the addition of deoxynucleotides. For this, a sequence of RNA (Primer) is synthesized on the template and the free 3'-OH end of RNA is extended by the DNA polymerase. This is the most commonly used method of initiating replication as seen in chromosome, plasmid and some viruses. However, a primer terminus can also be generated by introduction of a nick in duplex DNA (rolling circle) as shown in Fig. 5.17 or by pairing of a preformed RNA with the template (retroviruses) and even through a protein (certain viruses). Since the replication fork moves only in 5'-3' direction, the problem posed by antiparallel nature of the two strands is circumvented by replicating the two strands simultaneously but in two ways: One strand synthesized continuously in 5'-3' direction is the leading strand. On the other strand, however, DNA is synthesized in small segments in the reverse direction

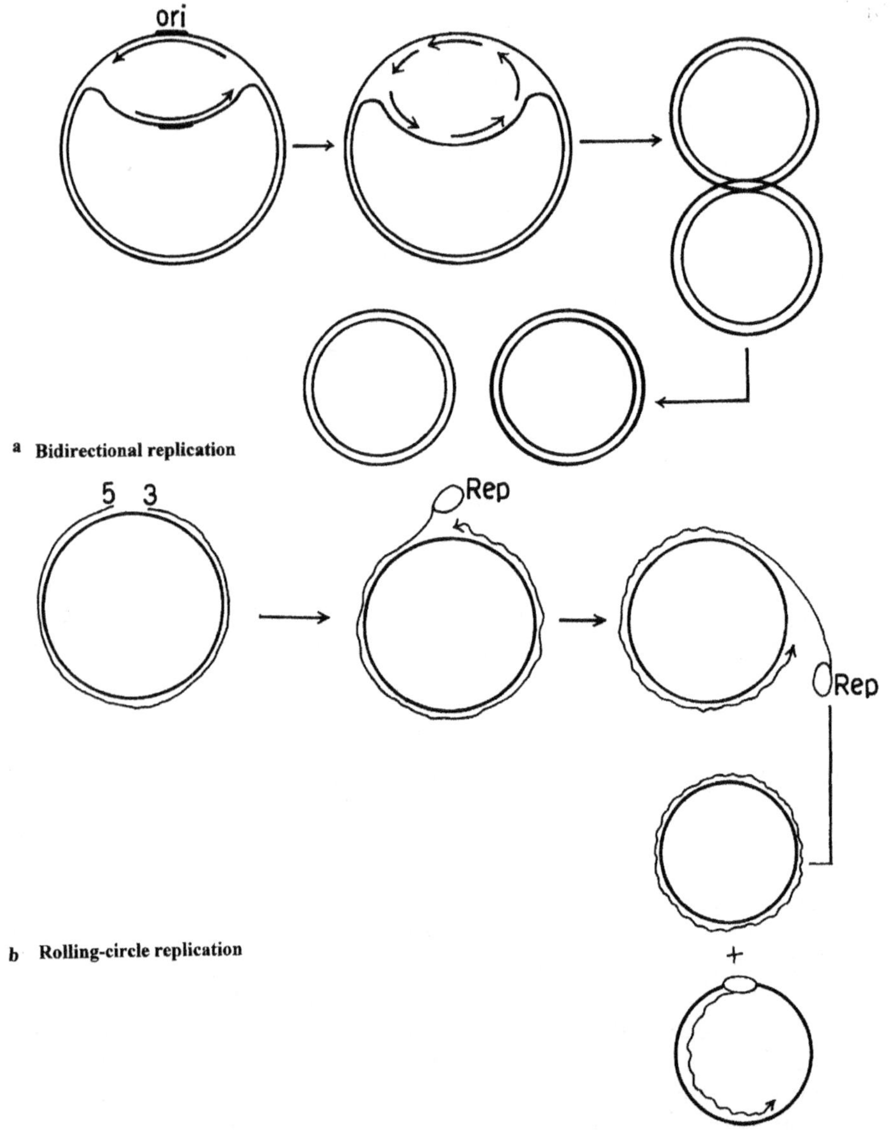

Figure 5.17a. Bi-directional replication in a circular replicon. The replication begins at origin region designated ori, and two replication forks move out simultaneously but in opposite orientation. Replication terminates when the two forks meet on the other side of the molecule. Note: In unidirectional replication, a single replication fork moves around till it returns to the origin. b. Rolling-circle replication. One strand is nicked at ori region to create a 3'-OH end which serves as a primer for the addition of dNTPs. Replication thus displaces the old strand which itself can act as a template for complementary strand synthesis.

(relative to fork movement) to maintain the customary 5'-3' polarity. Such a strand known as lagging strand is replicated discontinuously with each fragment consisting of about 1 to 2 kb and called Okazaki fragments. Such a mode has been called semi-discontinuous replication. While leading strand would require only one primer, in the lagging strand, each Okazaki fragment is primed by an RNA, ~10 bp long (Fig. 5.18). The RNA primers are subsequently removed by 5'-3' exonuclease activity of the Pol I which also simultaneously fills the gap by extending DNA sequence from 3'-OH of next fragment. Once this happens the nicks are sealed by the enzyme DNA ligase.

The initiation is brought about by binding of DnaA protein to AT-rich 9 bp region and 13 bp repeats in *oriC* that leads to DNA melting. In order to provide the template function DNA must first be unwound by a helicase, a step which requires the hydrolysis

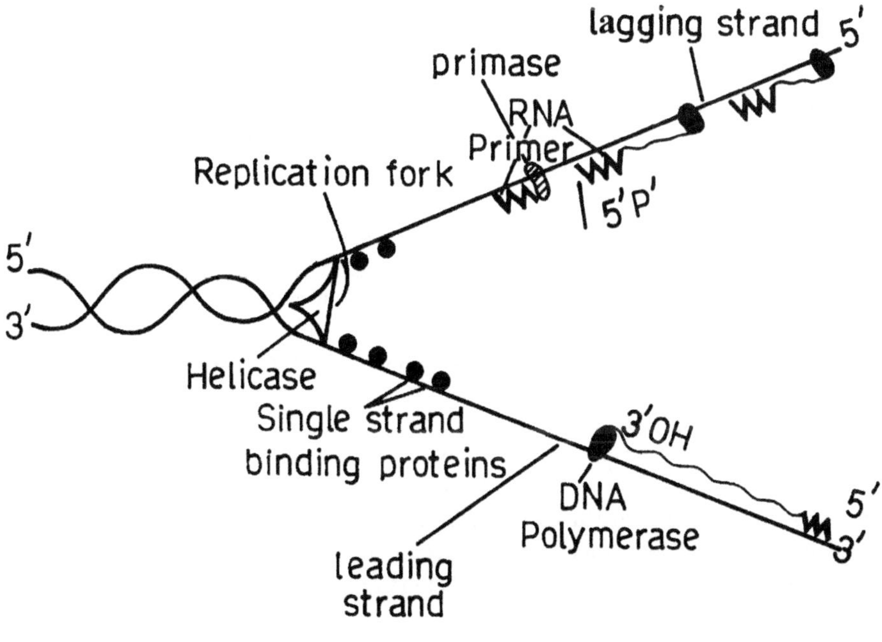

Figure 5.18. Semi-discontinuous replication of DNA. One template strand (leading strand) is synthesized in 5'→3' direction after initial priming continuously. The other, lagging strand, maintains overall 5'→3' direction by synthesizing DNA in short stretches or Okazaki fragments (discontinuous). All the necessary enzymes involved are shown in the figure.

of ATP to provide energy. Following this, the two single strands are bound by single strand binding protein (SSB) to prevent reformation of duplex. These activities essential for some of the priming reactions constitute a primosome. DnaB provides the 5'-3' helicase that unwinds DNA. DnaB binds to DNA along with DnaC and also DnaT. DnaG primase, which is an RNA polymerase smaller than the usual RNA polymerase gets activated by DnaB and is used to prime DNA synthesis. The primer is 11-12 bases long and starts with the sequence pppAG. SSB protein that keeps the unwound DNA in single stranded state is required at more than one step in replication. The primosome moves but in the anti-elongation direction (same direction as replication fork) and a single-stranded stretch is always available for primer synthesis. Primosome then moves to start the next Okazaki fragment.

The replication of a circular DNA results in a theta structure (see Fig.5.16). The bacterial DNA replication is closely adjusted with cell division. This is also an important requirement for unicellular organisms. However, in rapidly growing cells, a new replication fork may be initiated before the 'old' one reaches the terminus, thus creating a multiforked chromosome.

After termination of replication, the two daughter chromosomes must segregate into two daughter cells. The underlying mechanisms must ensure faithful nucleoid partitioning in the face of two problems: (i) the catenates resulting from the replication of DNA duplex must be unlinked (decatenation), and (ii) the dimeric chromosome, formed due to homologous recombination between daughter chromosomes, must be resolved.

Much of the unlinking function has been assigned to two type II topoisomerases, DNA gyrase and topoisomerase IV. Type I and type III topoisomerases are known to assist resolution of plasmid and gapped molecules. The resolution of dimeric chromosomes, on the other hand, is accomplished by the XerCD proteins, Arg R, and PepA proteins which require the *dif* site for their operation. Evidence for the involvement of structural proteins is beginning to accumulate suggesting that the nucleoid segregation may be an active process requiring protein synthesis. Chromosome segregation is effected by two genes *mukA* and *mukB*. While *mukA* product is similar to a outer membrane protein, *mukB* product provides the motor function that may physically partition the chromosome to two daughter cells.

For plasmids, a 'covalent extension' or 'rolling circle' mode of replication is employed. This allows rapid production of large numbers of such DNA molecules. In this mode of replication, a nick is made in one of the strands of circular DNA, thereby creating a free 3' -OH and 5'-P ends. The free 3' -OH provides DNA polymerase with a starting point to add deoxynucleotides. This produces a concatemer in which a genome length DNA sequence is tandemly repeated many times (see Fig.5.17).

5.9.2. Cell Division

Soon after the DNA replication, a bacterial cell divides into two through a process called binary fission. This is accomplished by the formation of a septum that divides the original cell into two daughter cells. The septum appears at the mid-cell in the form of an invagination. The septum formation is preceded by the organization of the periseptal annulus, where the inner membrane makes more close connection with the cell wall and outer membrane. In a new cell, the annulus occupies the central position. As the cell grows new annuli are formed on either side of the initial annulus, which itself defines the site of septum formation. The new annuli move to 1/4 and 3/4 position of the cell length and represent the mid-cell position for the next cell division (Fig.5.19a). The septum is composed of the same material as the cell envelope, and is initially formed as a double layer of peptidoglycan. The enzyme responsible for generating the peptidoglycan is the membrane-bound protein, PBP3 (Penicillin-binding proteins 3) and protein EnvA which splits the double layer, so that the two cells may separate. Mutants affecting the cell division may show the following phenotypes (i) long filaments where chromosome replication and segregation is not affected but septa formation does not take place, and (ii) mini cells are formed when septum formation occurs too frequently or in the wrong place with the result that a small sized cell lacking DNA is formed. Anucleate cells can also be formed due to aberrant segregation of the chromosome but as the septum formation is normal the size is unaltered.

Bacterial cell division is initiated through the formation of a FtsZ ring (Z ring) at the site of division (Fig.5.19b). The ring is composed of the tubulin-like FtsZ protein that has GTPase activity. Z-ring is thought to function in vivo as a cytoskeletal element that is analogous to the contractile ring in many eukaryotic cells. Evidences suggest that the Z-ring is utilized by all prokaryotic organisms for division and may also be by some eukaryotic organelles.

During each cell cycle, a division site is generated at midcell. Z-ring formation is initiated by the polymerization of FtsZ at this nucleation site. This process is rapid, requiring less than a minute. The discovery of associated GTPase activity suggested that polymerization involves GTP and might be analogous to tubulin polymerization into microtubules. During division, Z-ring is postulated to decrease in diameter at the leading edge of the septum.

The quantity of active FtsZ seems to be directly related to the frequency of cell

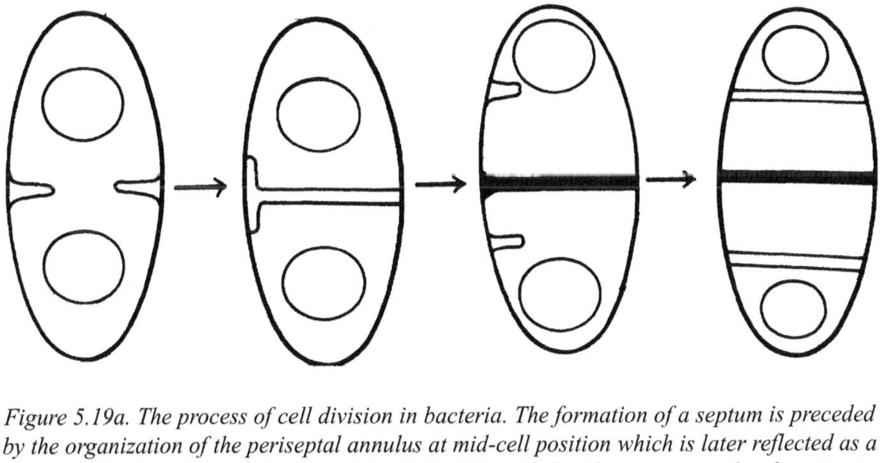

Figure 5.19a. The process of cell division in bacteria. The formation of a septum is preceded by the organization of the periseptal annulus at mid-cell position which is later reflected as a septum. New annuli are formed on either side of the annulus that move in polar direction to a position that identifies the mid-cell position at next cell division.

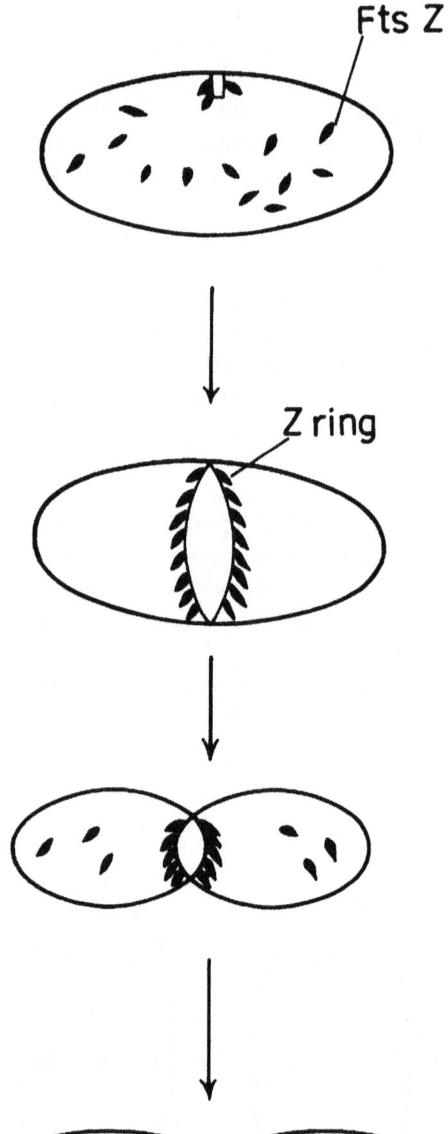

Figure 5.19b. FtsZ appears to be a cell division initiator protein. Early in the division cycle this protein is dispersed throughout the cytoplasm. As the cell elongates, FtsZ get organized in a Z-ring around the circumference that reflects the mid centre annulus position. This helps in the subsequent steps of division.

division. Mutation in this gene blocks septum formation leading to generation of filaments, whereas over expression induces mini cells by causing an increased number of septation events. The mini-cell mutation lies in the locus *minB*, deletion in this gene allows septation to occur at the poles as well as mid-cell. Thus, wild-type *minB* must suppress septation at poles. The *minB* locus consists of three genes, *minCDE*. The products of *minCD* leads to division inhibition. The expression of *minCD* in absence of *minE* or over-expression of same even in the presence of *minE* inhibits the growth leading to filament formation. The *minE*, on the other hand, inhibits the polar septa formation and protects the mid-cell sites from inhibition, thus restoring normal growth.

FtsZ has been found in *B. subtilis*, *Streptomyces coelicolor*, and even in *Mycoplasma* which not only lack cell wall but are also the smallest free-living cells with the smallest FtsZ. Moreover, *ftsZ* gene has been cloned from diverse members of archaea and has been detected in chloroplasts of *Arabidopsis thaliana*. Its presence in so many diverse bacteria with such different cell envelops suggests that its function is likely to be linked to the invagination of the cytoplasmic membrane (the septum formation). The universality of Z-ring in prokaryotes, make FtsZ an excellent target for antimicrobial agents.

A cell division gene in the halophilic archaebacterium, *Halobacterium salinarium* indicates its close relationship to tubulin, an important component of the cytoskeleton found in eukaryotic cells. FtsZ protein share several biochemical properties with eukaryotic tubulin, also involved in cell division, but they are supposed to be distant relatives according to amino acid and DNA sequence analyses.

DNA sequence of *H. salinarium* version of *ftsZ* puts it some where between the eubacterial *ftsZ* and eukaryotic tubulin gene. Archaebacteria, therefore, appear to provide the missing link between eubacteria and eukaryotes and supports the finding that tubulin may have evolved from FtsZ.

During vegetative septation in *B. subtilis,* FtsZ is present at mid-cell but as the cell enters sporulation FtsZ is localized to both cell poles. This switch in localization requires SpoA, the master transcriptional regulator for sporulation. The activation of one of the polar Z ring is likely to dictate the pole at which the spore will be formed.

For unicellular organisms like bacteria, cell division is the safest mean of cell multiplication. Complete coordination, therefore, exists between DNA replication and cell division so that the resulting daughter cells contain complete genome set. Infact, a fresh DNA replication is initiated as soon as the cell divides.

In many bacteria, the daughter cells do not separate, and depending upon the plane of subsequent divisions, this results in different associations (Fig. 5.20). For example, in chain-forming streptococci, subsequent divisions take place in the same plane, without the separation of resulting cells. However, if the successive divisions are in many planes, an irregular arrangement or cell clusters will be formed, as in *Staphylococcus*. If the second division is at right angles to the first, a tetrad is formed e.g., *Pediococcus* but if subsequent divisions take place at right angles to each other a planar sheet or 'windowpanes' of cells will result (e.g. *Lampromedia*). The cuboidal packet of cells in *Sarcina* and *Geodermatophilus* results from divisions at right angles in three planes.

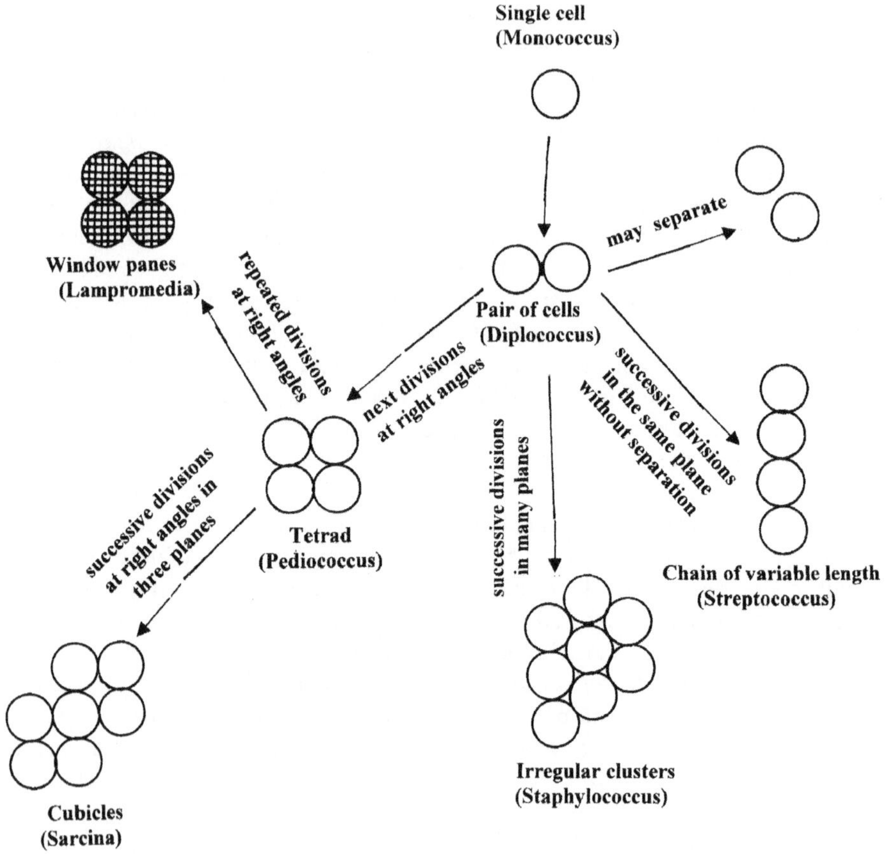

Figure 5.20. Cell divisions in specific planes can generate different types of bacterial cell associations.

5.10. Reproduction

As discussed earlier growth of unicellular organisms is intimately associated with an increase in the number of individuals in a population or reproduction. While reproduction can be broadly classified into asexual and sexual type, considerable variability exists in the overall process.

5.10.1. Asexual Reproduction

As cell division is a highly efficient mode of cell multiplication, most unicellular bacteria opt for binary fission (cell division) to multiply under optimum conditions. Asexual spores are commonly found in actinomycetes which may or may not be borne in sporangia and if latter, they may be produced singly or in short chains (Fig. 5.21). The fragmentation of the mycelia may provide another means of multiplication in actinomycetes. Several nocardioforms typically produce filaments that fragment into non-motile cells.

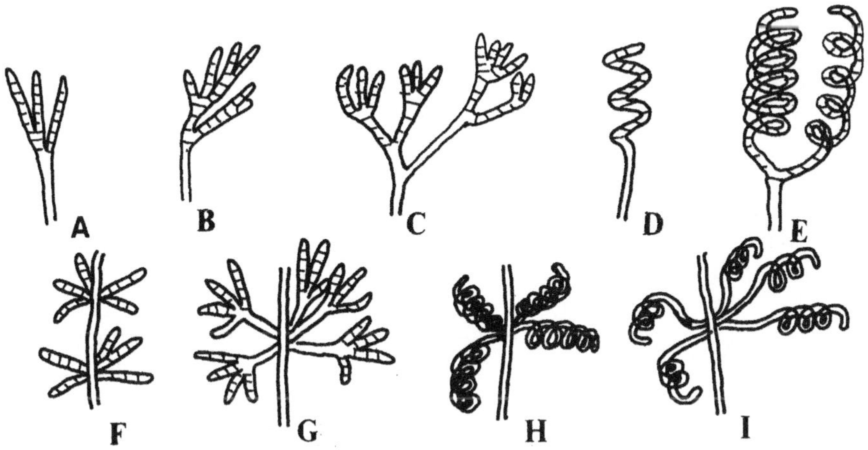

Figure 5.21.I. Different types of spores (conidia) bearing structures in Streptomycetes. A. Straight, B. Flexuous, C. Fascicled, D. Open loops, E. Open spiral, F. Monoverticillate without spirals, G. Biverticillate without spirals, H. Monoverticillate with spirals, I. Biverticillate with spirals.

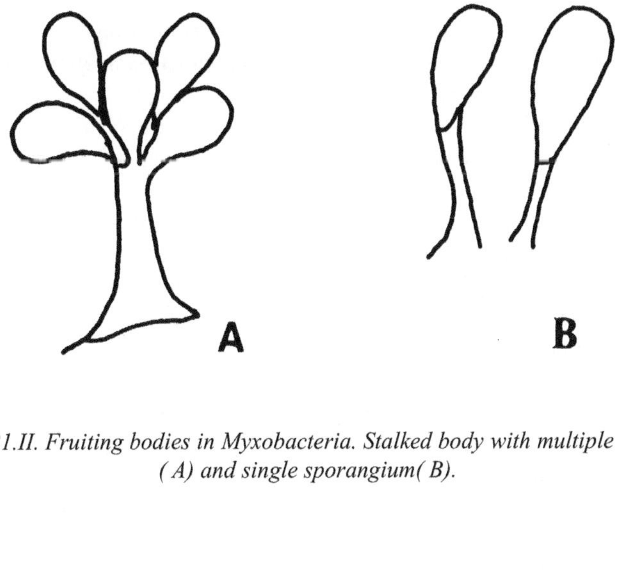

Figure 5.21.II. Fruiting bodies in Myxobacteria. Stalked body with multiple sporangia (A) and single sporangium(B).

The Chroococcacean cyanobacteria generally reproduce by binary fission although some members reproduce by budding. On the other hand, the Pleurocapsalean cyanobacteria undergo multiple fission to produce small coccoid reproductive cells, called baeocytes.

Myxobacteria under appropriate conditions produce fruiting bodies which contain cells of myxobacteria in dormant state and are called myxospores (Fig. 5.21). In some cases, myxospores can not be distinguished from vegetative cells, in others they may be refractile and encapsulated and are referred to as microcysts.

The budding and/or appendaged bacteria are so called as many members reproduce by budding which is carried out by variously shaped prosthecae or appendage. Some members, however, still prefer binary fission.

Some bacteria reproduce by growing their offsprings internally before releasing them. *Epulopiscium* spp., one of the largest known bacteria, releases litters of free-living baby cells. Based on 16S rRNA genes, this bacterium is closely related to *Metabacterium polyspora. Epulopiscium* cells are intestinal symbionts of surgeonfish, whereas *M. polyspora* lives amongst the microbiota found in the caecum of terrestrial guinea pigs.

The two bacteria resemble one another in several ways. For instance, both produce large inclusion bodies. While in *M.polyspora* these inclusion bodies are spores, in *Epulopiscium* they represent daughter cells that grow internally till they are released through a slit that opens in the depleted cell cortex. This led to the idea that daughter cell formation is a by-product of sporulation, a form of abortive sporulation in which the protective coating is not created. The evolutionary advantage of daughter cell formation is still not clear.

Although this phenomenon is rare, some other microorganisms are capable of releasing free-living offsprings. Certain cyanobacteria produce large numbers of internally held baeocytes, as described earlier, that are released as independent, sometimes motile cells when the parent cell lyses. Some other segmented filamentous bacteria, living anchored to the ileal walls of rodents such as mice and rats also sometimes produce live internal offsprings and other times, dormant spores. On phylogenetic tree, *Clostridium lentocellum* is closer to *Epulopiscium* than *M. polyspora*, but *Clostridium* produces only a single endospore.

The one characteristic spore produced by many Gram-positive rods is the endospore. However, the endospore is not meant for cell multiplication, instead it is used for tiding over the unfavourable environmental conditions, and for dispersal. Resistance to desiccation and high temperatures are the characteristic features of the endospores and are thus of adaptive significance in the survival of the microorganisms. Although only one endospore is produced per cell, its location within the cell is characteristic of a spore. Two basic types of locations are central and terminal with the latter sometimes be further classified as subterminal. The endospores are oval to spherical and are either completely contained within a normal cell or a cell may get distended to accommodate the spore (Fig. 5.22). Endospores are highly refractile and are resistant to basic aniline dye staining. Heating of spores, prior to staining makes them amenable to strong dyes such as malachite green or carbol fuchsin.

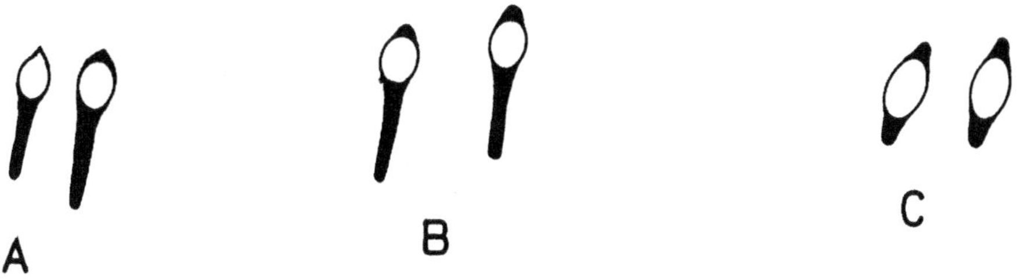

Figure 5.22. Intracellular location of endospores in different bacterial species. A Terminal, B. Subterminal, C. Central.

The general structural plan of an endospore, which may vary from species to species, is based on a multilayered structure (Fig. 5.23). The central core is made up of spore cytoplasm and nuclear material. The core also contains calcium dipicolinate which is responsible for high level tolerance to deleterious environmental conditions. The core is surrounded by a delicate membrane. This is followed by cortex, which in some cases may be distinguished into inner and outer cortex. The cortex in turn is enveloped by one or two spore coats that contain peptidoglycan. The coat may be smooth or show a variety of ornamentations. Finally, this whole structure is contained within a slightly wrinkled exosporium.

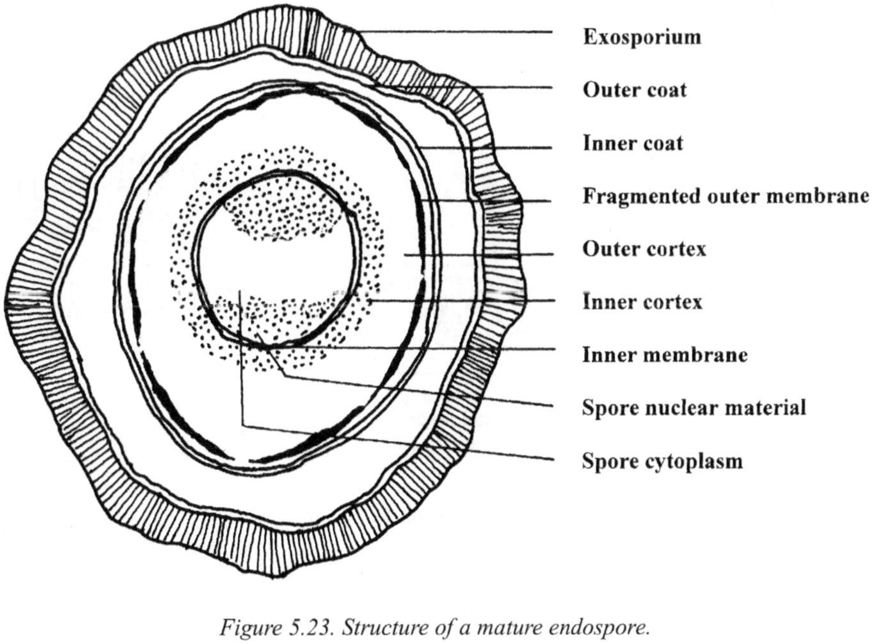

Figure 5.23. Structure of a mature endospore.

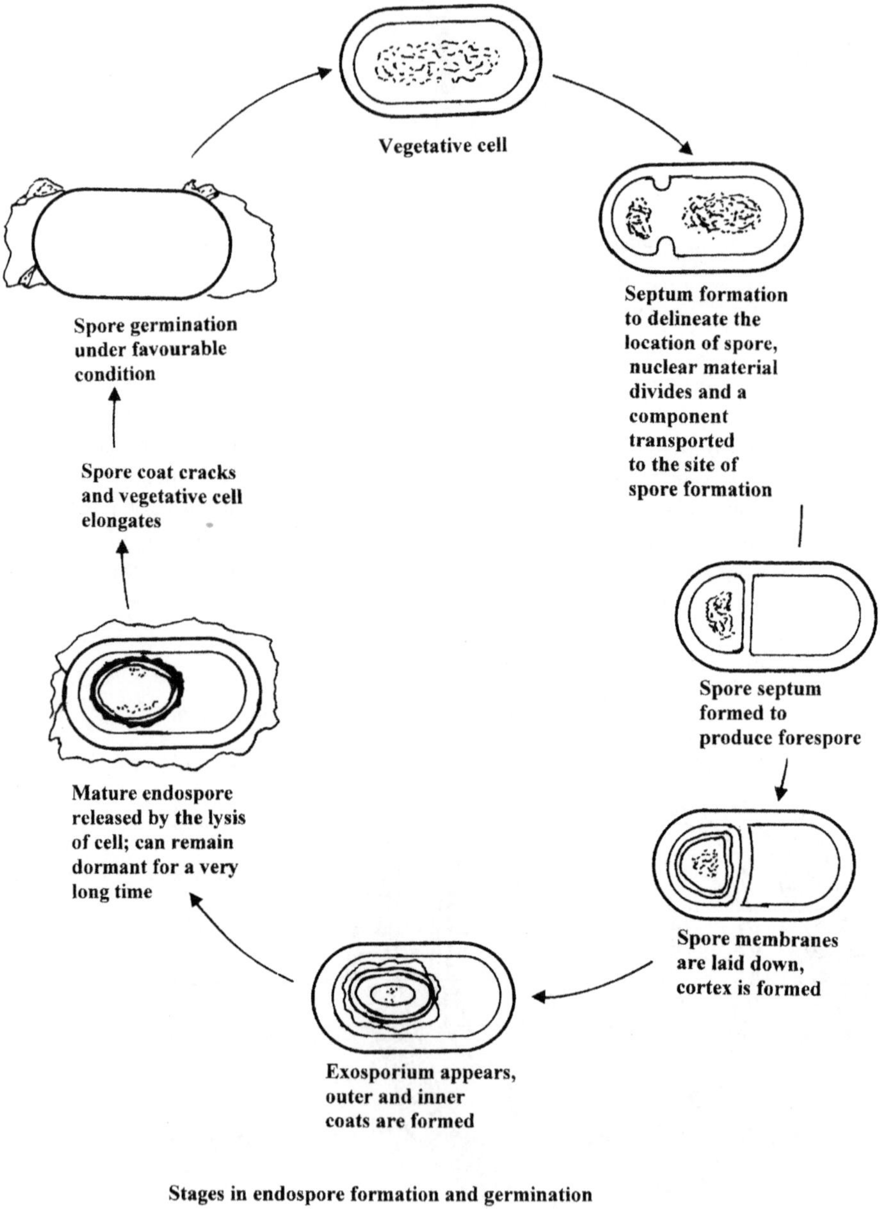

Figure 5.24. Stages in endospore formation and germination.

Besides the structural complexity, the biochemical uniqueness allows the endospores to attain remarkable physiological specialization. This includes the synthesis of dipicolinic acid and polypeptide consisting exclusively of one amino acid, such as cystine. Endospores can thus retain viability for a very long period. They have often been found in geological deposits lying dormant for thousands of years. Some can withstand boiling for more than one hour and are also resistant to several other toxic agents.

Endospores are formed when conditions are not conducive for growth. Such a stage of endospore formation called sporulation, is a sequential process consisting of some well-defined steps which are under strict genetic control (Fig. 5.24). The exhaustion of nutrients trigger sporulation. Initially, the nuclear contents get condensed and cytoplasmic membrane invaginates at a site which establishes the location of spore formation. A copy of bacterial chromosome is incorporated into this compartment that is referred to as "forespore" and the septum grows completely around the protoplast (engulfment). As the spore differentiates the multiple wall layers are laid down sequentially leading to the appearance of characteristic regions (maturation). Moreover, the characteristic chemical components specific to the spore are also synthesized.

On the return of favourable conditions, the endospores germinate to produce the metabolically active cell. During germination, the mother cell swells and lyses and the spore breaks out of the coat and elongates. There is a rapid shift from dormancy to metabolically active state. Phosphoglycerate, the principal substrate stored in the spore, is utilized, ATP is generated, and protein synthesis begins. The characteristic metabolic activities are then taken over by the cell as it resumes normal growth.

5.10.2. Sexual Reproduction

One of the most striking features of the biological evolution has been in the direction of the differentiation of sex. It is generally assumed that the fundamental purpose served by sex is reproduction, with microorganisms (along with many plants and simpler animals) providing the important exception. The second major function not performed by asexual reproduction, is the creation of genetic variability within families, populations, and species.

Bacteria virtually lacking morphological identity of male and female sexes led to the idea that they and many other lower forms do not reproduce sexually. However, the subsequent work clearly proved that the two mating partners in bacteria are physiologically different (mating types). If we go by the definition of the sexual reproduction as a process where the nuclear material between two different cells mix with a possibility of the exchange of their genetic material (recombination), sexual reproduction does occur in bacteria. We shall deal with the details of this process in **Chapter 7**. Essentially it involves the cell to cell contact between mating cells (conjugation) with male or donor cell unidirectionally transferring the genetic material and female or recipient receiving and processing the same to produce the progeny. Many Gram-positive bacteria produce a sex hormone, pheromone, to produce mating aggregates.

5.11. Control of Microbial Growth

The ubiquitous nature of bacteria, their rapid multiplication, pathogenic nature and other activities leading to material deterioration has resulted in finding ways by which their growth and multiplication can be controlled. Two types of ways, those which inhibit the growth (bacteriostatic) of the cells and those which kill (bactericidal) the cells can be envisaged. We have already seen that bacteria require a discrete set of physical and biochemical parameters for their growth. The same parameters, therefore, can be manipulated to get the desired effect.

5.11.1. Physical Control

Various physical factors required for optimum growth can also be used to control the growth.

5.11.1.1. Temperature. Exposure to high temperature is an age-old practice to preserve and prevent the spoilage of an article such as food. This is based on heat denaturation of proteins that affect the structure and metabolism leading to cell killing. The time required to get 90% killing at a temperature represents the D value (decimal reduction time). High temperatures would require relatively short time of exposure.

High temperature killing methods include incineration or direct combustion, dry heating, and moist heating. Heating of transfer loops, or decontamination of disposable materials and wastes like medical and household waste is done by incineration. Inert material such as glass and metal can be sterilized by dry heating in an oven at 180°C. Exposure to 100°C in boiling water or to 121°C under 15 lbs per square inch pressure in an autoclave is the most preferable method of killing microorganisms as even the heat-resistant endospores can be eliminated by this process. This combination of temperature and pressure is used to sterilize media and other implements and instruments. Cooking of food is an effective way of preventing microbial spoilage. The time of exposure, however, depends upon the nature of food. Ultra high temperature (UHT) sterilization consisting of 141°C for 2 seconds is effectively bactericidal and does not spoil the flavour of food.

Canning is used to preserve food by combining heat sterilization and hermetic sealing to prevent oxygen entry. A temperature of 115°C for 10-15 minutes is enough to ensure killing of spore forming *Bacillus stearothermophilus* and *Clostridium sporogenes* and to 121°C for 2.52 minutes to kill *C. botulinum* that causes fatal food poisoning.

Pasteurization is used to reduce the number of microorganisms in liquids like milk. Two different methods, viz., low temperature hold (LTH) involves exposure to 62.8°C for 30 minutes, and high temperature short time (HTST) at 71.7°C for 15 seconds, are used. Many pathogenic bacteria can be killed by pasteurization.

Low temperatures attained during refrigeration or freezing are also effective in food preservation. It not only reduces the growth rate but also the metabolic activities. Cold temperatures can increase the shelf-life of a food but it does not provide effective bactericidal activity. Many psychrophilic and psychrotrophic bacteria like, *C. botulinum* type E can grow slowly at 5°C. Since freezing at -20°C only precludes growth and does not kill, thawing can lead to the growth of surviving microorganisms and subsequent pathogenicity. Some killing effect can be exerted by ice crystal damage to cell membrane.

5.11.1.2. Desiccation. Bacteria like other living cells require water for growth.

Desiccation has been an ancient practice for preserving food and food products. Many bacteria like *Mycobacterium tuberculosis* and their spores, however, not only can withstand desiccation but can also be more easily disseminated in a dried state. Freeze drying, however, is more effective and is more preferrably used.

5.11.1.3. Oxygen. Oxygen is essential for some microorganisms and its removal can be used to prevent their growth. Packaging of food under anaerobic conditions or vacuum packing such as of coffee, eliminates spoilage by aerobic microorganisms and autooxidation. For anaerobic organisms, however, this packaging is done in concert with heat sterilization. Raising oxygen concentration under pressure (hyperbaric oxygen) is effective in controlling the growth of anaerobic, gas gangrene causing *C. perfringens*.

5.11.1.4. Radiation. Like all other living cells, bacterial cells are highly sensitive to electromagnetic radiations of both ionizing and nonionizing types. The short wavelength, highly penetrative ionizing radiations like X-rays and gamma rays exert their germicidal effect by generating highly reactive free radicals, such as OH^-. Their major effect is by destruction of DNA and RNA. Ionizing radiations have been used to sterilize the materials that are destroyed by heat, such as plastics. Pre-sterilized petridishes used in many laboratory experiments are sterilized by gamma radiation. Gamma radiation from ^{60}CO or ^{137}Ce have been used to enhance the shelf life of various foods like fruits, vegetables, and sea foods. Effectivity of radiation exposure is measured by killing of a highly resistant *Micrococcus radiodurans* (resists upto 1 Mrad of exposure), which is also effective against spores of *C. botulinum*. Although no residual radioactivity is detected in the exposed food, the usage of this process is still controversial. Under the law, such foods are supposed to be appropriately labelled for consumer's awareness.

The non-ionizing UV radiations are highly effective as at a wavelength of 260 nm they bring about thymidine dimerization in DNA that can prove lethal to a cell.

5.11.1.5. Filtration. Filtration is effective in eliminating bacteria from air. The important consideration is the pore size of the filter which should be about 0.2 μm to be effective. The air in the laminar flow hood is made microbe-free by passing it through high efficiency particulate air filters (HEPA).

5.11.2. Chemical Control

Various chemical compounds with antimicrobial properties are known. They once again either inhibit the growth (-static) or kill the microorganism (-cidal). Depending upon the target organisms the germicidal compounds are often referred to as virucidal (virus), bactericidal (bacteria), algicidal (algae) or fungicidal (fungi). Based on their uses they have been classified as disinfectants, sanitizers, antiseptics, antibiotics, and preservatives.

5.11.2.1. Disinfectants. Disinfectants are chemicals which are applied on inanimate objects and surfaces and are able to kill the microorganisms. Sanitizers are the disinfectant which reduce the number of microorganims. Ammonia and bleaching agents like hypochlorite are highly effective disinfectants. Ethylene oxide can be used to sterilize disposable plasticwares, surgical and anaesthesia equipments and implements. The toxicity of this gas, however, has limited its use and requires special sealed chambers. Household cleaning agents contain disinfectants which can inhibit the growth of microorganisms. The effectiveness of a disinfectant is determined by the phenol coefficient. It is determined by the ratio of the concentration (dilution) of a germicidal agent that kills the test bacteria in 10 minutes to the concentration (dilution) of the phenol that brings about the same effect. Another commonly used test recommended by the Association of Official Analytical Chemists (AOAC) uses the dilution method. In this, different disinfectants are compared with each other for their germicidal effect. An acceptable used dilution is one that kills all test organisms at least 95% of the time within 10 minutes. The test organisms generally used are *Staphylococcus aureus, Salmonella choleraesuis,* and *Pseudomonas aeruginosa*.

The traditional method of disinfecting water and water bodies is by chlorination. The incidental production of trace amounts of organochlorine compounds, such as carcinogenic trihalomethane (THM) has led to a strict regulation on THM level (100 mg/l) in drinking water. This makes the process impractical and expensive. In order to prevent the spread of pathogenic *Legionella pneumophila* through airconditioners and cooling towers, chlorination or bleaching is an effective method. Use of monochloramine instead of traditional chlorine is not only effective but least expensive in reducing THM. This can be generated by adding ammonia prior to or simultaneously with chlorine or hypochlorite. Ozone (O_3) treatment is also effective but is expensive and the effect is temporary.

5.11.2.2. Antiseptics. These are the antimicrobial agents which can be applied to living tissues but are safe for surface (topical) applications only. The concept of antiseptics was introduced by Joseph Lister. Before this, infections resulting from wounds and injuries often turned fatal. In determining the effectiveness of an antiseptic not only its antimicrobial activity but also its effect on living tissue is to be taken into consideration. This is demonstrated by toxicity index that is defined as the ratio of the greatest dilution of a test compound that can kill the animal cells (chicken heart tissue) in 10 minutes to the dilution that kills the bacterial cells in the same duration.

Alcohol, iodine, crystal violet as well as some heavy metal compounds are used as antiseptics. Alcohol is known to denature proteins, extract membrane lipids and functions as a dehydrating agent. All this can contribute towards its antiseptic action. Its easy evaporation and cracking effect on skin limits its use widely. Iodine is a strong oxidizing agent and, therefore, can kill all types of cells and spores. It is commonly used in topical applications. Mercuric chloride, copper sulphate, and silver nitrate are strongly antibacterial. Similarly, many phenolics and cationic and anionic detergents have effective antiseptic action.

5.11.2.3. Antibiotics. Antibiotics are the compounds that are produced by some microorganisms that can kill other microorganisms. This action of theirs have provided them with the property of active therapeutic agents. More so, because many of them have a far greater effect on the target microorganisms than on human cells. Today, the antibiotic production is a major pharmaceutical industry. Bacteria, specially the actinomycetes and fungi are the major producers of a large variety of antibiotics, though many of them are produced in modified form or entirely synthesized through chemical means.

Antibiotics can be classified on the basis of their mode of action (Table 5.3) or with respect to target microorganism. The latter may reflect a very selective target species or narrow spectrum antibiotic or those which may affect a wide-range of bacteria, also referred as broad spectrum antibiotic. Judicious selection of an antibiotic and/or its concentration is an important consideration in their usage in treating a particular disease. From a clinical point of view, it has become important as indiscriminate use can provide an easy selection pressure for the evolution of antibiotic-resistant strains of the pathogens and their rapid spread. The antibiotic susceptibility test is, therefore, important in deciding the type and the dose of an antibiotic. One of the commonly used laboratory tests is the Bauer-Kirby test. In this, the test organism is spread on nutrient agar surface and disc impregnated with a particular concentration of an antibiotic is placed on the surface. The antibiotic is able to diffuse in the medium and generate a concentration gradient. Inhibition of microbial growth can be seen as a halo or clear zone around the disc (Fig. 5.25). The diameter of zone of inhibition, which is different for different antibiotics, allows the classification of a microbe as sensitive (S), moderately resistant (I) or resistant (R) (Table 5.4). Many clinical laboratories now use automated light scattering device to assess the antibiotic sensitivity by monitoring the growth of a microbe in the presence of the desired antibiotic.

5.11.2.3.1. Minimum Inhibitory Concentration (MIC). The MIC test is useful in determining the lowest concentration of an antimicrobial compound that can inhibit growth. This is determined by monitoring the growth (turbidity) of a microbe in the presence of different concentrations of an antibiotic. This can help determine the effective dose of an antibiotic.

A slightly modified test determines the minimal bactericidal concentration (MBC) as the minimal concentration that can kill the microorganism. This is arrived at by

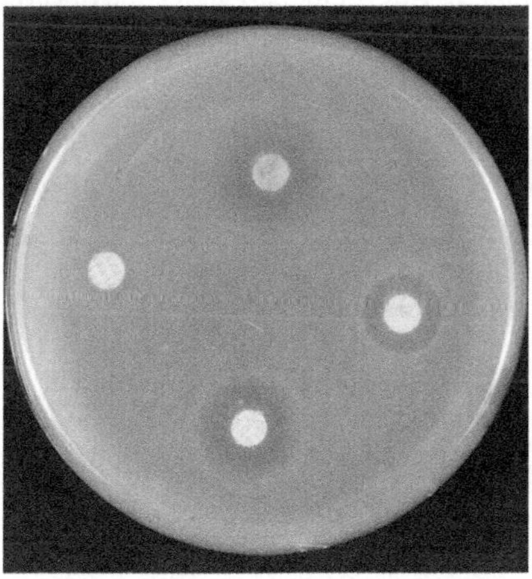

Figure 5.25. Bauer-Kirby Test. A simple plate test employed to determine the sensitivity of a strain to different antibiotics. The filter paper disc impregnated with the known quantity of an antibiotic is placed over the agar plate seeded with the culture. The size of the halo shown after an appropriate incubation is used to classify the strain.

plating a suspension showing no growth in MIC test.

The effectiveness of an antibiotic can also be determined by serum killing power. In this, a bacterial suspension is added to dilutions of the patient's serum treated with an antibiotic. No growth of bacteria at a critical dilution (1:8 or more) is considered effective.

Because of the antibacterial action, the discovery of antibiotics is considered to be an important landmark in medical sciences. In recent years, however, the medical practitioners and medical microbiologists are expressing a great concern about the development of resistance to many of the useful antibiotics among many of the pathogens.

5.11.2.3.2. Resistance to Antibiotics. As early as 1929, Alexander Fleming demonstrated that a contamination of *Staphylococcus aureus* culture by *Penicillium* (common bread mold) leads to bacterial cell lysis that was attributed to a substance, penicillin, secreted by the hypha. From the time the clinical effectiveness of penicillin was demonstrated by Florey's group in 1940, the antibiotics have been hailed as 'wonder drugs'. Rightly deserving this epithet, they provided succour to some of the worst infections in humans. Since 1940, many thousands of different antibiotics produced by fungi, actinomycetes, and other bacteria have been isolated. Of these, actinomycetes specially the species of *Streptomyces,* have dominated the scene contributing ~90% of the all known antibiotics. Many antibiotic-derivatives are either modified or synthetically produced now. Readers can find this information in **Chapter 9**.

With more and more antibiotics becoming available more and more human diseases could be treated and cured. For several years now, however, medical microbiologists have been observing a disturbing trend and raising alarm about it. Many antibiotics which could successfully kill bacterial pathogens are becoming ineffective. This may signal a return to the bad old days of "pre-antibiotics era" claims a researcher when even a simple infection could turn fatal.

Today, the clinical scientists and clinicians are facing the gruesome reality as bacteria are developing resistance with ease to literally any antibiotic medical research has thrown

Table 5.3. Some antibiotics and antimicrobial compounds, their mode of action and effective antimicrobial spectrum

Antibiotics/Compounds	*Mode of action*	*Effect*	*Action spectrum*
I.Cell Wall Inhibition			
1.Penicillins	Inhibit the enzymes involved in transpeptidase reaction and also penicillin binding proteins	Affected cell become susceptible to autolysis and osmotic shock	Most Gram$^+$ few Gram$^-$ bacteria
2.Cephalosporins	—do—	—do—	—do—
3.Semisynthetic penicillin, e.g., aminobenzyl penicillin or ampicillin	—do—	Easily penetrates the outer LPS layers of Gram$^-$ cell wall	Broad-spectrum Gram$^-$ and Gram$^+$ bacteria

Antibiotics/Compounds	*Mode of action*	*Effect*	*Action spectrum*
4.Others- Cycloserine	Prevents incorporation of D-alanine	Cell wall synthesis disrupted	Broad spectrum
5.Bacitracin	Prevents linkage between N-acetylglucosamine and N-acetylmuramic acid	Cell wall synthesis disrupted	Broad spectrum
6.Vancomycin	Binds directly to peptide inhibiting cross linking	Cells susceptible to osmotic shock	Gram$^+$ bacteria
II. Protein Synthesis Inhibition			
1.Aminoglycoside Antibiotics e.g. Kanamycin, Neomycin Streptomycin, Gentamicin and others	Binds to 30S subunit of ribosome causing misreading of mRNA. Some inhibit binding of aminoacyl tRNA	Inhibit protein synthesis leading to cell death	Gram$^-$ bacteria few Gram$^+$ bacteria
2.Other inhibitors- Tetracycline	Binds to 30S subunit of ribosome and prevents the binding of aminoacyl tRNA to A-site of mRNA-ribosome complex	—do—	Broad spectrum
3.Erythromycin	Binds to 50S subunit of ribosome, and prevents translocation of peptidyl tRNA	—do—	Mostly Gram$^+$ bacteria

Antibiotics/Compounds	*Mode of action*	*Effect*	*Action spectrum*
4.Chloramphenicol	Binds to 50S subunit and inhibits binding of aminoacyl tRNA to A, and P sites, thus peptide bond	—do—	$Gram^-$ bacteria few $Gram^+$ bacteria
5.Puromycin	Mimics tRNA, binds to A-site, forms a peptide bond with the growing polypeptide chain	—do—	Broad spectrum
6.Rifamycin or Rifampicin	Inhibits transcription	Inhibits protein synthesis	Broad spectrum
III. DNA Inhibition			
1.Intercalating agents like, Proflavine, ethidium bromide, mitomycin C	Disrupt DNA replication	Toxic to cell growth	Antibacterial Antitumorigenic
2.Griseofulvin	Inhibits DNA replication	—do—	Antifungal
3.Quinolones such as, Nalidixic acid, norfloxacin	Inhibits DNA gyrase by binding to β-subunit specially swivelase component	Cessation of DNA synthesis	$Gram^-$ bacteria especially those causing urinary tract infections
IV. Membrane Function Inhibition			
1.Polymixin B, Colistin	Interact with cytoplasmic membrane	Leakage of cell contents	$Gram^-$ bacteria
2.Amphotericin B (polyene)	Forms channels or pores in the cytoplasmic membrane of fungi	—do—	Many fungi and yeast

Antibiotics/Compounds	*Mode of action*	*Effect*	*Action spectrum*
V. Other Antimicrobial Compounds			
1.Sulphonamides,sulphones and para-aminosalycilic acid (analogues of vitamin, p-aminobenzoic acid)	Faulty functions. Inhibition of folic acid synthesis	Affect bacterial growth	Some Gram$^-$ bacteria
2.Trimethoprim	Inhibits dihydrofolate reductase thus blocks the synthesis of thymidine and purine	—do—	Wide spectrum especially those causing urinary and intestinal tract infections

Table 5.4. Susceptibility testing for some antibiotics by Bauer-Kirby test inhibition zone diameter (mm)

Antibiotic	*Disc conc.*	*Susceptible*	*Intermediate*	*Resistant*
Ampicillin	0.01 mg	14 or more	12-13	11 or less
Bacitracin	10 units	13 or more	9-11	8 or less
Chloramphenicol	0.03 mg	18 or more	15-17	14 or less
Erythromycin	0.015 mg	18 or more	14-17	13 or less
Kanamycin	0.03 mg	18 or more	14-17	13 or less
Neomycin	0.03 mg	17 or more	13-16	12 or less
Penicillin	10 units	22 or more	12-21	11 or less
Streptomycin	0.01 mg	15 or more	12-14	11 or less
Sulphonamides	0.3 mg	17 or more	13-16	12 or less
Tetracycline	0.03 mg	19 or more	15-18	14 or less

at them. This problem is further compounded by the fact that once a strain acquires a resistance strategy it does not feel shy to share it with others, even with those which may be completely unrelated.

The heavy use of antibiotics on humans or even farm animals or misprescription and underuse of drugs by physician or sometimes by patients have allowed the bacteria to evolve strategies to resist an antibiotic. This was best summed up by Flemming when he wrote in 1945, "the greatest possibility of evil in self-medication is the use of too small doses so that instead of clearing up infection, the microbes are educated to resist penicillin". Those that survive an antibiotic assault that kills even 99.9% of their fellow bugs are more likely to be superstrains, almost invincible to control for a long time to come. According to an estimate, in Europe until 1979, only 6% of pneumococci strains were resistant to penicillin, 10 years later this number rose to 44%. In US, more than 90% of the staphylococci are now resistant to penicillin. It is not difficult to envisage a similar situation for most other commonly used antibiotics.

In 1940, several years ahead of the introduction of penicillin into regular clinical practice, a bacterial enzyme that could hydrolyze ß-lactam ring was described. It is, therefore, clear that bacteria were ready with their defense strategies long before these toxic compounds came in regular use. This is the art that they have perfected over the years in the face of antibiotic onslaught. Most of the early studies on antibiotic resistance were carried out on laboratory strains of *E. coli* and *S. typhimurium*. A spontaneous mutation frequency of 10^{-9} or less per bacterial generation for streptomycin resistance gave no cause to worry for clinical applications. What was not anticipated at that time was the variety of mechanisms and the possibility of carrying the resistance gene/s on autonomously replicating transferrable plasmids, R plasmids, or even transposons.

Based on discrete modes of their antibacterial action, the resistance mechanisms are also few and distinct. These comprise, penetration barrier, target site modification, and active efflux to enzymatic modification/inactivation of the antibiotic. Though some minor mechanisms like sequestration of antibiotic, and metabolic bypass of inhibited reaction, can also play a role (Table 5.5).

Of the different mechanisms identified, enzymatic inactivation of the antibiotics is one of the most common biochemical mechanisms that bestow resistance to a whole variety of antibiotics. This is most intriguing because such a mechanism could not have evolved by mutations in an existing (often essential) bacterial enzyme that permitted it to catalyze the modification of a completely different substrate (antibiotic). It appears plausible that such a capability arose in few clinical isolates and the same was acquired by other bacteria through efficient gene transfer system. It is also clear, however, that once established in a pathogen, genes coding for such enzymes might have undergone mutations to remodel the active site of the enzyme to match the changing spectrum of antibiotic substrates. For example, a single base change in the gene for ß-lactamase (penicillinase) can change the substrate specifity of the enzyme leading to an extended spectrum.

The mutational changes in ß-lactamase gene occur rather frequently, possibly under the selection pressure of different ß-lactam antibiotics. Increased resistance may also come from increased expression of ß-lactamase through an up-promoter mutation or changes in transcriptional regulation of chromosomal ß-lactamase gene. These variations may have serious consequences as a single base change in ß-lactamase gene may render $ 100 million worth of research effort useless. One approach to counteract the destructive activity of ß-lactamase is to employ structural analogues of ß-lactams as enzyme inhibitors. For example, combination of amoxicillin and clavulanic acid can successfully circumvent ß-lactamase action. However, the chances of getting a mutation to mask the action of inhibitor must always be considered looming large on this approach. Many resistant strains may have a combination of mechanisms such as reduced affinity and thus reduced uptake due to altered penicillin binding protein and a mechanism to inactivate the drug.

Therapeutic use of a series of naturally-occurring and semi-synthetic aminoglycoside antibiotics led to selection and dissemination of a variety of resistance determinants.

Unlike ß lactamase, the resistance to aminoglycosides is brought about by different types of modifications such as those by acetyltransferases, adenylyltransferases, and phosphotranferases. The genes encoding these modifications do not undergo facile mutation to change their substrate spectrum. For example, when kanamycin was replaced by structurally-related gentamicin, an entirely new class of antibiotic inactivating enzyme was detected in gentamicin resistant strains.

Table 5.5. Mechanisms of antibiotic (antimicrobics) resistance in bacteria

	Mechanism	*Antibiotics*
I.	Restricted outer membrane transport	
	Penetration barrier	(i) Chloramphenicol (ii) Many β-lactam antibiotics (iii) Tetracycline
II.	Target site modification	
	(1) Altered penicillin binding proteins	(i) Penicillins
	(2) Ribosomal modification	
	(a) Alteration of a single amino acid in S12 protein on 30S subunit	(i) Erythromycin
	(b) Addition of methyl group to rRNA	(ii) Streptomycin
	(3) RNA polymerase modification	(i) Rifampin
	(4) Gyrase modification	(i) Quinolone
	(5) Modification of Peptidoglycan	(i) Vancomycin
III.	Active Efflux	
	Energy dependent transmembrane efflux	(i) Tetracycline
IV.	Enzymatic modification/ Inactivation	(i) ß-lactams
	(1) Hydrolysis-enzymatic cleavage	
	(2) Derivatization- adds acetyl, adenyl or phosphate group	(i) Aminoglycosides- kanamycin, neomycin, streptomycin (ii) Chloramphenicol
	(3) Target enzyme modification or metabolic bypass of inhibited reaction	(i) Sulphonamides (ii) Trimethoprim

Chloramphenicol acetyl transferase (CAT) genes are widely distributed among bacterial pathogens. The CAT genes (a few variants are known) appear to be of independent derivation like aminoglycosides modifying genes. Because of its toxicity, the use of this antibiotic has declined sharply in industrialized nations. However, its broad-spectrum action and inexpensive production leads to its extensive use in the Third World. This perhaps keeps the selection pressure on for CAT genes. The type I CAT has dual function. Besides acylating chloramphenicol, it also complexes steroidal antibiotic and fusidic acid.

Many microbes generate large quantities of glutathione but its use in providing antibiotic resistance has been demonstrated only in the case of fosfomycin, with which an adduct is formed. O-Phosphorylation of erythromycin has also been found in many resistant strains as also the hydrolysis of lactone ring of this class of antibiotics.

Another important mechanism of resistance comes from penetration barrier so that the cell appears intrinsically resistant. The outer membrane of Gram-negative bacteria offers one such permeability barrier, because lipophilic antibiotics and other agents cannot easily penetrate this layer. The porin channels may also be too constricted to allow the passage of such molecules. *Pseudomonas aeruginosa* has evolved a strategy by lacking the high-permeability porins thus even relatively hydrophilic antibiotics like ß-lactams also diffuse across the outer membrane very slowly.

Gram-positive mycobacteria have adopted a similar strategy. They are surrounded by a low permeability barrier that consists of a lipid bilayer of unusually high order and low fluidity. This has led to multi-drug resistance in many mycobacteria of clinical importance. Specific transport mutation such as in porin gene or other siderophore-like system can confer resistance.

Low permeability barrier alone has only little contribution in achieving significant level of drug resistance. A second contributor is always associated with the same either in the form of enzymatic inactivation or a membrane-associated energy-driven efflux.

Active efflux must have evolved as a major mechanism to avoid a number of obnoxious agents. Active drug efflux family systems have been described in bacteria. Four such families have been recognized: (i) "major facilitator (MF) family" which has a sequence homology to the glucose facilitators of mammalian cells, (ii) resistance nodulation-division (RND) family also used for exporting some heavy metal ions, (iii) Smr or Staphylococcal multidrug resistance family, consisting of small transporters, and (iv) ABC or ATP binding cassette transporters. The members of the first three families are single cytoplasmic membrane proteins that extrude drugs by using proton motive force or act as a H^+-drug antiporter. Many such efflux genes have been identified in antibiotic producing *Streptomyces* species that would be used to protect the producer strains. Many reduced uptake mechanisms described earlier are now shown to involve active efflux.

For example, the commonest mechanism of chloramphenicol resistance is the acylation of the drug. However, a non-enzymatic mechanism of efflux also exists. Active efflux has been demonstrated for several drugs, antiseptics and fluoroquinolones. In fact, multidrug efflux systems to handle wide variety of drugs, and other toxicants, such as cationic dyes, and organic cations etc., have been described.

Development of resistance to antibiotics by reduction in the affinities of their enzymatic targets is commonly encountered for antibiotics that inactivate a single target and are not analogues of the substrate. However, resistance due to target site alterations for the antibiotics that inactivate multiple targets irreversibly by acting as analogues of substrate is also common.

Penicillin inhibits transpeptidases by acting as structural analogue and forming an irreversible penicillolyl-enzyme complex. Bacteria possess multiple transpeptidases also called as penicillin-binding proteins (PBPs). The most common mechanism for penicillin resistance in clinical isolates is due to hydrolysis by ß-lactamase. However, resistance due to reduced permeability as in some Gram-negative bacteria or due to development of high molecular mass PBPs that have decreased affinity for the antibiotic are also common. The latter has been documented in *Haemophilus influenzae, Neisseria gonorrhoeae, Streptococcus pneumoniae,* various other streptococci and enterococci, *Statphyloccus aureus* and *S. epidermidis*. The penicillin-resistant gonococci and meningococci invariably possess mosaics of high-Mr PBP genes (PBPs), whereas penicillin sensitive isolates do not. PBPs genes of resistant isolates are essentially identical in parts to those in sensitive isolates except for 14-23% divergence. This mosaic structure is proposed to have arisen through interspecific homologous recombination. The PBP2 (*mecA*) gene of methicillin resistant *S. aureus,* on the other hand, produces normal (four) PBP2 but also contains an additional high-Mr low affinity PBP not found in sensitive isolates.

The rifamycins and quinolones both inhibit single enzymatic targets and alteration of these targets provide resistance. Rifamycin interacts with ß-subunit of RNA polymerase and numerous amino acid substitution or insertion and deletion result in decreased affinity for the antibiotic and thus resistance.

Quinolones like nalidixic acid or fluoroquinolone like ciprofloxacin inhibit DNA gyrase. Mutations in the *gyrA* gene encoding the A subunit of DNA gyrase provide resistance.

Resistance to few antibiotics occurs by synthesizing a new target enzyme that has much lower affinity for the antibiotic than the normal enzyme. Resistance to sulphonamides and to trimethoprim is usually achieved by this mechanism. In a few trimethoprim-resistant clinical isolates of enteric bacteria over-production of normal target enzymes also brings about the desired effect. More straightforward mechanism will be the development of the altered forms of the normal targets insensitive to the antibiotic. For example, macrolide antibiotics are resisted by methylating 23S ribosomal RNA.

5.11.2.3.3. Origin of Genes Conferring Antibiotic Resistance. The discovery of antibiotic modification as a self-protection measure in streptomycetes led to the proposal that antibiotic-producing strains provided the genes for resistance. Some support to this hypothesis came from nucleic acid and protein sequence comparisons of the aminoglycoside resistance determinants from producing organisms and clinical isolates. Many streptomycetes produce ß-lactamase though they do not synthesize ß-lactams. Similarly, chloramphenicol producing strain does not produce chloramphenicol acetyltransferase but many non-producing streptomycetes are known to make the enzyme.

The first tangible evidence of resistance gene transfer involving antibiotic producing streptomycetes came from the observation that tetracycline resistance gene (*tetK* and *tetL*) found in non-tuberculous mycobacteria and some *Streptomyces* species were similar to those identified in tetracycline-producing *S. rimosus.* These genes appear to be "foreign" because their G+C content differs from those of streptomycetes and mycobacterial chromosomal DNA. This is a clear case of gene pick-up and subsequent transfer among other genera. Thus, in nature, a substantial pool of antibiotic resistance genes or their close relatives exists. This gene pool becomes readily accessible when bacteria are exposed to strong selection pressure of antibiotic usage such as in hospitals, veterinary and agricultural set up, and animal and poultry husbandry.

The rapid spread of antibiotic resistance genes is assigned to their localization on plasmids such as R plasmids. The appearance of multi-resistance plasmids found in many pathogens is considered to be relatively recent event. When a new antimicrobial agent is introduced, plasmid-determined resistance is expanded by gene pick-up. The R plasmid itself is a clear example of a transmissible plasmid getting modified by picking up resistance genes.

The intermicrobial traffic of these genes and an associated marker pick up pathway got a further boost when many of these antibiotic resistance genes were found located singly or in small or large groups on transposable elements (see **Chapter 7**).

From the studies on transposable elements, a key structural constituent was identified and referred to as "integron". The integron is a mobile DNA element, consisting of two conserved segments flanking a central region that contains cassettes of antibiotic resistance genes. The integron is inserted with the help of a site-specific recombinase or integrase, the gene for which lies in the 5' segment along with a strong promoter or promoters for expression of antibiotics resistance genes. Integrase catalyzed insertion of resistance gene cassettes, and site specific deletion and rearrangement of the inserted cassettes have been demonstrated. The nucleotide sequence analysis of multi-resistant integrons show that the resistance cassette differ markedly in codon usage indicating that they are of diverse origins. Integrons may play a major role in gene mobilization by mediating the fusion between two R plasmids, or by interaction between the recombination hot spot of one and a secondary integrase target site on the other chromosome. The origin of R plasmid can also be explained by incorporating antibiotic resistance genes from various sources into an integron and then into mobile elements and finally into a replicon (Fig. 5.26).

The conjugative transposons represent a fortuitous combination of transposition and conjugative transfer and both are important for resistance gene flux in nature. The transfer of many such transposons, usually carrying tetracycline resistance is stimulated in the presence of the antibiotic. They perhaps represent a step in evolution when antibiotics not only controlled the expression of resistance genes but also promoted their own transfer. One may be tempted to consider these antibiotics as another group of bacterial pheromone.

Thus, the judicious use of antibiotics appears to be the only answer to the problems related to antibiotic resistance and its rapid, uncontrolled spread.

5.11.3. Preservatives

The use of chemical preservatives in controlling growth of unwanted microorganisms or pathogens in food or to prevent food spoilage is a common practice. However, strict regulations are to be exerted so as to ensure the safety of such additives or preservatives. Some of them have been traditionally used since ancient times.

The addition of salt (NaCl) to a food reduces the available water content and alters the osmotic pressure. Both actions will retard the growth of microorganisms as described earlier. Preservation in brine solution is an age-old practice and foods like fish, meat and vegetables can be effectively preserved by salting. The association of high levels of salt with high blood pressure and heart disease has made this process less preferable in recent times.

High sugar solution as a preservative also works on the same principle as salt. Fruits, candies, and others are preserved by high levels of sucrose. Maple syrup and

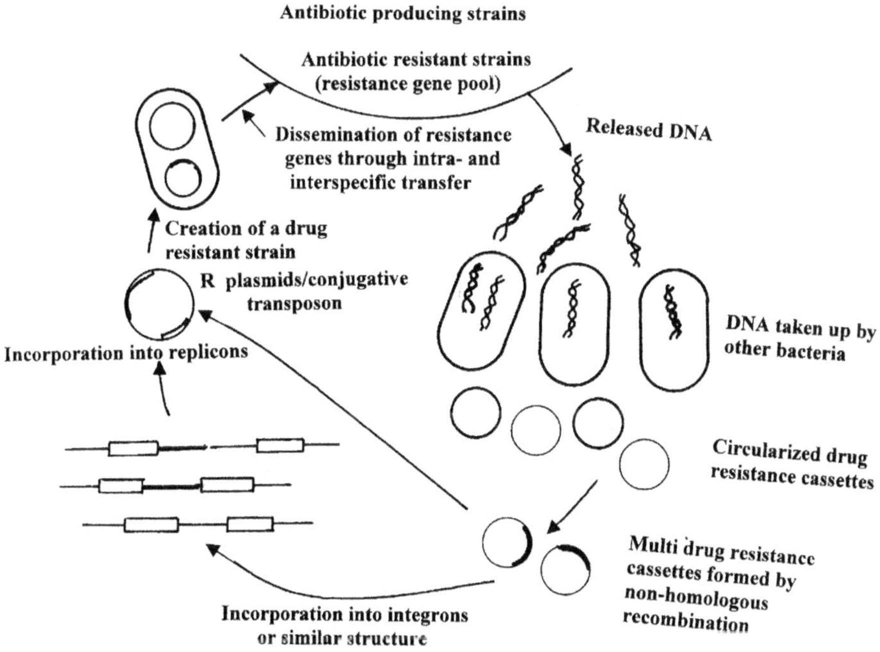

Figure 5.26. Possible route by which antibiotic resistance genes are acquired by bacteria. The process is facilitated under the increasing antibiotic selection pressure.

honey are naturally preserved due to their high sugar content.

Some organic acids or their salts such as citric, lactic, acetic, propionic, benzoic and sorbic acid are used in food preservation. They are able to inhibit bacteria that cause food spoilage. They have wide applications in bread, bakery products, cheese, pickles etc. Vinegar that contains acetic acid effectively inhibits bacterial and fungal growth. Sodium benzoate and methyl-and propyl-p-hydroxybenzoates are added to fruit juices, jams, jellies, soft drinks, salads and fruit dressings, ketchup and margarine as preservatives.

Nitrates and nitrites can produce strong oxidizing agents like nitrous acid and nitric oxide which can be used for preserving cured meats and meat products. Nitrates are effective against *Clostridium botulinum*. However, formation of nitrosamines, which are highly carcinogenic, by nitrates and nitrites with secondary and tertiary amines has restricted their use.

Sulphur dioxide and various sulphites as well as ethylene and propylene oxides possess antimicrobial properties. They are mostly used as fumigating agents on dried fruits, nuts and spices, although lemon juices and wine can also be preserved by using them.

CHAPTER 6
BACTERIA AND LIFE PROCESSES-II METABOLISM

A bacterial cell is a true example of an independent living cell, such that a single cell is capable of exhibiting all life processes. These life processes depend upon the flow of energy and materials during which the energy generated gets redistributed so that the cells can grow, multiply and carry out other cellular functions as described in **Chapter 5**. In this Chapter, the chemical reactions accompanying this energy flow and redistribution that collectively define the process known as cellular metabolism will be discussed. The simple organization of the bacterial cell and its amenability to grow on simple synthetic media has made them the organisms of choice to study different aspects of metabolism. Although simple, bacteria exhibit an extreme range of diversity in their metabolism that has helped them to live just about any where and feed on just about anything. There are three main components of metabolism: (i) A complex network of chemical reactions employed to generate and capture energy and raw materials from the environment (catabolism), (ii) Using these raw materials and energy to synthesize cell components that will allow cell sustenance (anabolism), and (iii) Regulation of the metabolic processes so as to exert utmost efficiency of the first two processes.

II-A-Generation of Cellular Energy

The formation and utilization of ATP and the proton motive force (PMF) is central to cellular energy transformations and forms the core of metabolism. Bacteria are highly versatile as to the choice of the pathway for generating energy. They have often been divided on the basis of metabolism that they employ (Table 6.1). Basically these include, (1) Chemoorganotrophic (chemoheterotrophic) metabolism in which organic compounds serve as the source of generating energy as well as carbon for cell constituents, (2) Chemoautotrophic (chemolithotrophic), on the other hand, rely on inorganic compounds for the same two purposes, (3) Photoautotrophic metabolism where light is used as energy source and CO_2 as a source of carbon, and (4) Photoheterotrophs that use light as energy source and an organic compound as carbon source.

Table 6.1. Classification of bacteria on the basis of energy-generating mechanism and principal carbon-source

Type	*Characteristic features*	*Examples*
1. Photoautotroph (photolithotroph)	Light is the principal energy source and CO_2 is the carbon source	Green and purple sulphur bacteria, cyanobacteria, green plants
2. Photoheterotroph (photoorganotroph)	Light as the energy source and an organic compound as carbon source	Purple non-sulphur bacteria
3. Chemoautotroph (chemolithotroph)	Energy obtained from the oxidation of inorganic compounds, such as NH_4^+, NO_2^-, H_2, H_2S,and Fe^{2+}, and CO_2 as the carbon source	Nitrifying bacteria, Thiobacilli, *Alcaligenes*
4. Chemoheterotroph (chemoorganotroph)	Organic compounds both as energy and carbon source	Most bacteria and animals

During both chemolithotrophic and photoautotrophic metabolism a proton gradient is established across the membrane followed by chemiosmotic generation of ATP (Fig. 6.1). In heterotrophic metabolism, on the other hand, an organic substrate is converted to end products with a consequent release of energy that is coupled to ATP formation. Since metabolism consists of different sequential steps, it is often referred to as metabolic pathways, where each step is mediated by a specific enzyme. There are two basic frameworks that define these pathways: In catabolic pathway, a substrate is broken down to smaller molecules (intermediary metabolites) with release of energy. The anabolic pathways begin with small molecules to generate larger molecules with consumption of energy. In a cell, these two types of pathways are not only interlinked but also precisely regulated. This brings about a better utilization of resources and prevents any wasteful activity.

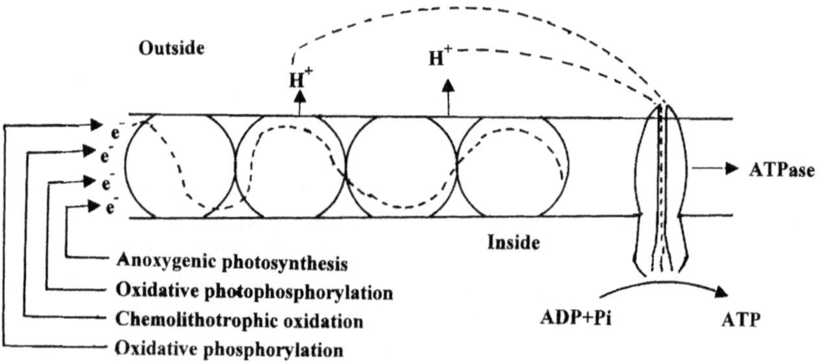

Figure 6.1. Schematic representation of the generation of proton motive force (PMF) and chemiosmotic synthesis of ATP.

As we shall see in subsequent discussion, glycolysis and tricarboxylic acid cycle (TCA) forms the core of metabolism and provides the interlinks with the metabolism of carbohydrates, lipids, proteins, and nucleic acid.

The proton motive force (PMF) is an electrochemical gradient that gets established across the cytoplasmic membranes of bacteria and archaeal cells (also mitochondria and chloroplast of eukaryotic cells), when cells, due to their metabolic activities, move protons (H^+) by specific proton translocating systems. This gradient consists of two components, the electrical one, $\Delta\Psi$ (delta Psi), and a chemical component, ΔpH. The PMF may provide energy for generation of ATP or may be harnessed directly for cellular work.

ATP is the "central currency of energy" of all living organisms. Much of the cellular metabolism is, therefore, directed towards the formation of ATP that stores chemical energy in the form of high energy phosphate bonds. When ATP is broken into ADP + Pi large amounts of free energy is released. In effect, bond breaking requires energy and bond making releases energy. In the reaction ATP$\rightarrow$ ADP + Pi (inorganic phosphate), the immediate formation of new bonds from the products releases considerably more energy than that was used to break the original bond. This leads to relatively large release of free energy during ATP breakdown.

Many metabolic pathways couple the energy-yielding reactions (exergonic) with the energy-requiring (endergonic) conversion of ADP + Pi to ATP. A cell, therefore, must have an efficient recycling of ATP and ADP to drive cellular functions. It has been estimated that a growing *E.coli* cell produces ~2.5 million molecules of ATP per second to support its energetics. Living cells possess two pathways for generating ATP. In substrate level phosphorylation, the generation of ATP is directly coupled with a chemical reaction. In oxidative phosphorylation, however, the protons translocated outside by the flow of electrons through membrane carriers first establish a PMF that is then coupled to ATP synthesis (see Fig. 6.10). These two processes can be separated from each other by chemicals like dinitrophenol (uncouplers).

The formation of ATP is catalyzed by a proton-conducting membrane-bound multicomponent enzyme, adenosine triphosphatase (ATPase), also referred as F_0F_1 ATPase, proton translocating ATPase, ATP synthetase, and ATP synthase. This enzyme not only hydrolyzes ATP but also synthesizes it. The energy of protons moving through ATPase is captured and transferred to form ATP from ADP + Pi by a process called chemiosmosis.

In the following sections, we shall discuss the various pathways utilized by bacteria to generate ATP and effect the flow of carbon to maintain the supply of various intermediates.

6.1. Respiration

Respiration is the process of conversion of carbohydrates and other substrates and consists of three distinct phases: (i) Glycolysis- a catabolic pathway in which an organic molecule is broken down to simpler molecules with the generation of some ATP and reduced coenzymes, (ii) the tricarboxylic acid (TCA) cycle- through this cycle, the small organic molecules produced during glycolysis are further oxidized to CO_2 and H_2O with the concurrent generation of more ATP and reduced coenzymes, and (iii) oxidative phosphorylation in which the reduced coenzymes are reoxidized and, thus, regenerated. The electrons and protons released are transported through a series of membrane bound carriers which finally establish a proton gradient or PMF. PMF is used for chemiosmotic generation of ATP and a terminal acceptor such as O_2 is reduced.

When molecular O_2 serves as the external electron acceptor, the process is known as aerobic respiration. Bacteria, however, may use a whole range of electron acceptors such as NO_3^-,SO_4^{2-}, CO_2, S° and Fe^{3+} and the process is referred to as anaerobic respiration. Some bacteria can use more than one electron acceptors. *Pseudomonas denitrificans*, for example, can use O_2 or NO_3^- as the terminal electron acceptor and thus can perform both aerobic and anaerobic respiration. The aerobic respiration, therefore, is an oxidation-reduction process, in which a carbohydrate combines with oxygen to form CO_2 and H_2O with a consequent release of ATP (38 molecules per molecule of glucose converted):

$$\mathbf{C_6H_{12}O_6 + 6O_2 \longrightarrow 6CO_2 + 6H_2O + 38\ ATP}$$

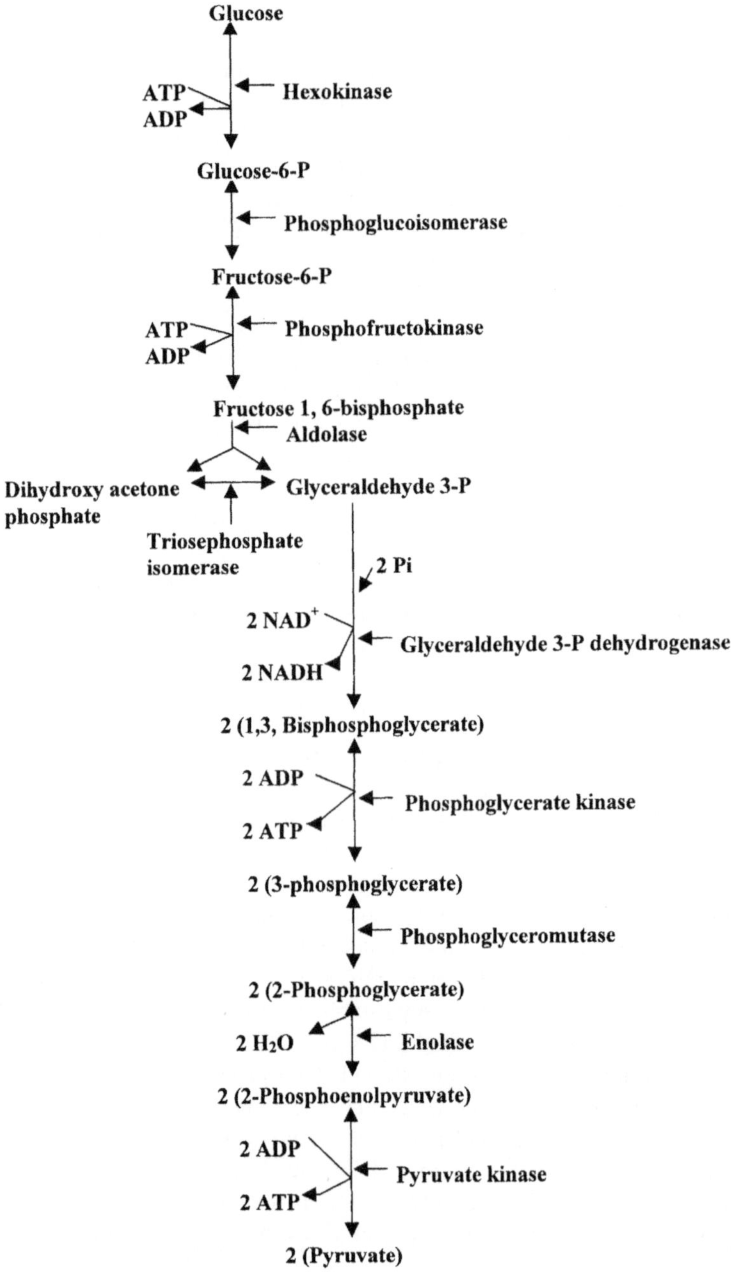

Figure 6.2. Embden-Meyerhof-Parnas Pathway of Glycolysis. Gains: 2 Pyruvate; 2 NADH; 2ATP.

6.1.1. Glycolysis

Carbohydrates are the most preferable substrates for the catabolic pathway called glycolysis. Almost all types of carbohydrates, such as polysaccharides, hexoses and pentoses, carbohydrate derivatives, like gluconic and glucuronic acid, and polyalcohols like mannitol and glycerol can be utilized by bacteria; glucose, however, is the most preferable substrate of many. Different glycolytic pathways are employed by different bacteria.

(i) Embden-Meyerhof-Parnas (EMP) Pathway: Glucose is converted to pyruvate by many anaerobic and facultatively anaerobic bacteria as well as eukaryotic cells by EMP pathway (Fig.6.2). The archaea are virtually the only exception.

The synthesis of pyruvate is accompanied by the net synthesis of 2 ATP molecules and 2 molecules of reduced coenzyme (NADH). The enzyme phosphofructokinase, is the key enzyme regulating the rate of glycolysis as it is allosterically inhibited by excess of ATP. Two substrate level phosphorylation steps, viz., conversion of 1,3-bisphosphoglycerate to 3-phosphoglycerate and of phosphoenolpyruvate to pyruvate, yield 4 ATP molecules. With the utilization of 2 molecules of ATP in the early part of glycolysis it results in the net availability of 2 ATP molecules.

Pyruvic acid is further converted to butanol or lactic acid or other end products depending upon the type of bacterium involved. In alcoholic fermentation, pyruvate is further broken down to ethyl alcohol, whereas lactic acid bacteria produce lactic acid from pyruvate.

In many bacteria like *E. coli* and related enteric bacteria, some species of *Clostridium* and *Pseudomonas*, a modified EMP-pathway operates under phosphate-limiting conditions. In this, 2-dihydroxyacetone phosphate is converted to methylglyoxal which in turn produces pyruvate through D-lactate. This pathway, thus, bypasses the phosphorylation of glyceraldehyde 3-phosphate but still yields pyruvate. However, 2 molecules of ATP are consumed rather than produced.

(ii) Entner-Doudoroff (ED) Pathway: Many bacteria and archaea employ a different glycolytic pathway known as Entner-Doudoroff Pathway (Fig.6.3). Many aerobic bacteria such as *Pseudomonas* species and others opting for this pathway lack the key enzyme, 6-phosphofructokinase and thus can not operate EMP pathway. In ED pathway, glucose-6-phosphate is oxidized to 6-phosphogluconate and then to 2-keto-3-deoxy-6-phosphogluconate (KDPG). KDPG is then cleaved to pyruvate and glyceraldehyde-3-phosphate which can regenerate pyruvate through later steps of EMP pathway. This pathway yields only one ATP molecule and, therefore, is only 50% as efficient as EMP pathway. Also, in this pathway reduced coenzyme NADPH is generated from NAD^+. NADPH and $NADP^+$ are used in biosynthetic pathways whereas NADH and NAD^+ are used in metabolic reactions that generate ATP. In an alternative pathway both glyceraldehyde 3-P and pyruvic acid are converted to acetaldehyde and then to ethanol.

Some bacteria (e.g., *Alcaligenes* species, *Clostridium* species) and some archaea (e.g., *Halobacterium* species, *Sulfolobus* species, *Pyrococcus* species) operate a partially phosphorylated or non-phosphorylated ED pathway. In this, glucose is converted to gluconate which is subsequently dehydrated to produce 2-keto-3-deoxygluconate. In

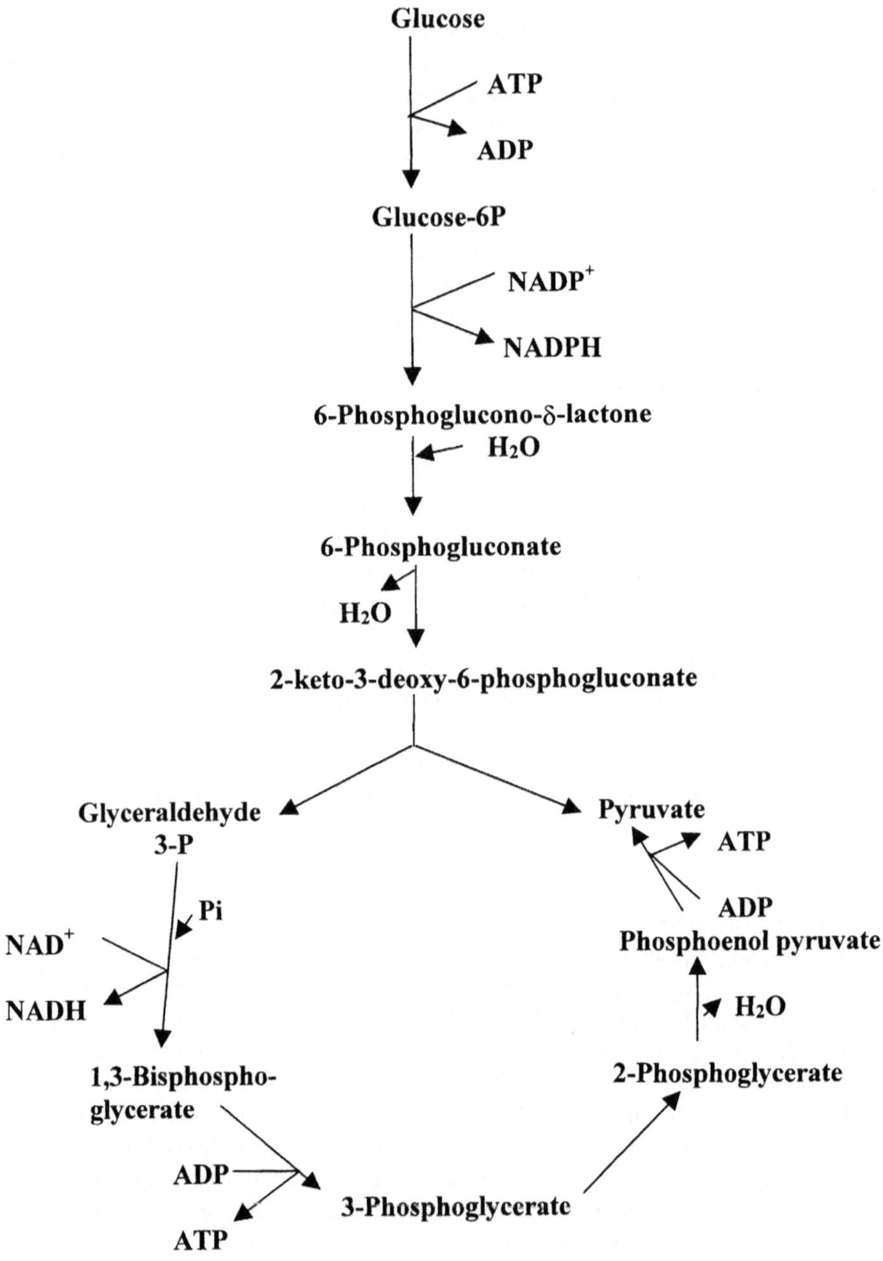

Figure 6.3. Entner-Doudoroff Pathway of Glycolysis.
Gains: 2 pyruvate; 1 NADPH; 1 NADH; 1 ATP.

one version, 2-keto-3-deoxygluconate is phosphorylated to KDPG that follows the normal ED pathway. In another, it is split into pyruvate and glyceraldehyde, with the latter then oxidized to glycerate, the same is then phosphorylated to 2-phosphoglycerate that is eventually converted to pyruvate as in normal ED pathway. ATP generation is restricted to further oxidation of pyruvate.

(iii) Pentose Phosphate Pathway: A third pathway or pentose phosphate pathway is very similar to ED pathway as glucose is converted to 6-phosphogluconate. However, 6-phosphogluconate is broken down to ribulose 5-phosphate which subsequently yields ribose 5-phosphate and xylulose 5-phosphate. This pathway, therefore, serves as a link between 6-carbon and 5-carbon carbohydrate metabolism and provides ribose required for DNA, RNA, ATP, NAD^+, and $NADP^+$ synthesis. Depending upon the species and metabolic requirements of some key intermediates several versions of this pathway are known (Fig. 6.4).

In some organisms, the oxidative pentose phosphate cycle accounts for complete oxidation of carbohydrates. This pathway involves a number of sugar interconversion steps with intermediates feeding into some or the other pathways. In this, glucose 6-phosphate is finally converted to ribulose 5-phosphate and CO_2. Ribulose 5-phosphate is maintained in equilibrium with ribose 5-P and xylulose 5-P by ribosephosphate isomerase and ribulose phosphate epimerase, respectively. The two can combine with the help of a transketolase to form sedoheptulose 7-P and glyceraldehyde 3-P. These two get converted to fructose 6-P and erythrose 4-P with the help of an aldolase and subsequently with a transketolase to glyceraldehyde 3-P and xylulose 5-P. While fructose 6-P can be converted to glucose 6-P and re-enter the cycle, glyceraldehyde 3-P can either give rise to pyruvate or with dihydroxyacetone phosphate forms fructose 1, 6-P and latter to fructose 6-P and then to glucose 6-P. Thus, repetitive turns of the cycle can account for the complete oxidation of glucose 6-P:

$$\textbf{Glucose-6-P} + \textbf{12 NADP}^+ \longrightarrow \textbf{6 CO}_2 + \textbf{12 NADPH} + \textbf{12 H}^+ + \textbf{Pi}$$

Some bacteria like *Acetobacter* and *Gluconobacter* may operate a part of this cycle under anaerobic conditions to yield acetate. In another variation, the formation of xylulose 5-P follows the pentose phosphate pathway but is cleaved into glyceraldehyde 3-P and acetyl phosphate by the action of phosphoketolase. While glyceraldehyde 3-P is metabolized to pyruvate and subsequently through triose phosphate pathway to lactate, acetyl phosphate is converted to acetyl-CoA and then to acetaldehyde and finally to ethanol. This important pathway (heterolactic fermentation) is carried out by bacteria such as *Leuconostoc*, and various *Lactobacillus* species. The pathway is also known as the Phosphoketolase or Hexose Monophosphate Shunt Pathway (Fig. 6.5).

(iv) Utilization of other Sugars: Besides glucose, many bacteria can use other mono-, di-and polysaccharides. Polysaccharides are broken down into monosaccharides that enter the glycolytic pathway at different points.

The polysaccharide mannans is broken down to mannose which are further metabolized. Common disaccharides, such as maltose, sucrose, and lactose are metabolized to produce glucose or fructose that can enter glycolysis. Galactose derived from lactose is converted to glucose 1-P which is then transformed to glucose 6-P and

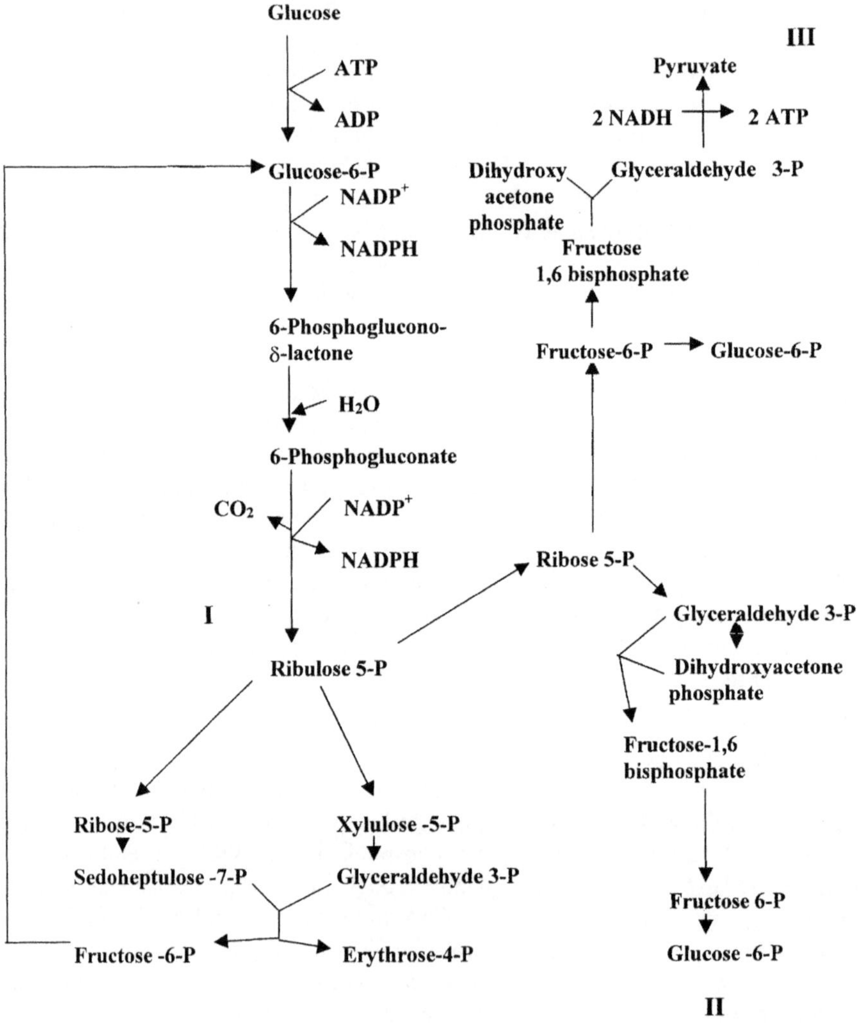

Figure 6.4. Different versions of Pentose Phosphate Pathway.
Gains: I= Ribulose-5-P; CO_2; 2 NADPH.
II= CO_2; 12 NADPH; Glucose-6-P broken down and resynthesized.
III= Pyruvate, 2 NADPH, 2 NADH, 2 ATP.

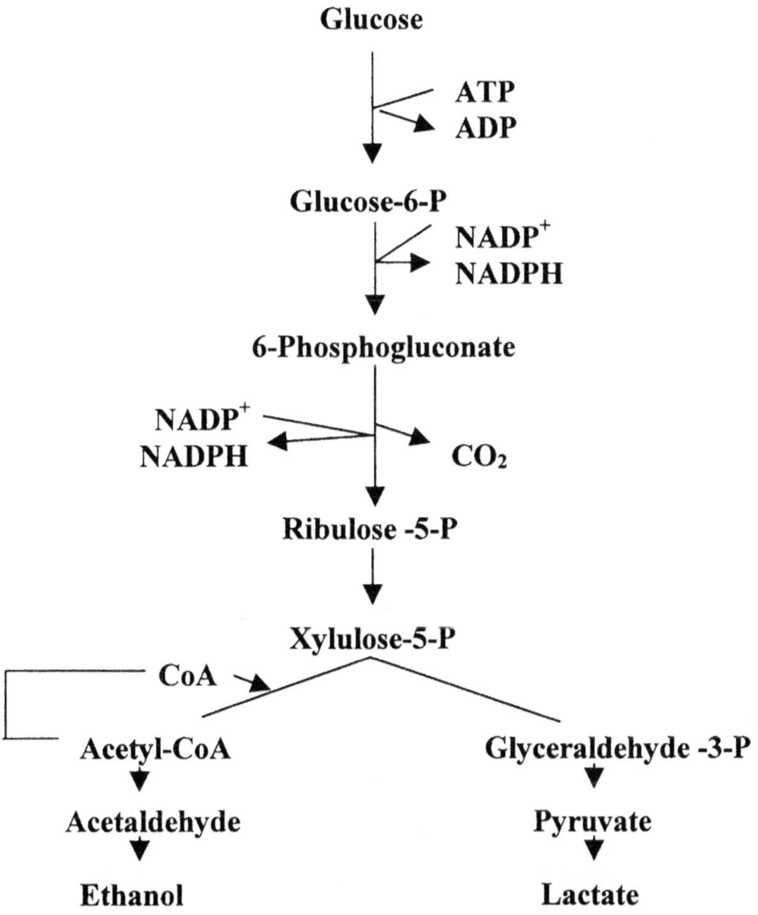

Figure 6.5. Hexose monophosphate shunt.

enters EMP pathway. Glucose 1-phosphate is also produced from glucose when sucrose or glycogen is metabolized and the same is transformed to glucose 6-P. This is a highly efficient reaction as glycolysis can be initiated without initial phosphorylation steps.

6.1.2. Lipid Conversion

Lipids can also serve as substrates for cellular production of ATP. Triglycerides can be cleaved by lipases into glycerol and fatty acids. Glycerol can be metabolized to dihydroxyacetone phosphate and then glyceraldehyde 3-phosphate which then enters glycolytic pathway. Similarly phospholipids can be cleaved by phospholipase, and

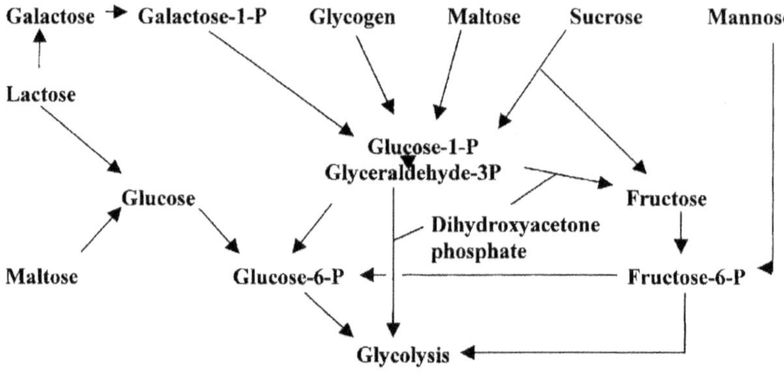

Figure 6.6. Conversion of higher sugars (di-and polysaccharides) to generate an entry point into EMP pathway. Note: Polysaccharides will be first broken down to these simpler sugars.

glycerol so formed can be handled as described above.

The fatty acid portion of lipids is metabolized by a pathway called ß-oxidation. In this process, fatty acids are broken down into small 2-carbon acetyl coenzyme A (CoA). Initially fatty acid molecule reacts with CoA to form a fatty acid-CoA using energy from ATP. Further reaction releases acetyl-CoA and forms a fatty acid-CoA molecule that is 2-carbon atoms shorter than the original molecule. Successive removal of 2 carbon atom units and production of acetyl-CoA continues until the original molecule is completely degraded. Each reaction generates reduced coenzymes, one molecule each of $FADH_2$ and NADH. The acetyl CoA is driven into the TCA cycle (Fig. 6.7).

6.1.3. Protein Catabolism

Many bacteria can utilize proteins as a source of energy to generate ATP. Proteins are first broken down by enzymes called proteases into smaller peptides and amino acids. Enzyme deaminase then removes amino group from amino acids. Removal of carboxylic acid by decarboxylases converts them into intermediates that either enter into TCA cycle or into different biosynthetic pathways. For example, alanine, glycine and serine can all be converted to pyruvate (Fig. 6.8).

6.1.4. Tricarboxylic Acid Cycle (TCA)

TCA cycle constitutes the second phase of respiratory metabolism. Alternatively known as Citric acid cycle or Krebs cycle, it is initiated by the reaction of pyruvate, generated during glycolysis, or protein catabolism, with CoA in the presence of NAD^+ to form acetyl-CoA, NADH and CO_2. This decarboxylation step is mediated by a multienzyme complex, pyruvate dehydrogenase (48 polypeptides). The acetyl-CoA can also be fed into TCA cycle through ß-oxidation of fatty acids as described earlier.

In the first reaction of TCA cycle, acetyl-CoA reacts with oxaloacetate to form citrate, regenerating CoA. Through a series of steps involving carboxylic acids, the Krebs cycle then reforms oxaloacetate (Fig. 6.9). During TCA cycle, CO_2 is released in two successive steps: conversion of isocitrate (6C-compound) to α-ketoglutarate (5C-compound) and of α-ketoglutarate to succinyl-CoA (succinate is a 4C-compound).

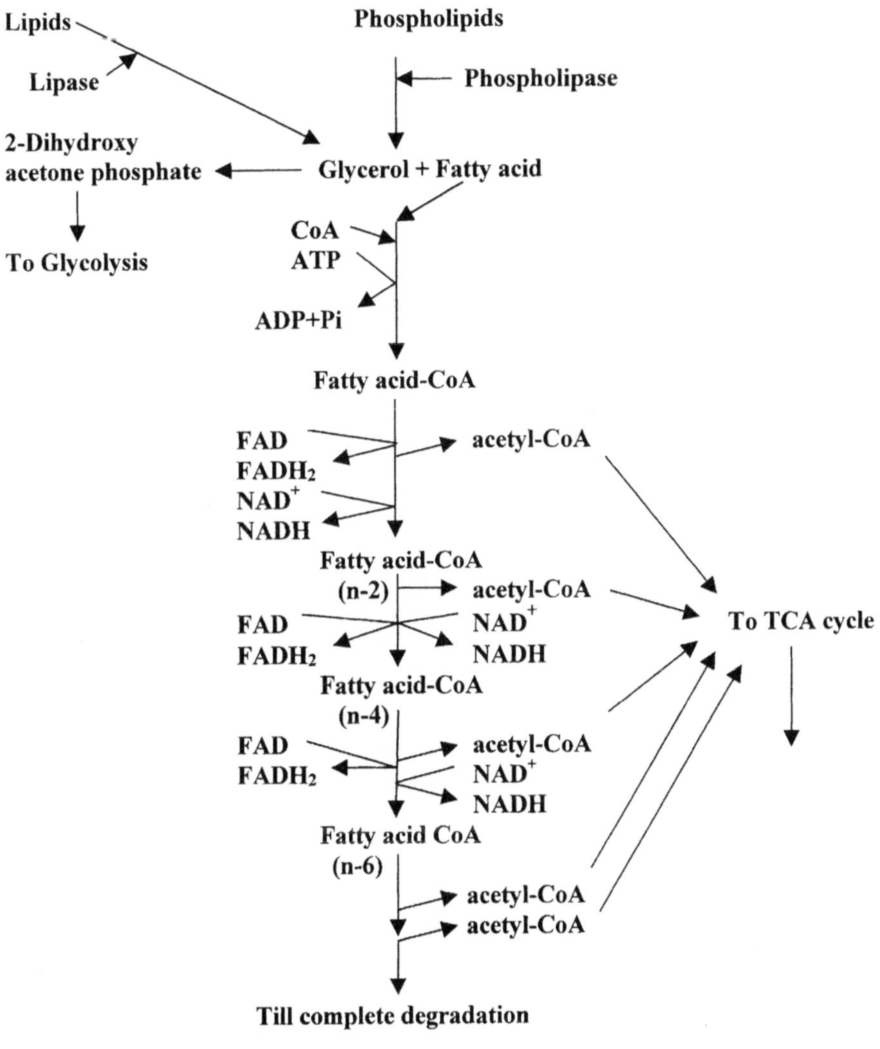

Figure 6.7. Utilization of lipids and phospholipids.

Also, reduced coenzyme NADH is formed at three different steps and an additional reduced coenzyme $FADH_2$ is formed from FAD (Flavin adenine dinucleotide) during conversion of succinate to fumarate. There is only one step of substrate level phosphorylation in TCA cycle that occurs during the conversion of succinyl-CoA to succinate. While in some bacteria ATP is formed during this step in others like eukaryotes, GTP is synthesized. GTP is an important energy source for some cellular reactions, such as, protein synthesis. GTP can also be converted to ATP when it transfers

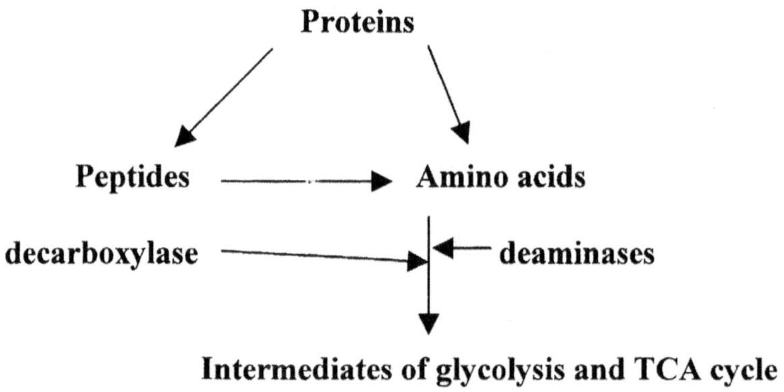

Figure 6.8. Utilization of proteins.

its high energy phosphate group to ADP (GTP + ADP → GDP + ATP).

The TCA cycle occupies a central position in cellular metabolism. It is highly efficient as its intermediates are not only reutilized but it also supplies precursor molecules to a large number of biosynthetic pathways as we shall discuss later in this **Chapter**. At the end of the TCA cycle, all the carbon from glucose is converted to CO_2.

2 pyruvate + 2 ADP + 2Pi + 2FAD + 8 NAD^+ → 6 CO_2 + 2 ATP + 2 $FADH_2$ + 8 NADH

If we sum up the EMP pathway with that of TCA cycle, there will be a net synthesis of 4 ATP molecules, 10 NADH and 2 $FADH_2$ molecule. The reduced coenzymes are used either for generating ATP (oxidative phosphorylation) or for the synthesis of NADPH required for biosynthesis.

6.1.5. Oxidative Phosphorylation

The final step in respiration consists of a process called oxidative phosphorylation. In this, the reduced coenzymes generated earlier in glycolysis and the TCA cycle are reoxidized. Oxidative phosphorylation is a sequence of oxidation-reduction reactions of the membrane-bound carrier molecules culminating in a terminal acceptor, such as oxygen (Fig. 6.10). While O_2 is reduced, ATP is synthesized through chemiosmosis. The final acceptor in anaerobic respiration may be an inorganic (e.g., sulphate or nitrate) or low molecular weight organic (e.g., fumarate) molecule. The members of a general series of electron transfers constitute:

NADH → flavoprotein → non-heme iron protein → quinone → cytochromes → O_2

Of these, flavoproteins and quinones are hydrogen atom (electron + proton) carriers and cytochromes and non-heme proteins are electron carriers. Metabolic poisons such

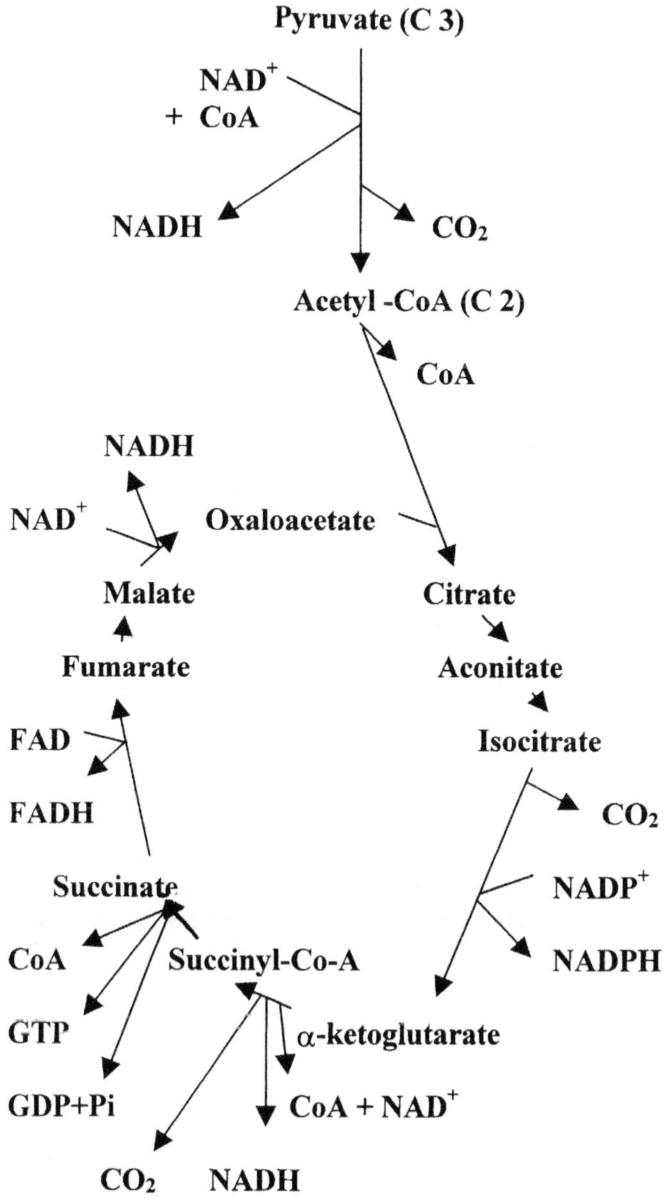

Figure 6.9. Krebs' Cycle/TCA Cycle. Through substrate level phosphorylation and electron transport phosphorylation: 15 ATP; I GTP. Gains: 3 CO_2; 4 NADH; FADH.

as cyanide and CO are strong inhibitors of electron transport. While cyanide binds to iron of certain cytochromes, CO binds to terminal cytochrome.

One of the important aspects of oxidative phosphorylation is the chemiosmotic generation of ATP. The sequential transfer of electrons and the coupled transfers of protons establishes a proton gradient across the membrane that drives the ATP synthesis (see Fig. 6.10). Compounds like 2-dinitrophenol can disrupt the linkage between two processes and are called uncouplers. The assymetric distribution of electron carriers leads to proton carriers transporting outside and electron carriers transporting towards inside. It has been estimated that as many as 10 protons may be transported across the membrane for each NADH molecule. A portion of the chemical energy released during electron transport is trapped in the form of a proton motive force that is used to generate ATP or perform other functions.

The chemiosmotic generation of ATP is based on the fact that the protons cannot diffuse back across the membrane but can only recross it via a specific proton channel, such as that provided by membrane bound ATPase. ATPase is a multicomponent enzyme system consisting of two major polypeptides, F_0 and F_1. While F_0 is membrane embedded, channel-forming component allowing the passage to the protons, F_1 lodged on the inner surface, catalyzes the reaction ADP + Pi $\rightarrow$ ATP. F_1 is also capable of catalyzing ATP hydrolysis to ADP and Pi. For every NADH molecule oxidized, 3 ATP molecules, and for every $FADH_2$, 2 ATP molecules are synthesized. Since 10 NADH and 2 $FADH_2$ molecules are synthesized during aerobic respiration (glycolysis + TCA cycle), a total of 34 ATP molecules will be available at the end. Most importantly, chemiosmotic generation of ATP is in addition to those formed during glycolysis and TCA cycle. Thus, the whole energetics of respiratory metabolism can be represented as:

Glucose + 6 O_2 + 38 ADP + 38 Pi $\rightarrow$ 6 CO_2 + 6 H_2O + 38 ATP

The generation of 38 ATP represents a theoretical maximal yield, a slight modification in the electron transport chain leads to fewer ATPs. For example, *E.coli* cell produces only 26 ATPs from respiration. Similarly anaerobic respiration also yields less ATP. Some bacteria like *Vibrio alginolyticus* couples the electron transport chain to translocation of sodium ions across the membrane, when growing in alkaline environment.

6.2. Fermentation

Many bacteria perform catabolic metabolism by a mechanism other than respiration, called fermentation. In fermentation, the organic substrate acts as an electron donor and an organic molecule derived from the process acts as electron acceptor. No external acceptor, such as oxygen, is involved and there is no net change in the oxidative state of the coenzymes involved. The synthesis of ATP in fermentation is due to substrate level phosphorylation only and is restricted to glycolysis.

A fermentative pathway begins with a substrate, goes through glycolysis and terminates with the end product whose nature depends upon the substrate. The end product may also be used to name the pathway (Table 6.2; Fig. 6.11).

Fermentation yields less ATP per substrate molecule and, therefore, is energetically

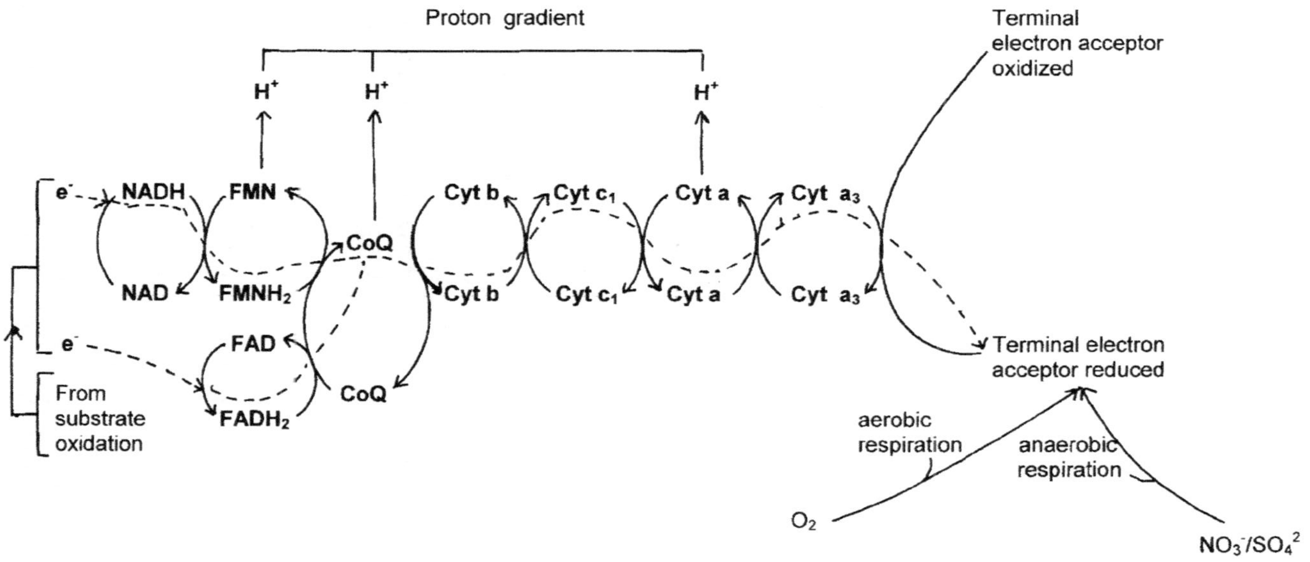

Figure 6.10. Membrane embedded electron transport chain that runs through a series of reactions. Pumping of H^+ across the membrane establishes proton gradient, the return flow of which ensures generation of ATP, and the reduction of a terminal electron acceptor.

less favourable. Those bacteria capable of undertaking both pathways switch from one to another depending upon the growth conditions.

Obligatively fermentative bacteria, such as, *Streptococcus* species do possess an ATPase but do not generate ATP by oxidative phosphorylation. Instead, ATPase is used to hydrolyze ATP to ADP and Pi, and the energy of this reaction is coupled with the export of proton from the cell.

6.3. Chemoautotrophy (Chemolithotrophy)

Chemoautotrophs (chemolithotrophs) perform respiration by metabolizing inorganic compounds. They oxidize an inorganic compound and couple it with the reduction of a coenzyme. The reduced coenzyme then participates in the electron transport and ATP is generated chemiosmotically. Although the terminal acceptor in majority of the cases is oxygen, some can carry out anaerobic respiration. Only a limited number of bacteria and some archaea are capable of chemoautotrophic metabolism. A number of inorganic compounds ranging from molecular hydrogen to reduced sulphur compounds, reduced iron compounds and nitrogen-containing compounds are used as substrates. These bacteria, therefore, possess special enzymes to utilize a particular inorganic compound (see Table 6.3).

6.3.1. Hydrogen Oxidation

Some hydrogen oxidizing bacteria, such as *Alcaligenes eutrophus*, produce hydrogenase that oxidizes molecular hydrogen to form water and reduced coenzyme, NADH:

$$\mathbf{H_2 + 1/2\ O_2 \xrightarrow{NAD} H_2O + NADH}$$

Some homoacetogenic bacteria, such as *Clostridium aceticum* and *Acetobacterium woodii* couple oxidation of H_2 to the reduction of CO_2 generating acetate in the process. Additional ATP is generated during this process when acetyl-CoA is converted to acetate. Special coenzymes, such as tetrahydrofolate and vitamin B_{12} are also employed.

Many extremely thermophilic archaea, such as *Pyrodictium* which grows at 110°C in deep sea thermal vents, oxidize H_2 with elemental sulphur acting as a terminal electron acceptor.

6.3.2. Sulphur Oxidation

Some bacteria, such as *Thiobacillus* and some archaea like, *Sulfolobus* utilize reduced sulphur compounds like H_2S, elemental sulphur S° and thiosulphate:

$$\mathbf{H_2S + O_2 \rightarrow S^\circ}$$

$$\mathbf{S_2O_3^{2-} + 2O_2 + H_2O \rightarrow 2SO_4^{2-} + 2H^+}$$

$$\mathbf{S^\circ + 1\tfrac{1}{2}\ O_2 + H_2O \rightarrow H_2SO_4}$$

Oxidation of H_2S leads to S° formation which is deposited either in the form of intracellular S° granules or extracellularly. NADH is not generated and therefore electrons derived from oxidation reaction enter electron transport at an intermediate point. Some thiobacilli generate ATP by combining sulphite with AMP to form ADP

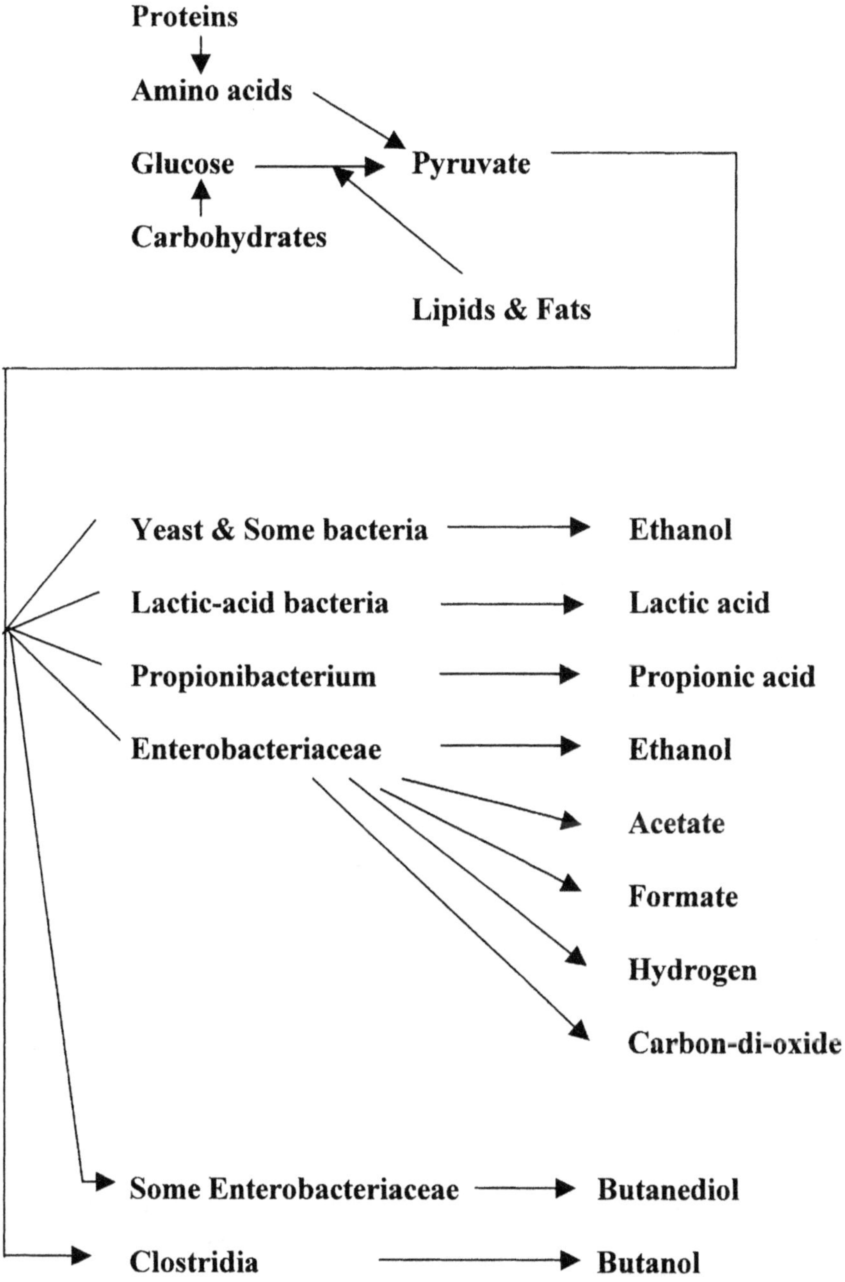

Figure 6.11. Fermentative products from different pathways.
Note: Pyruvate plays a central role in these reactions.

Table 6.2. Type of fermentative pathway in different bacteria

Fermentation Pathway	*End Products*	*Some Examples*
Homolactic acid	Lactate	*Streptococcus, Enterococcus,* and various *Lactobacillus* species
Heterolactic acid	Lactate+ethanol + CO_2	*Leuconostoc,* various *Lactobacillus* species
Ethanolic	Ethanol + CO_2	Commonly by many yeasts, rarely by few bacteria
Propionic acid	Propionate + acetate + CO_2	*Propionibacterium*
Mixed acid	Ethanol + acetate + lactate + succinate + formate + H_2 + CO_2	Members of Enterobacteriaceae such as *E.coli*
Butanediol	Butanediol + CO_2	*Enterobacter aerogenes, Klebsiella* species
Butyric acid	Butyrate + butanol + acetone + CO_2	*Clostridium* species
Amino acid	Acetate + NH_4^+ + CO_2	*Clostridium, Streptococcus, Mycoplasma, Peptococcus*
Methanogenesis	CH_4 + CO_2	Some Archaebacteria

Table 6.3. Components of chemolithotrophic metabolism

Primary electron donor	*Electron acceptor*	*Product/s*	*Example*
1. H_2	O_2	H_2O	*Alcaligenes eutrophus*
2. S_2O_3	O_2	SO_4	*Sulfolobus* species
3. S°	O_2	H_2SO_4	*Thiobacillus denitrificans*
4. S^{2-}	O_2	H_2SO_4 or S°	*Thiobacillus thiooxidans*
5. NH_4^+	O_2	NO_2	*Nitrosomonas* species
6. NO_2^-	O_2	NO_3	*Nitrobacter* species
7. Fe^{2+}	O_2	Fe^{3+}	*Thiobacillus ferrooxidans*
8. CO	O_2	CO_2	*Hydrogenomonas* species

and then 2 molecules of ADP combine to give ATP:

$$\mathbf{ADP + ADP \rightarrow ATP + AMP}$$

Formation of sulphate by bacteria like *Thiobacillus thiooxidans* and *T. ferrooxidans* has important ecological implications in acid mine drainage problems and their use in mineral biorecovery processes.

Sulfolobus, an acidophilic thermal archaea that grows at 90°C in sulphur-rich springs and at pH as low as 1.0 produces sulphuric acid by oxidizing H_2S or S° to H_2SO_4. These bacteria have played a significant role in ecosystems having extremely uninhabitable conditions.

6.3.3. *Iron Oxidation*

T. ferrooxidans generally found in acid mine drainage streams oxidizes both reduced sulphur and reduced iron:

$$2Fe^{2+} + 2H^{+} + \frac{1}{2}O_2 \rightarrow 2Fe^{3+} + H_2O$$

Interestingly, the electrons from Fe^{2+} directly enter the electron transport chain through an iron-sulphur protein and cytochrome C. Protons from a highly acidic environment enter the cell because of the less acidic interior, and generate ATP as they pass through ATPase. Oxidation of iron is linked to extrusion of protons that protects the cell from acidification.

6.3.4. *Ammonium and Nitrite Oxidation-Nitrification*

Some bacteria called nitrifying bacteria have the ability to oxidize ammonia or nitrite:

$$NH_4^{+} + 1\frac{1}{2} O_2 \rightarrow NO_2^{-} + H_2O + 2H^{+}$$

$$NO_2^{-} + \frac{1}{2} O_2 \rightarrow NO_3^{-}$$

Nitrosomonas oxidizes ammonia to nitrite in two steps, with hydroxylamine formed as the first product. Oxidation of hydroxylamine sets in an electron transport chain that leads to chemiosmotic generation of ATP.

Nitrification generates limited amount of ATP because some electrons are used to form reduced coenzymes for biosynthesis. These bacteria play an important role as they influence the soil fertility. While positively charged NH_4^{+} ions bind to negatively charged clay particles, negatively charged NO_2^{-} and NO_3^{-} cannot, and thus can be leached out.

6.3.5. *Methanogenesis*

Some chemolithotrophic archaebacteria, are able to convert CO_2 to methane (CH_4) through a specialized anaerobic respiration pathway. Such bacteria called methanogens use electrons from hydrogen to reduce CO_2. During this process, one molecule of ATP is synthesized for every molecule of CO_2 reduced in addition to NADPH. Methanogens appear to be the earliest inhabitants of the earth as they could grow autotrophically on H_2 and CO_2 under anaerobic conditions- a type of atmosphere prevailing on early earth.

6.4. *Photoautotrophy*

Many bacteria are capable of obtaining their energy directly from the energy of sun which they convert into chemical energy in the form of ATP. This process is known as photophosphorylation. These microorganisms possess pigments such as chlorophyll or bacteriochlorophyll which can absorb a particular wavelength of sunlight. This leads to excitation of the molecule causing the release of an electron which is then transferred through a series of membrane-bound carriers known as a photosystem. Electron transfer leads to the establishment of a proton gradient culminating in chemiosmotic generation of ATP (Fig. 6.12).

The general reaction flow of this process includes the trapping of light by light-harvesting 'antennae' pigments which then transfer the photons of light to a photochemical reaction centre (a particular pigment). At this reaction centre the electron flow begins leading to synthesis of ATP. Pigments of various colours, such as red, orange, and yellow carotenoids and green chlorophylls allow the bacteria to trap sun energy of different wavelengths. This characteristic leads to a stratification pattern of photoautotrophic microorganisms in water bodies.

Cyanobacteria and algae carry out oxygenic photosynthesis in which H_2O is split to serve as a source of electrons, liberating O_2 as a product besides ATP and reduced coenzymes. Other photoautotrophic bacteria are characterized by anoxygenic photosynthesis. They generate ATP but not oxygen because they use alternate electron donors such as H_2, S, and H_2S in place of water. The two types of photosynthetic bacteria differ in the nature and arrangement of their photosynthetic pigments, the electron donors used, and the number of photosystems as well (Table 6.4).

6.4.1. Oxygenic Photosynthesis

Cyanobacteria, prochlorobacteria, and algae contain chlorophyll a as the predominant reaction centre pigment and chlorophylls, carotenoids, or phycobiliproteins as accessory 'antenna' pigments. This group, like higher plants, possesses two photosystems through which they generate ATP and NADPH. Photosystem I has reaction centre with chlorophyll a P_{700} and of photosystem II with a slightly modified chlorophyll a P_{680}. The two photosystems are linked in a series by Z pathway of oxidative photophosphorylation (Fig. 6.13a). Of the two, photosystem II is a non-cyclic photophosphorylation pathway operating unidirectionally.

The electron flow begins by absorption of light energy by P_{680} chlorophyll that changes it into energetically excited state releasing an electron. This oxidation of P_{680} is balanced by splitting H_2O to generate oxygen, hydrogen ions, and electrons that reduces oxidized P_{680} back to its original state. The formation of oxygen makes this process known as oxygenic photosynthesis. The electrons are transferred from photosystem II through a number of membrane-bound carriers to P_{700} chlorophyll of photosystem I, which establishes a proton gradient sufficient to synthesize one molecule of ATP.

Light energy is also absorbed by P_{700}. The release of an electron due to its excitation is balanced by the donation of an electron through photosystem II. The electrons transferred through photosystem I is eventually utilized to produce co-enzyme NADPH from $NADP^+$, which is required for biosynthetic metabolism (Fig. 6.13a).

This pathway is generally non-cyclic with the electrons flowing from donor H_2O to accepter $NADP^+$. However, a cyclic photosystem I is also known in which ATP is formed but NADPH is not synthesized. Many cyanobacteria carry out a non-oxygen evolving photosynthesis at low light intensities in which cyclic photosystem I operates but photosystem II is inoperative. In such situations, cyanobacteria derive their reducing power by oxidizing H_2S in which elemental sulphur granules get deposited extracellularly.

6.4.2. Anoxygenic Photosynthesis

As the name suggests, in this process oxygen is not evolved but ATP is generated. In the anaerobic green and purple photosynthetic bacteria and the heliobacteria, a cyclic

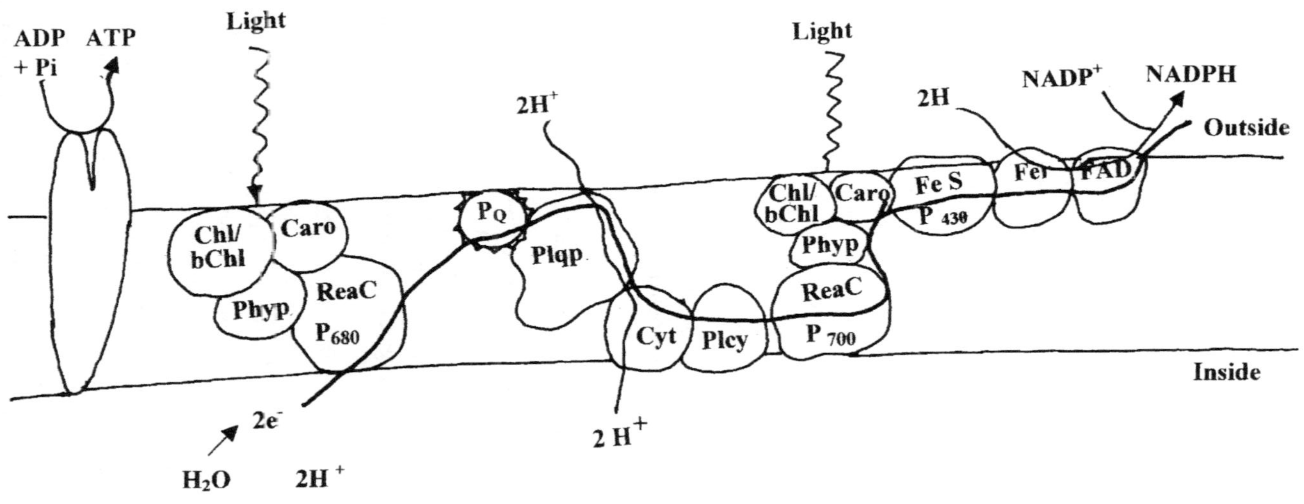

Figure 6.12. Photosynthetic membrane showing the components of oxidative photophosphorylation and chemiosmotic generation of ATP. The bold line indicates the route of electron transport. Caro= Carotenoid; Ferr= Ferredoxin; Phyp= Phycobiliprotein; Plcy= Plastocyanine; Plqp= Plastoquinone; Reac= Reaction centre.

oxidative phosphorylation or photosystem I alone operates. These bacteria contain bacteriochlorophylls (a or b) as the primary photoreaction centre (P_{870} in purple, P_{840} in green, and P_{798} in heliobacteria). The accessory pigments are bacteriochlorophylls c,d, and e and carotenoids. The ejected electron from the excited P_{870} passes through a number of membrane bound electron carriers to return to the same molecule to reduce it (Fig. 6.13b). During this cyclic electron flow, four protons are picked up from the cell cytoplasm; two are used to reduce the secondary quinone carrier and the remaining two are extruded outside the membrane. The latter establishes a proton gradient and generates ATP through a membrane bound ATPase.

As there is no mechanism to generate reduced coenzymes, anaerobic photosynthetic bacteria utilize reduced inorganic compounds such as H_2, H_2S, S or even organic acid as electron donor. In the latter case, the growth is known as photoheterotrophic or photoorganotrophic. The electrons from the initial oxidation of bacteriochlorophyll are used to reduce NAD^+ to NADH and the electrons from the external electron donor are utilized to re-reduce the oxidized bacteriochlorophyll.

Variation in different anoxygenic photoautotrophic bacteria exists in terms of different numbers of reaction centre pigment molecules, substitution of bacteriochlorophyll b for bacteriochlorophyll a and substitution of menaquinones for ubiquinones.

In the archean *Halobacterium,* purple coloured bacteriorhodopsin is lodged in cytoplasmic membrane (hence called red membrane) that also contains ATPase. When exposed to light or supplied with an organic substance, protons are translocated across the membrane establishing an electrochemical and hydrogen ion gradient. Enzyme ATPase is involved in ATP generation.

II-B- Biosynthesis of Cellular Building Blocks

As we have discussed in the last Section, a number of catabolic reactions lead to generation of energy and provide simpler components and other factors which the cell utilizes to synthesize the major cell constituents. These include carbohydrates, lipids, proteins, and nucleic acids. The metabolic pathways that build up macromolecules from simpler precursors, both organic and inorganic, is known as biosynthesis, or anabolism.

The first set of biosynthetic pathways are geared towards the synthesis of building blocks from which the macromolecules are organized. The cellular macromolecules generally occur in a more reduced state in comparison to starting substrate molecules. The anabolic pathways, therefore, require reducing power which is provided in the form of NADPH. The readers may recall that many catabolic reactions generate NADPH from its oxidized form $NADP^+$, though the latter may also be synthesized from NAD^+ through an ATP-involved reaction. Thus while catabolism is dependent on NAD^+/NADH, anabolism proceeds through the involvement of NADPH/$NADP^+$ system. Moreover, these biosyntheses also require ATP as they are generally endergonic as compared to catabolism which is exergonic.

Living cells have evolved an important link between anabolism and catabolism by operating a core metabolic pathway such as TCA cycle. This allows the flow of carbon

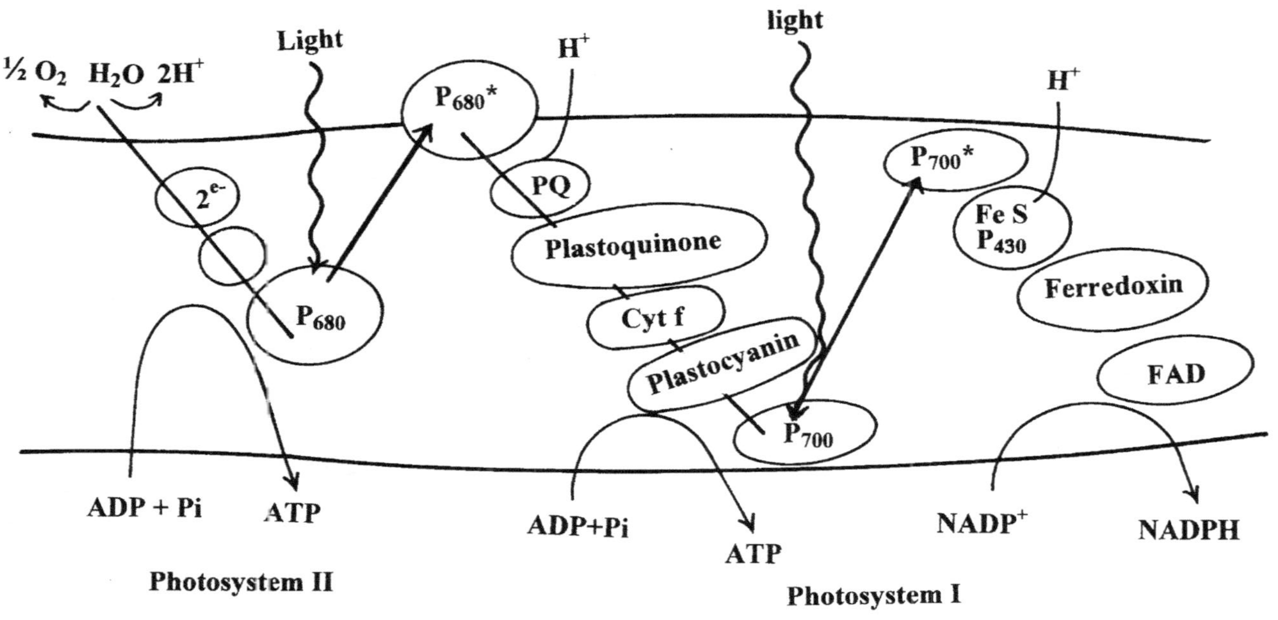

Figure 6.13a. Oxidative photophosphorylation or Z-pathway consists of two photosystems. Two separate photoreactivation steps i.e., excitation of P_{680} to P_{680} and of P_{700} to P_{700}* initiates a chain of electron transport with the net result of splitting of H_2O to produce O_2, generation of ATP molecules and reduced coenzyme (NADPH).*

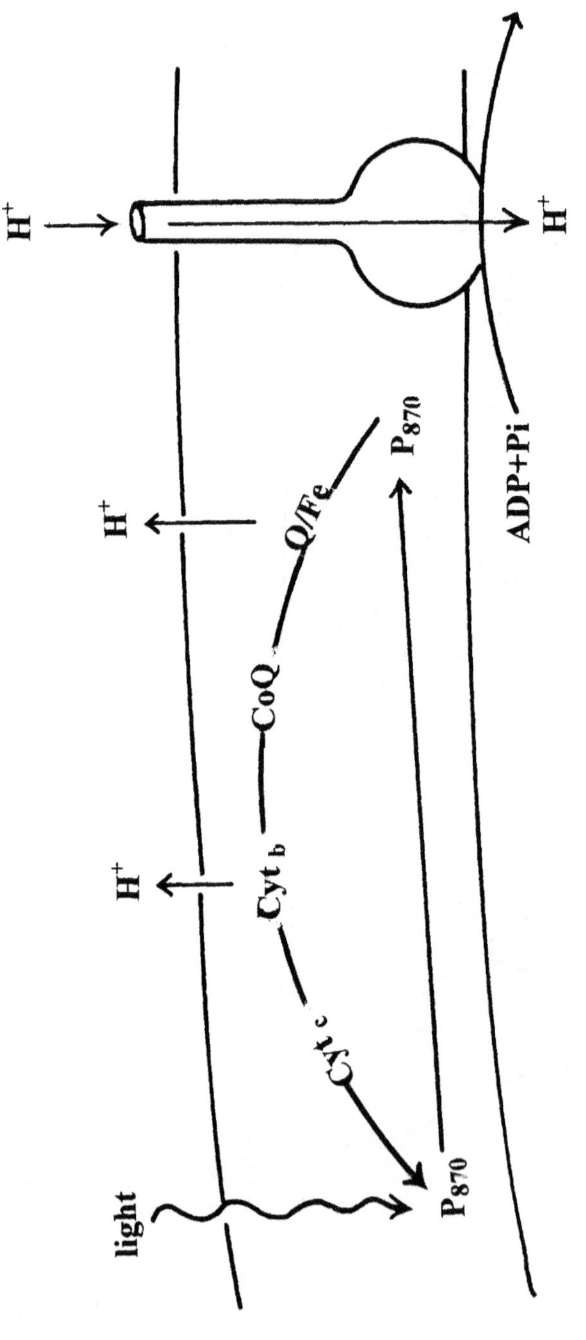

Figure 6.13b. Cyclic oxidative photophosphorylation of anoxygenic photosynthesis. In this system the chlorophyll molecule acts as an internal electron donor as well as acceptor (Photosystem I). Sufficient energy gradient is developed to generate ATP molecule.

Table 6.4. General features of photosynthetic organisms

	Photosynthetic Bacteria		*Cyanobacteria*	*Green Algae and Higher plants*
Features	**Purple**	**Green**		
Appearance	Brown, red, or purple	Yellow, green or brown	Bluish green, bright green, purple red, brown	Green
Photosynthetic apparatus	Specialized internal membranes	Chlorobium vesicles	Thylakoids	Thylakoids localized in chloroplasts
Participating pigments	Bacteriochlorophyll a and b, carotenoids	Bacteriochlorophyll a and b or c, carotenoids	Chlorophyll a with some containing chlorophyll b, phycobiliproteins	Chlorophyll a, b, xanthophylls, carotenoids
Absorption spectrum (in nm)	Infra-red; 850-910 nm, or even at 1000nm	Red to far red; 735-755 nm	Visible; 680-700 nm	Visible; 675-685 nm
Oxygen relationship	Mostly anaerobic	Anaerobic	Mostly aerobic	Aerobic
Principal electron donor	H_2S, H_2, S, thiosulphate, some organic compounds	H_2S, H_2, S	H_2O	H_2O
Photosystems	I	I	I & II	I & II
Photosynthetic products	Carbohydrates, organic acids, sulphur	Carbohydrates, sulphur	Carbohydrates, O_2	Carbohydrates, O_2

in both the directions so as to use many of the intermediates as precursors for biosynthesis as well as breakdown of the same to generate other chemicals and precursors. Such pathways have been referred to as amphibolic (dual purpose) pathways (Fig. 6.14). As the figure represents, while compounds are regularly siphoned to produce cellular macromolecules, they return to regenerate these precursors as well.

6.5. Carbon Dioxide Fixation

All living cells are built on a carbon skeleton. The flow of carbon, therefore, constitutes a major pathway in cellular metabolism. Although bacteria exhibit an extreme versatility as far as the carbon source utilization is considered, the fixation of carbon dioxide into organic molecules is a prime function and forms the basis of autotrophic metabolism.

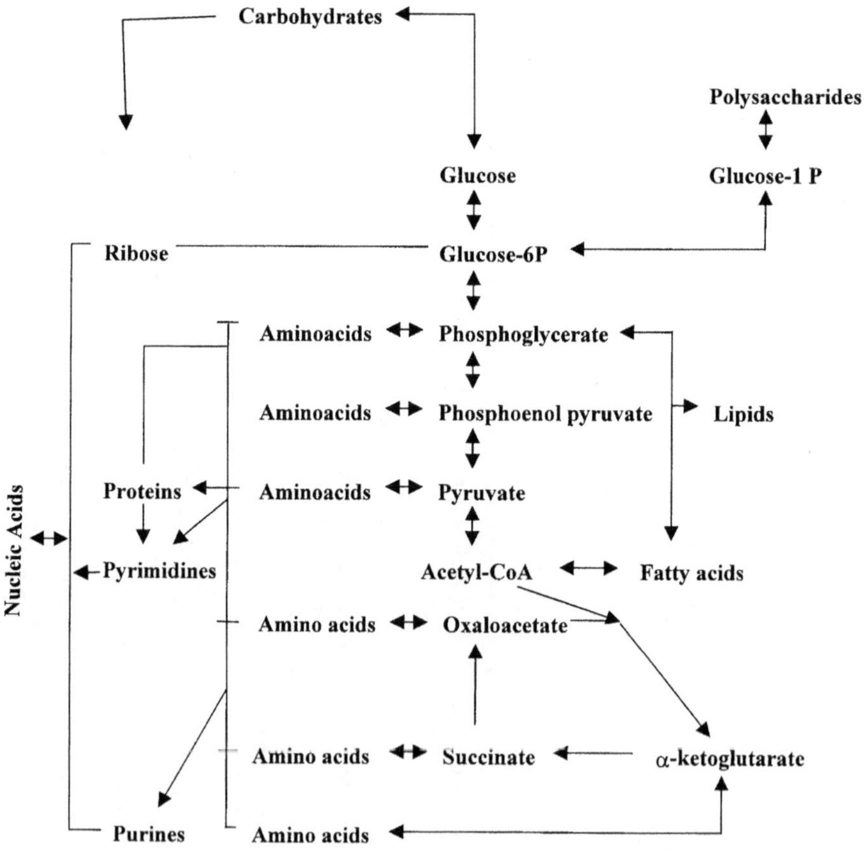

Figure 6.14. Interrelationship between catabolic and anabolic pathways, to generate all essential cellular components. Note: Only some essential intermediates have been shown.

6.5.1. Calvin Cycle

Like plant cells, many autotrophic microorganisms fix CO_2 via Calvin cycle. This fixation results in the synthesis of glyceraldehyde 3-P. Since the latter is a 3C compound, this pathway is often referred to as C_3 pathway.

CO_2 is the most oxidized form of carbon and its reduction to glyceraldehyde requires a plentiful supply of NADPH and ATP. Thus the overall reaction is:

$$3CO_2 + 9ATP + 6NADPH \rightarrow \text{Glyceraldehyde 3-P} + 9ADP + 6NADP^+$$

The supply of ATP and NADPH for photoautotrophs comes from light reactions of photosynthesis and for chemoautotrophs (chemolithotrophs) from oxidation of inorganic compounds as discussed in previous section. Since the reactions of cycles do not involve a direct input of light energy it is also referred as dark reaction. The first step constitutes the reaction of CO_2 with ribulose 1,5-bisphosphate to form an unstable 6 C compound that immediately splits into two molecules of 3-phosphoglycerate (Fig. 6.15). This reaction is catalyzed by ribulose 1, 5-bisphosphate carboxylase (RuBisCo), a key enzyme that regulates the rate of Calvin cycle. RuBisCo is the most abundant protein in the world as all autotrophic organisms contain it. Nitrifying bacteria, cyanobacteria and many sulphur-oxidizing autotrophic bacteria contain RuBisCo in insoluble crystalline form in polyhedral structures, called carboxysomes.

Three molecules of CO_2 combine with 3 molecules of ribulose 1,5-bisphosphate to form 6 molecules of 3-phosphoglycerate. Of the 6 molecules, 5 are recycled through sequential steps to regenerate 3-ribulose 1,5-bisphosphate while the remaining one molecule is reduced to form glyceraldehyde 3-P. The glyceraldehyde 3-P can further react to form glucose and subsequently other polysaccharides like starch and cellulose or fed into TCA cycle.

6.5.2. Reductive Tricarboxylic Acid Cycle

Green sulphur bacteria such as *Chlorobium* fixes CO_2 by a reverse (reductive) TCA cycle (Fig. 6.16). In this, oxaloacetate is converted to succinate sequentially through malate, and fumarate. Succinate is then activated to succinyl-CoA which in turn accepts CO_2 in the presence of reduced ferredoxin to form α-ketoglutarate. Another molecule of CO_2 is added to α-ketoglutarate to form isocitrate and then citrate. Citric acid is split to give oxaloacetate that re-enters the cycle and acetyl-CoA gets reductively carboxylated to pyruvate. Pyruvate is then activated to form phosphoenolpyruvate, which either picks up another CO_2 molecule to furnish oxaloacetate or is used for synthesizing cellular constituents. Most of the reactions are carried out in reverse order of TCA cycle except for the cleavage of citrate that is brought about by citrate lyase whereas citrate synthesis from oxaloacetate and acetyl- CoA during TCA cycle is mediated by citrate synthase.

6.5.3. Hydroxypropionate Pathway

The green non-sulphur bacterium, *Chloroflexus* uses H_2 or H_2S as electron donor but fixes $2CO_2$ molecules into one acetyl-CoA via the hydroxypropionate pathway. Although many steps have not been elucidated, acetyl-CoA is regenerated through an intermediate, such as hydroxypropionyl-CoA. The acetyl-CoA is reductively carboxylated to pyruvate which can be further utilized.

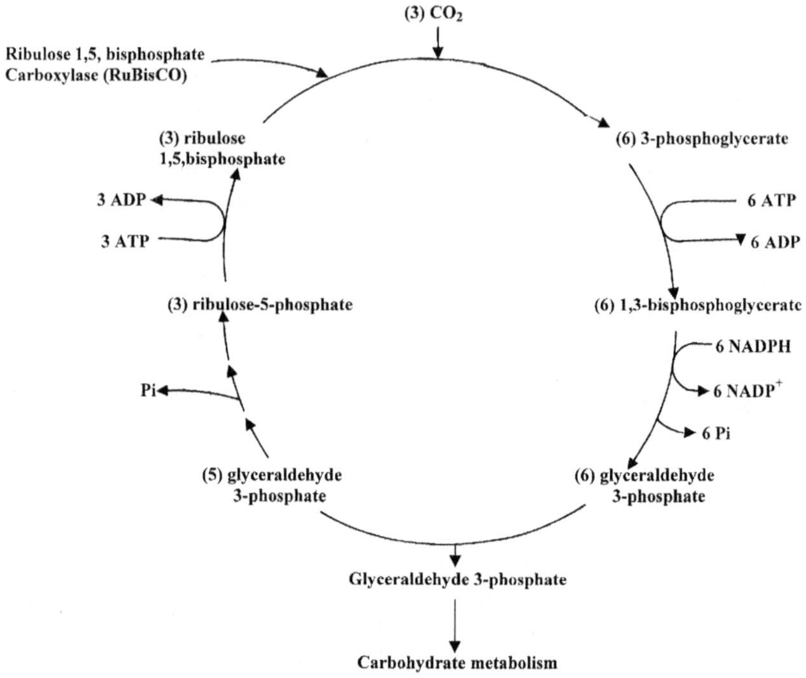

Figure 6.15. Calvin or carbon reduction cycle.

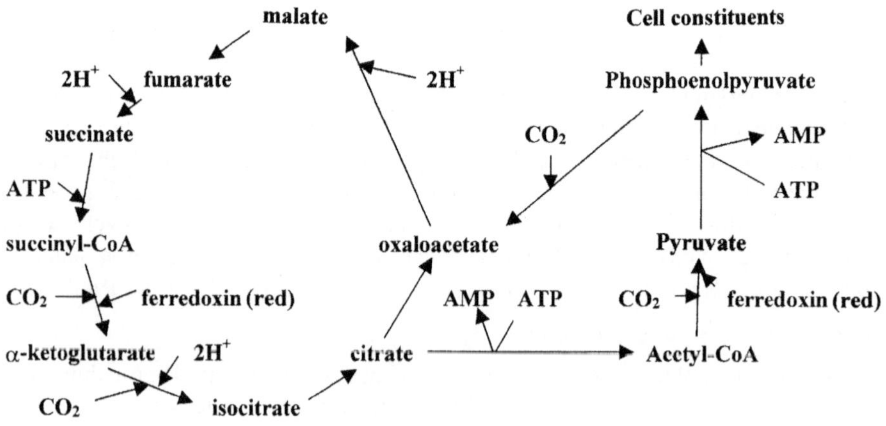

Figure 6.16. Reductive tricarboxylic acid cycle.

6.5.4. C_4 pathway

Many heterotrophs and autotrophs are capable of fixing CO_2 through a C_4 pathway. In this, pyruvate or phosphoenol pyruvate, formed during glycolysis, react with CO_2 to produce oxaloacetate, a 4C compound (Fig. 6.17). Oxaloacetate is then channelized through TCA cycle or used for amino acid and nucleic acid biosynthesis as we shall discuss later in this Section. This is not a very efficient pathway and many heterotrophs depend upon organic compounds as substrates for growth.

6.5.5. Assimilation of Organic C_1 Compounds

Besides CO_2, many bacteria have capability to utilize organic C_1 compounds through specialized pathways.

Methanotrophy- Obligate aerobic bacteria having the ability to use the most reduced form of carbon, methane (CH_4) have been called methanotrophs. Methanotrophs, such as, *Methylomonas*, *Methylococcus* and others can also grow on methanol as well.

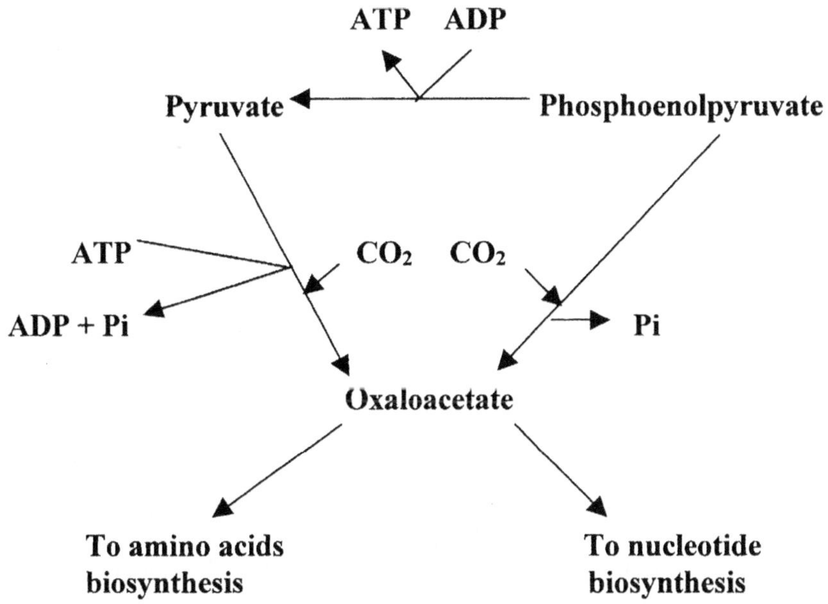

Figure 6.17. C_4 pathway used by heterotrophs and autotrophs to fix CO_2.

Methane is oxidized to methanol (CH_3OH) using methane monooxygenase as the enzyme. This enzyme has a broad substrate specificity as it can also act upon ammonium ions, ethane, chloro- and bromo-methane, propane, trichloroethylene and others, and can catalyze both oxidation and reduction:

$$CH_4 + O_2 + NADH + H^+ \rightarrow CH_3OH + NAD^+$$

$$CH_4 + O_2 + \text{Cytochrome C (reduced)} \rightarrow CH_3OH + H_2O + \text{Cytochrome C (oxidized)}$$

6.5.6. Ribulose Monophosphate Shunt

The methanol so formed is further oxidized to formaldehyde (HCHO). Formaldehyde is then fed into the ribulose monophosphate cycle by type I methanotrophs (Fig. 6.18). This pathway finally generates glyceraldehyde 3-P and also branches off to re-form ribulose 5-P through some rearrangement reactions. Glyceraldehyde 3-P can then be used in biosynthesis:

6 Formaldehyde + 2ATP → 2 Glyceraldehyde 3-P + 2ADP

6.5.7. Serine Pathway

Type II methanotrophs, on the other hand, metabolize formaldehyde through serine pathway (Fig. 6.18). Formaldehyde reacts with cofactor tetrahydrofolate to form methylene tetrahydrofolate which subsequently reacts with glycine to form serine and release the cofactor. Serine through deamination and subsequent reduction steps forms glycerate, which is then converted to phosphoenol pyruvate. Phosphoenol pyruvate reacts with CO_2 to produce oxaloacetate and the latter is converted to acetyl-CoA and glyoxylate via malate. Amination of glyoxylate to glycine helps run the cycle and acetyl-CoA is the final product:

$$HCHO + CO_2 + CoA + 2NADH + 2ATP \rightarrow \text{Acetyl-CoA} + 2NAD^+ + 2ADP + 2Pi + 2H_2O$$

6.5.8. Methylotrophy

A diverse group of bacteria, such as some *Pseudomonas, Bacillus,* and *Vibrio* can use other C1 compounds such as methanol, formate, or methylamine. These compounds are metabolized through serine pathway.

6.6. Carbohydrate Biosynthesis

6.6.1. Gluconeogenesis

Microorganisms show extreme versatility as far as the utilization of carbon source for growth is concerned. Each of these would require a pathway that will yield all the essential metabolites. For example, many bacteria can grow on L-malate, succinate or glycerol but they need to synthesize hexoses for cell wall mucopeptides and storage glycogen and as carbohydrate source for nucleic acid and glycoprotein synthesis. Biosynthesis of glucose from non-carbohydrate molecule is carried out through gluconeogenesis, and typically involves the formation of pyruvate or other intermediates (Fig. 6.19). Gluconeogenesis is achieved by a reversal of the flow of carbon from pyruvate in EMP pathway. However, many of the steps are insufficiently reversible,

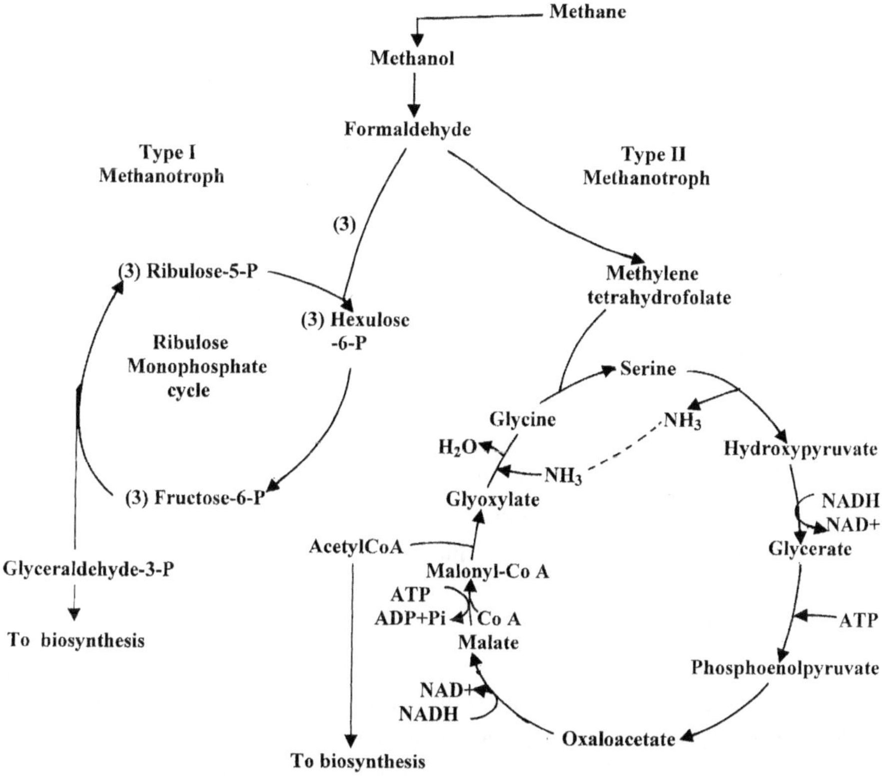

Figure 6.18. Utilization of organic C_1 compounds such as methane by methanotrophs. Methanotrophs essentially follow serine pathway.

leading to the utilization of alternative pathways. Pyruvate kinase, for example, is not reversible so phosphoenolpyruvate (PEP) is generated by pyruvate carboxykinase in the presence of GTP from oxaloacetate:

$$\textbf{Oxaloacetate + GTP} \longleftrightarrow \textbf{PEP} + \mathbf{CO_2} + \textbf{GDP}$$

The adequate supply of oxaloacetate is maintained by two reactions. In one, pyruvate carboxylase can synthesize oxaloacetate by CO_2 fixation:

$$\textbf{Pyruvate} + \mathbf{CO_2} + \textbf{ATP} \rightarrow \textbf{Oxaloacetate + ADP + Pi}$$

In the second, malate is oxidized:

Malate + NAD^+ → Oxaloacetate + NADH + H^+

Similarly phosphofructokinase which synthesizes fructose 1, 6-bisphosphate from fructose 6-P + ATP is essentially an irreversible reaction. Fructose bisphosphatase brings about the dephosphorylation of fructose 1,6-bisphosphate, to generate fructose 6-P:

Fructose 1, 6-bisphosphate + H_2O → Fructose 6-P + Pi

Synthesis of glucose from glucose 6-P is also carried out by an enzyme, glucose 6-phosphatase:

Glucose 6-phosphate + H_2O → Glucose + Pi

The biosynthesis of carbohydrate is favoured when cell has a good supply of ATP whereas carbohydrate breakdown is initiated when ATP level is relatively low.

6.6.2. Glyoxylate Cycle

Flow of carbon from fatty acids (lipids) or acetate to carbohydrates is carried out through glyoxylate cycle (Fig. 6.20). This pathway provides a shunt to TCA cycle and refurbishes oxaloacetate. The importance of this pathway lies in the fact that many key intermediates of TCA can be siphoned off for biosynthesis of important organic molecules, besides the synthesis of glucose. The pathway thus serves an example of anaplerotic sequences which replenish the key molecules for metabolism.

Isocitrate is broken down to succinate and glyoxylate by isocitrate lyase. Succinate continues through TCA cycle and glyoxylate combines with acetyl-CoA to produce malate. Malate is then converted to oxaloacetate, and the latter is then converted to phosphoenolpyruvate which can be utilized to form glucose through gluconeogenesis:

4 Acetyl-CoA → 2 Oxaloacetate → 2 Phosphoenolpyruvate + 2 CO_2

2 Phosphoenolpyruvate → Glucose

6.7. Polysaccharide Biosynthesis

A small but significant component of a cell is made up of polysaccharides. Polysaccharides are key components of the cell wall and also serve as store house of energy and carbon. Cells growing under nutritionally-rich conditions synthesize and store polysaccharides such as glycogen and starch which can be broken down when cell is deprived of energy or a key constituent.

While the most common sugar in polysaccharides is hexose, such as, glucose, two 5-carbon sugars, or pentoses like ribose and deoxyribose are key constituents of RNA and DNA, respectively.

Hexoses can be obtained from environment or synthesized from non-sugar materials, as we discussed in the early part of this **Chapter**. Glucose 6-P is converted to glucose1-P which is then used for starch or glycogen biosynthesis.

In glycogen biosynthesis, glucose 1-P reacts with ATP to give ADP-ribose (eukaryotes use UTP and form UDP-ribose). ADP-ribose is then added to the non-reducing end of an oligosaccharide that contains at least 4 glucose molecule by the enzyme glycogen synthetase. The chains so formed are composed of α (1-4) linkages; α (1-6) branched chain linkages are formed later by the action of a transglucosylase.

Starch is synthesized by a similar mechanism using UDP-glucose. Some bacteria,

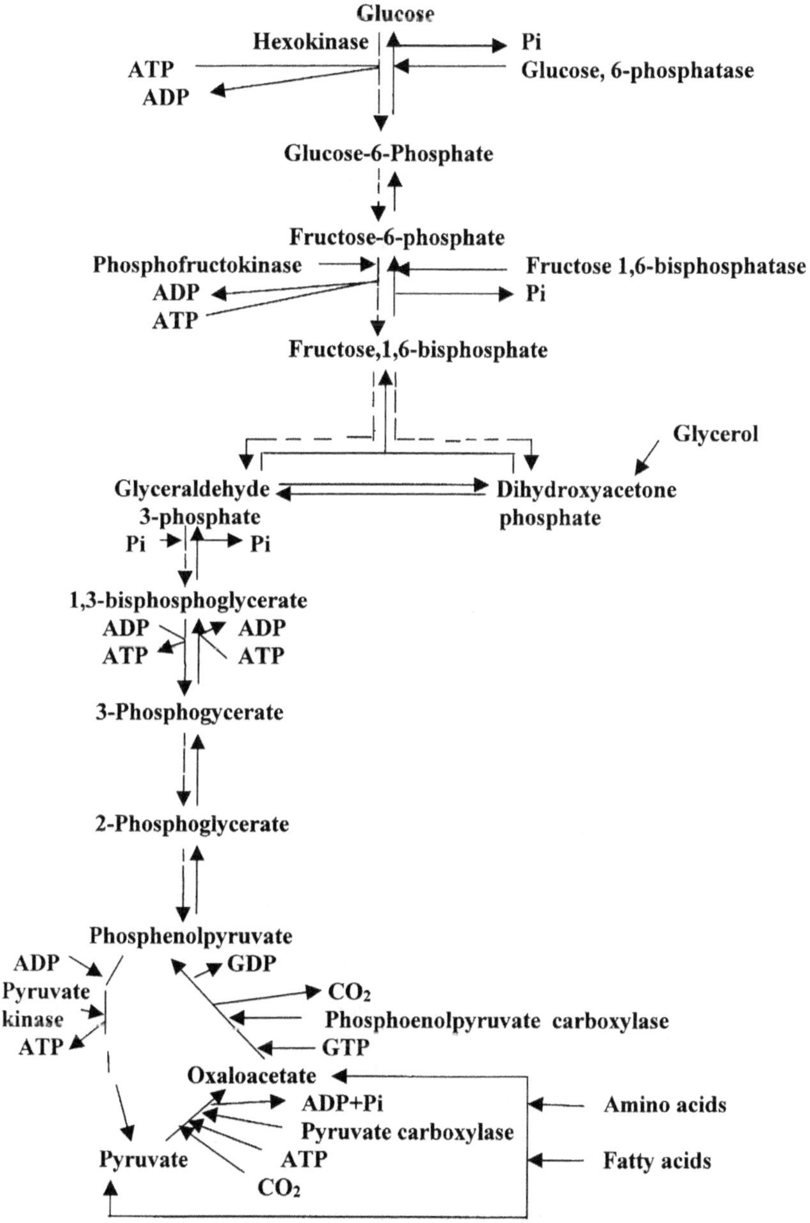

Figure 6.19. Gluconeogenesis. The conversion of noncarbohydrate substrates to glucose. Note: A major part of this pathway is the reversal of glycolysis (——). Only the enzymes different from glycolysis are represented here.

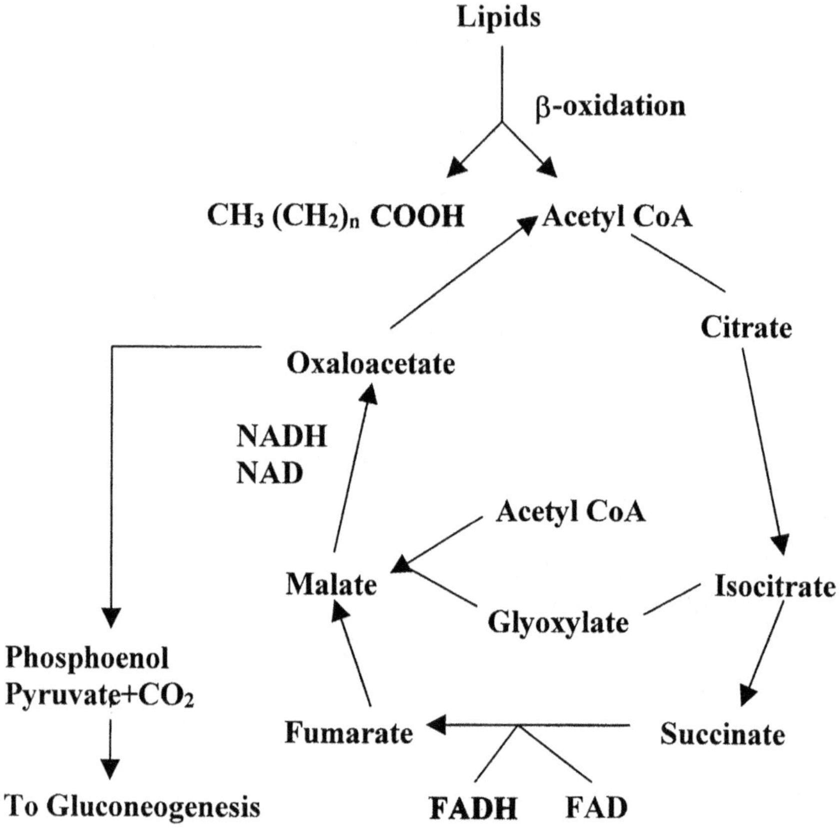

Figure 6.20. A shunt across the Krebs Cycle (The Glyoxylate Cycle) operates when lipids are metabolized. It yields the intermediates of Krebs Cycle as well as provides a pathway for conversion of lipids to carbohydrates.

like, *Streptococcus mutans* and *Leuconostoc mesenteroides* form sticky exteriors due to extracellular dextran production. The enzyme dextransucrase (glucosyltransferase) breaks sucrose into glucose and fructose and links glucose through α (1-6) and some α (1-3) branches into a dextran polymer.

6.8. Lipid Biosynthesis

The main constituent of lipids are fatty acids. These fatty acids impart interesting properties to the lipids as they are amphipathic in nature, i.e., they contain highly hydrophilic (water-loving) and highly hydrophobic (water-repelling) regions. Such regions are ideal to provide permeability barriers. Simple lipids (fats) consist of three fatty acids bonded to the C_3 alcohol glycerol. They are thus also referred to as triglycerides. When simple lipids contain additional elements such as nitrogen, phosphates, or sulphur, or small hydrophilic carbon compounds such as sugars, ethanolamine, serine, or choline, they constitute complex lipids. Phospholipids are a major class of complex lipids.

6.8.1. Biosynthesis of Fatty Acids

Fatty acid biosynthesis takes place by sequential addition of 2-C units derived from acetyl-CoA. Initially, acetyl-CoA combines with CO_2 to produce the key intermediate, malonyl-CoA. This step requires both ATP and vitamin, biotin and forms the basis of biotin requirement as growth factor by many organisms. Malonyl-CoA then contributes 2-C units to the growing fatty acid chain releasing CoA and CO_2 at each step. These reactions are carried out with the help of a protein known as acyl carrier protein (ACP) to which the substrates are bound (Fig. 6.21). Fatty acid biosynthesis requires energy in the form of ATP and reducing power in the form of NADPH.

C_{16} saturated fatty acid, palmitic acid, commonly found in membrane phospholipids is synthesized as follows:

8 Acetyl-CoA + 7 ATP + 14 NADPH → Palmitic acid + 14 $NADP^+$ + 8 CoA + 6 H_2O + 7 ADP + 7 Pi

Monounsaturated fatty acids like palmitoleic acid and oleic acid are synthesized from palmitic acid (C_{16}) and steric acid (C_{18}) precursors. The reaction is mediated by fatty acyl-CoA oxygenase and NADPH.

6.9. Biosynthesis of Amino Acids

Several amino acids serve as building blocks of cellular proteins. Amino acid biosynthesis requires the incorporation of inorganic nitrogen into organic molecules.

Molecular nitrogen (N_2) abundantly found in the atmosphere, remains largely unutilized as most organisms require a supply of fixed form of N_2 as NH_4^+, NO_2^-, NO_3^-, or nitrogen-containing organic compounds. Some bacteria and archaea possess the unique capability of nitrogen fixation in which molecular N_2 is converted to NH_4^+. Nitrogen fixation is important because majority of living organisms can derive their nitrogen supply for amino acid biosynthesis and other nitrogen-containing compounds indirectly through this process. We shall discuss this process in detail separately in **Chapter 8.**

An actively growing cell needs to synthesize 20 essential amino acids for producing different proteins. In fact, there will also be a number of other amino acids which may play important role in metabolism. An amino acid has to derive a carbon skeleton to which an amino group is incorporated. Amino acids have been grouped in different families based on the precursor used for carbon skeleton (Table 6.5). It should be interesting to note that these are derived from glycolysis and TCA cycle. The twenty amino acids originate from six different non-amino acid precursors, and that results in six different families of amino acids, as shown in Table 6.5.

Glutamate, the first amino acid of the glutamate family, is formed by direct addition of amino group from NH_3 to α-ketoglutaric acid (a pathway known as reductive

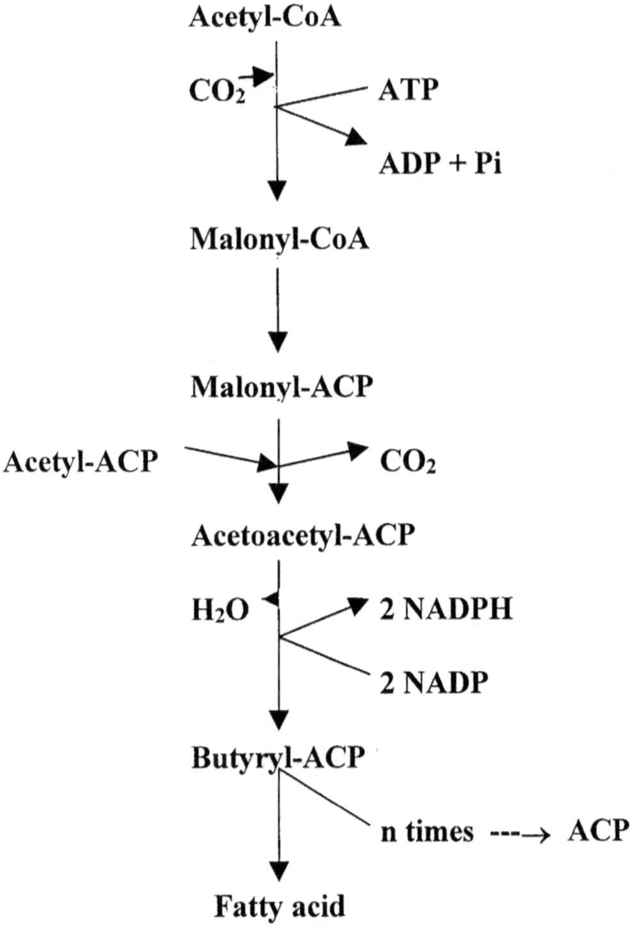

Figure 6.21. Fatty acid Biosynthesis
(Example) 8 Acetyl-CoA + 7 ATP + 14 NADPH $\longrightarrow$
Palmitic acid (C_{16}) *+ 14* $NADP^+$ *+ 8 CoA + 6* H_2O *+ 7 ADP + 7 Pi.*

amination), the key enzyme being glutamate dehydrogenase, and NAD (P)H supplying the reducing power. Other amino acids generally get their amino group from glutamate, or some other amino donor by a type of reaction called transamination. For example, glutamate can donate its amino group to oxaloacetate to form aspartate, regenerating α-ketoglutarate in the process (Fig. 6.22). Some bacteria may use other amino acid dehydrogenases in place of glutamate dehydrogenase and some may even use glutamine synthetase.

Table 6.5. Derivation of essential amino acids from the intermediates of TCA cycle and glycolysis. The three lettered and single lettered symbols are given in the parenthesis

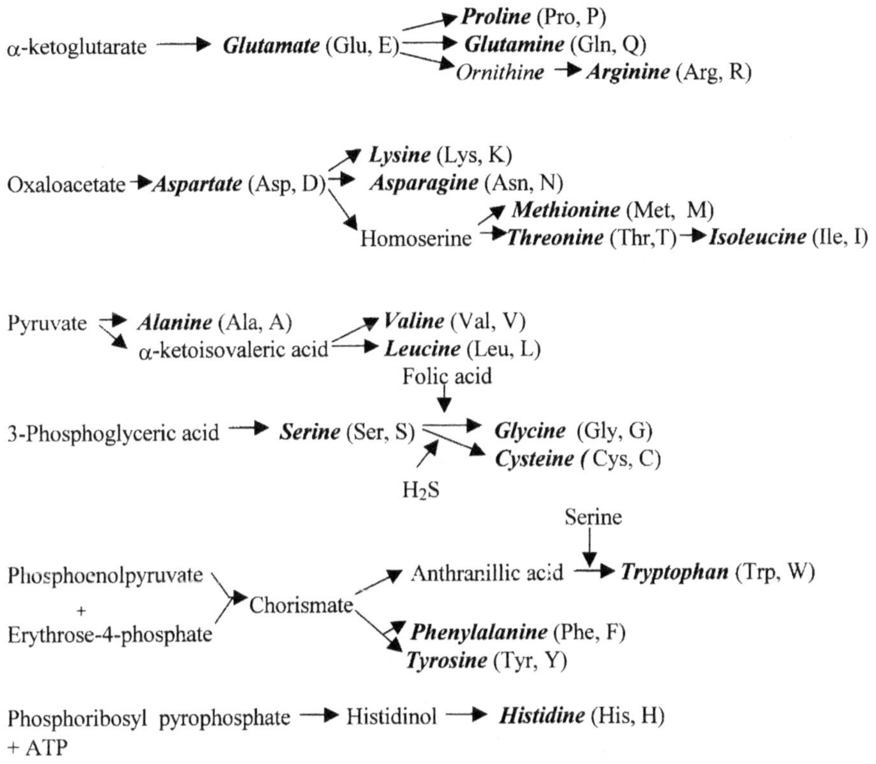

6.10. Purine and Pyrimidine Biosynthesis

The two nucleic acids, RNA, and DNA from a variety of biological sources are characteristically composed of ribose and deoxyribose, respectively, and phosphates and nitrogen-containing compounds called bases. The bases are classified into two groups purines and pyrimidines. The most commonly encountered purines are adenine and guanine, although a number of methylated purines with specialized functions are also known. The three pyrimidines are cytosine, uracil, and thymine. Thymine is found only in DNA and uracil only in RNA. Once again modified pyrimidines are also known.

Transfer of the amide nitrogen of glutamine to 5-phosphoribosyl pyrophosphate (PRPP) initiates the purine biosynthetic pathway. PRPP itself is synthesized from ribose 5-phosphate. The first step is critical as it controls the complete biosynthetic pathway. Addition of the entire glycine molecule (2 carbon + amino nitrogen) and sequential addition of a formyl group and the amide nitrogen of glutamine provides the first ring structure. The formation of second ring is contributed by the amino nitrogen from

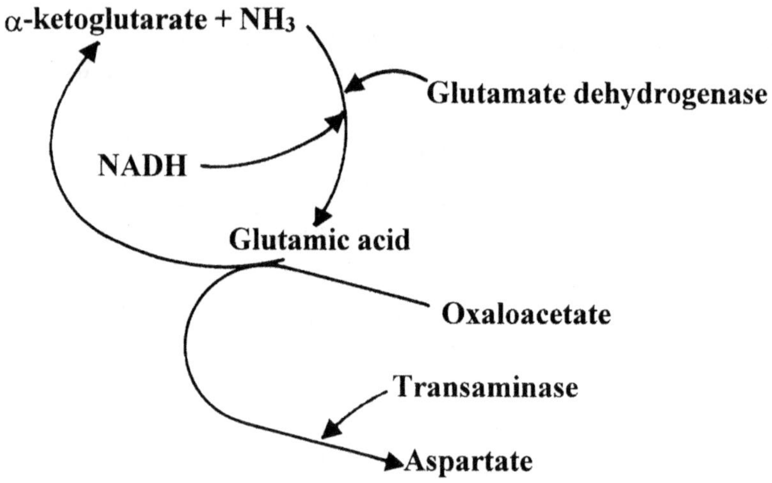

Figure 6.22. Transamination reaction for the biosynthesis of amino acids.

aspartate, addition of CO_2, and a formyl group. Ring closure generates the first complete purine structure of Inosine 5'-monophosphate (IMP). IMP is the branch point intermediate that is converted to AMP by addition of amino nitrogen from aspartate and to GMP by utilizing amide nitrogen of glutamine (Fig. 6.23).

The first enzyme specific to pyrimidine biosynthesis, the aspartate transcarbamylase brings about the condensation of carbamyl phosphate with aspartate to produce carbamyl aspartate with the release of inorganic phosphate. The cyclization of carbamyl aspartate forms a ring structured molecule, the dihydroorotic acid. The oxidization of dihydroorotic acid in the presence of NAD^+ forms orotic acid which accepts ribose 5'-phosphate in the presence of PRPP to form a nucleotide, orotidine 5'-phosphate. The latter is decarboxylated to yield uridylic acid (UMP). UMP can be converted to UTP which in the presence of ammonia leads to the formation of CTP (cytidine triphosphate). Moreover, UMP can also be converted to deoxyderivative, dUMP which on methylation gives rise to TMP (thymidylic acid). TMP is then converted to TTP in the presence of ATP (Fig. 6.24).

All the five trinucleotides serve as the basic building blocks for the two types of nucleic acids as we discussed in **Chapter 5**.

II-C- Biosynthesis of Cellular Macromolecular Constituents

From the essential macromolecules biosynthesis discussed earlier in this section, major cellular components are organized. These constitute:

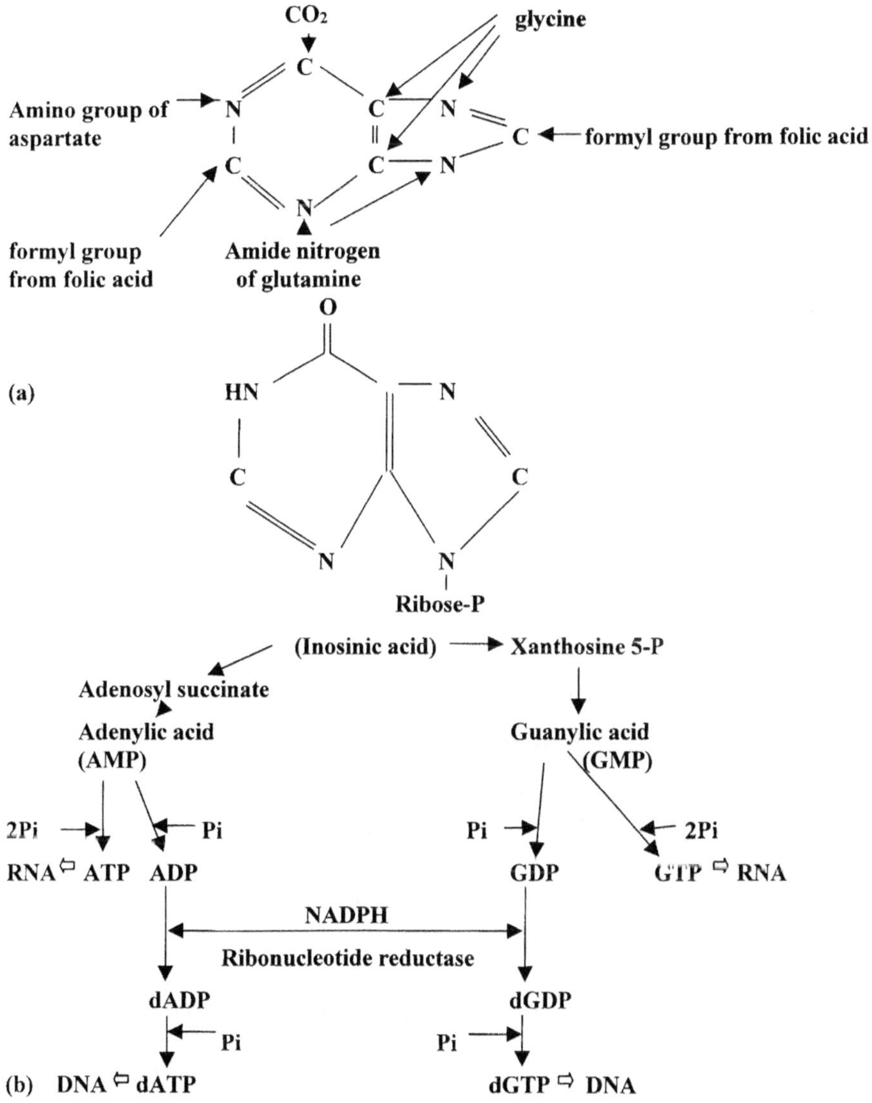

Figure 6.23. Biosynthesis of purines. a. The basic purine skeleton and component contributors, b. The pathway starting from inosinic acid, the first purine.

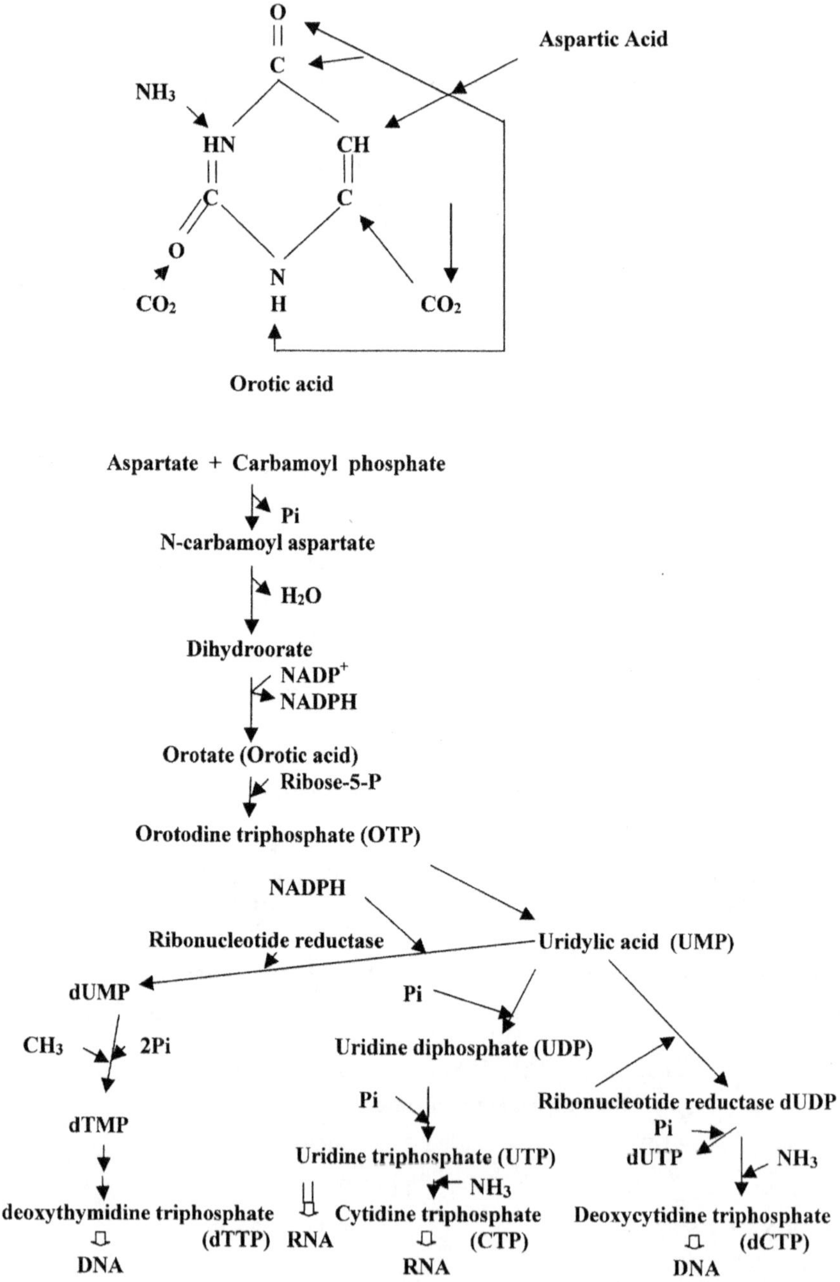

Figure 6.24. Biosynthesis of pyrimidines. a. basic pyrimidine skeleton and component contributors, b. the pathway of synthesis.

6.11. Peptidoglycan Biosynthesis

As we have seen earlier, peptidoglycans are required for cell growth and division as they constitute the cell wall polymers. The glycan portion of peptidoglycan consists of two disaccharides, N-acetylglucosamine and N-acetylmuramic acid. The former is synthesized from fructose 6-P, derived from glucose, and glutamic acid and acetyl-CoA. N-acetyl muramic acid is derived from N-acetylglucosamine by its reaction with UTP to give N-acetylglucosamine acid-UDP (G-UDP) which subsequently reacts with phosphoenolpyruvate to produce N-acetylmuramic acid-UDP (M-UDP). This last reaction is inhibited by antibiotic phosphonomycin (fosfomycin).

The peptidoglycan is synthesized in three stages (Fig. 6.25). In stage I, which takes place in cytoplasm, precursors are assembled in the form of UDP-N-acetylmuramic acid-pentapeptide. The latter consists of L-alanine—D-glutamate—diaminopimelic acid—D-alanine—D-alanine, of which the first four are found in the cell wall (in the form of tetrapeptide) and the last D-alanine is added at the carboxyl end during peptidoglycan formation. Each amino acid is added to M-UDP at the appropriate place by specific enzyme or ligase and requires ATP. Peptide bond formation takes place in the cytoplasm and not at the ribosomes. D-alanines are connected by D-alanine-D-alanine synthetase. Both these reactions are inhibited by the antibiotic cycloserine. In a separate reaction, UDP-acetylglucosamine is also made available.

In the second stage, M-UDP pentapeptide (M-UDP-pp) is translocated to the C55 carrier lipid, undecaprenylphosphate or bactoprenol, located in the cytoplasmic membrane. The C55 carrier lipid exists as a pyrophosphate that is dephosphorylated before it is employed as a carrier, a reaction that is inhibited by the antibiotic, bacitracin. At this stage, N-acetylglucosamine is transferred to this complex to form bactoprenol-N-acetylglucosamine-N-acetylmuramic acid-pentapeptide. This whole complex now moves across the membrane to lie on the outside of the cell. There it is associated with the cell wall precursors and called nascent peptidoglycan. The translocation step is inhibited by the antibiotics vancomycin, tunicamycin, and ristocetin.

In the final stage or stage III, nascent peptidoglycan is covalently bound to the existing cell wall by transpeptidation. In this reaction, cross-bridges are formed between the new and the existing peptidoglycan and the terminal D-alanine is cleaved. The energy released during cleavage is used to attach the fourth amino acid, D-alanine to diaminopimelic acid of the adjacent tetrapeptide. Several enzymes involved in peptidoglycan synthesis can bind penicillin and are collectively called as penicillin-binding proteins (PBPs). This binding can inhibit their activities, and forms the basis of penicillin action.

6.12. Lipopolysaccharide (LPS) Biosynthesis

Gram-negative bacteria contain an outer membrane consisting of LPS outside their peptidoglycan. LPS biosynthesis occurs at cytoplasmic membrane and bactoprenol or lipid carrier is required. To this lipid carrier, individual sugars of repeat units are sequentially added. Sugars are, however, activated by condensing with nucleotide triphosphates to form nucleotide diphosphate-sugar intermediates such as GDP-

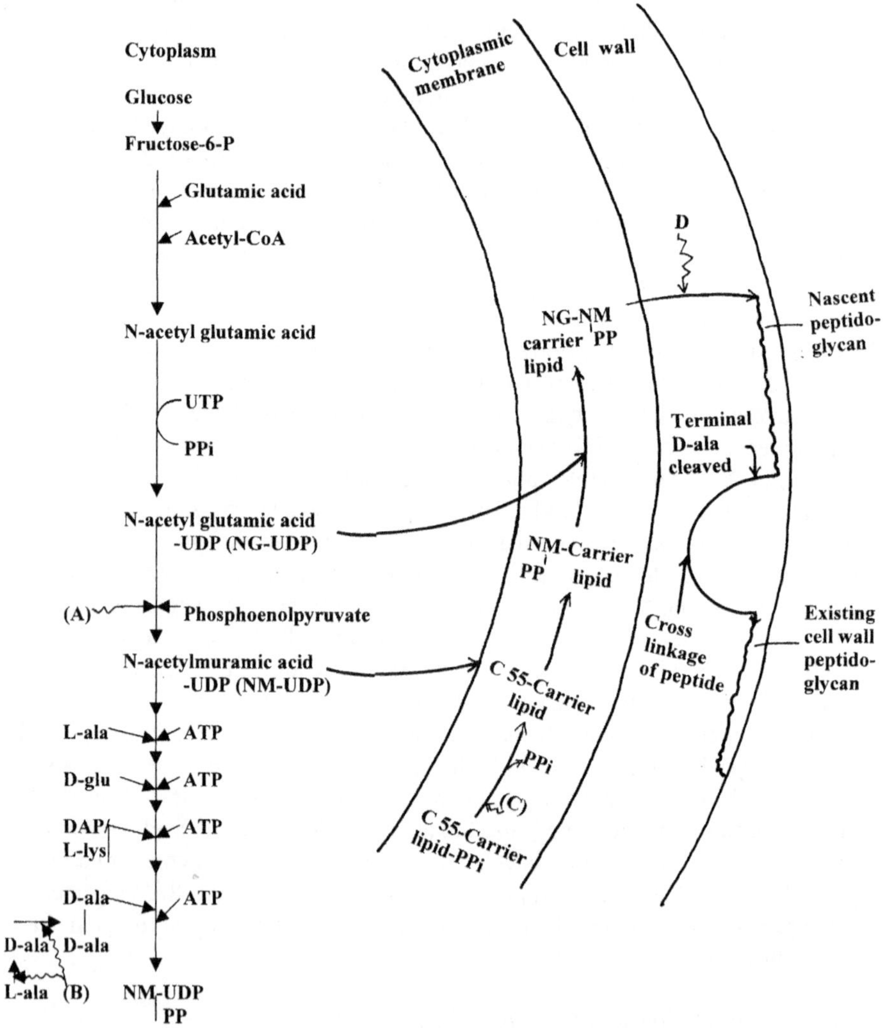

Figure 6.25. Peptidoglycan biosynthesis. ⇝ *represents the steps inhibited by different antibiotics. A. Phosphonomycin, B. Cycloserine, C. Bacitracin, D. Vancomycin. L-,D-ala= alanine, D-glu= glutamic acid, L-lys= lysine, DAP= diaminopimelic acid, PP= pentapeptide.*

mannose, UDP-galactose, or CDP-abequose. These activated sugars are sequentially added enzymatically to produce the repeat polysaccharide chain.

The lipid A fraction is organized in the cytoplasmic membrane by combining two molecules of glucosamine phosphate and then adding several fatty acid molecules notably ß-hydroxymyristic acid. Around the lipid A fraction, core polysaccharide is built by sequential enzymatic addition of sugars, such as KDO (2-keto-3 deoxyoctulosonic acid), heptose, galactose, and glucose (Fig. 6.26). The lipid A-core polysaccharide is transported to the outside of the cytoplasmic membrane after which the repeat polysaccharide is added on to produce the LPS molecule. Newly synthesized LPS molecule first appear at the adhesion sites (Bayer junctions) on the outer membrane suggesting that they are transported through them.

6.13. Biosynthesis of Phospholipids

Phospholipids are essential components of membranes. They are derived from the addition of fatty acids to glycerol 3-P. Glycerol 3-P is produced from reduction of dihydroxyacetone phosphate, an intermediate of glycolysis, with the help of NADPH. First, glycerol 3-P reacts with acylated ACP to form phosphatidic acid which then is activated by cytosine triphosphate (CTP) to form CDP-diacylglycerol. Phospholipids are synthesized by replacing CDP with serine, inositol, or glycerol to form a specific type of phosphatidyl series.

6.14. Biosynthesis of Poly-β-Hydroxybutyric Acid (PHB)

PHB is a common storage product of many bacteria. Its biosynthesis involves the formation of a 4C, acetoacetyl-CoA from 2 molecules of acetyl-CoA. This is then reduced with NADPH to form β-hydroxybutyryl-CoA, to which successive acetyl-CoA is added to effect chain elongation. Subsequent removal of CoA portion gives rise to PHB. These reactions require NADH rather than NADPH. Interest in PHB has been generated due to plastic-like characteristics of this compound. PHB is, therefore, recommended as biodegradable plastic alternative.

6.15. Biosynthesis of Proteins

Proteins are essential macromolecules of a cell with a multitude of functions. These range from structural, enzymatic, hormonal, as antibodies and others. Proteins are therefore synthesized through a number of biochemical reactions directly under the care of genes.

The proteins are organized from a primary sequence of essential amino acids that are arranged in a specific manner. It is the arrangement of amino acids that is directed by the gene and the whole process forms the basis of gene expression, and underlies the pathway of inheritance. By controlling the structure, and, therefore, the function of proteins, genes are able to control specific phenotypes.

A gene has often been defined by its function. Although the end product that determines the final phenotype is greatly different for different genes, the basic gene

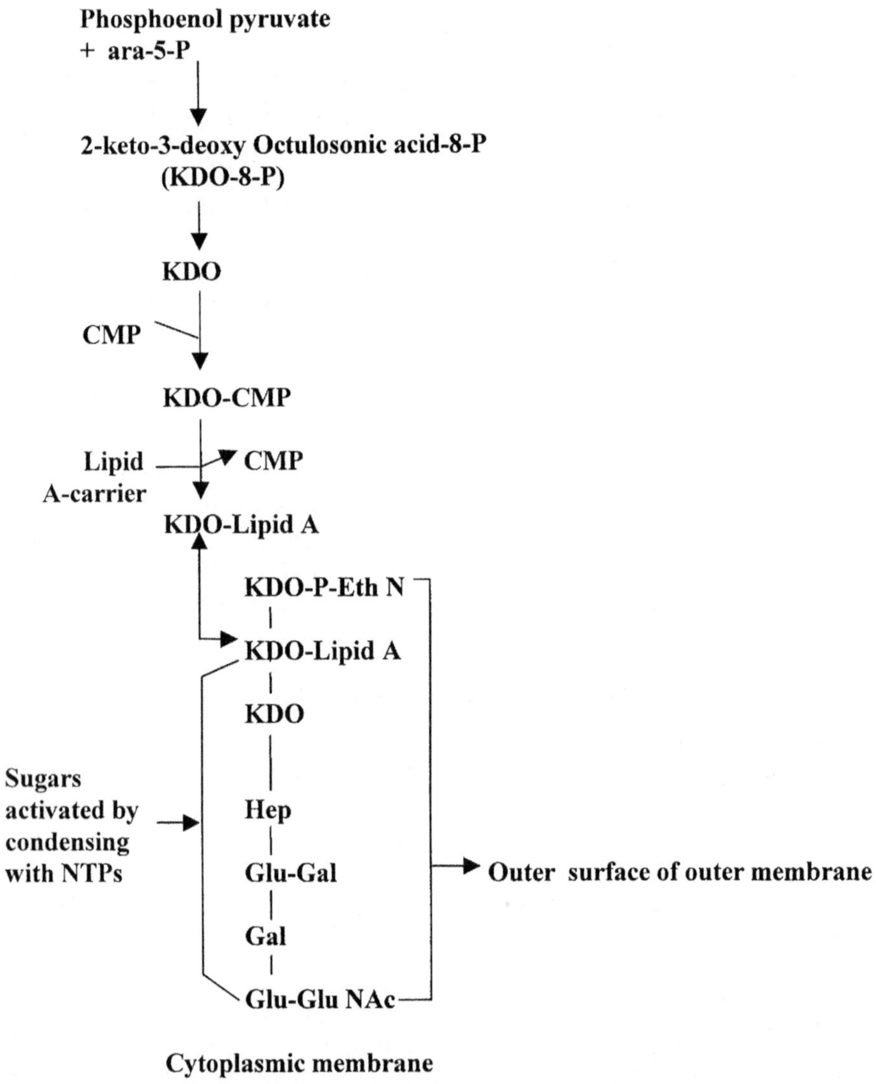

Figure 6.26. LPS biosynthesis in Gram-negative bacteria.

function is summed up by the central dogma (Fig. 6.27). Accordingly, the genes are perpetuated as sequences of nucleic acid but function by being expressed in the form of proteins. Two types of processes are required to cover the entire pathway of inheritance:

1) Genetic information contained in genes is perpetuated by replication that produces the identical copies (see **Chapter 5**).
2) The information is expressed by a two stage process that comprises the protein synthesis. These are transcription and translation.

Transcription generates a single-stranded RNA identical in sequence with one of the strands of duplex DNA. Three principal classes of RNA involved in protein synthesis are messenger RNA (mRNA), transfer RNA (tRNA), and ribosomal RNA (rRNA). Translation converts the nucleotide sequence of mRNA into the sequence of amino acids comprising the primary structure of protein. The primary structure may undergo further configurational modifications to generate the final product.

Molecular biologists equate a gene to a transcription unit comprising the DNA between a promoter, where transcription begins, and a terminator, where it ends. One DNA strand that directs the synthesis of RNA, including mRNA, is the template or antisense strand. The opposite DNA strand which bears the same sequence (except for having T in place of U) is called the sense or coding strand. In bacteria, mRNA can be synthesized for a single gene (monocistronic) or may be produced for the whole operon (polycistronic).

All bacterial genes are transcribed with the help of a multisubunit enzyme, RNA polymerase. This holoenzyme can be separated into two components, core enzyme, a multimer structure of α_2 β,β' subunits that is responsible for RNA elongation and sigma factor (σ) that adds specificity by recognizing the promoter at the stage of initiation. Bacterial promoters are identified by two short conserved sequences centered at -35 and -10 relative to the start point (or "upstream" of). The sequence TATAAT (or similar sequences known as the Pribnow sequence) are generally encountered at –10 region. RNA polymerase initially touches down at the -35 sequence and then extends to make

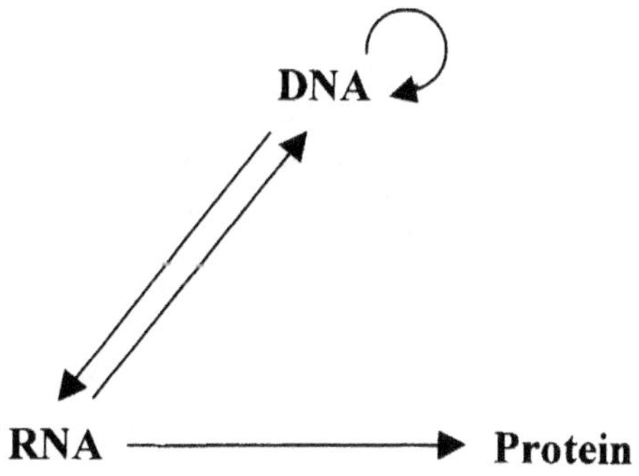

Figure 6.27. The central dogma of molecular biology that identifies the general route of information flow. While DNA, the master molecule, passes on the information contained in its base sequences to RNA, RNA uses the same to direct the synthesis of a specific protein. In some cases, RNA (as in retroviruses etc.) can provide the information to DNA.

contacts over -10 region. The enzyme overall covers ~77 bp of DNA. This closed binary complex is converted into open binary complex by melting an A-T rich 12 bp sequence that extends from -10 to start point. Incorporation of ribonucleotides converts it into a ternary complex. After the initial recognition is established sigma factor is released and the enzyme moves along DNA synthesizing RNA by incorporating complementary ribonucleotides (Fig. 6.28).

The "strength" of the promoter is described by the frequency at which RNA polymerase initiates transcription and is controlled by the conformity of the -35 and -10 region with the ideal consensus sequence. Many alternative sigma factors are known which can direct the core enzyme to recognize promoters with different consensus sequences. Both *E. coli* and *B. subtilis* contain alternative sigma factor that gets activated under adverse conditions such as heat shock or nitrogen starvation. The latter also contains a number of minor sigma factors, and also one that is active when sporulation is initiated.

RNA polymerase binds to the promoter site, a process called as template recognition. At this stage, DNA is locally unwound so that the template strand is available for base pairing with ribonucleotides. This structure is often described as transcription bubble. A small RNA transcript (~9 nucleotides long) is produced to initiate the process after which the enzyme (transcription bubble) moves along the DNA to extend the growing RNA chain (elongation) (Fig. 6.28). RNA polymerase terminates transcription at two types of sites. One GC-rich hairpin followed by a run of U-residues is the intrinsic terminator. Another one recognized by Rho factor is a stretch of 50-90 nucleotides, rich in C but poor in G, preceding the site of termination. Rho-dependent termination requires ATP and acts at a site where RNA polymerase has paused.

The process of transcription makes available all the three major classes of RNA viz., mRNA that carries the information coded in the gene for a polypeptide in form of three base codons, rRNA that is used for the organization and function of ribosomes, and tRNA which are the adaptor or the carrier molecule for different amino acids during translation.

Ribosomes are the ribonucleoprotein particles that act as tiny protein synthesizing factories. The bacterial ribosome, known as 70S type consists of two subunits, the small (30S) squashed-shape subunit and a large (50S), more spherical subunit (see Figs 4.11, 6.29a). Each subunit contains a single major rRNA called 16S and 23S, respectively, and minor rRNA, such as 5S rRNA in the large subunit. Each rRNA species is associated with several proteins that constitute a number of functional sites, such as P, A, and E sites, the EF-TU and EF-G elongation factor binding sites, peptidyl transferase, and the mRNA binding site. The major rRNAs have notable functions at some of these sites, specially mRNA binding site and P site on the 30S subunit, in which 3' end of rRNA seems to be involved. The rRNA also possesses the essential catalytic activity of peptidyl transferase. The rRNA shows extensive intrastrand base pairing to form short, imperfectly paired duplex stems and single stranded loops. Ribosomes normally exist as a pool of free ribosomes with separate small and large subunits. During translation small subunits bind to mRNA which is later joined by large subunit to generate an intact ribosome.

mRNA which is a string of codons, associates, with the 30S subunit that involves

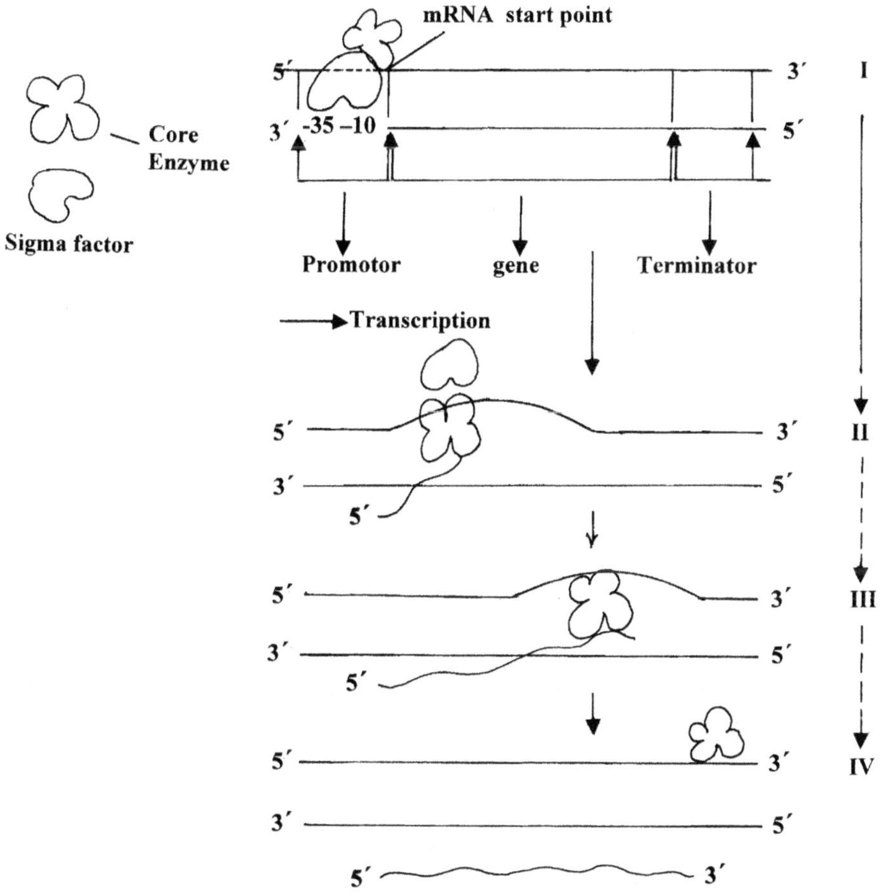

Figure 6.28. Transcription – Synthesis of mRNA from one (template/antisense) DNA strand. The gene is distinguishable in three regions: promoter, the coding region, and terminator. The enzyme RNA polymerase is a multisubunit enzyme and a sigma factor. I-Sigma factor is required for recognizing the promoter and attachment of enzyme to start point. II-Once transcription begins sigma is released. III-The process continues in 5'-3' direction and finally terminates at the terminator region, and IV-RNA transcript is released from DNA.

the binding of 3' end of rRNA to the Shine-Dalgarno motif which precedes the initiator AUG (or GUG) codon. The former is a polypurine (5' AGG AGG 3') consensus sequence lying within 10 bp region upstream of AUG. A complementary sequence is present close to the 3' end of 16S rRNA that is involved in base pairing with mRNA. A codon in mRNA, now associated with ribosome is recognized by an aminoacyl tRNA that has an anticodon complementary to the codon and an attached amino acid corresponding to the codon is brought at the ribosome site. A special initiator tRNA (fmet-$tRNA_f$) recognizes both AUG and GUG codons as initiator. The initiator AUG codon is recognized by f met-$tRNA_f$, but the same codon anywhere else in mRNA is recognized by Met-$tRNA_{m.}$ The fMet-$tRNA_f$ is first acylated with methionine and then an enzyme, found only in prokaryotes, adds a formyl group to amino group of the methionine, generating fMet-$tRNA_f$.

Two tRNA molecules are active in protein synthesis at any moment. An incoming amino acyl-tRNA binds to A site (acceptor site) and the codon representing the most recent addition of amino acid (with tRNA) lies in the P site (donor site). P site is occupied by peptidyl tRNA. Initially the fMet-$tRNA_f$ is lodged at the P site and the second codon exposed at A site directs the entry of second amino acyl-tRNA. A peptide bond is formed and the dipeptide now lodges at A site (Fig. 6.29b). This transfer generates a free deacylated tRNA at P site.

The ribosome now moves one triplet along the messenger, a stage called translocation, which also sees the exit of deacylated tRNA via another t-RNA binding site, the E site (exit site), and the transfer of peptidyl tRNA to P site. This frees the A site to accept the next amino acid. The continuation of this process involves the movement of the ribosome and the addition of amino acids one after another. The mRNA is read in the 5′ to 3′ direction and the peptide chain grows from amino end to carboxylic end. Once a termination codon, UAA, UAG, or UGA is reached, these are not recognized by any amino acyl-tRNA and the process of translation is terminated (Fig. 6.29). In *E. coli,* termination is catalyzed by two related proteins called release factors (RF). RF-1 recognizes UAA and UAG, and RF II, UGA and UAA termination codons. It has been clearly proved that the incorporation of amino acid is determined by the codon sequence of mRNA. Thus, after the completion of translation, amino acids are aligned in a specific manner that is determined by the codon sequence of mRNA or in effect by the gene. This generates the primary structure of the protein or polypeptide. For the detailed description of genetic code, the readers are advised to consult any book dealing with gene function.

The universality of genetic code is striking but exceptions do exist. These normally tend to relate to initiation and termination, and result from the production (or absence) of tRNAs representing certain codons. For example, the incorporation of modified amino acid selenocysteine takes place by using UGA codon (otherwise a chain termination codon) within the genes that code for selenoproteins. The incorporation of selenocysteine is controlled by four genes. The gene *selC* codes for tRNA (with the anticodon ACU) that is charged with serine. SelA and SelD modify serine to selenocysteine. The *selB* codes for a GTP-binding translation factor that recognizes selenocystyl-tRNA. This occurs because of the presence of a site called SECIS

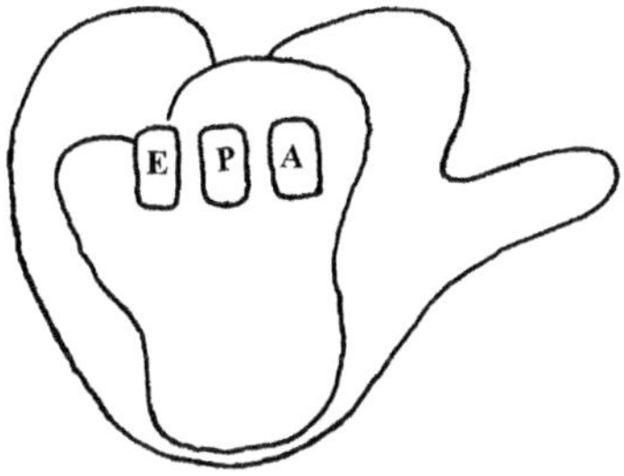

(a) Structure of the ribosome showing the position of the A- acceptor site, P- peptide site, and E- exit site.

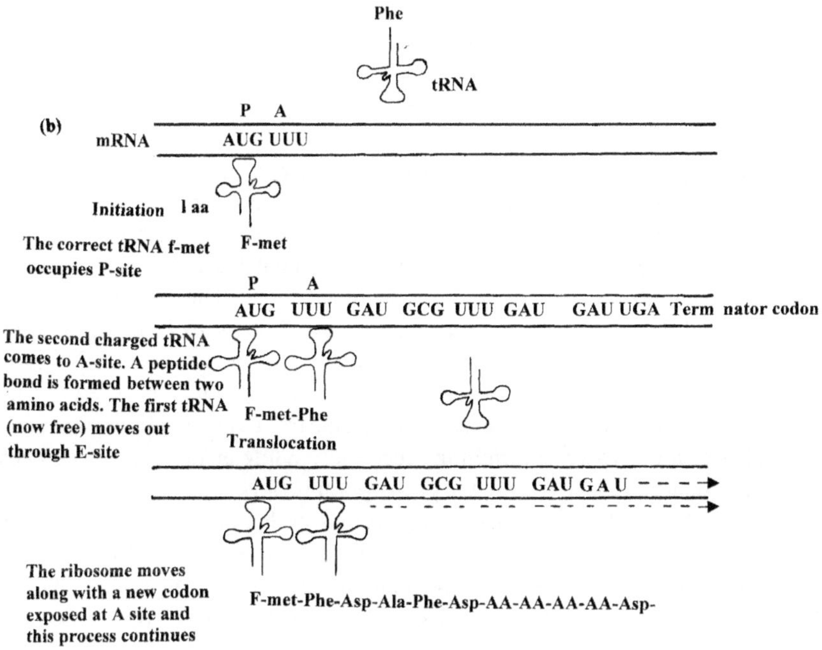

Figure 6.29a. Structure of the ribosome showing the position of the A- acceptor site, P- peptide site, and E- exit site. b.Translation: the process that leads to synthesis of protein on ribosomes under the message carried by mRNA (gene). Translation results in the alignment of amino acids in a specific sequence.

(selenocysteine insertion sequence) towards the 3′ end of UGA codon. SECIS forms a putative stem-loop structure that is used as a recognition site by SelB factor directing the selenocysteyl-tRNA to UGA codon on mRNA. In the absence of this structure, UGA acts as a normal stop codon.

Protein synthesis is an energy expending process. Upto 90% of all the ATP molecules synthesized by a rapidly growing bacterium are consumed at different stages of protein synthesis. ATP is used in the charging of tRNA and hydrolysis of GTP is required in the initiation and elongation, and termination stages. Additional factors are needed at each stage, that enter into cyclic association and dissociation, from the ribosome. Initiation requires initiation factors (IF). IF-3 is involved in binding of mRNA to 30S subunit as well as to keep the latter in a free form. IF-2 helps in bringing fMet-$tRNA_f$ to initiation complex at 30S subunit as well as excluding other amino acyl-tRNAs from the initiation reaction. IF-1 binds to 30S subunit and stabilizes the initiation complex. IFs are released before large subunit joins the initiation complex. EFs are involved in elongation. While EF-TU binds aminoacyl-tRNA to the 70S ribosome, EF-TU, and EF-G are required for translocation. As discussed earlier, release also require factors; RF1 and RF2 for recognizing the termination codon and RF3 for release of RF1 and RF2 from the ribosome. The reaction is completed by the release of polypeptide from the last tRNA and dissociation of the ribosome. The binding of these factors is mutually exclusive, ensuring that each step is carried out in correct order.

Many ribosomes may translate a single mRNA simultaneously, generating a polyribosome (or polysome). Bacterial polysomes are large, each consisting of tens of ribosomes. Bacterial mRNA has an extremely short life (a few minutes). The 5' end of mRNA starts getting translated before the completion of transcription. mRNA degradation is carried out by combined action of an endonuclease that cuts at discrete sites followed by an exonuclease that generates nucleotides. The prokaryotic rRNA and tRNA, like eukaryotic counterparts, are synthesized in the form of complex primary transcripts from which the mature molecules are released or processed. Processing reactions are mediated by particular ribonuclease and involve both endo- and exonucleases. Endonucleases release a mature molecule from a longer primary transcript and exonucleases are involved in trimming reactions. While eukaryotes also process their mRNA, prokaryotic mRNAs are the primary transcripts.

A typical mRNA contains besides coding regions both a non-translated 5' leader, and 3' trailer. Bacterial mRNA is usually polycistronic with non-translated regions between the cistrons. A growing *E.coli* cell contains ~20,000 ribosomes and ~200,000 tRNAs mostly in the form of aminoacyl tRNA. There may be 600 different messengers produced in 2-3 copies each.

The tRNA is a small molecule, 75-85 bases long and acts as an adaptor. This is an unique RNA species that contains a number of unusual bases that are generated by modification of the four standard bases after the synthesis of polynucleotide chain (Table 6.6) Some modifications are constant features, such as D and ψ. Others are specific for a particular tRNA or groups of tRNAs. The tRNA secondary structure is in the form of a clover leaf where complementary bases pair to produce stems with single stranded loops. The stem-loop structures are called the arms of tRNA. Four such arms have been identified: i) the acceptor arm that ends in an unpaired sequence whose free 2'-or 3'-OH group can be linked to an amino acid (ii) the T ψ C arm is named after the presence of this modified base in a triplet sequence (ψ = pseudouridine), (iii) the anticodon arm contains the triplet anticodon in the centre of the loop, and (iv) the D arm so called for its content of the base dihydrouridine (Fig. 6.30).

Table 6.6. Unusual or modified bases in tRNA

Normal base	*Modified bases*
Uridine	Ribothymidine (T), Dihydrouridine (D), Pseudouridine (ψ), 4-Thiouridine (S^4U)
Cytidine	3-methylcytidine, 5-methylcytidine
Adenosine	Inosine, N^6 methyladenosine (m^6A), N^6 isopentenyladenosine (I^6A)
Guanosine	7-methylguanosine (mG), Queuosine (Q), Wyosine (W)

When an amino acid is linked by an ester bond from its carboxyl group to 2' or 3'-OH of the ribose of the 5' terminal base (which is always A) of acceptor arm of tRNA, the latter is supposed to be charged and called aminoacyl-tRNA. This reaction is a two step process in which each amino acid reacts with ATP to form aminoacyl adenylate and then the activated amino acid is transferred to tRNA releasing AMP. Both these reactions are catalysed by a specific enzyme, aminoacyl-tRNA synthetase whose specificity is determined exclusively by the anticodon of the tRNA. There are at least 20 amino acyl-tRNA synthetases, each recognizing a single amino acid and specifically lodging it on a tRNA species.

Bacterial transcription and translation take place at similar rates in 5′ - 3′ direction. At 37°C, transcription of mRNA in *E. coli* occurs at ~40 nucleotides/sec which is very close to the rate of protein synthesis, roughly 15 amino acids/sec. By this analogy it takes ~ 2 min to transcribe and translate a mRNA of 5000 bp, corresponding to 180 kDa of protein. Several steps of protein synthesis are inhibited by a number of antibiotics (see **Chapter 5**).

Many newly synthesized polypeptides undergo post-translational modification. All newly synthesized bacterial proteins start with formylmethionine (the initiating amino acid) but archaeal proteins like those of eukaryotes begin with methionine. The formyl group is often removed by deformylase leaving methionine as the first amino acid and in some archaea (also eukaryotes) methionine is removed by methionine aminopeptidase, so that another amino acid is present in the beginning. In still other cases, an amino acid at carboxyterminal end may also be removed.

Amino acids like serine, tyrosine, and threonine may be phosphorylated through an ATP-dependent kinase. Extra carboxyl groups may be added to aspartate and glutamate. Similarly lysine may be methylated and some proline may be converted to hydroxyproline.

Many proteins are folded into tertiary and quaternary structures by the formation of covalent inter- or intrachain disulphide bridges involving cysteine residues. It was long thought that primary structure alone is sufficient to bring about spontaneous folding of the proteins. It is clear, however, that many proteins called as molecular chaperones

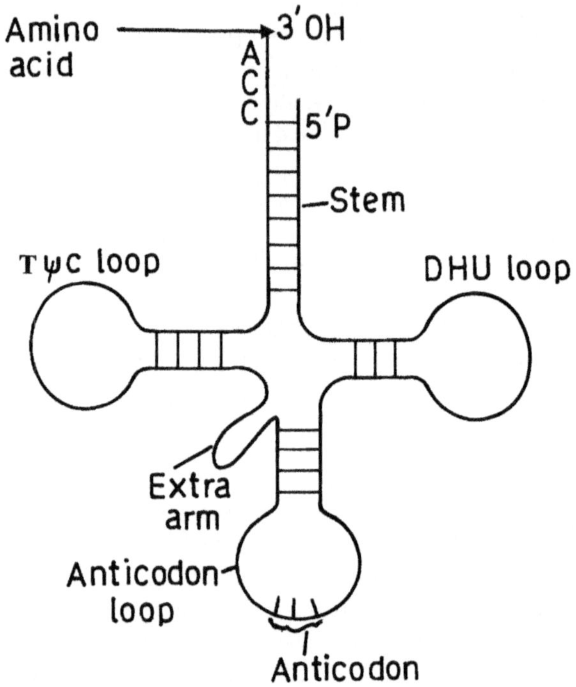

Figure 6.30. Diagrammatic structure of transfer RNA.

may assist in the proper chain folding. The chaperones do not become the part of the folded protein. They guide the unfolded, improperly folded or sometimes partially denatured proteins to fold properly. The chaperones are highly conserved and extremely wide-spread. The energy for their function is provided by ATP.

Modification of proteins through addition of carbohydrate side chain (glycosylation) is less prevalent in bacteria than eukaryotes. Removal of inteins is analogous to RNA splicing, that allows the reorganization of proteins with exteins. Inteins with a high degree of sequence homology to homing endonuclease have been reported in bacteria, archaea and eukaryotes. Removal of inteins by them thus is an autocatalytic process.

6.16. Analysis of Gene Expression

The most fundamental question in molecular genetics concerns the analysis of gene expression. While the classical route is to alter the gene by mutation and then follow the changed expression in the morphological or physiological state of the cell, it is not always the easiest of the options. Assaying the enzyme activity which is coded by a particular gene can be a more direct way of monitoring the gene expression. But, then, all gene products are not enzymes nor are the assay procedures simple for all of them. Some molecular techniques, therefore, have been devised to study the gene expression:

Western Blotting: If antibodies against the protein of interest are available, they could be used to monitor the expression. For this, cell extract is fractionated by SDS-PAGE and proteins are transferred to a nitrocellulose filter by blotting. The filter is then reacted with the specific antibody; this antibody is either linked with an enzyme directly or with a second enzyme-labelled antibody. The enzyme is then reacted with a chromogenic substrate that results in a visual band at the position of the antibody and, therefore, of the initial protein. By the analogy with 'Southern blotting' this technique is known as Western blotting (**see also Chapter 9**).

Northern Blotting: Alternatively, one could look at the production of the specific mRNA. This is done by isolating the RNA, fractionating it by gel electrophoresis, and transferring the RNA to the filter by blotting. This RNA can then be detected by hybridizing it with labelled DNA probe. By the same analogy mentioned above this technique is referred as Northern blotting.

PCR and RT PCR: Bacterial mRNA have a very small half life and thus are unstable. The isolated RNA, therefore, could be quite heterogeneous and may not run as a coherent band on a gel. Moreover, the amount of RNA synthesized under a particular set of conditions may be too small.

Large amounts of a specific DNA fragment can by produced by enzymatic amplification by employing a technique called polymerase chain reaction or PCR. Two synthetic oligonucleotide primers that will hybridize to the DNA of interest (DNA template) are mixed along with a heat stable DNA polymerase, such as *Taq* polymerase, and dNTP substrates. The mixture is then heated to denature the template, and cooled for primers to anneal to the template. The DNA polymerase now starts synthesizing the complementary strands. This cycle of heating (denaturation), cooling (annealing) and synthesis can be repeated with each cycle doubling the amount of target DNA. A single molecule of DNA in 20 cycles can be amplified to about a million copies (also refer **Chapter 7**).

Another version of this technique, RT-PCR employs mRNA and the enzyme reverse transcriptase, and an RNA-directed DNA polymerase to copy the same (cDNA). Then a standard PCR can be used to amplify the cDNA produced.

PCR based techniques have found diverse applications. These include: generation of DNA fragments for cloning, generation and amplification of gene probes, and molecular diagnostics, amplification of specific chromosomal segments for sequence analysis, in vitro mutagenesis and analysis of gene expression by detection of mRNA (RT-PCR).

Reporter Genes: The reporter gene is a gene whose expression can be easily assayed. Regulation of a specific gene can be studied by attaching the regulatory region to a reporter gene. The expression of the reporter gene will now be under the regulation of the selected gene. For example, regulatory region can be linked to the gene for β-galactosidase, the plasmid is transformed into cells and cells are selected on a chromogenic substrate like X-gal (5-bromo-4-chloro-3 indolyl-β-D-galctopyranoside). When the regulatory signal will be available, the reporter gene will be expressed. For example, under such a situation the colonies will turn blue.

Other commonly used reporter genes are those coding for chromogenic β-glucuronidase (Gus; *uidA*), green fluorescent protein (GFP), a protein that emits green light under UV-irradiation, and the enzyme luciferase (*lux*) that gives the cell a readily detectable luminescence.

The *phoA* gene of *E.coli* coding for alkaline phosphatase has also been developed as a reporter gene. The active enzyme is secreted into periplasmic space and thus the gene carries a signal sequence for the same. If the signal sequence is deleted, the enzyme can not be secreted and is thus inactive. An insertion of a DNA fragment may restore the enzyme activity if it contains transcriptional and translational start signals as well as signal sequence to facilitate the transport across the membrane. Thus the expression of *phoA* gene as a reporter will require in-frame translational fusion. The other examples cited above deal with transcriptional fusion.

The applicability of *phoA* as a reporter in identifying the gene coding for secreted protein has led to the creation of a Tn*phoA*. This transposon can be used for insertional mutagenesis. Transposition of this element in a chromosomal gene will give detectable alkaline phosphatase activity only if the gene codes for secreted protein.

Reporter genes are useful in studying gene expression under different conditions or virulence genes, or similar such differentially regulated genes.

6.17. DNA and RNA Synthesis

From the precursor molecules, both RNA and DNA are synthesized. RNA is synthesized through transcription as discussed earlier leading to all the three species of RNA, viz., rRNA, mRNA, and tRNA. RNA is copied from the template or antisense strand so that the resulting molecule has a base sequence similar to that of other or the sense strand of DNA, except that thymine is replaced by uracil. From the linear primary structure, many RNAs are derived through intrastrand base pairing and loop formation and in some cases modification of certain bases (as in tRNA).

DNA synthesis is an autocatalytic function or replication, where the two strands act as templates to synthesize the new strand. The resulting daughter DNAs consist of one old and one new strand (semiconservative). This process has already been discussed in detail in **Chapter 5**.

The important contribution of this process is to maintain the structure of gene and impart genetic continuation.

II-D- Regulation of Metabolism

For carrying out a multitude of metabolic reactions and for proper utilization of the available resources, cells must exert a fine regulation on its metabolic capabilities. This will prevent the wasteful synthesis of a compound if not required and at the same time will not allow the cell to get starved off an essential metabolite. As all essential steps in a metabolic pathway are driven by specific enzymes, they are the target molecules for this regulation. Morcovcr, majority of the enzymes are proteins, whose synthesis is directed by specific genes, the regulation can thus operate at gene expression level also. Two broad mechanisms of regulation operate in a cell:

1) Metabolic regulation exerted by a feedback inhibition of enzyme activity.
2) Genetic regulation that operates through the control of enzyme synthesis.

6.18. Feedback Inhibition

In this, the activity of an enzyme present in the cell is regulated by the end-product of the pathway also referred to as feedback inhibition. Thus, if an intermediate or end product is overproduced, it inhibits the first or any subsequent enzymes (Fig.6.31). Feedback inhibition is an efficient way to check the flow of material through a pathway so as to prevent any wasteful synthesis.

Enzymes are biological catalysts that lower the energy of activation required for a biochemical reaction to occur. This is effected by a proper binding of the substrate molecule to the enzyme at active site which actually alters the three dimensional configuration of the enzyme, thus inducing the proper fit. The ability to modify the three-dimensional shape of the enzyme molecule, and consequently its activity forms the basis of regulation of the rate of a reaction at biochemical level. Substances known as allosteric effectors may bind to the enzymes at sites other than the active site and may increase (activate) or decrease (inhibit) the activity of an enzyme (Fig. 6.31). In some cases, the end product of a pathway, which is not going to be structurally similar to the substrate may bind at these sites affecting the enzyme activity. This forms the basis of feedback inhibition or end-product inhibition. Many structural analogues may also employ this pathway for exerting their inhibitory effects. In others, this may lead to feedback activation. These two mechanisms are self regulating and control the rate of a particular metabolic reaction effectively.

6.19. Regulation of Gene Expression

Besides metabolic regulation, living cells possess regulatory mechanisms that operate at the gene level. Depending upon the environmental conditions, cells may "switch on" or "switch off" the genes so as to regulate the synthesis of enzymes and thus the metabolic reactions catalyzed by them. Such a regulation operates at the transcriptional level and responds to prevailing environmental conditions. Two such situations envisaged are: (i) when a new substrate is available, cells have to respond by effecting its metabolism, a process also called as induction and the enzymes so synthesized are known as inducible enzymes, (ii) when a cell is provided with an end product, it shuts off its internal biosynthesis by "switching off" the concerned genes, a process referred to as repression and the enzymes as repressible enzymes. It should be very clear that both induction and repression affect the synthesis and not the activity of enzymes and for both the cells must possess the necessary genetic entities. They, however, provide a strong regulatory pathway to avoid any wasteful metabolism and help cells economize on its resources. A cell may possess the genetic capability of synthesizing a large number of enzymes each in several copies under optimal growth conditions. However, if enzymes are estimated individually, it becomes clear that some enzymes are synthesized in much larger amounts in comparison to the other. This strongly suggests that a regulatory mechanism is invoked to control the levels of each enzyme and that all enzymes are not synthesized at all times.

6.19.1. Inducible Enzymes

Many bacteria possess biochemical pathways to metabolize a variety of chemical compounds as discussed in the earlier Sections of this Chapter. However, many of these pathways are not needed at all times. In other words, many enzymes are induced, and so are the concerned pathways only under a special set of conditions and exist only till such condition exists. Such enzymes known as inducible enzymes are commonly associated with catabolic pathways. As we have seen earlier, bacteria are extremely versatile as far as the utilization of a carbon source is concerned, which may either be available individually or in simple mixtures. A bacterium responds to the availability of a carbon source by inducing the enzyme/s necessary for its breakdown and these enzymes will be synthesized as long as that carbon source is available.

(i) *Lac* Operon: One of the classical examples of this type of regulatory mechanism is associated with the utilization of sugar lactose by *E. coli*. Lactose is broken down to glucose and galactose with the help of the enzyme ß-galactosidase, and is synthesized

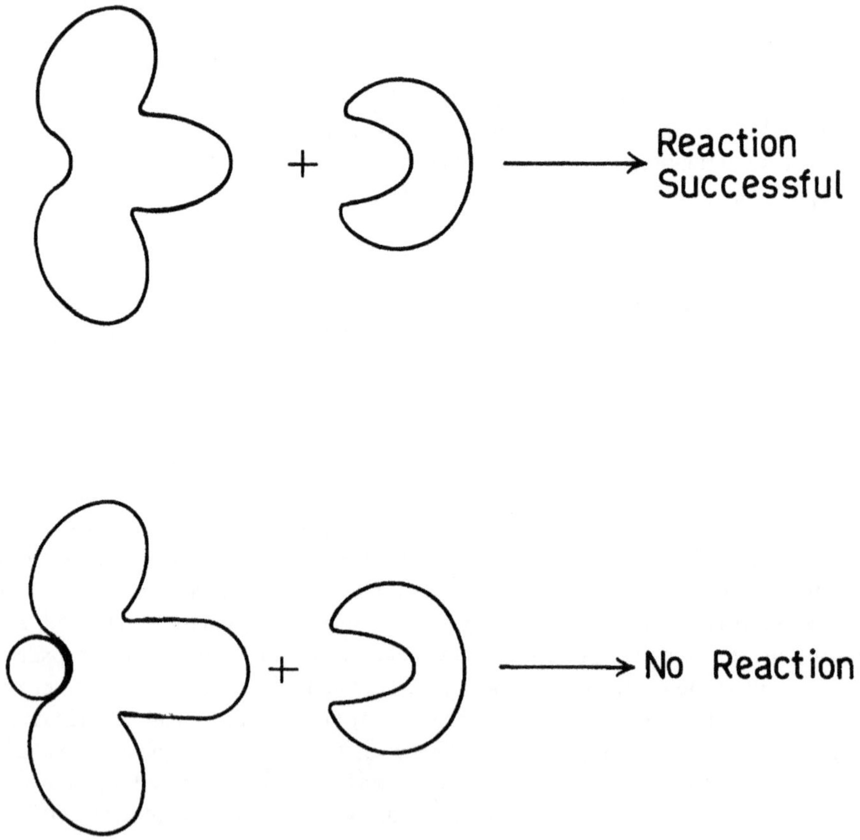

Figure 6.31. Feedback inhibition. A cellular control mechanism in which the end product controls the activity of an enzyme in a metabolic pathway. It may inhibit by binding to an allosteric site.

only when the cell has to utilize lactose as sole carbon source. In its absence, the enzyme is synthesized in very low (basal level) amounts. These enzymes were referred to as adaptive enzymes till in 1953 Monod *et al.* renamed it and many such other enzymes, as inducible enzymes.

Lactose utilization by *E.coli* requires three enzymes, viz.,β-galactosidase, β-galactoside permease, and β-galactoside transacetylase which are concurrently induced in the presence of lactose (more correctly allolactose, a derivative of lactose). Through extensive mutational analysis in *E.coli*, Jacob & Monod (1961) demonstrated that the genes controlling their synthesis are *lacZ, lacY, lacA,* respectively, located in the same order contiguously on *E. coli* genome. Jacob & Monod named them as structural genes and proposed that their close association indicates that they function as one unit and transcribed in a single polycistronic mRNA (RNA derived from more than one gene). In another mutation, the regulatory system was altered from inducible to constitutive (where the enzyme is synthesized continuously and does not respond to regulation). They labelled this gene as *lacI* and called it a regulator gene, whose product, a repressor, is responsible for regulating the transcription from structural genes. In order to effect regulation by *lacI* on three structural genes they visualized and later confirmed the presence of a target site or *lacO* gene. This gene was mapped, as expected, close to structural genes, on the left of *lacZ*. The structural genes along with this common operator (*lacO*) defined a functional unit named operon by them. The operon concept was subsequently elaborated by adding a promoter gene *lacP*, left of *lacO* to original *lac* operon proposed by Jacob & Monod. The promoter identifies the site where RNA polymerase binds to initiate transcription. Both the promoter and operator define the cis-acting sites. The regulation mechanism of *lac* operon is explained in Fig. 6.32.

Under normal conditions, i.e., in the absence of lactose, the repressor, synthesized constitutively by *lacI*, binds to operator *lacO* and prevents transcription. Thus, the three enzymes are not synthesized, when not needed (absence of lactose). When lactose (=allolactose) is added it acts as inducer and inactivates the repressor through binding. The inactive repressor leaves the operator thus allowing the transcription to start and proceed through the structural genes. The enzymes needed for metabolizing lactose thus become available. The presence or absence of lactose thus provides the regulatory switch that controls the synthesis of the necessary enzymes at transcription level.

Since the product of the regulator gene functions by "switching off" the operon, inducible regulation is often cited as an example of negative control. *lacI* gene comprises an independent transcription unit with its own promoter and terminator. Its product, the repressor, is a tetramer protein consisting of identical subunits of 38 kDa each. The dual response of repressor is due to presence of two binding sites, one for operator and another for inducer. In the absence of inducer, operator binding site is used to occupy the operator, preventing the passage of RNA polymerase through the operon. In the presence of inducer, however, inducer binds to repressor effecting an allosteric modification, leading to inactivation of the same. A mutation in *lacI* that may inactivate the repressor will lead to unhindered expression (constitutive) of the structural genes. The repressor can function in *trans,* i.e., even when it is present on a different genetic element than the *lac* operon. Several other types of mutations have been isolated in *lac*

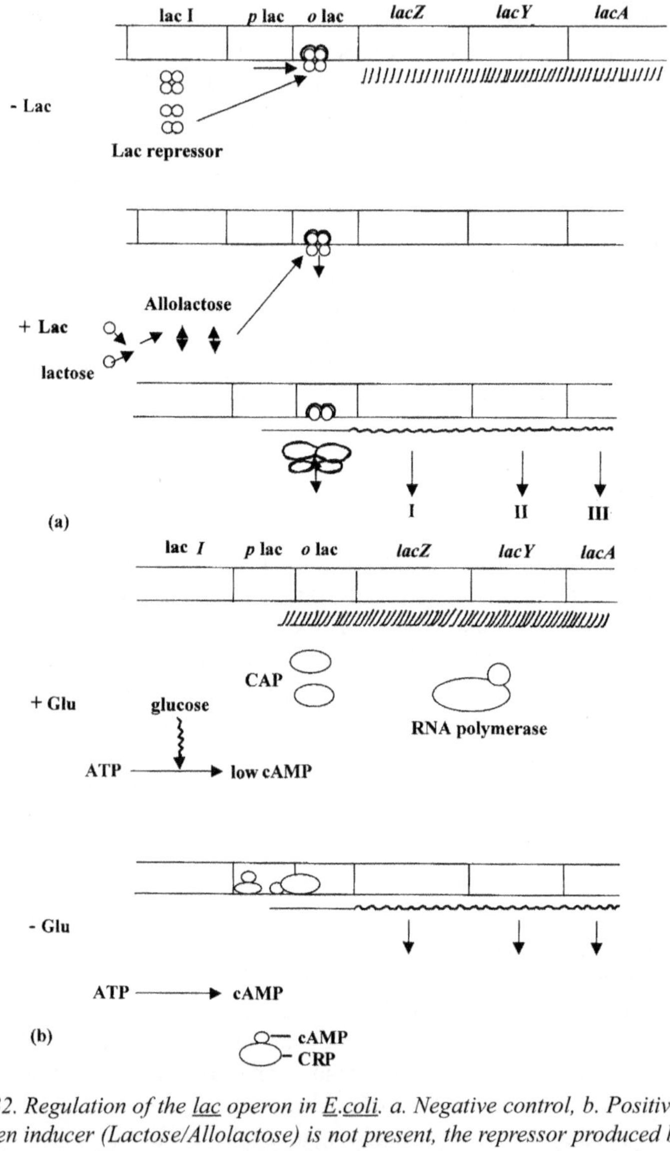

Figure 6.32. Regulation of the lac operon in E.coli. a. Negative control, b. Positive control. a. When inducer (Lactose/Allolactose) is not present, the repressor produced by the regulatory lacI gene binds to operator disallowing any transcription through the operon and thus no enzyme synthesis. When lactose is present, however, it is converted into allolactose that acts as the inducer. The inducer binds to the repressor and dislodges the same from the operator. The cell now produces the enzymes necessary for metabolism of lactose, b. The promoter activity of the operon is stimulated by a combination of cAMP and a specific protein known as the cAMP receptor protein (CRP). The cAMP-CRP complex binds to a site at the promoter and stimulates RNA polymerase binding. When cells are grown in the presence of an easily metabolized sugar such as glucose, catabolic repression takes place and lac operon is not switched on.

genes which have given further insight into the repressor structure and function and other constituents of *lac* operon (Fig. 6.33). Interested readers can find the details in any book dealing with molecular genetics.

The *lac* operon is also a case of positive regulation, that is, the operon is activated instead of being shut off (Fig. 6.32). There are some promoters at which transcription initiation by RNA polymerase takes place only by the assistance of an ancillary or the activator protein. These proteins such as CAP (catabolite activator protein) or CRP (cAMP receptor protein) are in turn regulated by small molecules such as cAMP. All CAP responding promoters have atleast one copy of the target sequence or CAP binding site- a pentamer ($\frac{TGTGA}{ACACT}$), which may be present at different locations relative to start point in different operons. When *E.coli* cell is grown in the presence of glucose and another C-source such as lactose, galactose, and arabinose, it preferentially utilizes glucose keeping off the expression from other operons. This effect is called glucose effect or catabolite repression.

It is generally believed that catabolite repression is set in train by the ability of glucose to reduce the level of cAMP that is synthesized from ATP by the enzyme adenylate cyclase. Expression of catabolite-regulated operons is inversely related to the levels of cAMP. This is so because the stimulatory effect of CRP is brought about only when it associates with cAMP and then this complex binds to promoter facilitating RNA polymerase binding. CRP-cAMP complex affects both RNA polymerase binding and can also induce DNA bending. CAP interaction, therefore, serves one general purpose and that is to switch off the alternative metabolic pathways, when they become unnecessary.

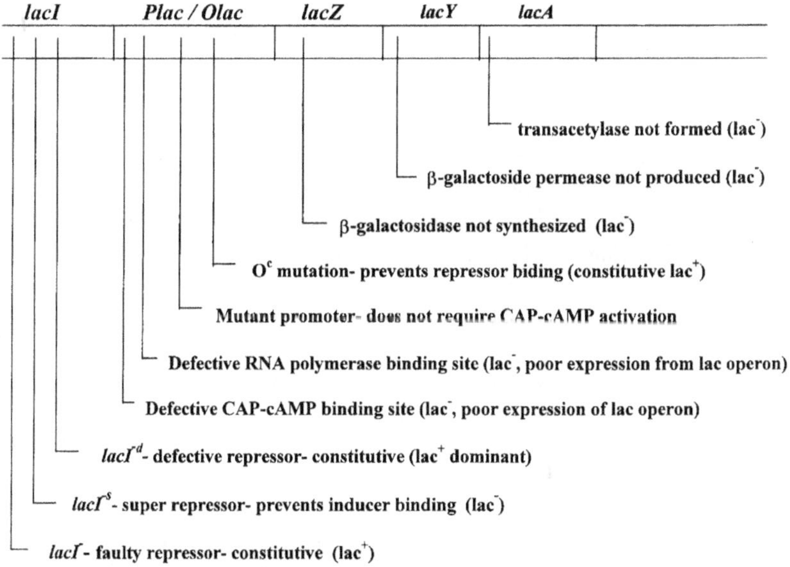

Figure 6.33. Depicts mutations which allowed the understanding of role of different components in lac operon regulation.

Upon binding, it is suggested that the CRP-cAMP bound complex will promote helix destabilization, facilitating RNA polymerase binding and increasing the efficiency of open promoter formation. It has been established that CRP causes DNA bending of around 90° or more. The bend enables CRP to directly interact with RNA polymerase. This interaction increases the RNA polymerase promoter binding as well as the transcript elongation activity.

Many operons involved with carbohydrate utilization such as *gal, ara,* and *pts* operons are similarly affected by cAMP-CRP. These, therefore, can also be referred to as members of carbon/energy regulon. Positive regulation by cAMP of operons such as *tna, ilv,* and *nadA* and negative effector regulation of *cya* and *ompA* by cAMP-CRP is also reported.

The catabolite involved in catabolite repression is an inducible component of the phosphotransferase system (PTS), III^{glu} (coded by the *pts-crr* gene) which is critical for this control system. It appears that III^{glu} controls the activity of pre-existing adenylate cyclase rather than its synthesis. The PTS system employs several proteins, some of which are specific for a given sugar. A phosphate group is transferred from phosphoenolpyruvate (PEP) to a carbohydrate during its transport across the membrane. The phosphate group is transferred from protein to protein until, in case of glucose, it reaches III^{glu}. When glucose is absent, phosphorylated III^{glu} accumulates and in turn interacts with adenylate cyclase and activates the enzyme. This results in a significant rise in the levels of cAMP. When glucose is present, it is transported across the membrane. During this transport III^{glu} transfers its phosphate to glucose to form glucose 6-P. Non-phosphorylated III^{glu} binds to inhibit adenylate cyclase causing intracellular cAMP levels to go down and thus catabolite repression. Cells grown on non-PTS sugars or sugar derivatives (e.g., glycerol), cAMP levels are high and operons like *lac* are fully active.

However, it became clear subsequently that glucose effect may have an alternative explanation as both CRP and cAMP are required for glucose effect also. This is because the major glucose transporter gene, *ptsG* is under the control of CRP-cAMP. Glucose under such a situation appears to prevent the entry of inducer (e.g., lactose) into the cell resulting in an increase in the concentration of free Lac repressor and thus switching off of the operon. The underlying mechanism is known as "inducer exclusion". Non-phosphorylated III^{glu} has also been implicated in inducer exclusion. It appears that it inhibits the accumulation of carbohydrate by non-PTS uptake systems. It has been shown that non-phosphorylated III^{glu} directly interacts with lactose permease inhibiting its transport. In glucose-lactose medium, *lac* transcription is induced quickly after the depletion of glucose. It is proposed that CRP-cAMP may confer on the cell an effective way to switch from glucose to lactose utilization.

Since the discovery of *lac* operon, a number of bacterial genes involved in similar functions have been found to be organized in operons. This appears to be a demand for effecting proper regulation as all the genes needed for one function can be controlled simultaneously or coordinately. Moreover, it also leads to effective utilization of relatively smaller genomes like that of bacteria. Several other negatively inducible but catabolite-repressed operons have been described. They, of course, differ in the number of structural genes, position of regulator gene in relation to operon, and number and positions of promoter and operator.

(ii) *Ara* Operon: The utilization of the sugar L-arabinose by *E.coli* is another example of inducible operon, which like *lac* operon is also regulated by catabolite repression. In fact, three operons are involved in metabolizing L-arabinose. The *BAD* operon encodes three catabolic enzymes that converts arabinose to D-xylulose 5-P which is further utilized through other pathways; the *FGH* operon codes for three proteins involved in L-arabinose uptake from low external concentrations, and an unlinked *araE* encoding a transporter of low affinity that operates at high concentration of the sugar. Upstream of the P_{BAD} lies the region where activator proteins AraC, and CRP-cAMP bind. These are referred to as *araI* and CRP site, respectively. The regulator *araC* is located upstream of the *BAD* operon and is transcribed in the opposite direction that is from the opposite strand of DNA. There are two operators, *ara* O_1 and *ara* O_2 upstream of *ara BAD*. The operons normally remain repressed but get activated (induced) in the presence of arabinose.

Although in general sense, the regulation of *ara* operon appears very similar to that of *lac*, mutation in *araC* suggested that *ara* operon is regulated by positive induction rather than negative inducible nature of *lac* operon. The mutations in *araC* that inactivates its protein product are non-inducible or "superrepressed". Readers may recall that a similar mutation of *lacI* is constitutive.

The regulation of *ara* operon has been explained on the basis of two alternative states of AraC, P_1 and P_2. In the absence of arabinose, AraC protein exists in P_1 form which acts as an antiactivator. This is because it does not repress transcription like a classical repressor but rather prevents activation of the operon that can be carried out by the P_2 state of protein. In the P_1 state (absence of arabinose), AraC protein preferentially binds to the operator *ara* O_2 and another site *ara* I_1 and that results in bending of DNA and inhibition of transcription. When arabinase is present, it associates with AraC converting the same into P_2 state. This results in binding of AraC to both *ara* I_1 and *ara* I_2 and in turn activation of the operon (Fig.6.34). The AraC protein not only controls *ara* operon but also negatively autoregulates its own synthesis so that less AraC is synthesized in the absence of L-arabinose than in its presence.

The *ara* operon like *lac* and *gal* operons is also subjected to catabolite repression. Thus, in the presence of glucose or other better carbon sources, *ara* operon does not express. The CAP-cAMP is a positive activator like AraC and binds at the CAP-binding site to facilitate RNA polymerase binding. Its action has been explained by the proposal that CAP-cAMP complex may help in opening the loop of DNA created by binding of AraC to *ara* O_2 and *ara* I_1. This may prevent the binding of AraC to these sites and instead facilitate the binding of AraC to *ara* I_1 and *ara* I_2 and activating the transcription from P_{BAD}. The transcription of the gene *E* and *FGH* operon is also activated concurrently.

The single control of three functional units (two operons and the gene *E*) by AraC is an example of a regulon where genes dispersed about the genome are under the control of a single repressor or activator protein.

(iii) *Mal* Regulon: Another positively regulated operon in *E.coli* involving the utilization of sugar maltose also falls under the definition of regulon (Fig. 6.35). This consists of four clusters of genes located at different sites on *E.coli* chromosome. One operon consists of genes *malP* and *malQ* which code for enzymes that break maltose and other polymers into glucose and glucose1-P. The third gene is the regulatory *malT* gene. Another gene *malS* located at a different site also codes for the enzyme that breaks down polymers of maltose. The third cluster consists of two operons and its six genes code for products that transport maltose into the cell. These two operons consisting of three genes are transcribed in opposite direction. Of these, *lamB* forms a large channel in outer membrane for passage of polymers of maltose, called malodextrins. Malodextrins are synthesized by the breakdown of starch. The LamB protein also serves as the receptor for phage lambda. The fourth cluster consisting of three genes is also involved in maltose polymer catabolism.

The first three clusters are regulated by an activator encoded by the *malT* gene. The MalT activator protein specifically binds the inducer, maltotriose which is synthesized from maltose. This complex activates the transcription by changing the DNA conformation.

This regulatory pathway is complex because maltose polymers are natural constituents of cell, playing many roles including protection against osmotic stress.

6.19.2. Repressible Enzymes

The inducible operons that we discussed in the earlier section all belong to catabolic pathways. However, the enzymes, needed for carrying out biosynthetic reactions, such as those for amino acids, vitamins, nucleosides, are often regulated by repression. The need for repressible enzyme systems is just the opposite of inducible enzymes. Here, the enzymes of a biosynthetic pathway should not be synthesized if the end product is available. This allows the cell to conserve energy as wasteful synthesis is checked. The overall pattern of regulation of inducible and repressible enzyme is same as both are negatively regulated. Thus, mutational inactivation of regulator gene would lead to constitutive expression from the operon even when the end product is available. In repressible operons, the product of regulator gene is inactive and called aporepressor. The end product acts as corepressor and once it associates with aporepressor, an active repressor is formed which negatively regulates the operon.

(i) *Trp* Operon: One of the classical examples of repressible regulation is the biosynthesis of amino acid, L-tryptophan. This is an essential amino acid as it is a constituent of many proteins and thus must be synthesized. The products of five structural genes of the *trp* operon provide the enzymes to make tryptophan from chorismic acid. This operon is transcribed from a single promoter *Ptrp* and regulated by a single regulator *trpR* (Fig.6.36).

As shown in the figure, *trpR* codes for an aporepressor that cannot repress the operon allowing the synthesis of tryptophan. Tryptophan, once available in plenty, can act as corepressor which binds to the aporepressor, and changes its conformation to form active repressor. The active repressor binds to operator switching off the expression and thus further synthesis of tryptophan. The TrpR repressor negatively autoregulates itself.

In addition to the promoter-operator complex, another site involved in regulating the *trp* operon is a region that lies between the operator and *trpE*, the first structural

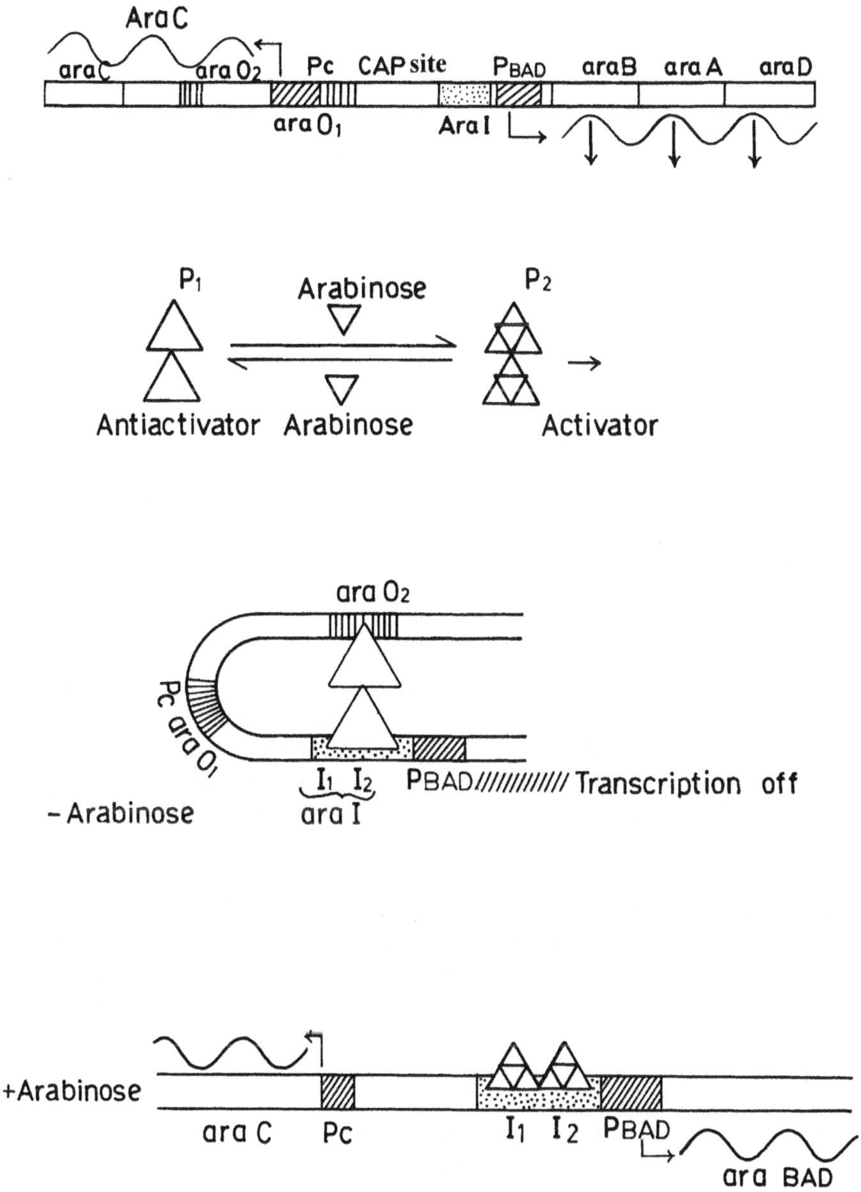

Figure 6.34. Structure of ara operon and regulation of operon by the presence or absence of arabinose. In the absence of arabinose, AraC dimer binds to two sites and represses the operon. In its presence, arabinose binds to AraC altering its conformation and operon is activated.

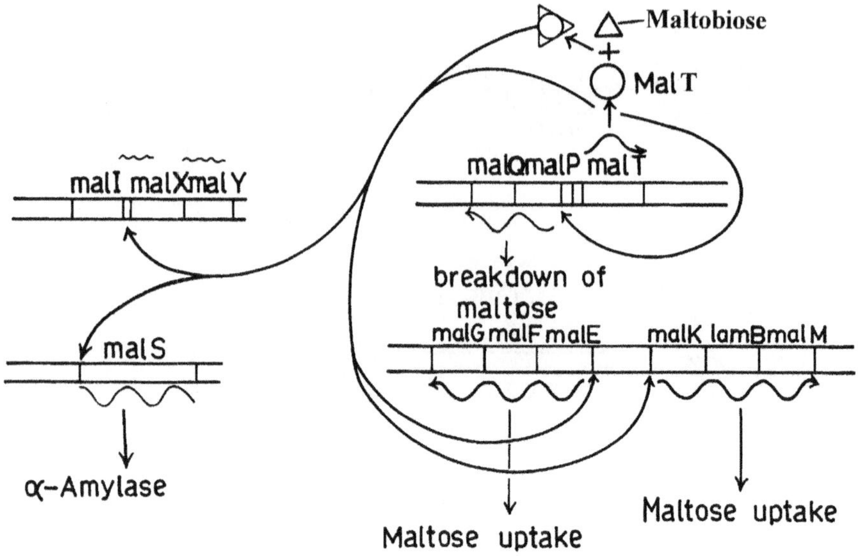

Figure 6.35. Components of mal regulon. The genes required for metabolism of maltose are distributed and are controlled coordinately by a common regulator.

gene of the operon. This *trpL* or leader region consists of a short coding sequence (162 bp long) that represents a leader peptide of 14 amino acids and the attenuator (intrinsic terminator) that provides a barrier to transcription into the structural genes. The attenuator controls the progression of RNA polymerase into the *trp* genes as a response to the level of tryptophan. When tryptophan is present in adequate amounts, this terminator is efficient so that *trp* genes are not transcribed (repressed= ~10% of RNA polymerase allowed to proceed). When tryptophan is removed, RNA polymerase has free access to the promoter and is also not compelled to terminate in the leader region. The sequence of the leader region suggests that it contains a ribosome binding site whose AUG codon is followed by a short coding region that contains two successive *trp* codons (in the region designated 1). When the cells run out of tryptophan, the ribosome that initiates translation of the leader region gets stalled at *trp* codons and this influences termination at the attenuator. The leader sequence has several complementary regions that can be organized into alternative base-paired hairpin loop-like structures (Fig. 6.37) influenced by the process of translation. In this, pairing between region 3 and 4 is the essential signal for intrinsic termination followed by a string of U residues. Such a situation would arise when tryptophan is available in plenty. In the absence of tryptophan, however, ribosome gets stalled in region 1 allowing regions 2 and 3 to pair, preventing the formation of termination hairpin, and RNA polymerase moves into the structural genes. This form of control, dependent on a mechanism that is responsive to the concentration of a particularly charged aminoacyl-tRNA in the cell is known as attenuation. Attenuation has a ~10x effect on transcription and together with ~70x influence on initiation of transcription in classical regulator-promoter-operator system allows ~700 fold regulation of Trp operon.

The biosynthesis of several amino acids (e.g. Phe, Thr, Leu, Ile, Val) is under very similar control with leader DNA sequences that encode short polypeptides containing multiple codons (usually adjacent) for the amino acid, whose synthesis the operon directs.

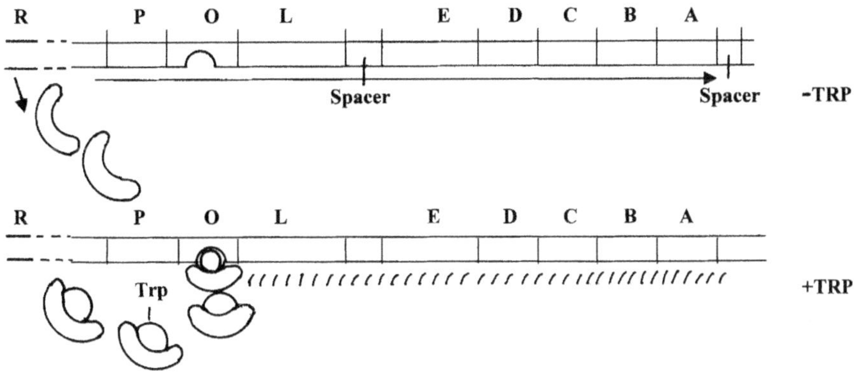

Figure 6.36. Overall structure of the trp operon and its regulation. The repressor produced by trpR on its own is unable to interact with the operator when tryptophan is not present. Thus, the operon is switched on to synthesize the necessary trp biosynthetic enzymes. When enough tryptophan is available, it binds to TrpR aporepressor and forms the active repressor. The latter now interacts with the operator and switches off the operon.

6.19.3. Transcriptional Attenuation

A new type of attenuation involving RNA polymerase rather than ribosome is used by both *E.coli* and *S.typhimurium* to regulate the synthesis of uridine monophosphate (UMP), the precursor of all pyrimidine nucleotides. One *pyrBI*, of the six unliked genes and operons involved, is regulated over 100 fold by the intracellular concentration of UTP. In the regulatory region of *pyrBI*, a RNA polymerase pause site is present flanked by U-rich (T-rich in DNA) area. When cellular UTP levels are low, RNA polymerase takes a long pause so that translating ribosome catches up the stalled polymerase. When the polymerase finally escapes the pause region and reaches transcriptional attenuation region, close proximity of ribosome prevents rho-independent terminator hairpin and thus permits transcription through the *pyrBI* structural genes. When UTP levels are high, polymerase takes a short pause and then quickly transcribes the attenuator. As the ribosome is still away, transcription termination loop is formed and structural genes are not transcribed.

6.20. Translational Control

The natural control of gene expression may also be exerted at the level of translation.

As is expected, this plays a minor role as it would be rather wasteful to produce large amounts of all mRNA and then not translate some of them. A few such mechanisms may operate in the cell.

The ribosomes bind to the mRNA at a specific sequence called Shine-Dalgarno sequence that is complementary to the 3′ end of the 16 S rRNA. This sequence lies approximately 7 bases upstream of translation initiation AUG codon. Most known sequences match the consensus ribosome binding site in atleast 4-5 bases. As such different sequences will act with different levels of efficiency, i.e., they can be weak or strong ribosome binding sites. The distance separating the ribosome binding site form the initiation codon can also have a strong influence on gene expression.

In regulation of *pyrC*, transcription can start at different points in the promoter region depending upon the availability of the pyrimidine. When pyrimidine is present in plenty, it starts at a specific point that facilitates the hairpin loop formation. Such a loop preferentially sequesters the Shine-Dalgarno sequence inhibiting ribosome binding, and thus translation. Alternatively, when pyrimidine level is low, initial transcription generates a shorter transcript that cannot incorporate SD sequence and the latter is available for ribosome binding.

The genes in the *arg* regulon are controlled by a conserved upstream 18 bp sequence called ARG box. This sequence shows a dyad symmetry and is used for repressor binding. The base composition and degree of homology between the ARG box sequences, as well as their number and their position relative to the promoter forms the basis of the differential effectiveness of the repressor.

In polycistronic operons, the ribosome after translating the first gene may dissociate and then reassociate with the next initiation codon. The efficiency of this may vary leading to differential expression of genes within an operon, an effect known as polarity. In many cases, however, the second initiation site may be very close. Thus, after the release of the first polypeptide and 50S subunit, 30S subunit without dissociation contacts the next initiation codon to reinitiate the translation process.

Another process called as ribosomal frame-shifting may take place in some cases. Here, the ribosome after reaching a specific site on the mRNA (usually a string of adenine residues) may slip back by one base (a-1frameshift) and then continue polypeptide synthesis. This will lead to the reading of mRNA from a different reading frame and results in two proteins, one of which will contain portion read in two different reading frames. This achieves further down regulation of gene expression. The best examples are some insertion sequences.

Translational control may also operate at the level of 'codon usage'. Most amino acids can be coded for by more than one codon some times with equal efficiency as they are recognized by the same tRNA. But in some cases, different tRNA species may recognize different codons, and some of these tRNAs may be present in the cell at quite low levels. If a gene contains a large number of such codons that require these 'rare' tRNA molecules, translation will suffer delays and the amount of end product will be affected. Study of a large number of *E.coli* genes suggests that highly expressed genes have a high degree of 'codon bias' or in other words they contain the codons that can be recognized efficiently by the common tRNA species. Genes expressed at moderate or

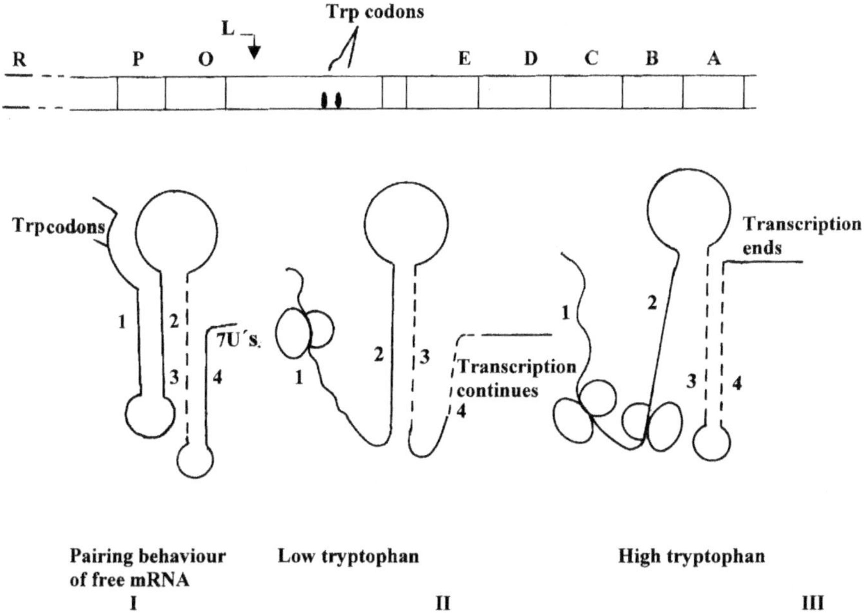

Figure 6.37. Attenuation in Trp operon regulation. I- The intrastrand pairing in mRNA that may involve 1-2 and 3-4. II- At low levels of tryptophan, ribosome gets stalled at Trp codon in the leader region. This allows 2-3 pairing and continuation of transcription. II- At high levels of tryptophan, ribosomes are in contact with region 2 that allows the 3-4 pairing, and termination of transcription.

low levels will have different codon bias.

Translational control can also occur by the production of 'antisense RNA'. Transcription of the gene in both the direction from the region specially in the vicinity of ribosome binding site and translation initiation point will yield both sense and antisense RNA. The hybridization of two RNAs will block the binding of ribosomes and initiation of translation. Though, more common in eukaryotes, some bacterial genes like the outer

Table 6.7. Some global regulatory systems in E. coli

Type of signal	*Response*
A. Nutrient limitation	
i) Carbon	a) Catabolite repression b) Control of fermentative vs oxidative metabolism
ii) Nitrogen	Response to ammonia limitation
iii) Phosphorus	Starvation for inorganic orthophosphate (Pi)
B. Growth limitation	
i) Stringent response	Response to lack of sufficient aminoacylated tRNAs for protein synthesis
ii) Stationary phase	Switch to maintenance metabolism and stress protection
iii) Oxygen	Response to anaerobic environment
C. Stress	
i) Osmoregulation	(a) Response to abrupt osmotic upshift (b) Adjustment to osmotic environment
ii) Oxygen stress	Protection against reactive oxygen species
iii) Heat shock	Response to abrupt temperature increase
iv) pH shock	Tolerance of acidic environment

Category of mechanism	*Gene/operon-regulated*
DNA binding activator or repressor —— do ——	a) Numerous C-source operons such as *lac, ara, gal, mal* and others b) Genes involved in glycolysis and Krebs cycle
a) Sigma factor (σ^{54}) b) Two component system	Glutamine synthetase gene and operons for amino acid degradation
Two component system	38 genes including alkaline phosphatase and phosphate uptake genes
(p) ppGpp metabolism	rRNA and tRNA genes
Sigma factor (σ^{s})	>40 genes including those for trehalose synthesis and DNA binding protein
a) CAP family of DNA binding protein b) Two component system	a) >31 transcripts including nitrate reductase genes b) >20 genes including cobalamine synthesis
a) Two component system b) Sigma factor (σ^{s}) c) Antisense RNA	a) K^+ uptake system b) > 16 genes including an outer membrane lipoprotein gene and genes regulated in stationary phase as above c) Porin gene
a) AraC family of DNA-binding proteins b) LysR family of DNA-binding proteins	a) > 10 genes including those for superoxide dismutase and antisense RNA regulator of porin b) > 10 genes including of catalase
a) Sigma factor (σ^{32}) b) Sigma factor (σ^{24})	a) > 20 heat shock proteins (Hsps) including (DnaK, DnaJ), GrpE chaperones, some proteases and others b) >10 proteins including σ^{32} and a periplasmic protease Several genes including amino acid decarboxylase
ToxR-related DNA binding protein	

membrane protein gene, *ompF*, and the replication of some plasmids is regulated by this process.

6.21. Developmental Regulation- Quorum Sensing and Cross Talking

Although bacteria are generally unicellular, the significance of cell-cell interactions in the regulation of some aspects of their physiology is getting increasingly recognized.

One such phenomenon referred to as quorum sensing confers the cells the ability to respond differently to the environment based on cell density. Such a response is mediated by the secretion of the signaling molecules and thus more are the number of cells higher is the level of signal.

The best-studied example of quorum sensing is the luminescence emitted by some *Vibrio* species inhabiting the light organs of squids and some fishes. These bacteria secrete acylated homoserine lactones which interacts with LuxR protein, a transcriptional regulator which in turn activates the expression of luciferase gene and the genes required for the production of its substrate. The induction varies with the levels of the lactone and will thus be higher at high cell density.

Similar quorum sensing has also been observed during the spread of plasmids in *Agrobacterium tumefaciens,* production of some antibiotics by species of *Streptomyces,* development of transformation competence and sporulation initiation in *Bacillus subtilis*, and cell aggregation to form fruiting bodies in *Myxococcus xanthus*. The nature of signal molecules in Gram-negative and Gram-positive bacteria have been found to be different. While in the former it has been found to be a lactone, the latter produces a type of peptide for the same purpose.

6.22. The Stringent Response

Besides the individual regulation of functional operons, the bacterial cells may face a situation that will demand a co-ordinated series of events to take place. Such a situation is encountered under conditions of amino acid starvation and is called as stringent response. These include the cessation of production of rRNA, tRNA, and ribosomal proteins and inhibition of synthesis of carbohydrates, lipids, and peptidoglycans. However, transcription from some genes may also be simultaneously stimulated. The stringent response is mediated by a pair of unusual nucleotides- guanosine tetraphosphate (ppGpp) and guanosine pentaphosphate (pppGpp) also known as magic spot nucleotides I and II, respectively.

Mutants defective in the enzyme phosphoryl transferase (*relA*) that synthesizes ppGpp do not exhibit this response and are known as relaxed. The enzyme is associated with the ribosome and gets stimulated by the binding of uncharged tRNAs at A site on the ribosome. This leads to a dramatic (>10 fold) rise in the concentration of ppGpp by stimulation of RelA protein. The intracellular levels of ppGpp is also regulated by SpoT (magic spot) protein. This protein normally degrades ppGpp but under amino acid starvation this inhibitory activity is suppressed and concentration of ppGpp rises. In fact, *spoT* gene product may even synthesize ppGpp during growth rate regulation. Most evidences suggest that ppGpp blocks the initiation of transcription of tRNA and rRNA genes by acting directly on their promoters. Other evidences point out that ppGpp does not block the initiation, but the elongation of the transcripts. This molecule has also been implicated in the transcription of some biosynthetic operons and growth rate regulation.

6.23. Global Regulatory Mechanisms

The bacterial cells must be able to adapt to a wide range of environmental conditions. Unlike the conditions available in the laboratory, the external environment not only varies widely but that these changes can be abrupt. For example, nutrients are usually limiting and highly variable, amount of water and other solutes may suddenly change, temperatures may fluctuate and the cell may face stresses of various kinds like pH, oxygen and several others. In order to respond to such changes, the bacterium must make adjustments in the rate of synthesis of its cellular components so that the cell does not make more of a component than it needs at the cost of others for which it may starve. Thus, a cell will have to invoke regulatory systems that will simultaneously regulate numerous operons. These systems are called global regulatory mechanisms wherein a single regulatory protein controls a large number of operons, which are then said to be the members of the same regulon (Table 6.7).

In bacteria, a signal transduction pathway also exists as a response to a wide-variety of environmental stimuli. This pathway is mediated by pairs of proteins that communicate with each other by a conserved mechanism involving protein phosphorylation. The pair comprises a sensor and a response regulator (modulator/effector) component. The two component systems from diverse sources belong to a large superfamily, in which sensor proteins show similarity in C-domains also called transmitter domain and regulator proteins share homology in their N-domain, the so called response receiver domain. While the sensor protein is an autokinase and is usually lodged in the membrane, the response regulator protein receives the phosphoryl group from the former. The N-domain of the sensor receives an environmental signal and transduces the same to C-domain, which causes the transmitter domain to phosphorylate itself.

The phosphoryl group is subsequently donated to the receiver domain of the response regulator. Phosphorylation of the receiver domain causes a conformational change in the output domain of the response regulator that in turn translates out the appropriate response in terms of controlling a particular set of genes. The similarities between these systems suggest that the diverse environmental signals can be transduced through a common mechanism to regulate a wide range of processes in bacteria, including response to nutrient deprivation, osmolarity changes, chemotaxis, sporulation, symbiosis and bacterial pathogenesis (Table 6.8).

Table 6.8. Some members of superfamily of signal-transducing regulators

Environmental signal	*Sensor*	*Regulator*	*Target*
1. Nitrogen status	Ntr B	NtrC	Nitrogen assimilation genes
2. Medium osmolarity	Env Z	OmpR	Outermembrane porin genes
3. Phosphate status	Pho R	Pho B	Phosphate regulon
4. Chemotactic signal	Che A	Che Y/Che B	Chemotaxis proteins

The structural-functional similarities also suggest that there might be a cross-talk between the component members of different kind. This will provide a means for integrating many types of environmental stimuli.

CHAPTER 7
BACTERIA AS MODEL SYSTEMS IN GENETIC ANALYSIS

Microorganisms are ubiquitously present, but the role played by them in human affairs overshadowed for a long time their importance in the study of some fundamental biological, including genetic, problems. Although, Louis Pasteur brought them into the realm of the experimental science, it was in the field of Genetics that microbes had the greatest difficulty in finding a foothold. Microbial Genetics is, therefore, one of the youngest branches of Genetics.

The entire edifice of classical genetics is based on transmission of characters from one generation to another. Such ideas originated from controlled hybridizations conducted between races and varieties of the same species of plants and animals which differed in certain distinct hereditary characters. For a long time, both microbiologists and geneticists conveniently agreed that laws of Genetics can not be applied to microorganisms, especially bacteria, as they do not seem to reproduce sexually and that they are inconveniently small and simple to provide any morphological/phenotypic distinctions. It was only after the classical work of Beadle and Tatum, in 1943, with the fungus *Neurospora crassa* that microorganisms like bacteria were seriously looked into. The discovery of sexual reproduction in *E.coli*, just three years later, provided the necessary impetus for the creation of Microbial Genetics as a science. Since then, there has been no looking back for bacteria, as they became the source material for one exciting discovery after another. Their morphological simplicity which appeared to be a serious bottleneck initially has now been exploited to the hilt by geneticists who coupled it with their metabolic versatility to generate a wealth of information.

7.1. Why Microbes?

Bacteria and other microorganisms, offer some very unique features that can be utilized for genetic analyses. Most importantly they are haploid, making the isolation and expression of a mutant very convenient because unlike the diploid organisms, the recessive mutation can not be masked by the wild-type counterpart as the two will not exist together. The growth requirements of most culturable bacteria are very simple, such that they can be grown on a simple synthetic culture medium. For the same reason, effect of any physical and chemical factor can be easily demonstrated. Their simple morphological and genetic organization make them amenable to structure-function as well as detailed genetic analysis. Their rate of reproduction is high and life span is short, so that a statistically significant population can be raised and analyzed in a reasonably short time.

Genetics is the branch of science that deals with generation, expression, and transmission of biological variation. In this Chapter, we shall discuss how microbes/bacteria have emerged clear winners in contributing significantly towards all the three aspects. Readers, however, must note that the expression of biological variation has been dealt with under the heading gene expression and regulation in **Chapter 6** and will find only a cursory mention here.

7.2. Generation of Biological Variation (Mutation)

Availability of variability is a hallmark of biological populations and bacteria are no exception to this. Although this variability was discernible since the time they could be viewed through microscopes of all levels of refinement, their genetic connotation could only be clarified by the classical work of Beadle and Tatum (1943) in the fungus, *Neurospora crassa*. They were able to induce and characterize several mutations in *N. crassa* by treating the haploid conidia with X-ray. The haploid nature of microbes led Tatum to introduce *E. coli* in Microbiology Laboratory Course at Stanford University. Tatum was able to induce and stock a number of mutants in this bacterium. Studies done by Luria and Delbrück (1943) helped in distinguishing environmental adaptations from spontaneous mutation. The technique for demonstrating spontaneous mutation by Newcombe (1949) and of replica plating (see Fig 7.2b) by Lederberg and Lederberg (1952) further facilitated this line of study.

Induction of mutations in bacteria was first achieved by X-ray and UV irradiation; several chemical agents and other molecular means were subsequently applied to cause mutations with high efficiency and precision.

A mutant refers to an organism whose DNA base sequence has undergone a change but which may or may not cause a changed phenotype. As we know, each gene controls the synthesis of a protein (in some cases RNA is the final product) by determining its amino acid sequence. The protein which may contribute a structural or a functional (enzymatic) component is thus responsible for a final product that leads to a particular phenotype. The effect of a mutated gene is percolated through a changed amino acid sequence that may form the "faulty" proteins incapable of performing the assigned function. When an amino acid substitution does not affect or only partially affect the protein, a silent and a leaky mutation, respectively can be produced. For details of mechanism of mutation, readers are referred to any book of Genetics. Once again the haploid nature of microbial systems, offer the unique advantage of screening for a mutant, spontaneous or induced by a mutagen. For the sake of screening mutants can be classified into four categories:

1) Phenotypic- Although abundantly popular in eukaryotic organisms, morphological simplicity of bacteria offer a very few such characters. However, mutants affecting colony shape and size and pigmentation can be isolated sometime.

2) Nutritional- These are a very common class of mutants in bacteria. As we discussed in **Chapters 2 and 6**, bacteria have very simple requirements as they are capable of synthesizing all their essential metabolites through various enzyme-mediated pathways. Such bacteria are called prototrophs (autotrophs/phototrophs). Mutation in one of the steps of such pathways may lead to non-production of that essential metabolite, leading to the requirement for the same that is referred to as auxotrophy. Auxotrophic mutants have been valuable tools in molecular genetics.

3) Degradative- Based on their metabolic versatility, bacteria are also capable of utilizing a large number of organic molecules as we have seen in **Chapter 6**. For example, bacteria can utilize many types of C-sources ranging from mono-, di- or polysaccharides as well as complex hydrocarbons, amino acids, proteins, fats and even some xenobiotic compounds. Mutation in any one of these pathways will lead to non-utilization of that particular C-source.

4) Resistance - Most bacteria are extremely sensitive to antimicrobial agents such

as antibiotics, antimetabolites, metabolic inhibitors, and phages. Mutation may lead to the acquisition of resistance to such agents.

One of the commonest ways of classifying mutants is to identify the nature of events. In nature and so also in the laboratory, mutations sometimes arise without any provocation, whereas at other times one could generate mutations by applying certain agents called mutagens. They are, thus, referred to as spontaneous and induced mutations, respectively. Similarly mutations can also be classified on the basis of conditions in which the mutant phenotype is expressed. Accordingly, a non-conditional mutant will display the mutant phenotype under all conditions, whereas the expression of a conditional mutant will depend either on environmental conditions or on the presence of other mutations. For example, a temperature-sensitive (Ts) mutant will behave as mutant only at a certain temperature (restrictive) but as wild-type (nonmutant) at another temperature (permissive). This class of mutants have contributed significantly towards understanding the functions of some essential genes. Similarly, suppressor-sensitive mutant will express mutant phenotype only if the cell does not carry the products made by the suppressor genes (*sup*°). These have been designated as Am (for amber or UAG mutations), Oc (ochre or UAA mutations), and Op (opal or UGA mutations). The suppressor gene product either substitutes for the defect in the mutant or enables the altered gene product to be utilized normally. It is also customary to classify mutations on the basis of the nature of changes that may have occurred. Accordingly mutations that are caused by a change (substitution/addition/deletion) of one base are referred to as point mutations (Fig.7.1) in contrast to a change that may remove a part or whole of the gene which are then called deletion mutations. Addition of a chunk of DNA into the base sequence of a gene leads to the generation of insertion mutation.

Isolation and Characterization of Mutants: The mutations have played a central role in all genetic analysis. They are the main cause of creating allelic variation, thus contributing significantly towards biological variations. They not only reveal a function at phenotypic level but also at biochemical level. Mutations aid in the elucidation of a metabolic pathway and its regulation. They are required to understand the interaction between genes and proteins. They are also employed to identify and assess the action of external agents. Because mutations are so useful but rare, they need to be isolated from a population of cells. The haploid nature of microorganisms can be very handy in isolation, screening, and characterization of mutants.

7.2.1. Induced Mutagenesis

Although mutations may arise spontaneously, they are so rare that various agents, called mutagens, are employed to introduce them. The mutagens range from physical (e.g., UV, X-ray, etc.) to chemical (base analogues, nitrous acid, alkylating agents, acridine dyes, etc.) agents. For details of mutation induction, readers can refer to any book on Genetics.

7.2.1.1. Detection. After mutagen treatment, the mutations can be detected either through a selection or a screen.

7.2.1.2. Selection. This requires a condition that will allow the growth of a specific mutant but not the non-mutant parental cells. This is a very powerful tool and allows the selection of a rare mutation from a large population of cells. For example, the isolation of antibiotic resistant mutants can be done by directly plating a mutagenized (treated with a mutagen) sensitive population on agar-medium containing that specific antibiotic, as only the resistant mutants will grow.

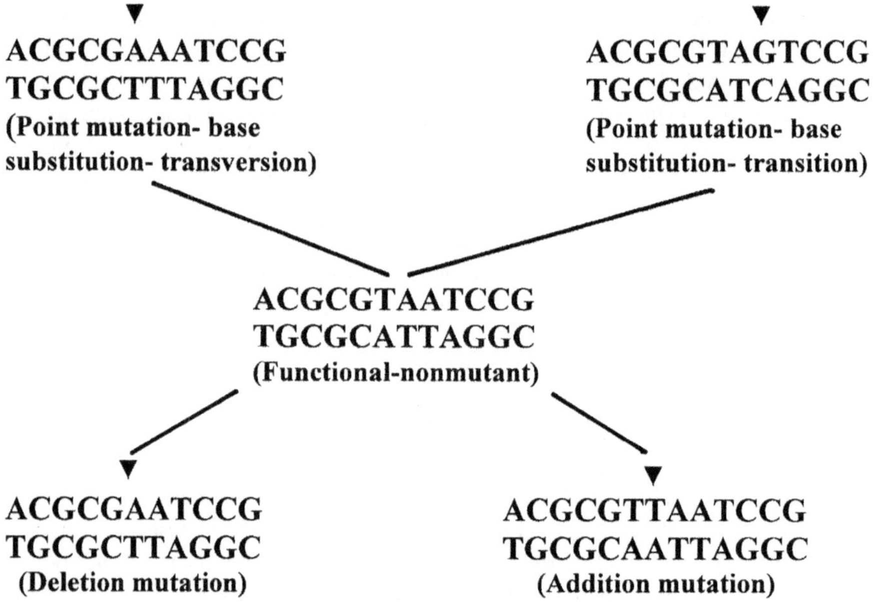

Figure 7.1. Classification of mutations: A part of a functional gene is shown in which mutational changes have been depicted. Such changes can take place anywhere in the gene. The deletion and addition can lead to frame-shift.

7.2.1.3. Screening. In case a direct selection is not possible, mutants will have to be detected by screening. In this, the medium to be employed supports the growth of both mutant and non-mutant cells. The mutants are then screened either by a changed phenotype (For example, sugar-utilization mutants on indicator agar; **see Chapter 2**) or by their failure to grow on a simple nutrient medium (for example, auxotrophs). The latter is generally carried out by two ways: i) in patch replication, the mutagenized colonies are arranged on a master plate and then manually transferred to a nutritionally-defined medium, or ii) the mutagenized colonies are replica plated on a defined medium plate (Fig.7.2a, b). Replica plating is an useful technique to screen for different types of mutants such as auxotrophic and temperature-sensitive conditional mutants.

The induced mutagenesis increases the frequency of mutations so that the desired mutation can be obtained with a reasonable number of replica plates. Further enhancement can be brought about by introducing an enrichment procedure. The enrichment brings about an increase in the proportion of mutants in a population.

7.2.1.4. Reversion. Most point mutations are revertible i.e., they can give rise to a wild-type allele restoring the function. The process of regaining the original phenotype is called reversion and an organism that has reverted is referred to as a revertant.

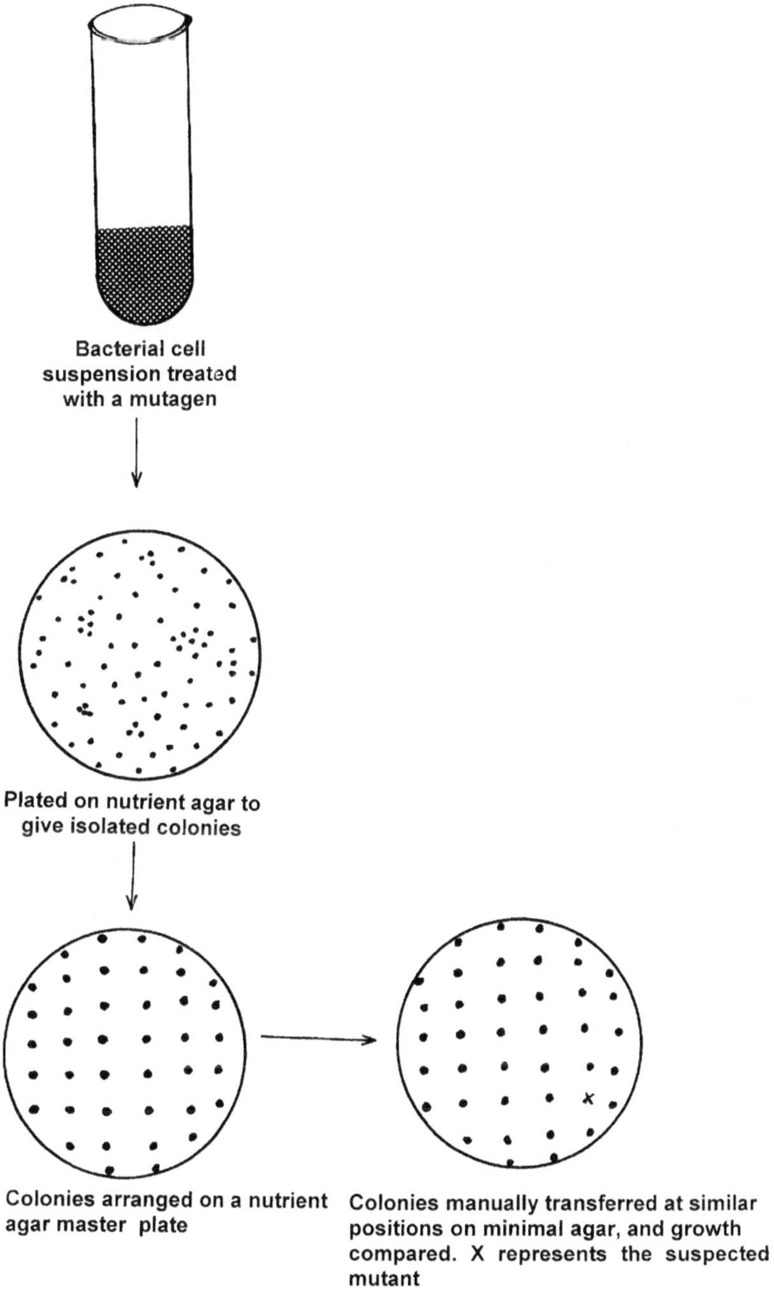

Figure 7.2a. Patch replication for screening of mutants.

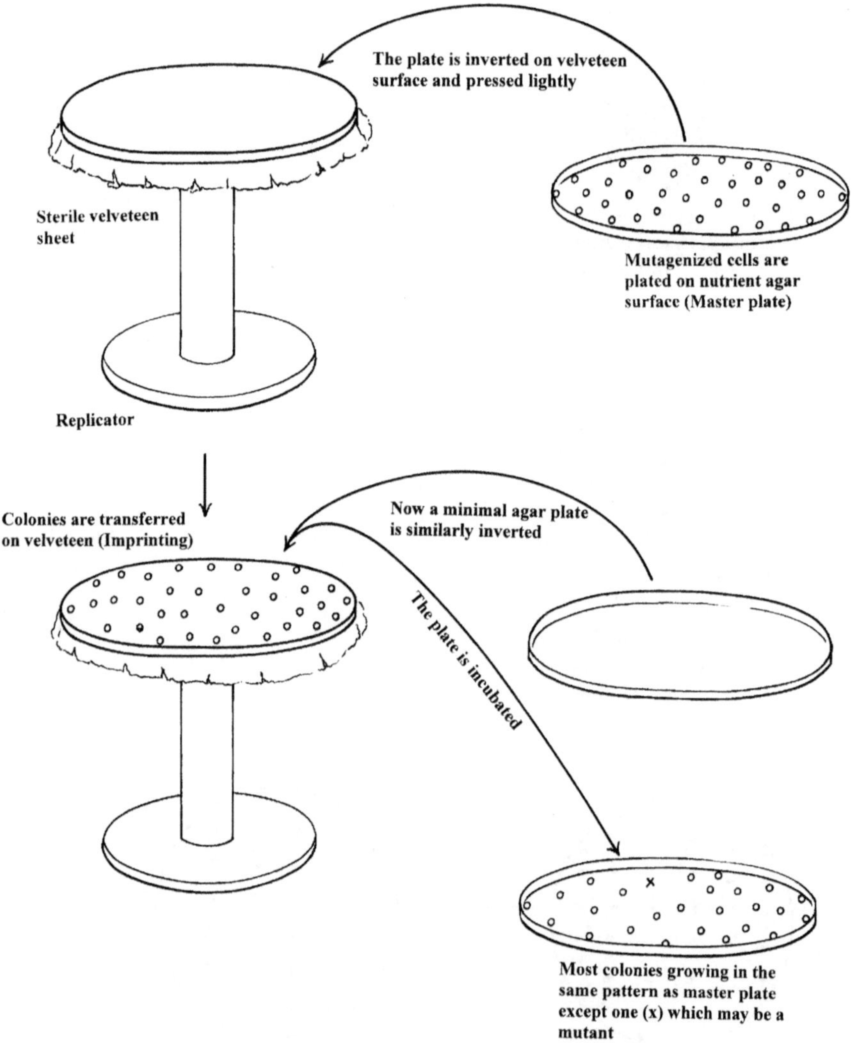

Figure 7.2b. Replica plating for screening of mutants.

Reversion can also result spontaneously, or it could be stimulated by mutagenesis, often by the same mutagen that initially introduced the mutation. Reversion has been employed to differentiate between different mutations. For example, a single point mutation will revert at a comparatively higher frequency than double point mutations and a deletion mutation will not revert at all. The reversion of known mutations has also been used to identify potential carcinogenic agents.

7.2.2. Insertion Mutations

Another class of mutations that can be introduced in bacteria is through insertion of genetic mobile or transposable elements. Three different classes of such elements, viz., insertion sequences (IS), transposons (Tn) and phages like Mu have been described in bacteria. Though many of them are resident members of bacterial genome and/or plasmid, the others can be introduced either through a plasmid or a phage. Such elements have the capacity to insert into genome or plasmid rather randomly and their insertion within a functional gene can lead to disruption of the reading frame and thus to mutation. The mutations are introduced at a relatively high frequency. In fact, majority of the spontaneous mutations in bacteria have now been assigned to the insertion of these elements. For further details about these elements readers are referred to later sections of this Chapter.

7.2.3. Mutagenicity Testing

The susceptibility of microorganisms to chemical mutagens has led to their deployment in testing the mutagenicity of various chemicals. Their usefulness is further promoted as an extremely large population (10^9 cells) can be exposed to the test chemical and analyzed on a single petriplate. Such an experiment with higher organisms would not only be impossible but also impractical.

Bruce Ames designed a test (named after him as Ames Test) using histidine-requiring (*his*) mutation in *Salmonella typhimurium*. Some of the later tests employed *E. coli* and yeasts. The test is based on the capability of a test chemical to induce prototrophy (positive selection). For this, a population of *his* mutant was spread on a nutrient agar plate not containing histidine (or containing traces of histidine). A well was scooped out in the centre and the same was filled with test chemical. The diffusion of chemical created a gradient of concentration. While in the centre, where concentration is very high, the colonies did not develop because lethal mutations might have been induced, but towards the periphery, some prototrophic colonies may appear if the chemical is mutagenic. If the chemical is not mutagenic no colony would form on the agar plate (Fig. 7.3). This has become a routine test in determining the toxic nature of a chemical.

With a high correlation between mutagenicity and carcinogenicity (cancer-causing), Ames test can also be applied for testing a carcinogen. In this test, the potential carcinogen is first incubated with a rat liver enzyme preparation to simulate what normally occurs in the liver. Following this, various concentrations of transformed chemical are mixed with *his* auxotroph and plated to determine if reverse mutation is induced. Once tested positive, the carcinogenic nature of the chemical is further tested.

7.2.4. Analysis of Mutations

With the advent of microbial systems in genetic analysis, isolation and characterization of mutations became an important area of study. This led to a large collection of mutations not only affecting different phenotypes but also the same phenotype. Two

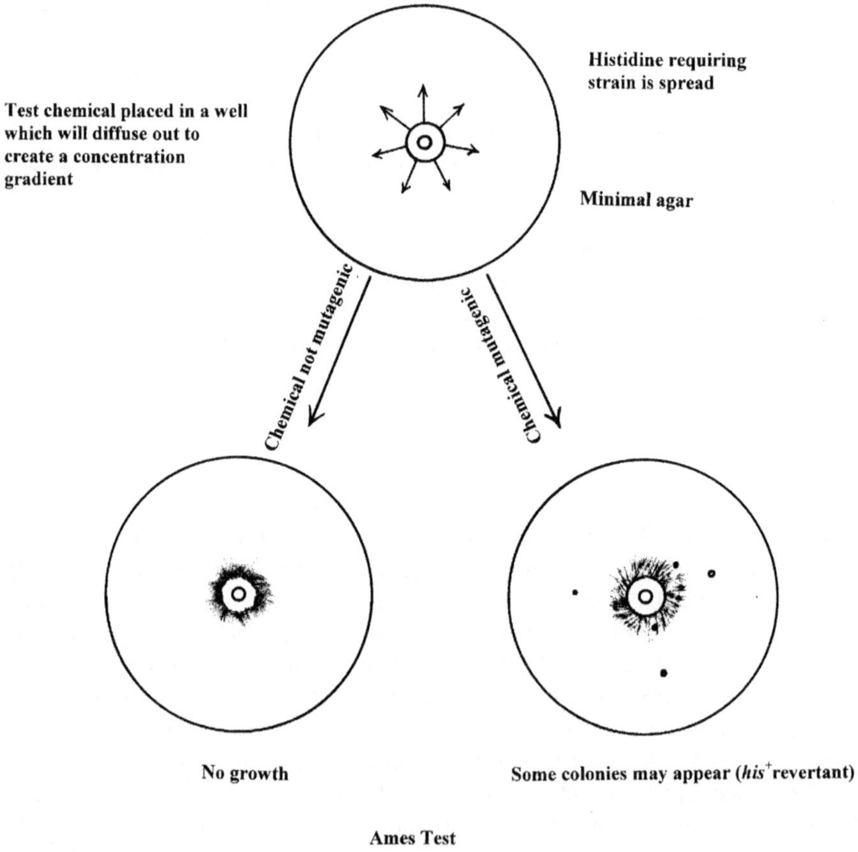

Figure 7.3. Ames Test. A simple mutagenicity assay employed for monitoring the possible hazardous nature of a test chemical.

most important tools in distinguishing the mutations of the latter category are complementation and recombination. While complementation allows you to determine the number of genes responsible for a particular phenotype and to distinguish regulatory genes from regulatory sites, genetic recombination results in physical exchange between two genetic loci to create new combinations.

7.2.4.1. Complementation. A particular phenotype frequently results from the activity of many genes. It, thus, becomes important to determine the number of genes and regulatory elements constituting a phenotype. Although recombination can be used to know such a number, genes affecting the same phenotype are expected to be located very close making recombination a difficult/inconvenient tool. Instead, complementation can be used for this purpose, especially when the mutations are recessive. The basic requirement for this genetic test is that the cell should be partial diploid or merodiploid for the two sets of genes (genes in *trans*). As you will see later in this Chapter, in haploid bacteria this can be done by placing these genes, one set on chromosome and another on plasmid. For example, if a particular phenotype is controlled by three genes, viz., *a, b,* and *c*, the partial diploid could be a^+, b^-, c^-/ a^-, b^+, c^+ or any combination in which atleast one wild-type allele is present on either of the two components. Such partial diploids will exhibit wild-type properties as atleast one wild-type allele of the concerned gene is present in them. In other words non-allelic mutations will always complement each other. By the same analogy, if the two copies of any of these three are in mutant form, such as a^+, b^-, c^-/a^-, b^-, c^+, the phenotype can be of two types. If the two b^- mutations are functionally independent they will give rise to:

i) a wild-type phenotype, if the mutations are in two loci making *b* a multiple locus

ii) a mutant phenotype, if the mutation is in the same locus in both the cases.

In the case (i), the mutations are said to be complementing each other, thus defining separate functional genes. In the case (ii), they identify the same gene and thus do not complement each other. This is a very convenient test as any time a new mutation is isolated, it can be put through a complementation test with the known mutations and placed either along with the earlier known mutations (if no complementation) or to identify a new locus (if it complements the known mutations). If two mutations do not complement each other, they may also mean that one gene may affect the expression of other or its product may inhibit the other gene product.

Let us consider a complementation test involving 10 (1-10) independent mutations affecting a single phenotype. First, the two mutations are placed in partial diploids (= in *trans*) pairwise and their response for complementation (+) or no complementation (-) is recorded, and tabulated (see Table 7.1).

If we now analyze these results, following conclusions can be drawn:

(i) No mutations of the same type (1-1, 2-2,—etc.) complement each other (-), as seen along the diagonal.

(ii) Mutations 1 and 5; 2, 4, and 9; and 3, 6, and 7 do not complement each other but complement all others (1-2, 2-3,—etc.).

(iii) These constitute three complementation groups or cistrons and in effect identify three different functional genes, say, *X, Y, Z*. In other words, members of same cistron do not complement each other but members of different cistrons complement each other.

Table 7.1. Results of a complementation test carried out between the mutants affecting the same phenotype

Mutation No.	*Mutation number*									
	1	2	3	4	5	6	7	8	9	10
1	-									
2	+	-								
3	+	+	-							
4	+	-	+	-						
5	-	+	+	+	-					
6	+	+	-	+	+	-				
7	+	+	-	+	+	-	-			
8	-	-	-	-	-	-	-	-		
9	+	-	+	-	+	+	+	-	-	
10	±	-	±	-	±	±	±	-	-	-

(iv) If we consider mutation 8, it is not complementing any of the mutations, suggesting thereby that it may either be a deletion mutation or a *cis* dominat mutation that may prevent the expression of related genes.

(v) If we consider mutation 10, it is certain that it fails to complement 2, 4, and 9 and thus forms an additional member of the cistron *Y*. However, this mutation weakly complements (±) the others that is 1, 3, 5, 6 and 7. The simplest interpretation can be that the mutated *Y* gene product inhibits the activity of a good copy of *Y*. Because the *Y* activity is weak it does not give a full (+) response with other cistrons. This mutation is said to be negative dominant.

Many multimeric polypeptides consisting of identical subunits may express intragenic complementation. In this, two different mutant polypeptides may aggregate to form a functional or partly functional proteins. Since this happens only with some mutations, it is a very rare phenomenon.

7.2.4.2. Recombination. Recombination is the process by which the genetic material from two related organisms is physically exchanged. This leads to reshuffling of the genes initially present in two different members of a chromosome pair and create new combinations. This activity richly contributes towards generating the biological variations and along with mutation has played a central role in evolution.

Generally speaking, recombination involves the alignment of two related DNAs followed by a cut in one of the strands at homologous point and rejoining of the broken ends in such a way that new combination of genes can be formed (Fig. 7.4). Because of the requirement for fairly strong homology between the participating DNA, this process is also known as homologous recombination. As this recombination mechanism is found universally in all sexually reproducing organisms, it is also referred to as general recombination.

Many microorganisms, including bacteria, do not have well-defined sexual reproduction and have evolved different parasexual mechanisms to bring about genetic variations. Bacteria also offer the cases of non-homologous or illegitimate recombination, in which the participating molecules are not similar in their DNA sequence. This is exemplified by a very limited homology shared by two molecules such as in some phage genome integration in bacterial genome or the case of transposition where this homology requirement may be further diluted. Such a recombination mechanism is also known as site-specific recombination.

7.3. Mechanism of Gene Transfer

The one critical requirement for recombination involves the two participating genomes to come together. Bacteria being haploid have evolved three major mechanisms for gene transfer so that recombination could be carried out.

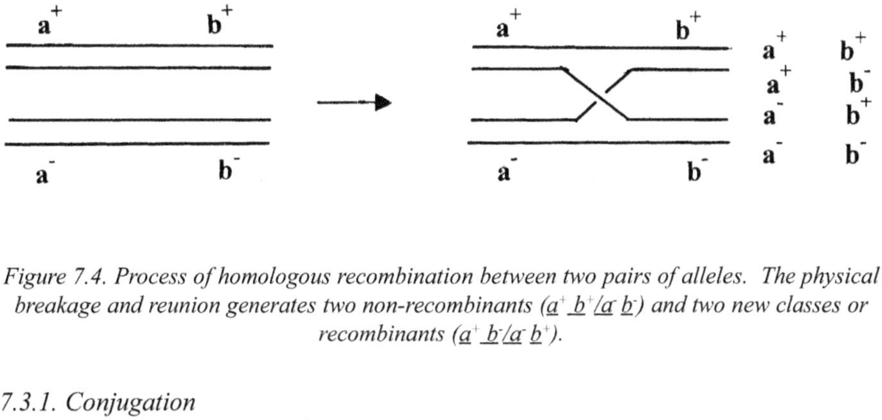

Figure 7.4. Process of homologous recombination between two pairs of alleles. The physical breakage and reunion generates two non-recombinants $(\underline{a^+ b^+}/\underline{a^- b^-})$ *and two new classes or recombinants* $(\underline{a^+ b^-}/\underline{a^- b^+})$.

7.3.1. Conjugation

In order to demonstrate the process of recombination (in effect sexual reproduction), Lederberg and Tatum employed multiple auxotrophic strains of *E. coli* K-12 and set about doing the cross between them. While one of the strains required methionine (*met*) and biotin (*bio*), the other one showed requirement for threonine (*thr*), leucine (*leu*) and thiamine (*thi*). Lederberg and Tatum showed, in 1946, that when cells of these two strains were mixed and spread on minimal medium, prototrophic colonies appeared at a low frequency of 10^{-5} to 10^{-6} (Fig. 7.5). These colonies continued to grow on minimal agar suggesting that some sort of reassortment of genetic material (recombination) between the two genotypes had occurred in the following way:

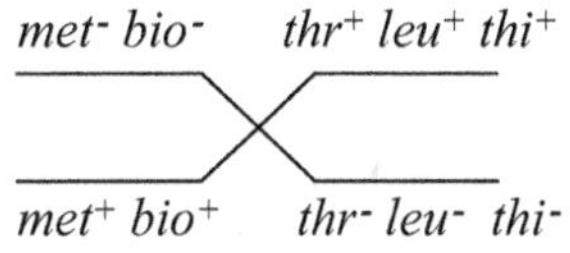

Of the different genotypes that can be generated through recombination between any two genes, one prototrophic type could grow on minimal medium. Since the process involved coming together of two single-celled organisms, it came to be called as conjugation.

That conjugation indeed involves cell to cell contact was demonstrated by Davis (1950) using an U-tube set-up (Fig.7.6). In this, the two participating strains were placed in two arms of the U-tube separated by a sintered glass filter. This filter completely inhibited the passage of the cells and thus effectively the contact between them. When some material from two arms were plated separately no recombinants could be detected. The removal of filter led to the reappearance of prototrophic recombinants.

Lederberg had earlier (1947) demonstrated that some unselected markers in the earlier cross showed a tendency to remain associated with a selected marker. Thus, *E.coli* appeared to have the linkage phenomenon similar to that of higher organisms.

Lederberg, Cavalli and Lederberg (1952) also showed that bacteria have clearly evolved mating type differences, though may lack morphological differentiation of sex. Accordingly, the strains were referred as F^+ or fertile and F^- or infertile. For any cross to be successful, the mating should involve one F^+ and one F^- cell. The experiments conducted by Hayes (1952) further clarified the functional differences between F^+ and F^- cells. He showed that treatment of F^- cells with UV irradiation or antibiotic streptomycin strongly inhibited the recombinant production but the same treatments of F^+ did not show any effect. He, therefore, suggested that F^+ acts as donor or male and F^- as recipient or female with the viability of F^- being crucial as this strain produces the recombinants. Moreover, there is only unidirectional transfer of genetic material i.e., from donor or F^+ to recipient or F^-. The analysis of recombinants suggested that male genetic material is not transferred completely and, therefore, the recombinants may show a bias towards the markers of F^- strain. In the absence of complete transfer of male genetic material, the zygote so formed is partial diploid or merozygote. In other words, the portion that is transferred appears in diploid condition and the recombinants would be restricted to the same.

Lederberg *et al.* (1952) were able to demonstrate that F^+ transmits the mating type infectiously converting all the F^- cells with whom it sponsors mating, into F^+, but its efficiency as a genetic donor is low. This characteristic of F^+ was assigned to a sex factor or F factor by Hayes that is absent in F^- cells (Fig. 7.7).

A new class of donor strain was subsequently isolated by Lederberg and his group and Hayes from the original F^+ strain. This donor gave a high frequency (10^3 times more than F^+) recombination with an F^- strain and thus named Hfr. In compatibility test, it behaved like a donor, but did not convert the recipients as was shown with F^+. Moreover Hfr could also revert to F^+ condition. The genes which are transferred early are characteristic of an Hfr and limited. Thus, while some markers are transferred with high efficiency, others still exhibit a gradient of high to low, typical of $F^+ \times F^-$ crosses.

The actual nature of chromosome transfer by Hfr donor was revealed by Jacob and Wollman (1955) using "interrupted-mating" procedure. In their experiment, mating was disrupted at different times by vortexing the mating population vigorously. The cells were then plated to reveal their genotypes. Jacob and Wollman showed that each Hfr

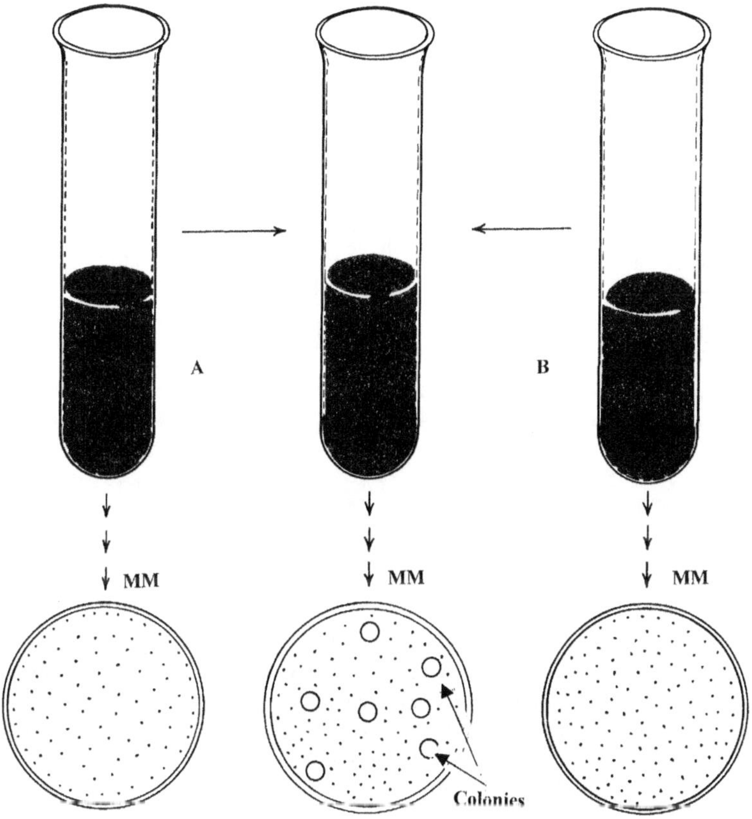

Figure 7.5. Lederberg & Tatum's experiment that demonstrated the occurrence of sexual reproduction in the bacterium, E. coli, based on generating genetic recombinants. Two multiple auxotrophic strains, A and B fail to grow on an unsupplemented nutrient medium (MM). However, when they are mixed for a few hours and then plated, a few colonies appear on the agar plate. These represent the recombinants in which the exchange of genetic material has taken place enabling them to synthesize all the required nutrients.

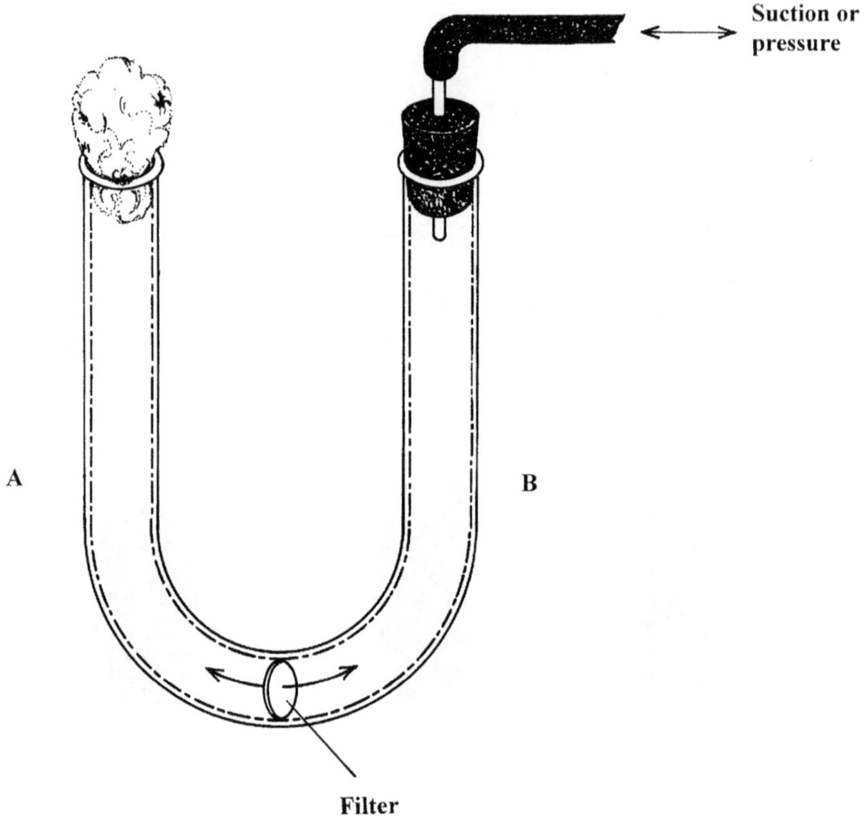

Figure 7.6. Davis' U-tube to demonstrate that cell to cell contact is necessary for producing recombinants (conjugation). The two strains, A and B, described above, were placed in the two arms separated by a sintered glass filter. On applying the suction or the pressure, liquid can pass through the filter but not the bacterial cells. The cells plated after several hours of incubation from either arm did not give any recombinants. However, if the filter is removed so that the cells are free to move, recombinants reappear.

marker appears at a characteristic time. In general, the transfer begins at a specific point (origin of transfer or *Ori*T) and proceeds linearly till the time two cells remained together (Fig.7.8). The observed kinetics suggested that there is a distinct relationship between the time at which a given marker is transferred from Hfr to F^- and its map location on the chromosome. Moreover, the closely linked markers appeared at the same time and distantly located at different times. In subsequent experiments, the time of entry of several other donor markers was ascertained that led to the construction of a temporal map of *E.coli* where the genes are located on a time scale (in min.) (Fig.7.9). It also became clear that chromosomal transfer by Hfr is never complete and is broken leading to only a

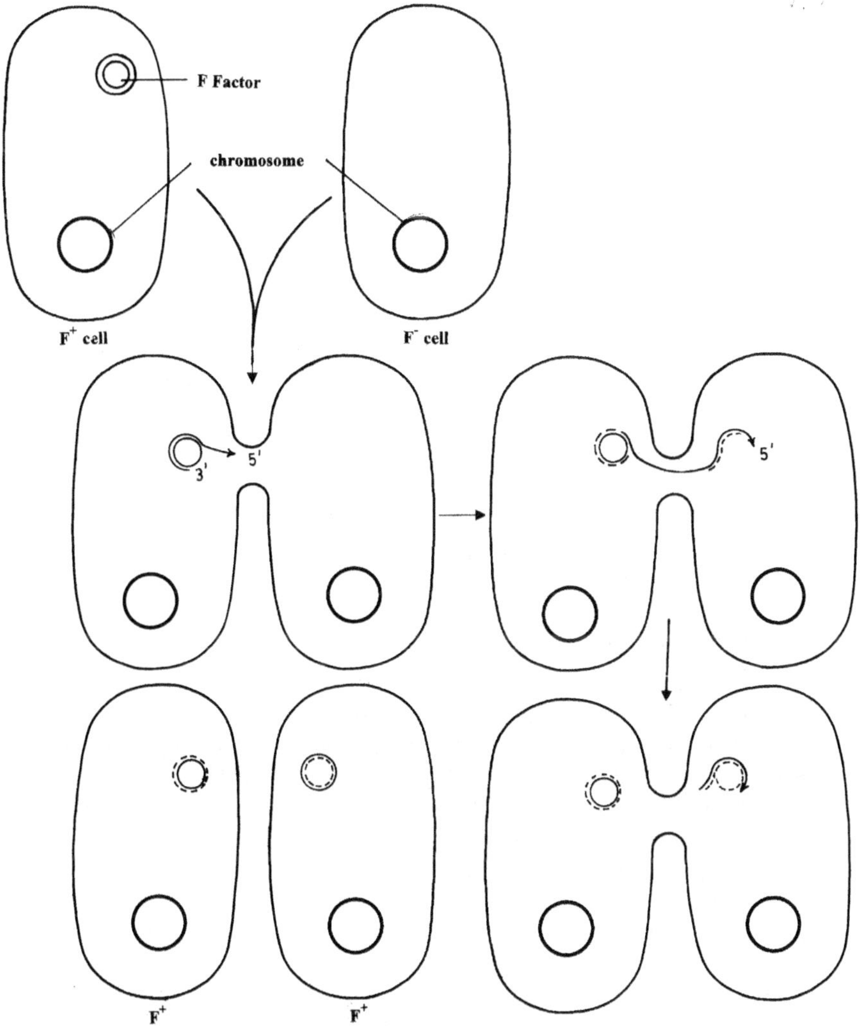

Figure 7.7. An F^+ (donor/male) x F^- (recipient/female) cross in bacteria. The circular F factor (plasmid) is nicked and transferred in a single stranded state to the F^- cells, where the complementary DNA strand is synthesized. The resulting cells (exconjugants) are F^+ or functionally male. In other words, F^- is converted to the F^+ state (infectious transfer).

part of the same being received by the recipient. This explained why only few markers are transferred efficiently as they must have been close to *OriT* (proximal or early marker), whereas those which are far (distal or late markers) do not get a chance to get transferred and, thus, to appear among recombinants. The resulting structure, therefore, is a merozygote.

In subsequent years, many other Hfr strains were isolated from different F^+ strains. Each Hfr had a characteristic order of transferring the markers which in all cases was sequential and linear and once again limited (Fig.7.10). On comparing these gene orders, it could be established that all markers of *E. coli* are linked in a single group and that the bacterial chromosome is circular (Fig.7.10). Consistent with the later experiments and to provide an uniformity, the current *E. coli* map has been standardized for 100 temporal map units. It coincides with the time taken for complete transfer of Hfr chromosome to F^-.

While on a 100 minute temporal map, the genes are ordered on the basis of time units in relation to a specific *Ori T*, conventional recombination maps can also be generated. The most popular method is to select for Hfr marker that transfers late and then testing these recombinants for markers that must have entered ahead. If these recombinants showed the proximal recipient marker that can be generated only when a crossing over occurred between them. The recombination frequency can be calculated by the following example:

$$\textbf{Hfr}\ (a^+ b^+)\ x\ \textbf{F}^-\ (a^- b^-)$$

select for b^+ (distal marker) and check for segregation of *a*

$$\textbf{RF} = \frac{\textbf{Total no. of recombinants}\ (a^- b^+)}{\textbf{Total no. of population}\ (a^- b^+) + (a^+ b^+)}$$

From the behaviour of the F^+ and Hfr strains of *E. coli*, it is clear that in F^+ cells, the F factor/plasmid exists in autonomous state, independent of the bacterial chromosome. But in Hfr, it is integrated at a specific but limited site, characteristic of each Hfr. Taking clue from Campbell's model of phage genome integration in bacterial chromosome, it was suggested that both bacterial chromosome and F plasmid are circular so that a single crossover between the two is sufficient for integration. Implicit in this model is the fact that reversal of this event will lead to the release of the F plasmid. Both genetic as well as physical evidences are available now to confirm this model. The sites of integration were subsequently identified to be as IS with F plasmid carrying two copies of IS3 and one copy each of IS2 and ϒδ (Tn 1000). With bacterial chromosome carrying several copies of these homologous elements, they can be employed by F plasmid for integration (and also excision). The location and orientation of these elements infact determine the nature and orientation of the chromosome transfer by an Hfr. It is now clear that chromosome transfer by Hfr is initiated by breaking (nicking) the chromosome at *OriT* of integrated F plasmid. The *OriT* along with a part of the F plasmid (leading region), followed by the chromosome is transferred to F^- cell. The main body of the F plasmid comes at the tail end, rarely gets transferred and,

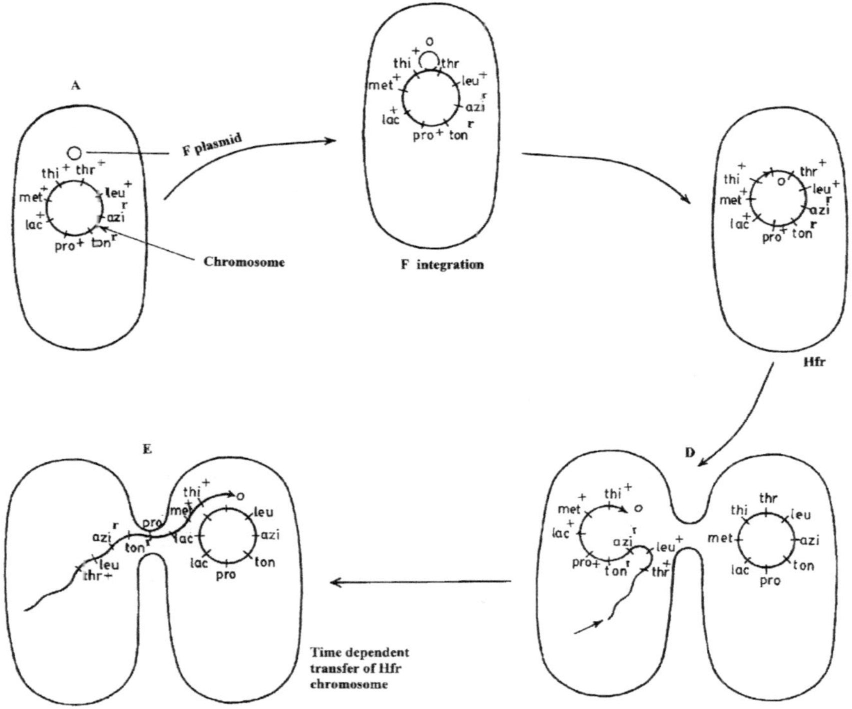

Figure 7.8. The sequential linear gene transfer carried out by the Hfr donor. This involves the integration of F factor into the chromosome (A-B) to generate Hfr (C) which conjugates with an F- cell. The Hfr chromosome is nicked at *oriT,* *located on the F factor (D) and then transfer begins with* *oriT* *entering first and subsequently dragging the chromosome along (E). The extent of chromosome transfer is governed by the time that the two cells remain in contact. The main body of the F factor comes at the tail end, which seldom enters the recipient and thus no conversion takes place.*

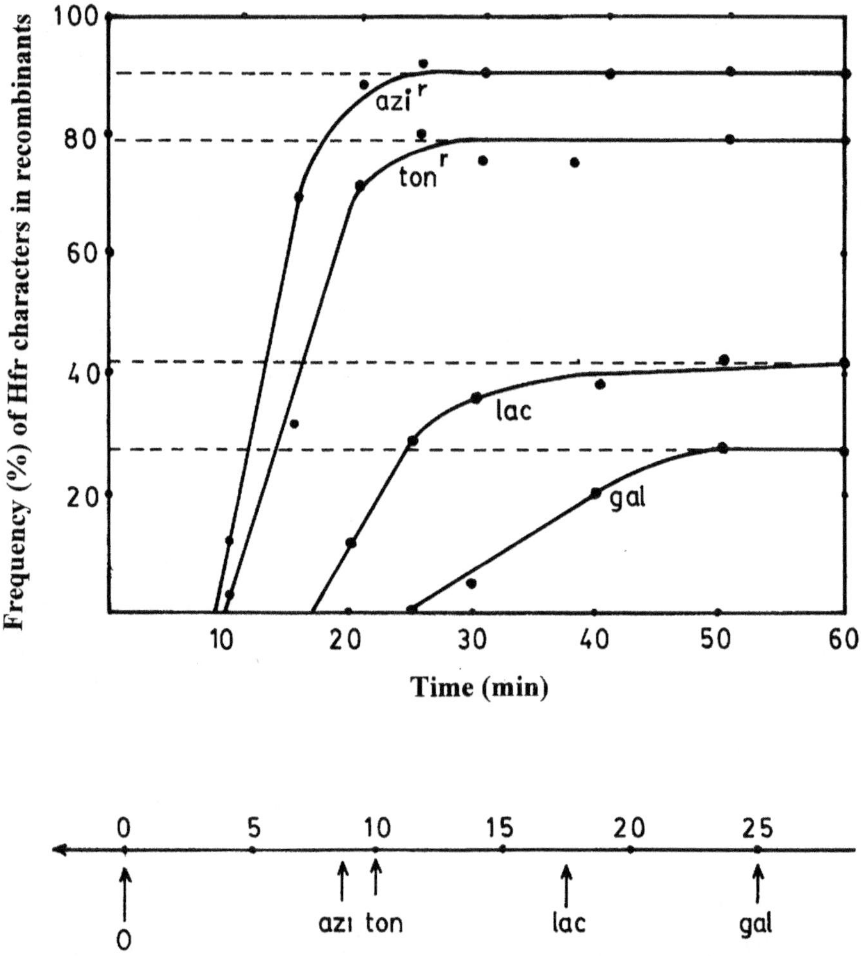

Figure 7.9. Temporal gene mapping in E. coli. The genes are mapped on the basis of time (in minutes) at which they enter the recipient cells. This technique was devised by Wollman, Jacob & Hayes on the basis of the "interrupted mating experiment". The time at which each of the donor markers begin to appear amongst the recombinants forms the basis for the construction of a linkage map.

therefore, no conversion of F⁻ usually takes place (Fig. 7.11).

In 1960, another type of donor strain was isolated by Adelberg and Burns from an Hfr strain. This strain conferred donor ability to those recombinants gaining the early marker, something that is not possible with Hfr. Thus, F⁻ strain was not only infectiously converted but also gained bacterial marker/s. This modified sex factor, referred to as F-prime (F′) was explained to have arisen through an illegitimate genetic exchange during excision that took place between a site located on F factor and another on chromosome (Fig.7.12). This led to an F factor that carries flanking marker (proximal

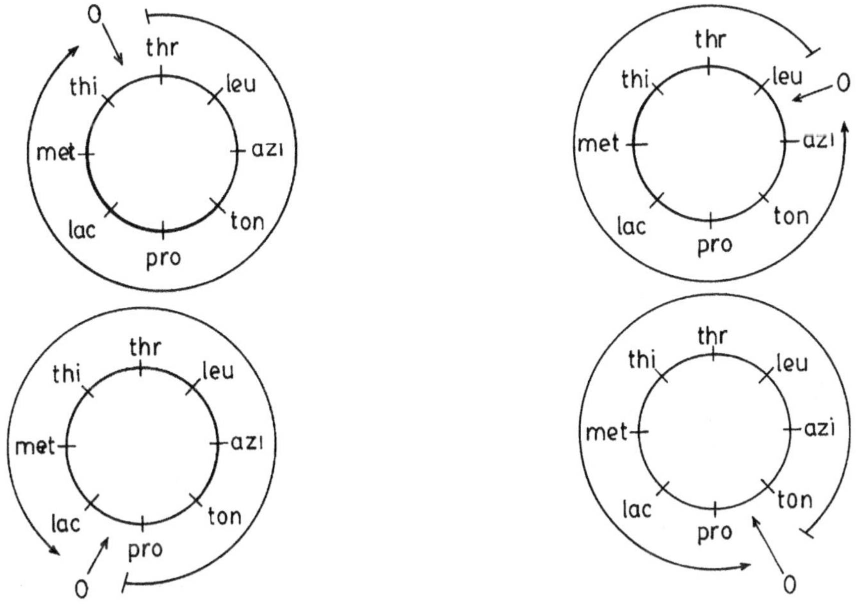

Figure 7.10. F plasmid can integrate at different points (→0) and orientation in (outer broken circle) the bacterial chromosome, generating Hfrs with characteristic orientation of gene transfer. The differences in the orientation, however, do not affect the established linkage except for those which get separated by the F insertion. These results suggested that bacterial chromosome is circular. The figure depicts only hypothetical sequence.

or distal) of the chromosome but leaves a part of its own into the chromosome. Besides this type of exchange often called as Type I, another or Type II exchange will lead to F′ picking bacterial markers from both the flanks along with complete F plasmid (Fig.7.12). When an F′ is transferred to a recipient cell, the plasmid is integrated by using the homology between the bacterial marker carried by F′plasmid and recipient genome.

F-prime factors have played important roles in genetic analysis of bacteria, and, therefore, have been sought after by the geneticists. This is because, when F′ factor is introduced in an F^- cell carrying a copy of the similar chromosomal genes, the bacterium

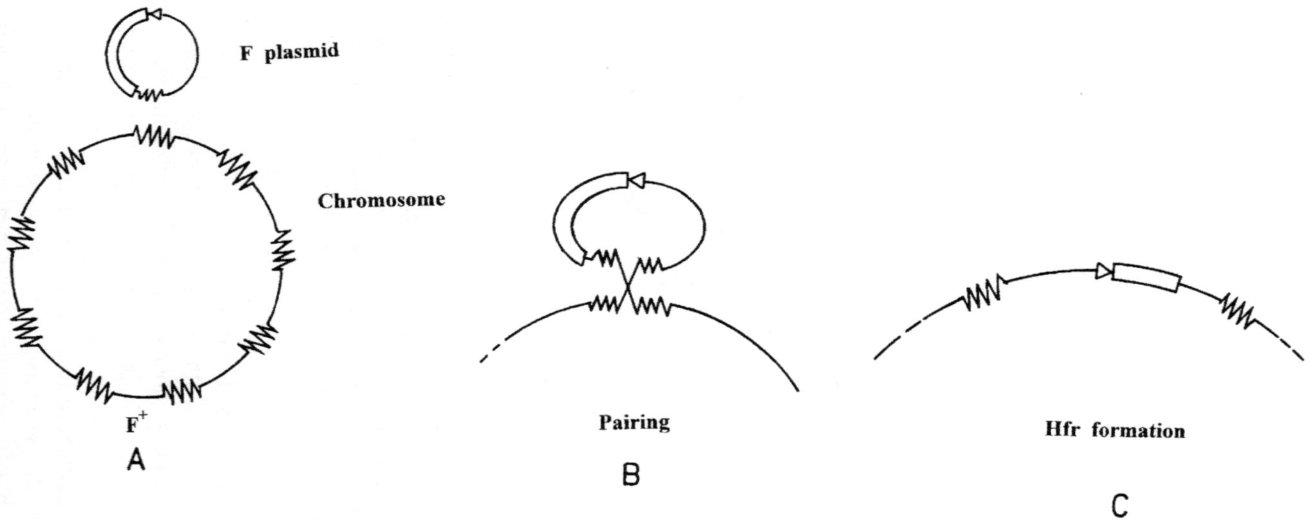

Figure 7.11. Insertion of F factor into the chromosome involves homologous regions (⌃⌃ or IS). Such regions are used for pairing and then crossing over takes place to result in insertion. If both the structures are circular single crossover is sufficient for this event Note: excision will also involve these regions.

becomes partial diploid for this region. Thus, heterozygosity for different alleles can be created and dominance relationship can be worked out. Similarly, the requirement for *cis*-acting and *trans*-acting components for gene expression can be clarified by placing them on these two genetic entities.

Various genetic and physical analyses have confirmed that F-factor/plasmid like bacterial chromosome is covalently closed circle of ~100 kb in size. The structural plan of F plasmid (Fig.7.13) consists of (clockwise): (i) the 20kb region that contains 2 copies of IS3 and one each of IS2 and Tn 1000 (ϒδ), (ii) there are three replication initiation regions; Rep FIA consists of both unidirectional (*Ori S*) and bidirectional (*Ori V*) replication origins, Rep FIB can sustain plasmid replication in the absence of RepFIA, and Rep FIC represents an incomplete remnant of a replication system functional in some other related plasmids, (iii) the *Ori T* is located in a region called leading region, and (iv) a large region, called as transfer region, consisting of ~over two dozen genes that control pilus biogenesis, mating aggregate stabilization, DNA metabolism, surface exclusion, fertility inhibition, and regulation.

The ability of the Hfr, F^+ or F' to act as donor by having F plasmid in free or integrated state has also been traced to characteristic hairy appendages possessed by the donor. *E. coli* cells are surrounded by appendages called type I pili, but the male cells contain another type of pilus called F-pilus or sex pilus. Many bacteriophages showing preference for male cells also adsorb to F-pilus. Each male cell may carry 1-5 pili which are 1-2 μm long and grow perpendicular to the cell. They are hollow rods of an external diameter of 85 Å and an axial hole of 20-25 Å diameter consisting of only one protein, pilin. Many other plasmids like R and I also produce pili of their type. From the observation that inhibition of F pilus leads to inhibition of conjugation, it was inferred that they provide the conduit for cell to cell contact (Fig. 7.14) and chromosomal transfer. While, the experimental evidences support the first function, DNA transfer through pilus has not been established so far. Later studies have suggested that they bring the two cells together, help in making cell aggregates, which triggers the conjugation specific events. Surface proteins of the recipient cell may also play an important role.

One convenient method of distinguishing the donor strains as Hfr or F^+ (or F') is to effect plasmid curing. Many agents like acridine dyes are known to affect plasmid replication by stacking themselves into the DNA. While this action leads to the loss of free plasmids (F or F'), no effect is seen on Hfr.

As mentioned earlier, the transfer of donor bacterial chromosome commences at *ori T*, located in the F plasmid that is under the control of integrated sex factor. Two classes of models were initially proposed to explain the chromosome transfer. In one, the donor chromosome undergoes replication prior to its transfer. However, in the absence of the nature of the signal that will be required for replication, this model was not considered tenable. In the other, the replication, was considered concomitant with transfer. In its original form, this model relied on the transfer of one of the replicas of the donor chromosome, later it was also applied to the situation where only one strand of donor DNA is freed for transfer. Under both conditions, cell to cell contact provided the necessary signal for replication. Various genetic, biochemical and biophysical evidences clearly suggest that donor DNA in all cases is transferred in a single stranded

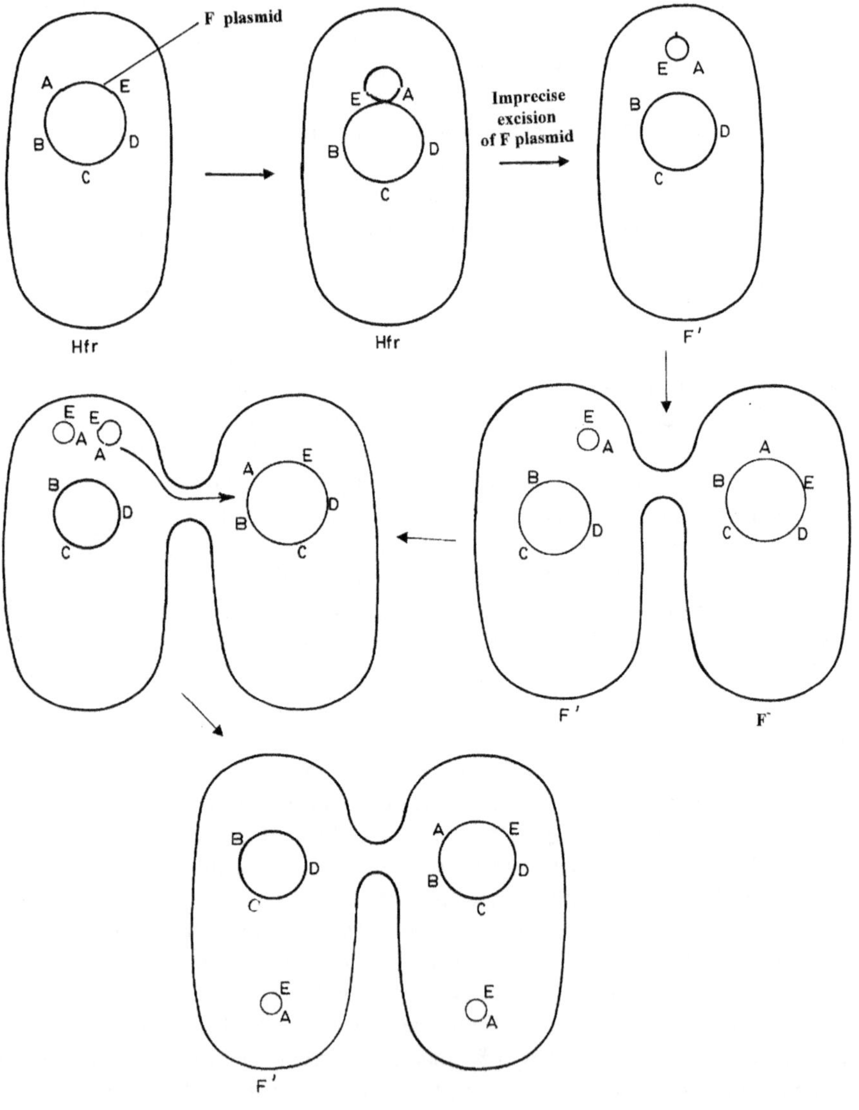

Figure 7.12. The formation of F′, a new class of donor strain. The F plasmid may be excised from the Hfr (A), and if this excision is not precise it can pick up one or few bacterial genes and leave some of its own in the chromosome (B,C). When such a cell with modified F or F′ mates an F⁻ (D), the latter becomes a partial diploid or merozygote for the genes carried by F′.

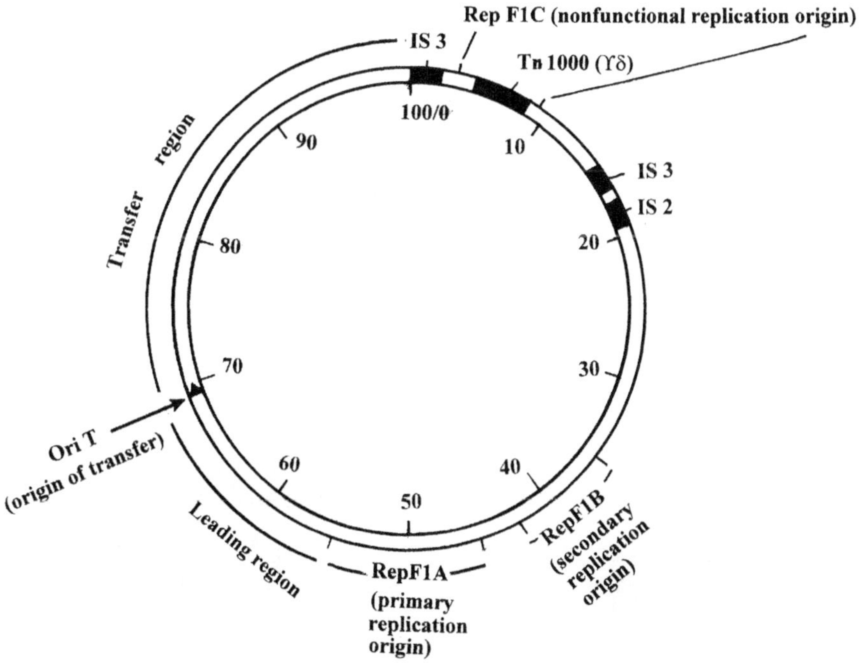

Figure 7.13. Structure of F plasmid (100 kb in size).

state. Such a single stranded DNA is generated by a nick at *ori T* site in F and 5' end enters the recipient first. The 3'-OH end is used for adding deoxynucleotides by a rolling-circle mechanism. This single stranded DNA in F^- cell provides a template for lagging strand synthesis. The leading 5' end is attached to DNA helicase I (*traI* product) in the intercellular junction so that in effect a growing loop of partially duplex DNA is presented to the recipient. When mating terminates, recipient will contain a linear fragment of F plasmid DNA (leading region) and a single strand overhangs of variable length at the distal 3' end.

In order to integrate the donor fragment recombination must take place. One major requirement for this is to have a single stranded region in one of the two DNAs, to which RecA protein can bind and initiate strand exchange. These single stranded regions are created by RecBCD enzyme, a multifunctional protein that loads onto DNA with flush or nearly flush duplex ends and provides both helicase and nuclease activity. Rec BCD, therefore, unwinds and degrades both strands as it tracks along the duplex molecule until it encounters a *chi* (5'-GCTGGTGG-3') sequence. The nuclease activity is then modulated so that as the enzyme proceeds beyond *chi*, the strand ending 3' is no longer degraded and could be recruited by RecA to initiate exchange. Alternatively, recombination could also initiate by making use of a transient single stranded gap in the loop of Hfr DNA being transferred. If RecBCD enzyme initiates precisely two exchanges, one near each end, the recombinants formed will integrate the Hfr DNA as a single fragment. Some 80% of the recombinants have been suggested to be derived from this type of event often called as "long-chunk integration". In contrast, the remainder seem to arise by both crossovers located near the selected marker or "short-chunk integration" (Fig. 7.15). After RecA-mediated strand exchange is completed, a Holliday joint or junction is created. The RuvAB and RecG proteins catalyze branch migration of the Holliday junction creating a heteroduplex. RuvC protein resolves Holliday junction by endonuclease cleavage across the point of strand exchange to give either "patch" (noncrossover) or "splice" (crossover) products.

The type of conjugation mechanism described above is not found in all bacteria. The F plasmid has a limited host range and can replicate only in enteric bacteria. Many conjugative plasmids may have self-transfer property but cannot mobilize the resident

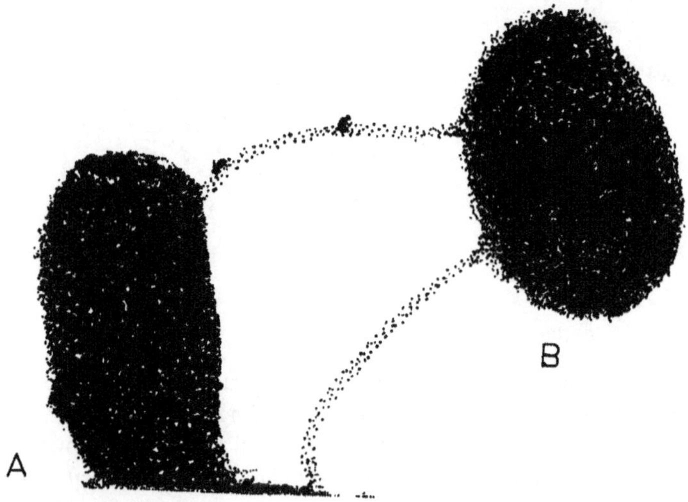

Figure 7.14. The diagrammatic representation of an electron micrograph showing two conjugating E.coli cells connected with long appendages or sex pili.

chromosome. Some such plasmids can, however, acquire this property by gaining insertion sequences. For example, plasmid R68.45, a broad-range mobilizable plasmid acquired this feature, once it gained a copy of IS21. Hfr-like strains have also been described in *Pseudomonas aeruginosa. Agrobacterium tumefaciens* that causes crown gall tumors in plants, transmits the plasmid (Ti) DNA during infection. This plasmid is not replicated in the plant cell and thus maintains itself by integrating a portion

(T-DNA) in plant genome.

Conjugation is less documented in Gram-positive bacteria. In *Enterococcus faecalis*, plasmid-carrying cells form clumps with plasmid-free cells and plasmid transfer occurs within these clumps. The plasmid-containing donor and plasmid-free recipient cells produce a binding substance on their surfaces. The donor cell additionally produces an aggregation substance (adhesin), when it comes in contact with the recipient. In effect, the recipient produces a small peptide hormone, pheromone, that diffuses out of the cells, enters the donor, and induces the plasmid encoded adhesin production. The aggregation substance helps in making the cell clumps.

Self-transmissible plasmids have been found in many types of Gram-positive bacteria, such as *Bacillus, Streptococcus, Staphylococcus,* and *Streptomyces*. The *tra* regions in most cases is much smaller. Many Gram-positive bacteria also possess self transmissible conjugative transposons (refer later section). They also encode Tra functions but do not mediate chromosomal transfer. Many plasmids of Gram-positive bacteria are promiscuous. For example, the plasmid pAMb1, of *E. faecalis* can transfer into *S. aureus* and other Gram-positive bacteria.

7.3.1.1. Application of Conjugation. Conjugation has provided bacteria one important means of homologous recombination. This mechanism, therefore, has contributed significantly to bacterial evolution, as in other sexually reproducing organisms by creating the necessary variability. Also this method is used for mapping exercise as described above.

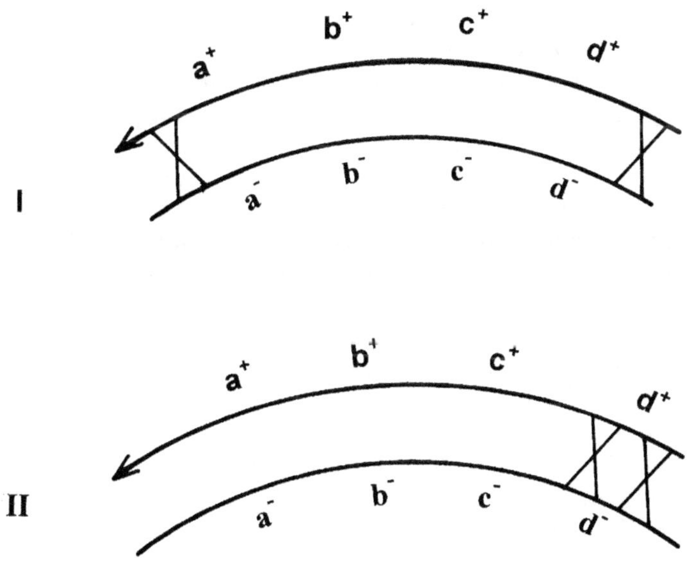

Fig. 7.15. Two types of integration in conjugational recombination. I. long chunk integration-the crossover points occur near the ends of the fragment leading to the integration of whole fragment, II. short chunk integration- the crossover points occur around the selected gene only, e.g., gene $\underline{d}^+$, that results in recombinants inheriting $\underline{d}^+$ only.

While conjugation is not popular amongst bacteria, two parasexual pathways have been evolved by them to effect gene shuffling. They are transformation and transduction.

7.3.2. Transformation

Transformation was the first mechanism of gene transfer described in bacteria. In 1928, Fred Griffith found that pathogenicity of the pneumococci, *Pneumococcus* (=*Streptococcus*) *pneumoniae* could be correlated with the presence of a capsule around the cell. Such virulent strains produced smooth glistening colonies and were referred to as S strain. In contrast, the avirulent, rough-colony (R) strains lacked the capsule. Many serotypes, SI, SII, SIII etc., were identified amongst S strains based on the differences in the composition of the capsule. S strains could give rise to R strain by mutation and R may revert to S type. During these mutational changes, serotype specificity is not lost. For example, SI will give rise to RI and SII to RII but never to RIII and so on. Griffith had also observed that heating a S type bacterial suspension at 80°C leads to the loss of virulence. Thus, when he injected mice with either live RII or heat-killed SIII type bacteria, mice did not show any signs of disease. However, when they were injected with a mixture of live RII and heat-killed SIII bacteria, the mice died (Fig. 7.16). From the blood of dead mice, Griffith could isolate live SIII type of bacteria. He, therefore, concluded that dead pathogenic bacteria, released a 'transforming principle' that transformed the live non-pathogenic RII into SIII virulent bacteria. This change appeared to be a permanent one as such SIII bacteria continued to produce the same type of cells.

Later, many researchers refined Griffith's experiment by designing an in-vitro transformation system and demonstrated transformation of R to S type bacteria by cell-free extracts from S bacteria. Then about 16 years after Griffith's work, Avery, MacLeod and McCarty (1944) demonstrated that the transforming principle is nothing but DNA. Griffith's experiment is thus considered one of those classical experiments that have revolutionalized biology. Not only it led to the chemical identification of the genetic material as DNA but also settled the controversy that DNA and not protein (or any other chemical constituent) is the hereditary material. In the meantime, transformation was extended to some more bacterial genes but the real impetus came when penicillin-sensitive pneumococci could be transformed to penicillin resistant state by treating the former with DNA derived from the latter, basically because, an easily scorable phenotype was available in comparison to complex pathogenicity-related character.

Transformation is basically classified into two types: natural transformation and artificial transformation.

7.3.2.1. Natural Transformation. Many bacteria have evolved transformation as a natural mechanism of gene transfer. The three sps. viz., *S.(=P.) pneumoniae, Bacillus subtilis* and *Haemophilus influenzae* are the best known examples of this category, and of the three, *B. subtilis* proved exceptionally useful as it provided a number of markers for experiments and in being genetically better known than the other two.

Transformation involves three main steps:

1) the release of transforming DNA by donor
2) the uptake of donor DNA by recipient

3) the processing of donor DNA by recipient to produce a changed phenotype (transformant).

I - Donor DNA- In all naturally-transforming systems, the donor cells release DNA at a particular time that can be used as donor DNA. For laboratory investigations, this will demand the isolation of DNA from appropriate donors.

II - Uptake- The uptake of donor DNA is a multi-step process:

(i) Competence: It was observed from the beginning that recipient cells are not able to take up DNA at all times. It thus appeared that recipient cells acquire a state called competence when they could take up and process the donor DNA. Two major changes are expected to take place during competence. These are permeability changes and/or the synthesis of the receptor sites. In the three naturally transformable systems, competence reaches at the end of the active growth phase (late exponential phase), suggesting that a critical cell concentration is important. While in *Pneumococci* and *Haemophilus* almost 100% cells become competent, this phase lasts only 15-30 minutes, in *Bacillus* only ~20% cells become competent but the competence phase remains for several hours. One interesting aspect of competence in *Pneumococci* is that it is transferable, and that the transfer rate is higher than growth rate. The spread of competence has been assigned to a proteinaceous competence factor (CF) synthesized by the cell. Inhibition of protein and RNA synthesis during treatment with CF prevents acquisition of competence. Inhibition of DNA synthesis, on the other hand, has no effect. Pneumococcal cells also fail to respond to CF if cells are deprived of choline, a component of the cell wall. Also, autolytic enzyme degrades the cell wall that contain choline. It is, therefore, suggested that CF acts upon the cell by causing a limited release of autolysin into the periplasmic space that weakens the cell wall and/or exposes the receptor sites on the membrane so that they become accessible to donor DNA.

In *B. subtilis*, the physiological difference between competent and non-competent cells was observed in a variety of ways. The competent cells appear to be smaller and less dense and are penicillin resistant. The latter suggested that they are non-growing. While protein synthesis is normal, DNA synthesis is greatly reduced. *B. subtilis* has also been employed to identify the genes whose products are required for competence and to understand its regulation. The regulation of competence is brought about by a two-component regulatory system (as described in **Chapter 6**). ComP, a sensor protein lodged in the membrane perceives the signal from exhaustion of nutrients and high cell density. This causes the ComP to autophosphorylate and then transfer phosphate to ComA, the response regulator component. Phosphorylated ComA acts as transcriptional activator of several genes including *srfA* which is required for the expression of other competence genes (Fig. 7.17).

B. subtilis cells also produce a small peptide called competence pheromone. The effect of this pheromone is concentration-dependent and the desired effect is produced when cell density is high (late exponential/early stationary phase). This pheromone peptide is derived from a longer peptide, the product of *comX* gene. The *comQ* gene, located upstream is proposed to provide the protease for this trimming reaction. The pheromone triggers the phosphorylation of ComP. There is another interesting correlation between acquisition of competence and sporulation in *B. subtilis*. When the critical

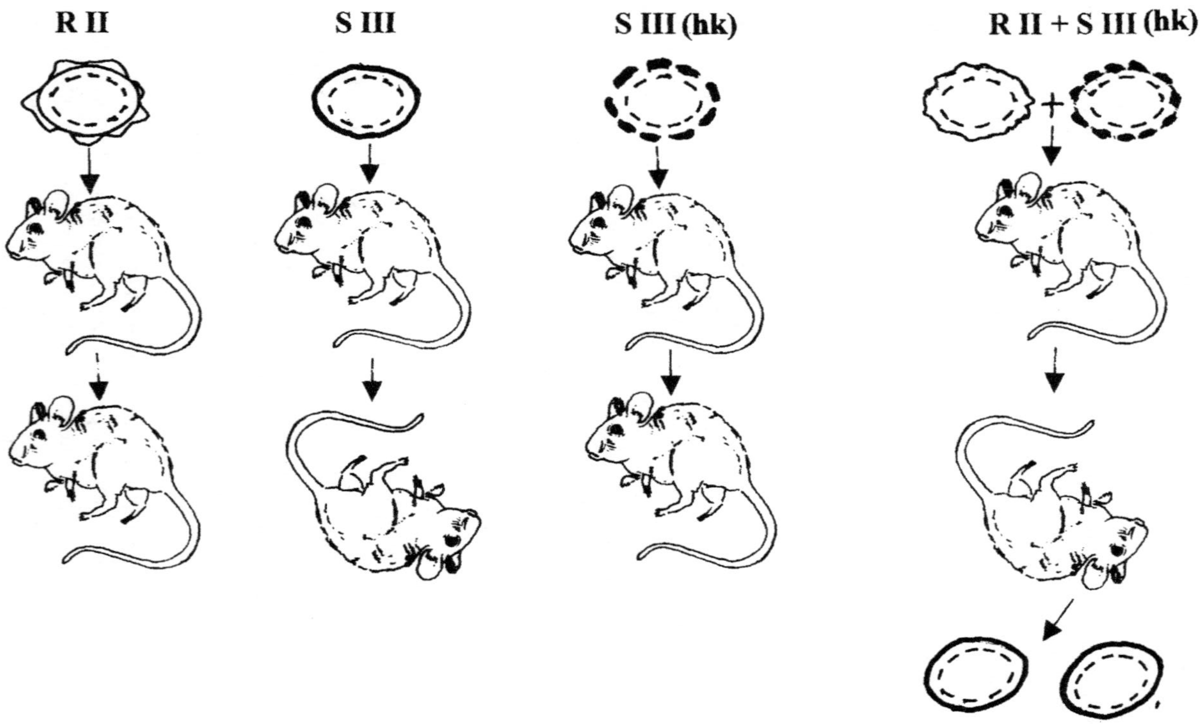

Figure 7.16. Griffith's experiment on transformation in <u>*Pneumococcus*</u>*. The avirulent RII strain injected in mice does not produce disease symptoms, but virulent SIII strain can cause fatal infection. The virulence of the SIII strain can be lost by heating the bacterial suspension at ~80°C. A mixture of RII and heat-killed (hk) SIII bacteria, however, becomes virulent and live SIII bacteria could be recovered from the blood stream of the infected dead mice. The last set showed that a material released from hk SIII (a transforming principle) transformed the live RII bacteria into SIII type.*

cell density reaches, the bacteria also begin to sporulate. The two processes appear to share some regulatory genes. Some of the gene products are expected to modulate cell permeability and some that may bind DNA. Polyhydroxybutyrate (PHB), a storage polymer, has been shown to form some channels in the membrane for DNA entry.

(ii) Uptake of donor DNA: In transformation, the donor cell is only required to release the DNA. Since recipient will take up naked DNA, experiments can be designed to understand parameters affecting the same. For example, it could be possible to show whether DNA should be in double or single stranded state, is there any minimum or maximum size of DNA, is any host specificity required during uptake, and what is the fate of donor DNA upon its entry into the recipient cells?

Transformation starts with a random collision between the competent recipient cells and donor DNA. Donor DNA is first bound reversibly and then is converted into an irreversible bound state. While in the former state DNA can be released by a variety of means, the latter is identified by the acquisition of DNAase resistance. The uptake of DNA is concentration dependent upto a certain level. Thus, increasing the DNA concentration causes a proportional increase in the number of transformants till it reaches a plateau, which reflects the saturation of the recipient's ability to take up any more donor DNA. The duplex (ds) state of the DNA is most preferable form for uptake. Although the cells may take up single stranded (ss) DNA, the conditions required may be nonphysiological and, therefore, deleterious. DNA uptake is energy-requiring and is therefore inhibited in the presence of metabolic poisons. Similarly near neutral pH is optimum for duplex DNA uptake.

The requirement for specificity of source of DNA is variable. For example, *Pneumococci* and *B. subtilis* can virtually take up any DNA, others like *H. influenzae*

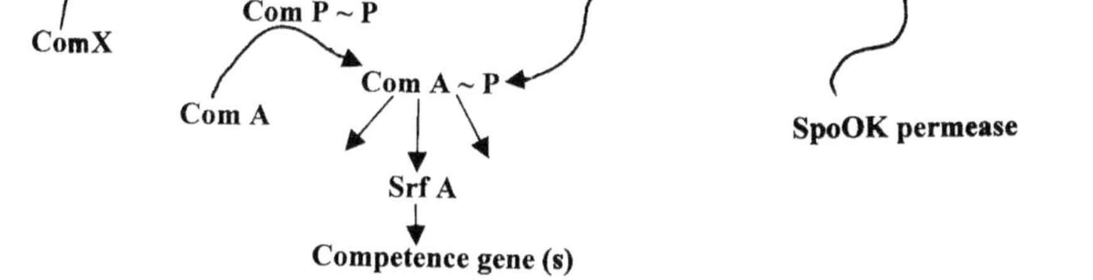

Figure 7.17. Development of competence in Bacillus subtilis. The figure depicts the possible pathways. It consists of a two component regulatory system and is also interrelated with sporulation.

and *Neisseria gonorrhoeae* demand strict homology. In the latter two, the uptake sequences, which are short, ~10-12 bp long, have been identified. Donor DNA must possess these sequences in order to gain entry. Though some bacterial species do take up non-homologous DNA, they do not process them to produce transformants. A specific size limit has not yet been found out, but very small DNA may offer difficulties both during uptake as well as integration. Naked phage DNA can also be used to transform a recipient cell. This process, known as trasfection, results in successful completion of phage life cycle.

Soon after its entry, the donor DNA may enter into an eclipse phase. This phase is of short duration (5 min. in *S. pneumoniae*) and DNA is modified in such a way that it loses its capacity to transform in a fresh reaction. After this eclipse phase, donor DNA regains its transforming ability which increases with the increase in the total cell number. During eclipse phase dsDNA is converted into ssDNA, which when recovered from the recipient cell does not sponsor transformation. But later, this ssDNA becomes a part of the recipient DNA that brings back its transforming ability. Several physical, biochemical, and genetic evidences are available to show that DNA is converted into ss state. *H. influenzae*, however, is an exception, as no distinct eclipse phase is observed in this bacterium. It has been proposed that in *H. influenzae* the dsDNA is taken up in subcellular compartments called transformasomes. This DNA is not converted into ss state (and, therefore, no eclipse) until it gets associated with the recipient genome.

There has been some debate as to the stage when conversion of ds DNA into single strands take place. One model suggests that it happens during uptake. An enzyme located close to the cell surface "pulls in" one strand while the other one is degraded. The other model explains that the single strands represent the strand that gets associated with the recipient genome. In pneumococcal cells, the bound DNA is degraded initially with single strand breaks. Only the competent cells, however, have the ability to get a part of this DNA in an irreversibly bound state. Pneumococcal cells having residual (~3% of the wild-type level) endonuclease and exonuclease activity behave normally in transformation assay. Mutants isolated implicated the endonuclease in this reaction. Thus, endonuleolytic cleavage is responsible for generating ssDNA during transformation.

(iii) Integration of Donor DNA: From the discussion mentioned above, it is clear that donor DNA is integrated in single-stranded form. Thus, a strand replacement step is required during integration. In *H. influenzae*, as we have seen earlier, donor DNA does not pass through a single- stranded intermediate state. It is taken up in a duplex form and appears to be converted into single strands concomitant with its integration into the recipient genome. A number of physical and genetic evidences suggest that donor DNA is integrated in a single stranded form in all cases. The ss fragment, protected by some protein like SSB protein and then aided by a protein equivalent to RecA, synapses and assimilates in dsDNA of recipient at homologous region. Subsequent nicking and trimming of recipient strand continues till the donor fragment is completely assimilated. The sealing of nicks by ligase creates a heteroduplex region (Fig.7.18). The appearance of the donor marker in the progeny cells depends upon whether or not the mismatch repair occurs. In case of many high efficiency markers, the correction preferentially takes place in the direction of donor or does not take place at all. In the latter case, replication will lead to the formation of two cells, one containing donor and another recipient genotype, and with the selection favouring donor phenotype, transformants would be selected. This has been supported by the observation that *hex* mutants of *S. pneumoniae*, defective in mismatch repair, when transformed, expressed all markers at high efficiency.

7.3.2.2. Mapping by Transformation. In many bacteria, transformation has been used as a tool for mapping genes and in some to confirm the mapping done by other means. This procedure is based on the principle that if two genes are located so close to each other (linked) that they can be carried on the same DNA fragment, transformation for two markers (cotransformation) can be achieved (Fig. 7.19). If the two genes are far from each other and, therefore, present on two different DNA fragments, the frequency of cotransformation, at the same concentration of DNA, would be roughly the square of the two single transformation frequencies. Whereas for linked genes, the cotransformation frequency will be nearly the same as single transformation frequency. Thus, cotransformation at a frequency substantially greater than the product of the two single transformation frequencies implies physical proximity of the genes. This can be used for determining the gene order. For example, if genes *a* and *b*, and *b* and *c* could be cotransformed but *a* and *c* could either not be cotransformed or cotransformed at exceedingly low frequency, the order must be *a b c.*

The linkage can also be confirmed by another method. Since the transformation frequency depends upon DNA concentration, any decrease in concentration of DNA will bring about a parallel decrease in transformation for e.g., *a, b,* and *ab* linked situation. However, if the two genes are unlinked the curve relating transformation frequency with DNA concentration will be steeper.

7.3.2.3. Artificial Transformation. The natural transformation system occurs only in some bacteria. The discovery by Mandel and Higa in 1970 that *E. coli* cells treated with $CaCl_2$ at low temperature can be transformed with non-infective phage DNA and other workers showing that under same conditions *E. coli* cells can preferentially take up plasmid DNA, opened up an exciting field of research. Transformation thus, became an important tool in genetic engineering research, where the transfer of a foreign DNA in an appropriate recipient is desired. Two different methods of making cells competent for transformation commonly practised are: chemical treatment and by an electric shock.

(i) Chemical Treatment: In this method, the cells are made competent by chemical treatment. Although $CaCl_2$ treatment may still be the most popular method, it can be replaced by other multivalent cations like Mg^{2+} and Mn^{2+}. Many other compounds like hexamine $Co_{(III)}Cl_2$, organic solvents and sulphahydryl reagents, simple freeze-thaw cycles, growth in rich medium, can also improve the transformation frequencies.

It has been known that multivalent cations can form relatively stable association complexes with phosphates. They can thus shield the phosphate on the cell surface as well as on the DNA, thereby creating conditions for favourable DNA transfer across the cell envelope. Temperatures near 0°C in conjunction with high concentrations of divalent cations will crystallize the more fluid membrane structure. This is supposed to make them more susceptible both to stable phosphate shielding and to DNA association

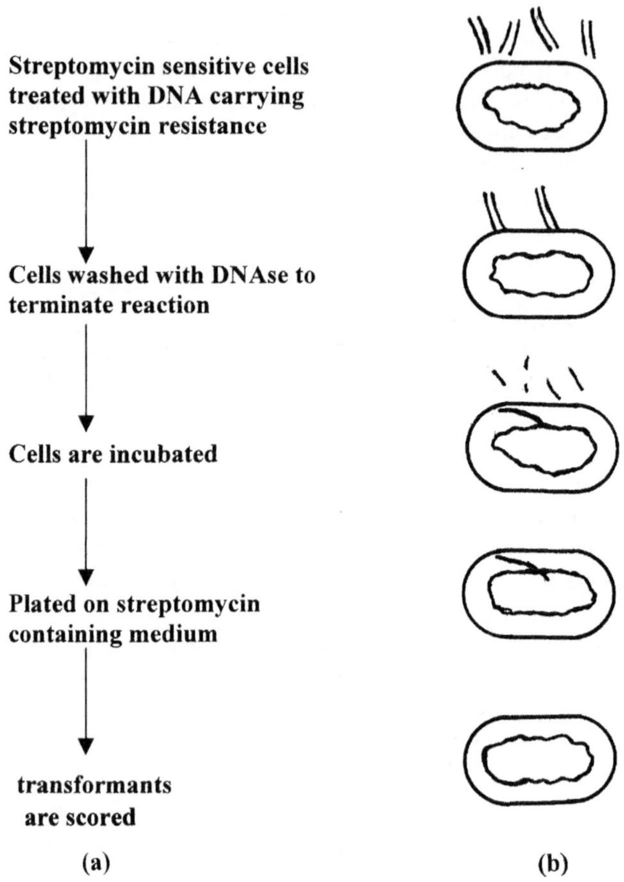

Figure 7.18. A typical transformation protocol for streptomycin-resistance (a), and possible steps in the formation of transformants (b). For details see the text.

with the cells. Some experiments have suggested that DNA enters through large channels forming the zones of adhesion. Others have implicated channels composed of poly-ß hydroxybutyrate (poly-HB), and polyphosphate [poly (P)] in addition to Ca^{2+} channels for DNA uptake.

A brief temperature shock (approx 90 sec.; 0°C → 42°C → 0°C) appears to facilitate uptake, although the effect may not be very significant. The final step in all cases is to transfer the cells to 37°C which also melts the membrane structure and allows the cells to grow. Transformation by whole plasmid DNA has been effectively demonstrated, where the size of the plasmid may play a significant role.

(ii) Electroporation/Electrotransformation**:** The treatment of both prokaryotic and eukaryotic cells with a brief pulse of high voltage electricity has been found to permeabilize them towards entry of a variety of macromolecules. The voltage potential transiently depolarizes the membrane and induces pores that can be used for entry of macromolecules. Electroporation or electrotransformation is the most efficient method available for transformation.

An electric shock generated by the discharge of a high voltage capacitor is passed through a mixture of bacterial cells and DNA. The field strength and pulse time is optimized. Another important parameter is to grow the cells in a low Mg^{2+} medium

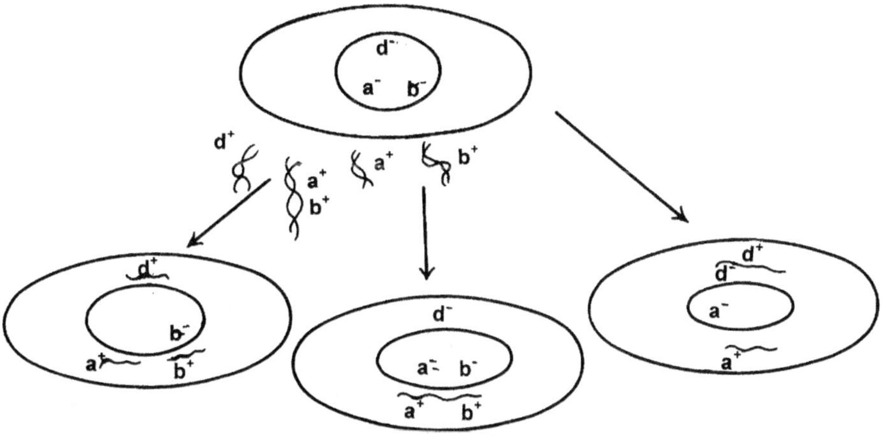

Figure 7.19. Cotransformation for linked and unlinked genes. Transformation based mapping is done by selecting for cotransformants as linked genes are likely to be carried on the same piece of DNA. Note: One should be cautious about classifying these cotransformants (see the text).

and the cells should be suspended in low salt buffer. A very high cell density is used as loss of cell viability is expected during electroporation. The uptake of transforming DNA and expression of the transformed phenotype follow the same step as in chemical treatment method.

Transformation has served as an important tool in introducing genetic material in an appropriate recipient cell. With the methods of inducing competence available now, literally any type of bacterial cell can be transformed.

7.3.3. Transduction

Besides transformation, another method of gene transfer came to light when Zinder and Lederberg (1951) conducted experiments to study genetic recombination in *Salmonella typhimurium*. Being closely related to *E. coli*, a process analogous to *E. coli* recombination was expected in this bacterium also. They employed the same strategy and mixed two multiple auxotrophic strains, one requiring phenylalanine, tyrosine, and tryptophan (*phe, tyr, trp*) and the other requiring methionine and histidine (*met, his*). When this mating mixture was plated on amino acid-free minimal medium, prototrophic colonies appeared at a frequency of $\sim 10^{-5}$. Taking clues from the inference drawn from

E. coli crosses, they suggested that genetic exchange can take place in *S. typhimurium* also. In further experiments, Zinder and Lederberg found that many other characters can also be exchanged between two *Salmonella* strains. These were cells which are unable to utilize galactose (Gal^-), xylose (Xyl^-), mannitol (Mte^-), or maltose (Mal^-) or those that were resistant to streptomycin (Str^r). However, on placing the two strains in two arms of Davis U-tube (refer conjugation), it was observed that cell to cell contact is not necessary for these recombinants to arise. It was, therefore, clear that the underlying process is different from *E.coli* conjugation and was suggested to be mediated by a filterable agent (FA). This FA was small enough to cross through the pores of sintered glass filter that allows the mixing of donor and recipient genetic material. Subsequent studies were then directed towards the identification of FA and it turned out to be *Salmonella* phage P22. This conclusion was drawn from the observations that- i) P22 and FA have the same host range, ii) the dimension and the mass of the two are the same, iii) both can be inactivated by heat or treatment with anti-P22 serum, (iv) both are protected from inactivation by hydrolytic enzymes such as DNAase (thus, transformation is ruled out) and trypsin, and (v) phage-resistant strains of *Salmonella* that no longer adsorb phage P22 also do not interact genetically with FA.

Zinder and Lederberg thus discovered a third mechanism of genetic exchange, which differed radically from either transformation or conjugation. This process mediated by a phage was referred to as transduction. Lederberg's group in 1956 was also able to observe the transfer of *gal* marker in *E.coli* through phage lambda (λ). However, there was an important difference between *Salmonella*-P22 and *E.coli*-λ systems. In the former, virtually any gene can be transferred and thus the process was called generalized transduction, in the latter where only *gal* (and a few more) could be transferred came to be known as restricted or specialized transduction. The two transduction pathways are related only in having a phage as a vector for the donor DNA transfer. Otherwise, the mechanism by which the transducing particles arise, the way by which the part of bacterial genome is carried by the phage and the incorporation of transducing DNA into the recipient genome are quite different. We shall discuss the generation of these differences in the subsequent sections. The recombinant cell formed through either of the pathways is referred to as transductant.

In order to understand the transduction mechanism, we should go through the various steps in phage growth in a bacterial cell first. Bacteriophage, phage, or bacterial virus like any other virus are tiny living objects which can thrive only inside a living cell, where they are capable of multiplying themselves. They are intracellular parasites of non-cellular organization, consisting of only nucleic acid and proteins and are basically classified by their hosts and express unique host specificity. Bacteriophage or phage live on bacteria (Fig.7.20).

In the first step of the host infection, they adsorb on the host cell wall (specific receptors) and then, by creating a through passage, inject their DNA or RNA into host cell cytoplasm. Once inside, their nucleic acid modulate the host cell metabolism to synthesize the viral components and the latter are then organized into virus particles. These particles are released by the breakdown of the host cell wall and the viral progeny find another host cell to restart this cycle. The phages which opt for this life style are called virulent phage or lytic phage as they tend to essentially kill the host cell at the end of their growth/multiplication cycle.

There are other types of phages (P22, P1, λ) known as temperate phages which may follow a slightly modified pathway. After injecting the phage DNA, viral components are not immediately formed, instead phage DNA integrates into host genome (phage P1 is an exception as it remains in free form called as P1 plasmid). Phage genome which remains quiescent is known as prophage and the bacterial cell carrying a prophage is referred to as lysogen. The whole process is called lysogeny. Many temperate phages may spend a long time in prophage state and may be distributed to all the daughter cells formed by the division of the host cell. At some point of time, however, the prophage gets induced. During induction, the integrated prophage excises out and starts the process required for viral multiplication. The latter then culminates into host cell lysis and release of the progeny. Thus, the host cell killing is delayed in case of temperate phages. The cell lysis in both types of phages is observed as clear area or plaques appearing on a bacterial lawn growing on agar medium.

7.3.3.1. Generalized Transduction. Many bacteriophages are capable of bringing about generalized transduction. Besides the phage P22, the first to be discovered, *E.coli* phage P1, described in 1955, are the two well worked out systems. Both phage P1 and P22 are temperate phages. P1 is a prototype bacteriophage having an icosahedral head, an intricate tail complete with tube, contractile sheath, base plate, and tail fibres. P22 in contrast is much simpler, the tail being so reduced as the head appears to be resting on a base plate. The DNA in each P1 particle amounts in size to ~2% of its host and is also approximately 2.4 times that of P22 in size. The most significant feature of the genomes of P1 and P22 are the terminal redundant ends, which is extensive in phage P1 (~12%) as compared to P22 (~2%). The ends are not only repetitive, they are also different in different particles (circular permutation). One can envisage that P1 and P22 DNA are packaged from concatemers of several genome lengths and that the length of the DNA packaged in phage head is greater than a genome equivalent. As a consequence, the DNA in each phage has a sequence repeated at the two ends. Moreover, because subsequent molecules are cut out from the same concatemer, their ends will be circularly permuted. The limited nature of the circular permutation, more strongly exhibited in phage P22, suggests that there is a preferred site or region, where sequential packaging begins (*pac* site) and the packaging must stop after a few cycles. While phage P22 has a single chromosomal attachment site between *proA* and *proC* genes, phage P1 genome acquires a plasmid form and remains free even in a lysogen.

Transducing phage formation takes place when lytic growth ensues and the packaging mechanism seizes upon host DNA as the packaging substrate (see Fig. 7.21a,b). The formation of P1 transducing particles was elegantly explained by Ikeda and Tomizawa in 1965 (Fig.7.21a). Their experiment was based on density labelling of the DNA of donor cells with 5-bromouracil (BU) before infection, and replacing BU with thymidine (T) at the time of infection. In fact, three different experimental set-ups were used. In one, T was used all through, i.e., before and after infection, in the second, donor cells were grown with BU and shifted to T-medium at the time of infection, and in the third BU was used throughout. From this, the fate of host DNA synthesized before infection

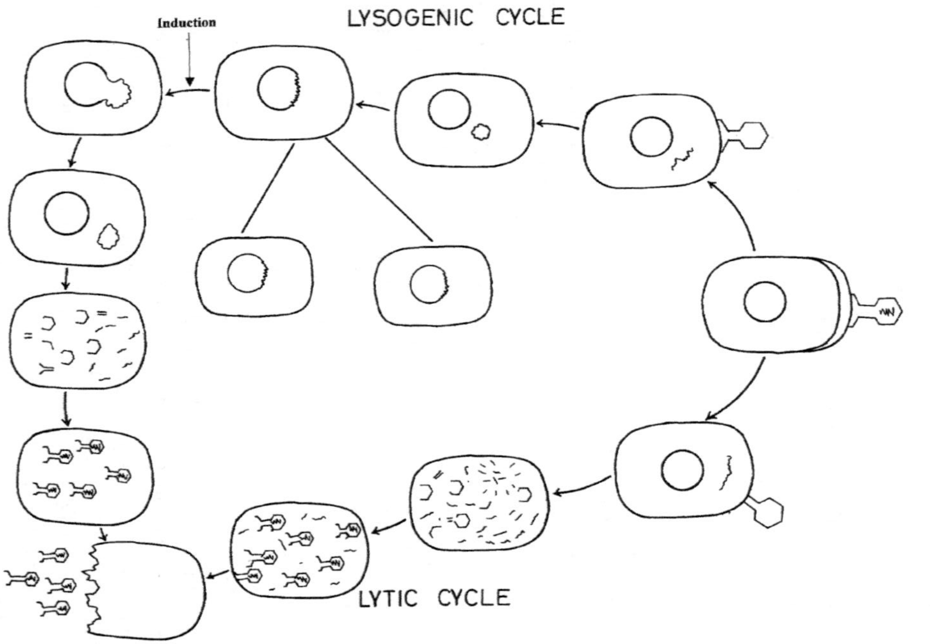

Figure 7.20. Life cycle of bacteriophage. Based on their life style phages can be classified as (a) virulent or lytic, and (b) temperate phage. The lytic cycle consists of phage (i), adsorption to host cell (ii), injection of nucleic acid (iii), replication of phage nucleic acid (iv), synthesis of phage proteins (v), packaging of nucleic acid in phage head (vi), phage assembly, and (vii) lysis of the host cell and release of phage progeny. In the lysogenic cycle the first two steps are common and then phage nucleic acid may integrate in the host chromosome and maintained and distributed in this situation for a long time. The host cell is called lysogen and phage as prophage. Prophage may get induced whereby it gets excised and follows the later steps of lytic cycle to multiply and release the progeny.

could be followed separately from that of the phage DNA (or host DNA) made after infection. The phages obtained after lysis were centrifuged in CsCl and analyzed for transducing activity and plaque-forming ability (infectious particles). It was observed that these two types of phages form separate classes. As shown in Fig.7.21a, in set up I and III, the transducing and infectious particles exhibited densities reflecting, as per the medium used, normal or heavy DNA, respectively. However, in set up II, while the transducing particles showed heavier density, the infectious particles contained normal density DNA. This clearly showed that all the DNA in transducing particles is likely to be derived from the host DNA made before infection, whereas infectious particles contain only light DNA, made after infection. Thus, the two types of DNA are not mixed up in a transducing particle (Fig. 7.21b).

In another experiment, where host DNA was labelled with ^{3}H and phage DNA with ^{32}P it was estimated that transducing particles of P1 represent 0.3% of the total particles released. P22 also showed similar behaviour, with transducing particles amounting to 1-5% of the total population. Highly transducing derivatives of both P1 and P22 have been isolated where their representation may range from 2% to up to 50%, respectively. It has now been established that *Salmonella* chromosome offers sites resembling *pac* of P22 that can be recognized by phage system as substrate. In phage P1, however, the packaging appears to occur in a non-specific manner either from DNA ends or by non-specific cutting by packaging enzymes.

Once packaged within phage particle, transducing DNA is delivered to recipient cell as if it were phage DNA. In recipient cell, three possible fates await the donor DNA: degradation by nucleases, recombination with homologous recipient DNA, or persistence as a non-replicating stable structure. The latter may well constitute the most predominant form (~90%). Of the 10-15% of the DNA which gets associated with the recipient only 1-2% may really be involved in recombination to give rise to complete transductants (Fig.7.22).

The persistence of donor DNA in a recipient cell without recombination and replication is referred to as abortive transduction. This was first recognized when transductants were selected for colony growth. It was seen that among the normal-sized colonies, some very small colonies also appeared. Analysis of individual cells in such colonies revealed that each colony contained one cell that could give rise to another small colony. It was interpreted to mean that the donor DNA either recombines with the recipient DNA to yield a complete transductant or fails to do so when it is destined to abort. Abortively transduced gene may be expressed and can complement mutations in the recipient bacterium. When the transduced cell will divide, one daughter cell will not receive the donor DNA as the same is not replicating. Protein product, however, will be distributed between the two cells, allowing the growth of even that cell which does not inherit the donor DNA. However, since the cell lacks that DNA, the protein product will be gradually diluted leading to the cessation of the growth and the colony will remain small. The cell that retains the transduced DNA will be able to reinitiate this process and can give rise to another small colony when division takes place (Fig.7.23).

Support for this mechanism has been obtained by labelling the donor DNA (the bacterial cell from which transducing phage particles are produced) with BU and

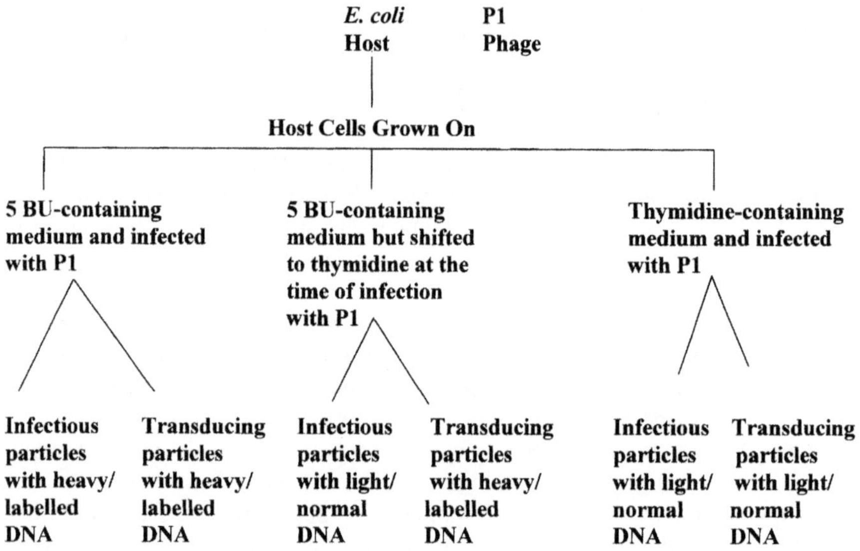

Figure 7.21a. Representative diagram showing the basis of Ikeda and Tomizawa's experiment. The experiment clearly suggests that transducing particles carry the host DNA synthesized before infection.

examining its fate by re-extracting it from recipient. It was observed that major proportion (70-90%) of this DNA remained in fully heavy DNA form suggesting that it is not associated with the recipient genome. This condition persisted for as long as 5 hours, indicating that it is neither integrated nor degraded. The sedimentation behaviour and gel mobility suggested this to be in a circular supercoiled form. This was surprising because the linear ds transducing DNA does not have any mechanism to circularize. These circular molecules could be disrupted by treatment with proteases or detergents implying that circularity is brought about by association of ends with a protein. Its absence in infectious particle was explained as this to be cellular rather than a phage encoded protein. However, many proteins both in P22 and P1 seem to have some effect on abortive transduction suggesting their involvement in this pathway.

In conclusion, it can be said that the transducing DNA must undergo recombination with recipient genome so as to produce the complete transductants. Alternatively, the DNA is circularized, and protected from degradation, and the action of recombination enzymes. Once destined to abort, this DNA does not yield complete transductant.

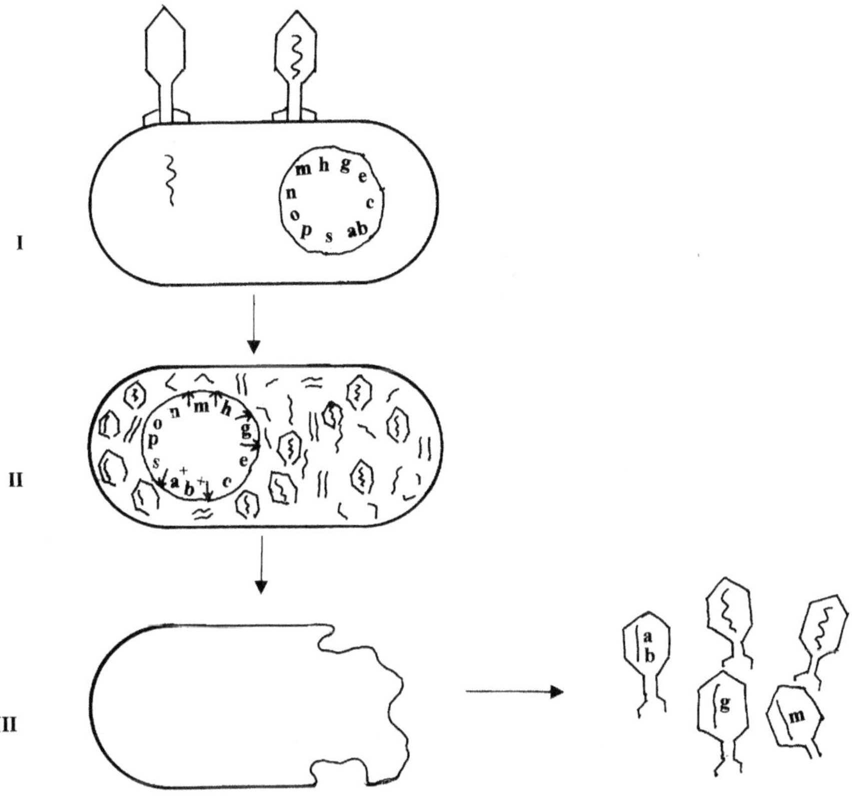

Figure 7.21b. Formation of generalized transducing phages. I. Phage infects a bacterial cell and injects its DNA. II. Phage DNA undergoes replication and capsid components are synthesized. While DNA is packaged into phage head, some heads get stuffed with bacterial chromosome fragments. III. When the bacterial cell is lysed, two types of particles are formed: infectious, carrying phage DNA and transducing carrying bacterial DNA.

Two types of DNA integration events have been detected in Rec$^+$ but not in Rec$^-$ cells. Some DNA is found in the continuous stretches as expected by recombination. In some there is a patchy integration (~10%) which comes from reuse of the component nucleotides of degraded donor DNA.

Differences normally observed in transduction frequencies of different markers by P1 and P22 have been ascribed to differential recombination and differential marker pick-up, respectively.

7.3.3.2. Restricted or Specialized Transduction. As mentioned earlier, phage lambda, a temperate coliphage behaved very differently in a transduction assay. While it followed a life cycle very similar to P22 it did not transduce the host genes with the exception of *gal* (later *bio* also).

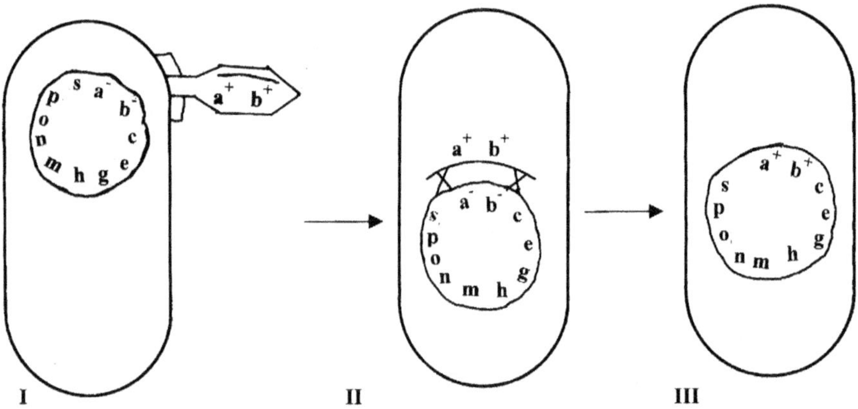

Figure 7.22. Formation of transductant. I. When a phage carrying $\underline{a}^+$ $\underline{b}^+$ *bacterial genes, for example, infects a bacterial cell that is* $\underline{a}^-$ $\underline{b}^-$*, II. The injected DNA pairs with the homologous region and, III. Through recombination the donor DNA is inserted.*

The restricted transduction was discovered in 1956 by Lederberg's group in the course of an experiment designed to test whether phage λ can transfer *E.coli* genes from donor to recipient cell. For this, a prototrophic wild-type *E.coli* cell lysogenic to λ [*E.coli* K-12 (λ)] that carried the λ prophage was induced by UV treatment to generate phage progeny (phage lysate). A variety of non-lysogenic mutant *E.coli* cells were infected with these phages and plated on different selective agars to test whether any wild-type allele of the K-12 donor cells had been picked up by λ and transferred to the mutant recipient. The result of this survey was mostly negative except in case of *gal*, with about 10^{-6} of lambda-infected *gal*$^-$ cells being converted into *gal*$^+$ character of the donor. It was thus clear, that lambda phage has transducing power like P1 and P22 but is restricted to a few bacterial marker/s.

When these *gal*$^+$ transductants were examined they appeared to carry a λ prophage and were genetically unstable. It can be inferred that along with *gal* character, the transductants also acquired the λ prophage and each *gal*$^+$ colony had 1-10% of the cells losing *gal*$^+$ character and returning to *gal*$^-$ nature of the recipient. This suggested that during transduction, phage λ along with donor (*gal*$^+$) marker is added onto the recipient genome rather than exchanged, creating a *gal*$^+$/*gal*$^-$ partial heterozygote. The recombination between two copies of *gal* gene leads to descendants without the donor

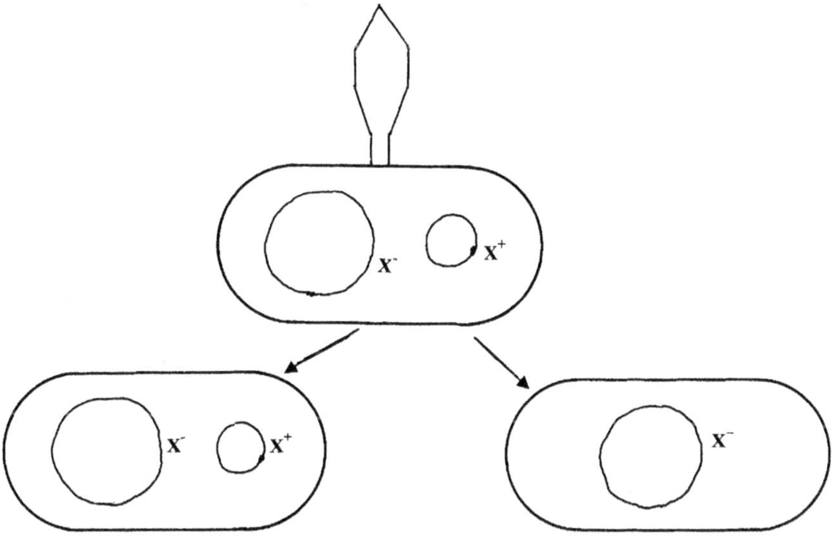

Figure 7.23. Abortive Transduction. Sometimes the transducing DNA does not integrate but acquires a circular form. Such a cell does provide the transduced phenotype (X^+). When such a cell divides, the transducing DNA is transmitted to only one cell. This can continue for several generations every time giving one transduced and one non-transduced cell.

marker (Fig.7.24). Later studies showed that λ can also transduce *bio* marker of *E. coli* donor which like *gal* flanks the integrated λ.

The generation of λ transducing phage can be explained on the basis of its life cycle inside the *E.coli* cell. Lambda phage injects its DNA into host cells upon infection. This DNA which is principally double stranded carry complementary single stranded overhangs or cohesive ends (cos sites) and circularizes by the pairing between these single stranded ends, and finally the ends are sealed to give a covalently closed circular (ccc) DNA. The circular phage genome is integrated into bacterial genome at a site called *att*λ located between *gal* and *bio* genes (Fig.7.25). Once integrated it lies there as a prophage, a copy of which would be transferred to all the cells that are produced by the division of the lysogen, and can even be conjugationally transferred to appropriate recipients. Upon induction, prophage genome loops out and is excised to a free-form. In this form, λ initiates and completes a lytic cycle and releases a burst of phage progeny. On rare occasions, λ looping out may be imprecise when a part of the bacterial genome is included in the loop and a part of λ genome is left out in the host chromosome. The excision of such a loop leads to the generation of λ genome with bacterial gene/s, and since only the flanking bacterial genes get a chance to be included, it creates λ *gal,* or λ *bio* or in very rare case λ *gal, bio*. The readers may recall a similar event taking place during F plasmid excision and generation of F′s. On the basis of the distribution of λ genes in the prophage genome, λ *gal* phages usually lack a number of essential genes. Such phages are referred to as defective or λ d*gal*. By the same analogy, in the creation of λ *bio*, the essential genes are not lost, generating λ p*bio*. When such phages infect host cells at low multiplicity of infection (MOI) so that a single phage particle attaches to a single host cell, λ p*bio* but not λ d*gal* can form infective particles. In order to reproduce particles like λ d*gal*, a host cell will have to be infected at high MOI so that each gets infected with one wild-type (or any other mutant phage that can provide the missing functions) along with λ d*gal*. The wild-type or any other appropriate phage acts as a helper phage and allows the multiplication of λ d*gal*.

A specialized transducing phage can transduce a bacterial gene essentially in two ways (Fig.7.26). In Type I mechanism, the transductants are haploid, stable and nonlysogenic. These result from two crossovers that replace the chromosomal, for example, gal^- gene with the donor gal^+ gene. In Type II, a single crossover between the recipient and phage-carried allele leads to insertion of phage as well as the donor gene. Such transductants are heterogenotes or partial diploids, unstable and lysogenic as described earlier. They, however, are incapable of getting induced and produce λ d*gal* particles unless helped by a helper phage, but expression of many phage-specific genes can occur.

If cells are infected at an MOI such that many cells are infected by both λ d*gal* and a normal λ^+, a dilysogen can be formed. This happens by the insertion of normal λ at attachment site *attB*, followed by insertion of λ *gal* by using the left attachment site. When these heterozygous lysogens were induced by UV treatment the released phage particles had an exceptionally high transducing power. In such a lysate, ~50% of the phage carried the transducing *gal* gene leading to it being referred to as high frequency transducing lysate (HFT) in comparison to original low frequency transducing lysate or LFT shown in Fig.7.24. Both LFT and HFT could only transduce *gal* marker (Fig.7.27).

7.3.3.3. Applications of Transduction. Transduction has been used both for genetic mapping and strain construction. Although it is not a very reliable and strong tool in mapping, cotransduction frequencies can be used to ascertain physical proximity. One of the important contributions of transduction is in effecting gene transfer in bacteria. As the amount of bacterial DNA carried by a phage is limited it can also be utilized for introducing small but defined changes in the recipient genome. The specialized transducing phages also provide the paradigm for cloning experiments and many of them or regions their of have been utilized in developing cloning vectors, as discussed later.

7.4. Transposable Elements

Geneticists all along believed that genes have a fixed position on the chromosome. This idea was reflected in the generation of extensive linkage maps of both prokaryotic and eukaryotic organisms. It was only in 1940s that Barbara McClintock discovered the presence of genetic mobile elements while studying the mottled phenotype of kernels of maize. Some thirty years later, such mobile elements were recognized in bacteria too and are now known to be ubiquitous in occurrence, from bacteria to human beings.

The bacterial transposable elements were identified by their property of insertion into an active gene that leads to gene inactivation (=mutation). On the basis of the absence or presence of recognizable gene/s, they were classified as insertion sequences

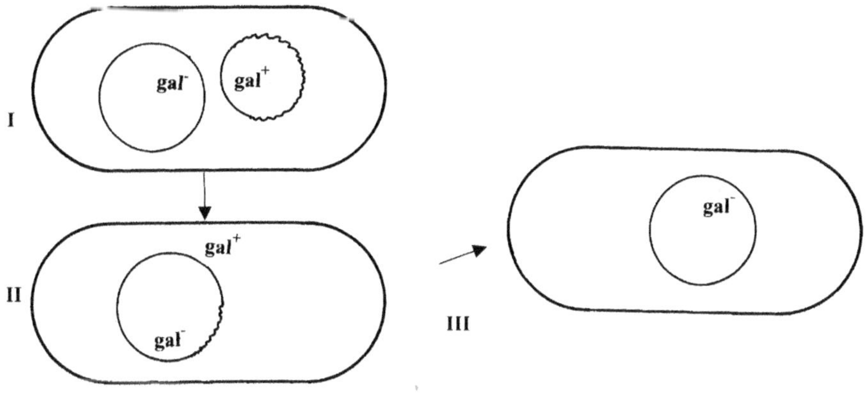

Figure 7.24. Specialized Transduction. I. A bacterial cell is infected by λ gal. The injected DNA circularizes and pairs with gal region of recipient, II. A single crossover leads to insertion of gal⁺ gene along with phage DNA, creating a gal⁺ / gal⁻ partial diploid, III. The donor gene (along with λ) can be segregated out by recombination.

or insertion elements (IS) (Table 7.2) and transposons (Tn) (Table 7.3), respectively. All isolated transposable elements are designated this way with abbreviation followed by a number (e.g., IS1, Tn5 etc.). When it is necessary to refer to the genes carried on the transposon, the standard genotypic designations, for example, Tn3 (*amp^r^*), in which ampicillin resistance gene is carried by the transposon, is used (see Table 7.3). Transposons, commonly carrying antibiotic and/or metal resistance genes, are the ones most frequently studied because they can be identified by a simple plating test. Occasionally, a transposon may be designated in a non-standard way, such as Tn 1000 found in F plasmid is called ϒδ. A transposable element can be frequently located in a particular gene, creating a mutation. Such mutants are designated by the gene and the allele number followed by two colons and then the name of the transposon. For example, 22nd mutation in *lacZ* due to insertion of Tn5 will be designated as *lacZ* 22::Tn 5.

7.4.1. Identification

1) The simplest procedure used for identification of prokaryotic transposable element is the inactivation of gene in which they insert that can be scored by a mutant phenotype. Since in bacteria, the genes are organized in operon, this could lead to the creation of polar mutations. For example, a mutation in *lacZ* by insertion of a transposon may lead to *lacZ, lacY, lacA* phenotypes (Fig. 7.28)

2) These mutations can not be reverted by known chemical mutagens, suggesting they are not base substitutions or single base additions or deletions.

3) If a plasmid is transferred into a cell containing such polar mutations, similar polar mutations occasionally appear in the genes carried by the plasmid, although the location of these mutations are quite different. A similar situation can also arise if the same cell is infected with a phage.

I

II

III

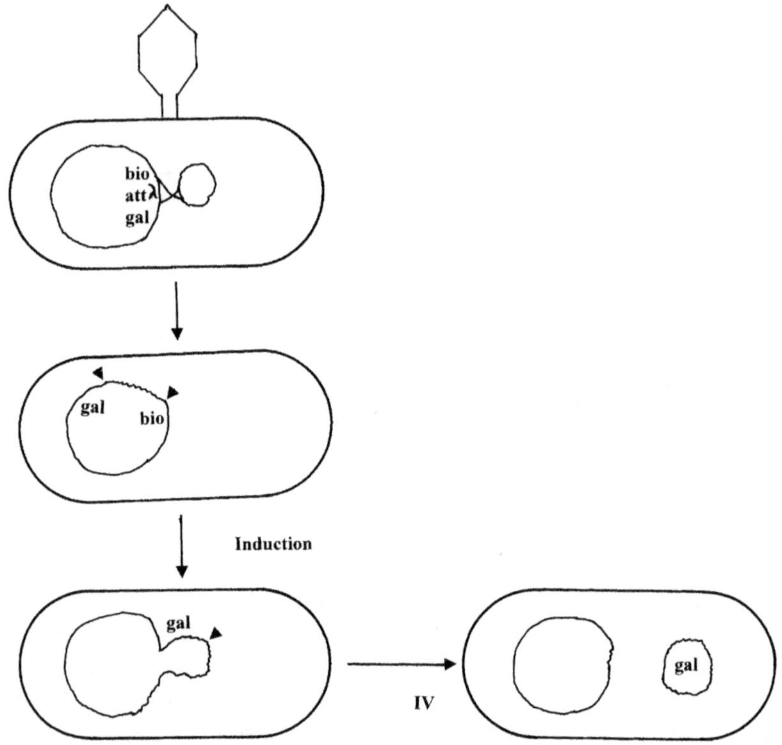

Figure 7.25. Formation of specialized transducing phage. I. Phage (e.g., λ) injects its DNA which circularizes and pairs with the bacterial chromosome at a specific site (att λ), II. A crossover leads to λ DNA integration into host chromosome, III. When phage DNA loops out during induction, a part of the bacterial chromosome (gal) is included and a part of its own is left in the chromosome, IV. The transducing phage picks up (gal) (or bio) gene of host.

4) Physical characterization of such plasmids (with polar mutation) shows the size of the plasmid larger than the original plasmid. The size of this extra segment is specific and varies from element to element (Fig. 7.29). Moreover, if such plasmid DNA is denatured and cooled rapidly to allow intra-strand reannealing, stem-looped structure can be observed on a single strand of DNA (Fig. 7.29).

5) When a phage DNA carrying the polar mutation is hybridized with the DNA of a normal phage lacking the mutation and the hybrid molecule is viewed through electron microscope, a loop like structure is formed. The length of the loop indicates the length of the insertion. However, if both the DNA carry mutation, the loop may not be seen. Alternatively, if the wild-type phage DNA is hybridized with various phage DNA carrying different mutations but derived from the same host cell, loops of similar sizes but at different locations can be seen.

Such analyses revealed the existence of several different IS elements. A similar approach can also lead to the detection of transposons. Although most transposable

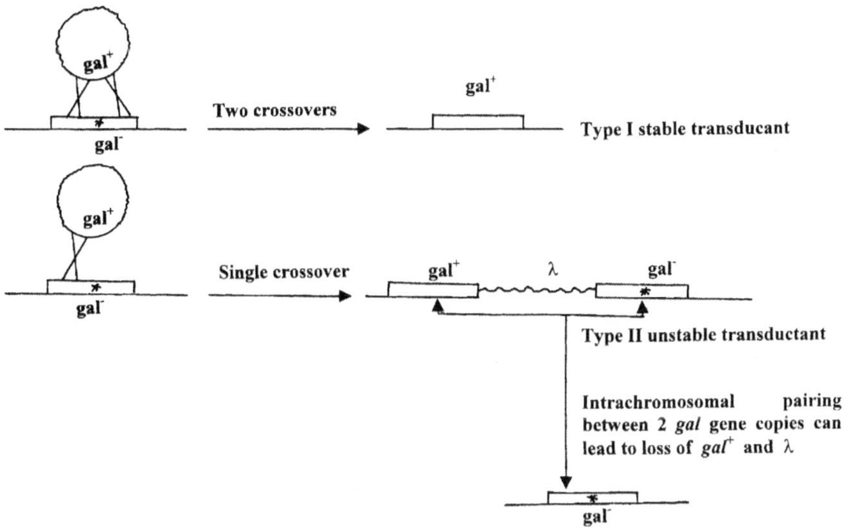

Figure 7.26. Two pathways of getting transductants based on the number and position of the crossovers. With two crossovers flanking the gene to be transduced leads to the replacement of the recipient's counterpart and a type I stable transductant; a single crossover, on the other hand, results in the addition of the transducing gene along with λ to the recipient genome creating a type II unstable tranductant. The partial diploid so created can indulge in intrachromosomal recombination and loss of transducing gene along with λ will bring back the recipient genotype.

elements get detected by spontaneous mutations, they can also be used to create mutations by transferring them into an appropriate cellular environment. Both plasmids and phages can be used as vehicle for their intercellular transmission. Most IS elements besides showing mobility have got adjusted with their hosts, so much so that several indigenous copies of them may be located both on chromosomes and plasmids. Tns, on the other hand, are widely distributed on plasmids (e.g., R plasmid) and can be freely disseminated between different bacteria. New copies of Tn can also be acquired by transposition from other locations (Fig.7.30). Both the events can be easily traced by the associated antibiotic resistance markers. They are principally classified on the basis of antibiotic resistance marker/markers they possess (Table 7.3).

IS elements which produce polar mutations contain transcription stop (except for IS1) and also chain termination signals in all possible reading frames. The termini of each IS element are characterized by inverted-repeat (IR) sequences, ranging from 16-44 bp (Fig. 7.31). The intervening region is variable in size ranging from little less than 1kb to several kilobases (Table 7.2). This region consists of atleast two apparent coding sequences.

Transponsons have been broadly classified into two classes:

(i) Composite transposons: These constitute a variety of transposons carrying antibiotic resistance markers flanked by two identical or nearly identical copies of an

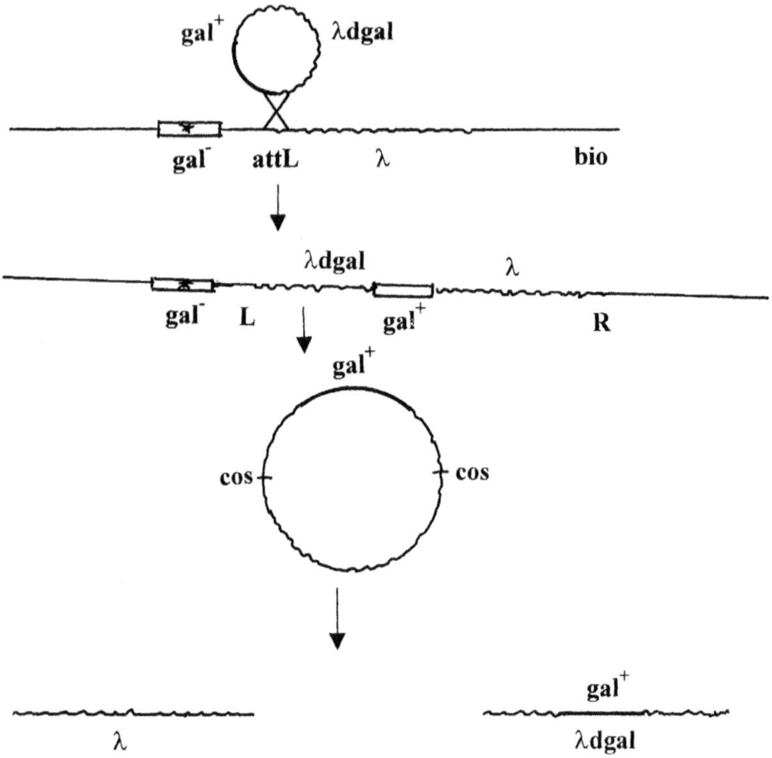

Figure 7.27. Phage λ dilysogen formation and its induction to produce an HFT lysate. Such a lysate will contain 50% λ and 50% λ dgal particles.

IS element (Table 7.3). The IS elements in these composite units, called as type-I transposons can either be in an inverted or direct repeat orientation (Fig.7.31). However, since each IS element itself is flanked by IRs, the whole unit remains in inverted repeat configuration. These elements were first identified by their capacity to transpose from one plasmid to another or to a phage. When such a plasmid was denatured and then renatured, single-stranded circles containing a stem-and-loop structure could be seen by electron microscope (Fig.7.29).

Composite transposons have two interesting features: (i) sometimes the terminal IS element can transpose independently, and (ii) if located on a small plasmid, it can act as a composite unit of two transposons (Fig.7.31).

(ii) Noncomposite transposons: These are flanked by short inverted repeats and the antibiotic resistance gene/s as a part of the transposable unit. They belong to a number of families, the members of which share sequence and the structure. The Tn3 family of transposons are about 5 kb in length and contain short (38 bp) inverted repeats and none are flanked by IS-like element. The transposon carries three genes, one coding for ß-lactamase (which confers ampicillin resistance) and two others required for transposition as described later (Fig.7.31). Other examples are: Tn 21, Tn 501, and Υδ.

Table 7.2. Properties of some commonly found insertion sequences in E.coli strains

Name	Size (in bp)	Number of copies and location	Target repeat (in bp)
IS1	768	5-8 in chromosome	9
IS2	1327	5 in chromosome; 1 in F	5
IS3	1258	5 in chromosome; 2 in F	5
IS4	1426	1 or 2 in chromosome	11-13
IS5	1195	unknown	4
IS10R	1329	few on chromosome	9
IS50R	1533	few on chromosome	9
Tn1000(Υδ)	5980	1 or more in chromosome; 1 in F	-

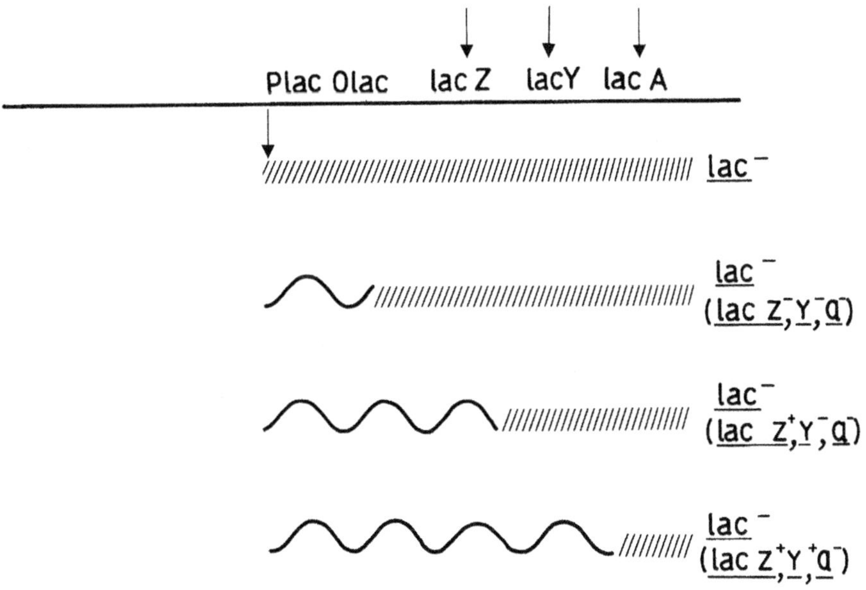

Figure 7.28. The introduction of polar mutations due to insertion of a transposable element (→). Since many of the bacterial genes are transcribed in a polycistronic mRNA, any insertion would lead to disruption of the reading frame distal to the point of insertion. The stappled line represents the portion of the operon where transcription is lacking and the wavy line the portion which is transcribed.

Table 7.3. Properties of selected composite type-I transposons reported in E.coli

Name	*Genetic Marker*	*Size (in bp)*	*Type and size of flanking element (in bp)*	*Relative direction of IS element*
Tn 5	*kan*	5818	IS50 (1533)	Inverted
Tn9	*cam*	2638	IS1 (768)	Direct
Tn10	*tet*	9300	IS10 (1329)	Inverted
Tn204	*cam,fus*	2457	IS1 (768)	Direct
Tn1681	*ent*	2088	IS1 (768)	Inverted

Genetic markers represent resistance to *kan*= kanamycin; *cam*= chloramphenicol; *tet*=tetracycline; *fus*=fusidic acid; *ent*= enterotoxin

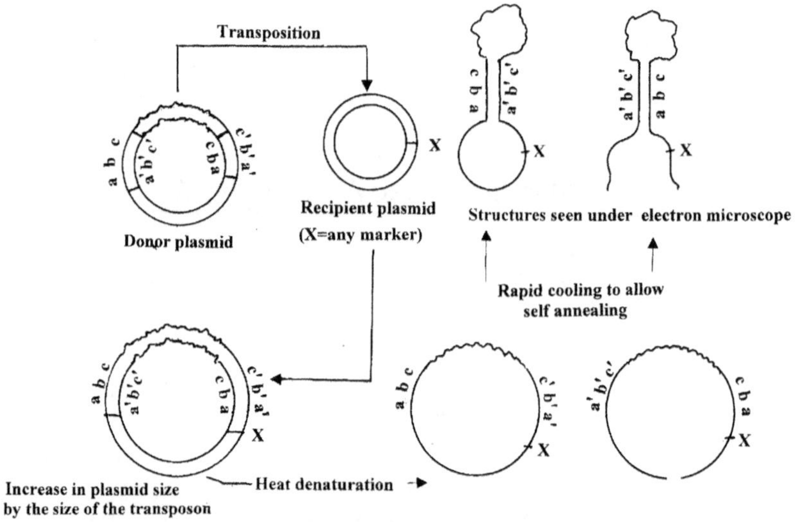

Figure 7.29. Consequence of transposition in a circular replicon and the detection of a transposable element by a characteristic stem-loop structure through heat denaturation and intramolecular reannealing.

The temperate phages Mu and D108 also behave as transposable elements as their whole life cycle is based on transposition event.

7.4.2. Transposition

The result of transposition is the insertion of one of these elements in a recipient DNA. Base sequence analysis of many transposons and their insertion sites reveal no extensive sequence homology suggesting that insertion is random. This is further supplemented by the lack of a requirement for *E.coli recA* system. However, localized homologies have been detected in several cases. IS4 and Tn7, on the other hand, show extreme specificity in their site selection. For example, Tn7 preferentially integrates into one chromosomal site referred as *att* Tn7. Insertion of an IS or Tn is always accompanied by the duplication of a short sequence (3 to 13 bp) in the recipient DNA (Table 7.2 and Fig.7.32). This sequence is present in one copy in the recipient before transposition and is not derived from the transposable element. Thus, inserted element appears sandwiched between these target repeats. The length of these target repeats but not the sequence is characteristic of an element. The transposition requires the enzyme transposase, which is coded by the transposable element. Transposase creates a staggered cut in the target DNA which is followed by the joining of the element to the two respective ends. This leaves a gap, one across from each strand of the target which is then filled up and the

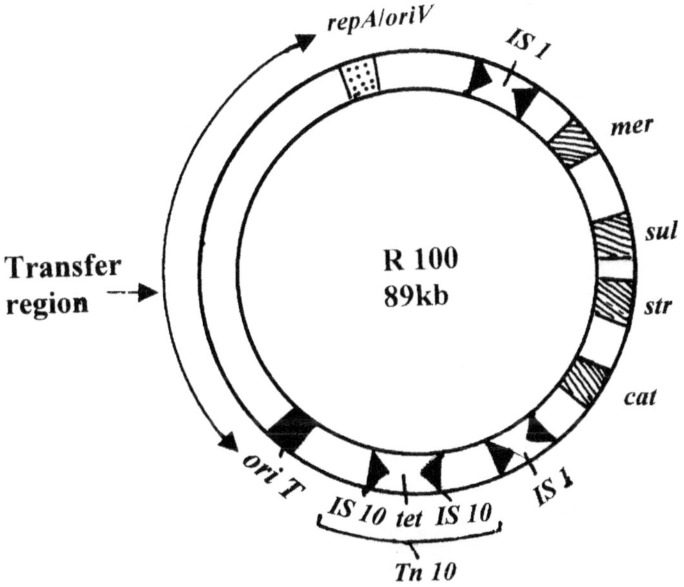

Genetic map of a conjugative plasmid R100 of *E.coli*

Figure 7.30. Structure of an R plasmid, R100 showing the location of resistance genes. mer= mercuric ions, sul= sulphonamide, str= streptomycin, cat=chloramphenicol acetyl transferase, tet=tetracycline. OriT= origin of transfer, tra= group of genes needed for conjugation, repA/ oriV= replication functions and origin of replication, IS1, IS 10= insertion sequences, Tn10= transposon.

nicks are sealed, leading to target site duplication (Fig.7.32).

The transposable element is cleaved clear from the donor element also by transposase. Two types of transposition mechanisms are known: In replicative transposition, joining of the transposon to target creates a replication fork. The transposable element is duplicated that results in the formation of a cointegrate, in which donor and recipient are joined by a copy of transposable element at each joining point. The transposable element also codes for a resolvase that catalyses a site-specific exchange to resolve this structure leading to a copy of the transposable element both in donor and recipient. The classical examples are Tn3 and phage Mu (Fig. 7.33). In non-replicative transposition, a "cut and paste" pathway is followed, in which donor loses the transposable element while it moves to a new site in the recipient. The examples of this category are Tn5 and Tn10 (Fig. 7.33).

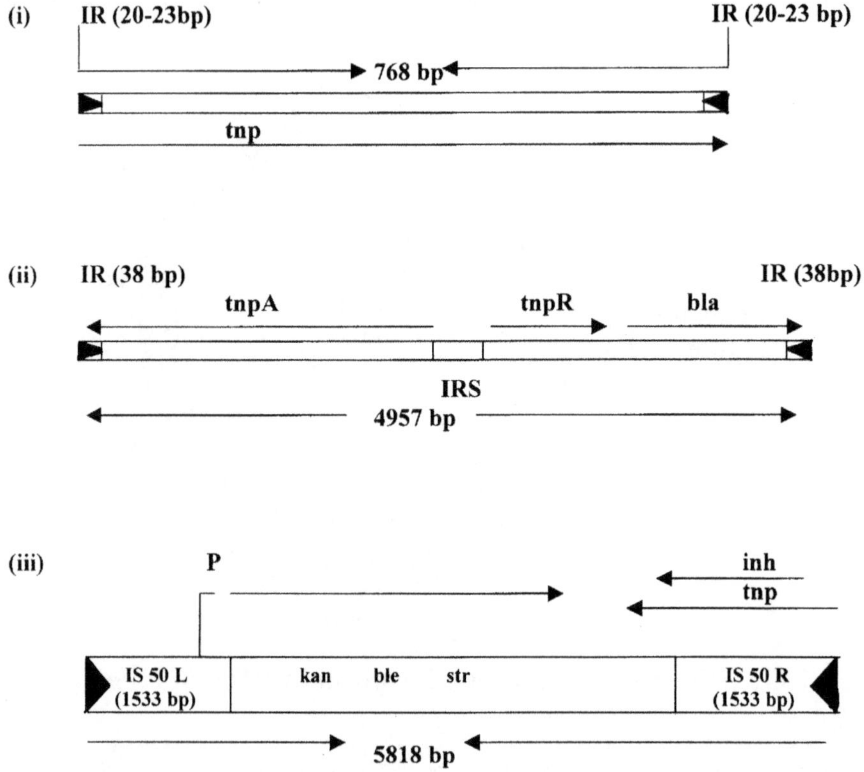

Figure 7.31. Structural plan of three representative transposable elements (i) IS 1, (ii) Tn 3, (iii) Tn 5. IR= inverted repeat, tnp/tnpA= transposase, tnpR= repressor and resolution protein, IRS= internal resolution site, inh= transposase inhibitor, IS 50 L and IS 50 R= the left (non-functional) and the right (functional) flanking inverted copies of insertion sequence, IS 50, bla= gene coding for ß-lactamase that confers ampicillin resistance, kan, ble, str= genes coding for resistance to kanamycin, bleomycin and streptomycin, respectively, P= promoter for drug-resistance gene.

Transposable elements can also excise from a DNA molecule by a *recA*-independent mechanism. In case the excision is precise, the gene function can be restored. Transposable elements besides inducing random mutations are also capable of introducing chromosomal rearrangements such as deletions, inversions and replicon fusion. They have also provided selective advantage to the host cells. Each one of them have contributed significantly in generating variability in prokaryotic populations. They have thus come out of the bracket of "selfish DNA" as they were usually christened.

7.4.3. Conjugative Transposon

Many bacteria belonging to both Gram +ve and Gram -ve group possess another class of transposon, called conjugative transposon. They have been so named as they have the capacity to promote their own transfer and do not require a carrier such as a plasmid or a phage. They are neither flanked by perfect terminal inverted repeats nor do they generate target site duplication upon insertion. They all carry tetracycline resistance gene but some of the larger transposons may carry additional antibiotic resistance markers. Their transfer and transposition are *rec* independent and are mediated by the formation of a ccc intermediate. The transposon is excised from the donor and integrated randomly. Excision and integration involves Xis and Int protein exhibiting sequence homology with similar proteins of temperate phages.

7.5. Genetic Engineering

The genetic analysis of an organism is dependent upon the availability of variation in the population. As we have discussed in the preceding Sections, two main mechanisms of generating variability are mutation and recombination. The usefulness of mutations and their rare nature led to the methods which can enhance the frequency of mutation. Through controlled hybridization amongst plants and animals some useful traits could be transferred into economically important crops and animals. These methods, however, suffered from the fact that members to be employed for crossing have to be closely homologous, that is, the genes from distant organisms could not be brought together. Also, the transfer of 'harmful' genes along with the 'beneficial' ones could not be prevented.

Since the early 1970's, new techniques have been developed that allow the modification of the genotype of an organism in a predetermined and directed way. This approach has been called recombinant DNA technology, genetic engineering, gene cloning, or gene technology. The fundamental aspect of this technology is based on the creation of new combinations of genetic material from DNA molecules of different origin. These recombinant molecules are introduced into appropriate, albeit unnatural, host organisms, where they can be multiplied and selected for conferring a particular phenotype. The protocol for a recombinant DNA experiment essentially consists of following steps (Fig. 7.34):

1) The "foreign DNA" to be cloned is isolated by an appropriate technique
2) The vector or the carrier molecule is similarly chosen and processed
3) The "foreign" and vector DNA are joined covalently
4) The recombinant molecule is introduced in a suitable host cell by transformation

or transfection.

5) Cells harbouring recombinant molecules are recognized and isolated by suitable selection procedures

6) Individual positive colonies, now called clones, are propagated and grown in bulk

7) The cloned DNA or its product may be isolated from the host cell in large amounts

7.5.1. Availability of Gene of Choice

One of the pre-requisites for a recombinant DNA (rDNA) experiment is the availability of the gene of choice on a suitable-sized DNA fragment. Following methods can be employed to generate them: i) mechanical shearing (ii) cleavage by restriction enzymes

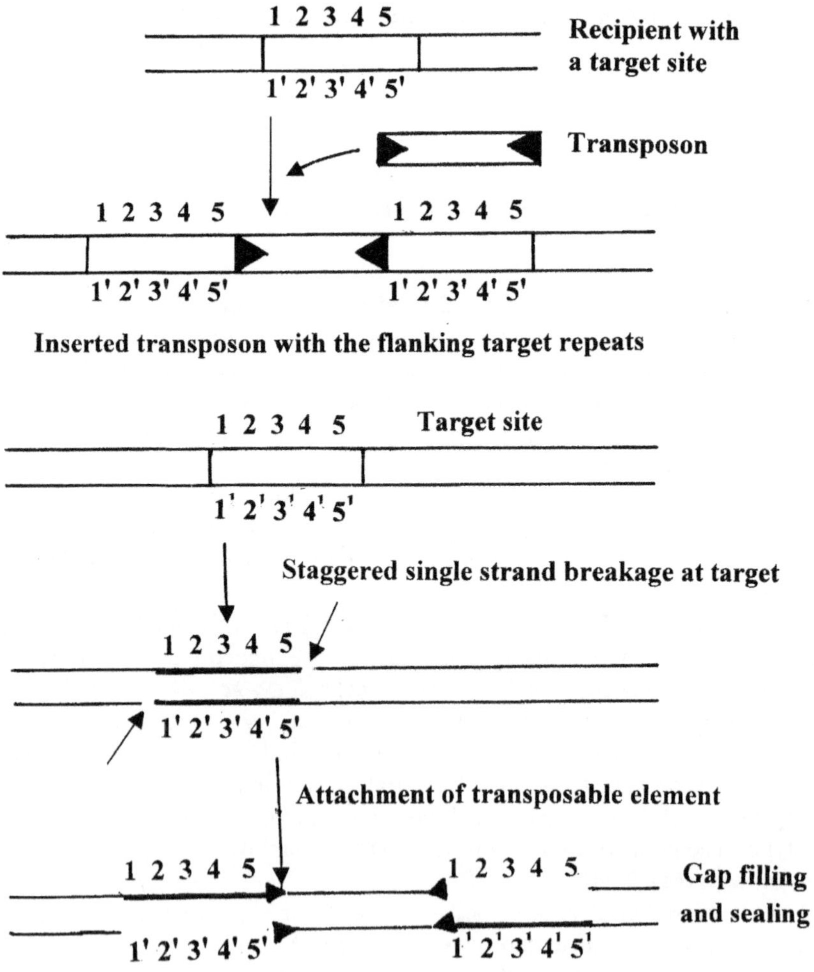

Figure 7.32. Mechanism of target site duplication during transposition.

(iii) RNA directed DNA (cDNA) synthesis (iv) chemical synthesis, and (v) PCR amplification.

7.5.1.1. Mechanical Shearing. Although the mechanical shearing is the simplest way of getting the DNA fragmented, the loss of some parts and difficult recovery of suitable sized fragments limits the use of this approach. Such shearing can be brought about by hydrodynamic forces, such as passing the DNA solution rapidly through a small orifice or by vigorous stirring.

7.5.1.2. Enzymatic Cleavage. The technique for the dissection of DNA molecules into discrete fragments by specific enzymes became known only in late sixties and laid the foundation of rDNA experiments. This approach was derived from the observation that the behaviour of a phage may differ in different strains of host bacterium. Phage λ, for example, grown on *E.coli* B, grows very poorly (3-5 orders of magnitude) on *E. coli* K. Those particles which succeed to grow on strain K thrive very well on this strain in consecutive infections. One growth cycle in *E.coli* K will, however, result in their poor growth on strain B. It was explained by suggesting that *E. coli* K is the restricting host which reduces the biological activity of λ DNA. However, some of the phage DNA is effectively modified so that it starts growing efficiently in strain K. The former process was termed as "restriction" and the latter as "modification". Experiments of Arbor and Linn (1968) and Meselson and Yuan (1968) revealed that restriction is caused by an enzymatic cleavage by restriction endonucleases and modification is the result of DNA methylation. Three different classes of restriction enzymes are currently known:

Type I - These are multifunctional enzymes, acting simultaneously as an endonuclease and a methylase and require Mg^{2+}, ATP, and S-adenosylmethionine as a cofactor. The cleavage site of the enzyme is located away from recognition site.

Type II - Unlike type I enzymes, the binding sites and cleavage sites of type II enzymes coincide and are specific. These are remarkably stable and only require Mg^{2+} as a cofactor. They generate complementary staggered ends in most cases.

Type III - These nucleases also cleave DNA at specific sequences, which lie in the vicinity of the recognition sites. Although they generate protruding ends, they differ from each other and, therefore, can not be recombined.

Following recommendations of Smith and Nathans (1970), each enzyme is represented by a three-letter code derived from the latin name of the bacterium from which the enzyme is isolated. For example, enzyme *Eco* stands for *Escherichia coli*, the source of the enzyme; similarly *Pst* is obtained from *Providencia stuartii*. These names are, therefore, italicized. A fourth letter may be added to identify either the serotype of the source organism or the strain. *Hinf*, for example, is derived from f serotype of *Haemophilus influenzae* and *Eco*R from the strain R of *E. coli*. Sometimes, more than one enzyme may be isolated from the same strain. These are usually distinguished from each other by roman numerals, such as *Hae* II or *Hae* III, *Eco* RI or *Eco* RII (Table 7.4).

Of the three, type II enzymes are most popularly used and, therefore, will be discussed in some detail. The type II restriction enzymes may recognize tetra-, hexa-, or octanucleotide sequences on the DNA. The number of times, on an average, that these sites are likely to occur in a genome is easy to calculate. As any of the four bases, A, T,

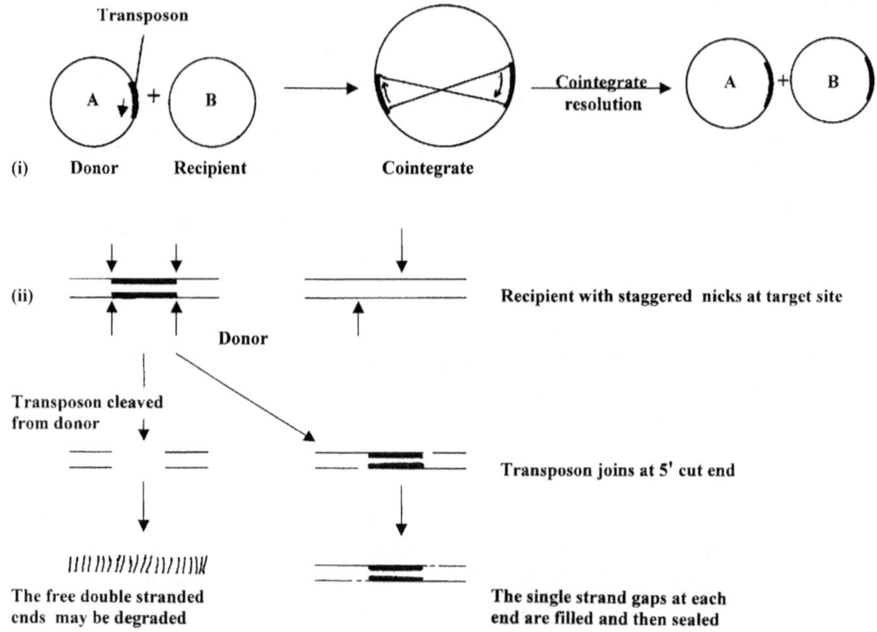

Figure 7.33. Two modes of transposition. (i) Replicative, (ii) Non-replicative.

C, or G is possible at each position in the recognition sequence, a four-hitter enzyme should have sites every: $4^4 = 256$ bp; six-hitter every $4^6 = 4096$ bp, and a eight-hitter $4^8 =$ every 65,536 bp. The latter, therefore, are called as rare cutters. This is, however, an approximation and actual number of sites are dependent upon many factors. The recognition sequences are conventionally written from 5'-3' direction. For example, site for *Eco* RI is:

5'- G ↓ A A T T C - 3'
3'- C T T A A ↑ G - 5'

The majority of the known recognition sequences have an axis of rotational (dyad) symmetry. Thus, the sequence reads the same in opposite strands when read in 5'-3' direction, e.g.,

5'- G A A ¦ T T C - 3'
3'- C T T ¦ A A G - 5'

Type two restriction enzymes cleave DNA at their recognition sequences in two ways. While some create protruding, staggered, single-stranded 5' or 3' ends, others make even cuts to generate flush or blunt ends (Fig. 7.35).

The symmetry of the sequence and the generation of staggered ends which are complementary in antiparallel allows two DNAs, which are completely non-homologous but cleaved by the same enzyme, join end to end. This is regarded as one of the most important milestones in the development of rDNA technology.

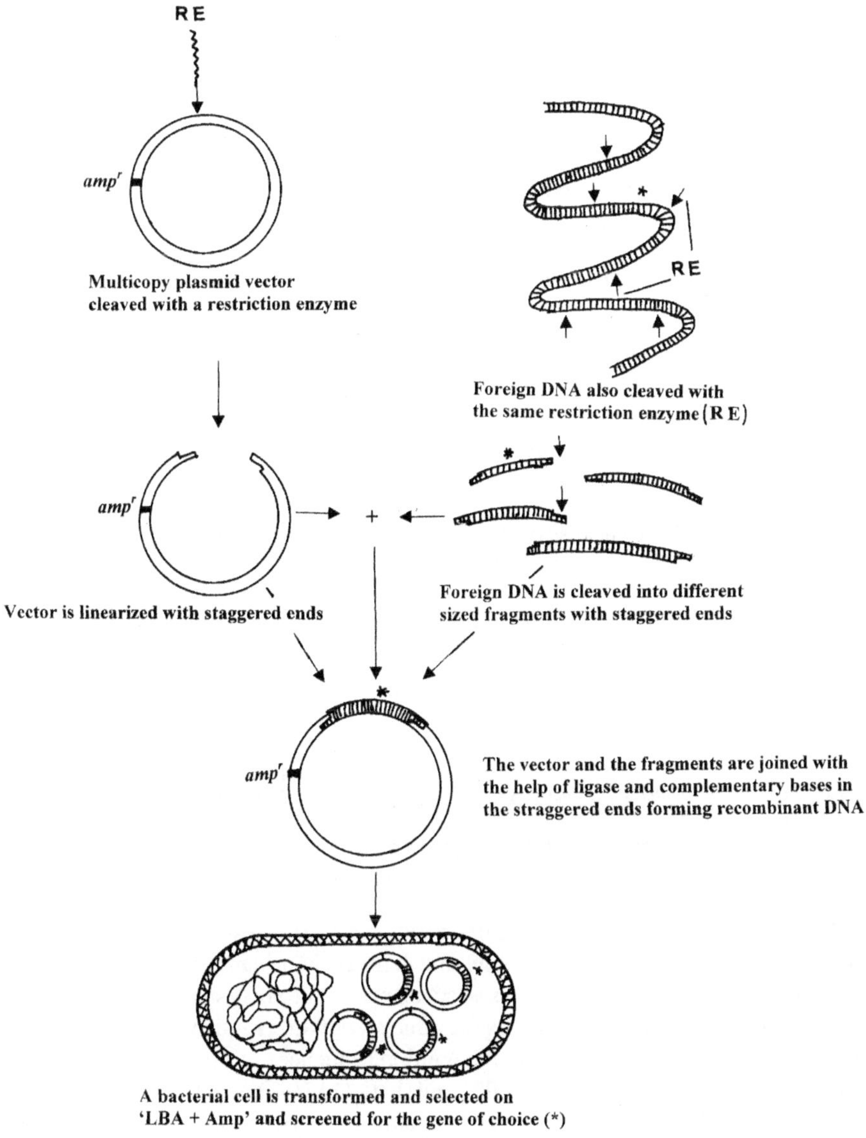

Figure 7.34. Schematic representation of a typical recombinant DNA experiment.

Table 7.4. Some commonly used restriction enzymes with their source, recognition, and cleavage sites starting with 5' end

Enzyme	*Source*	*Restriction site*
Bam H1	*Bacillus amyloliquefaciens*	G ↓ G A T C C C C T A G ↑ G
Bgl II	*B. globiggi*	A ↓ G A T C T T C T A G ↑ A
Eco RI	*Escherichia coli*	G ↓ A A T T C C T T A A ↑ G
Hind III	*Haemophilus influenzae*	A ↓ A G C T T T T C G A ↑ A
Hae III	*H. aegypticus*	G G ↓ C C C C ↑ G G
Hpa II	*H. parainfluenzae*	C ↓ C G G G G C ↑ C
Pst I	*Providencia stuartii*	C ↓ T G C A G G A C G T ↑ C
Not I	*Nocardia otitidis-caviarum*	G ↓ C G G C C G C C G C C G G C ↑ G
Sma I	*Serratia marcescens*	C C C ↓ G G G G G G ↑ C C C
Taq I	*Thermus aquaticus*	T ↓ C G A A G C ↑ T

Note: The target sites are 4-, 6-, or 8 nucleotides long and have a typical dyad symmetry

Some restriction enzymes' recognition sequences are not symmetrical. These may recognize up to four different sequences and generate non-identical termini. There are still other enzymes which are known as isoschizomers. They may share the same cleavage and recognition sequences, or possess the same recognition sequence but differ in their cleavage site or some may differ in recognition sequences but still generate identical termini. For example, digestions with *Bam*H1 (G/GATCC), *Bgl* II (A/GATCT), *Bcl* I (T/GATCA) and *Sau*3A (/GATC) will generate 5'-GATC-3' cohesive ends. Thus, overhangs generated by *Bam*H1 and *Bgl* II can recombine to give a sequence 5'-GGATCT-3' that will now be resistant to digestion with both the enzymes. Such recombinant molecules, however, can still be cleaved by enzymes such as *Dpn* I, *Sau*3A, and *Mbo* I as their recognition sequence is 5'-GATC-3'.

Thus, using an appropriate enzyme, fragments can be generated from any DNA genome (both prokaryotic and eukaryotic). These fragments can be separated by agarose gel electrophoresis and their sizes can be determined by running a size marker (such as λ *Hind* III digest or 1kb DNA ladder) along with (Fig. 7.36).

7.5.1.3. cDNA Synthesis. Cloning of an eukaryotic gene in a prokaryotic vector and background is often fraught with the problems of gene expression. This is based on different plan of gene structure in these two groups. While eukaryotic genes are split (consist of exons and introns), the processing of hnRNA to mRNA by RNA splicing is alien to prokaryotes. Under such circumstances an enzyme, first described from some animal viruses (retrovirus), called reverse transcriptase is used. This enzyme can make a complementary DNA strand from an RNA template. The RNA from RNA-DNA hybrid is chewed either by RNAseH or ribonuclease in the presence of alkali. Klenow polymerase or T4 polymerase is employed to convert ssDNA through a self priming mechanism into dsDNA and hairpin is digested with S1 nuclease. The single stranded DNA so converted into dsDNA is called complementary DNA (cDNA).

Such a cDNA will not possess the intron sequences of the original gene (Fig. 7.37a). These mRNAs are usually isolated from the tissues which may overproduce the particular species of mRNA and the purified mRNA is then used for producing cDNA.

7.5.1.4. Chemical Synthesis of Gene. A gene for cloning can also be made available by chemical synthesis. For such an exercise, either the base sequence of the gene should be completely known or the same should be derived from the base sequence of its RNA or amino acid sequence of protein product. With the pioneering work of H. G. Khorana, it became possible to chemically synthesize a gene from its known sequence. The one limitation of this approach, however, is that the gene should be reasonably small so that it could be handled conveniently.

7.5.1.5. PCR Amplification. With the genome sequences available of the target or related organism, primers can be designed and gene of choice can be PCR-amplified. PCR involves heat denaturation of double stranded target DNA into single strands. The primers are short oligonucleotide sequences complementary to the flanking regions of target DNA/gene. The primers pair with heat-denatured target DNA (primer annealing) and in between DNA is replicated by a DNA polymerase (such as *Taq* DNA polymerase) using target DNA as template (primer extensions). The whole process is repeated for the required number of cycles (Fig. 7.37b). So strong is this technique, that with right kind of primers a few micrograms of a DNA region can be synthesized from a few molecules. This DNA can then be used for cloning and several other applications.

Of the different methods described above, restriction enzyme cleavage, cDNA synthesis and PCR amplification are the popular approaches. In rDNA experiments this will constitute the "foreign" gene.

7.5.2. Vector

The "foreign" gene needs a carrier molecule that will facilitate its transmission into and replication/expression in the host cell. The usefulness of a vector depends upon certain features, like: (i) it must be a small molecule, (ii) it should have the ability of replication, (iii) it should preferably carry a single site for a number of restriction enzymes, (iv) it should be easily transferable into a host cell, and (v) there must be a simple selection or screen for the vector.

The most commonly used vectors are plasmids, phage λ, phage M13, and structures derived from these.

7.5.2.1. Plasmids. A large number of plasmid vectors are available that may fall into low copy, medium copy, or high copy number classes. The plasmid must contain at least one origin of replication along with its *cis*-acting control elements. Plasmid vectors can be introduced into a bacterial host cell by transformation, for which the cells are made competent as described in the earlier Section. In order to select for successful transformants, selectable markers are required. The most commonly used selectable markers are genes that confer resistance to antibiotics like ampicillin, chloramphenicol, kanamycin (neomycin), and tetracycline.

In order to create an efficient vector, a large number of plasmids have been modified to incorporate the characters as described above (Fig. 7.38). In one such approach, the vector size was reduced by removing ancillary regions. This enables the vector to accept larger segments of foreign DNA, improves its transformability and characterization by restriction mapping, and increases its copy number which in effect improves the yield of foreign DNA. In another approach, the number of useful restriction enzyme cleavage sites within the plasmid vectors are expanded and their distribution is rationalized. This is achieved by adding synthetic cloning sites termed "polylinkers", "restriction site banks" or "multiple" or "polycloning site (MCS)" that consist of sequences that are recognized by commonly used restriction enzymes. Such vectors are pUC, pGEM, pBluescript etc.

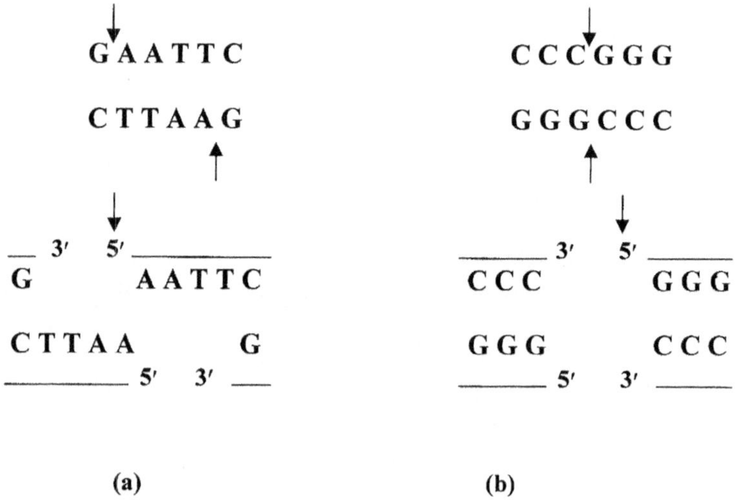

Figure 7.35. Restriction enzyme generated cuts at the target sites in a double stranded DNA. a. Staggered cut or cohesive-ended molecule (e.g., Eco RI), b. Flush cuts or blunt-ended molecule (e.g., Sma I).

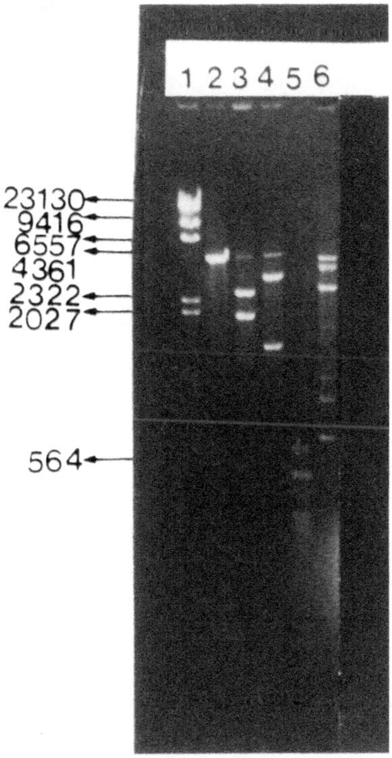

Figure 7.36. Restriction digestion of DNA and determination of size through a marker (λ Hind III digestion- lane 1). DNA sample digested with different enzymes (lanes 2-6).

In many of the earlier vectors, the plasmid carried two or more different selectable markers, and a restriction site located within one of the markers was used for cloning. In such a case, the insertion of foreign DNA could be identified by loss of that marker (insertional inactivation) (Fig. 7.39). In many of the modern vectors this facility is provided by having the MCS into a gene that codes for a biologically active fragment of an enzyme, such as ß-galactosidase, in a specific host cell. This is based on the fact that the part of the *lac* operon carried by the vector codes for the amino terminal fragment of ß-galactosidase, which in a host cell that provides the carboxy-terminal portion of the enzyme, leads to functional association and forms enzymatically active protein (α-complementation). Such cells when induced with IPTG produce blue colonies on a medium containing chromogenic substrate such as X-gal. Insertion of a foreign DNA fragment in MCS will disrupt α-complementation and the recombinant clones will appear white. Plasmids of pUC series and 'Blue-Script' belong to this category and recombinant selection is referred to as "blue-white" selection (see Fig. 7.46).

In some vectors, the origin of replication is derived from single stranded filamentous bacteriophage such as M13. These vectors are constructed in a duplicate set (+ and -) where M13 *ori* sequence is inserted in opposite orientation (plasmid Bluescript). This

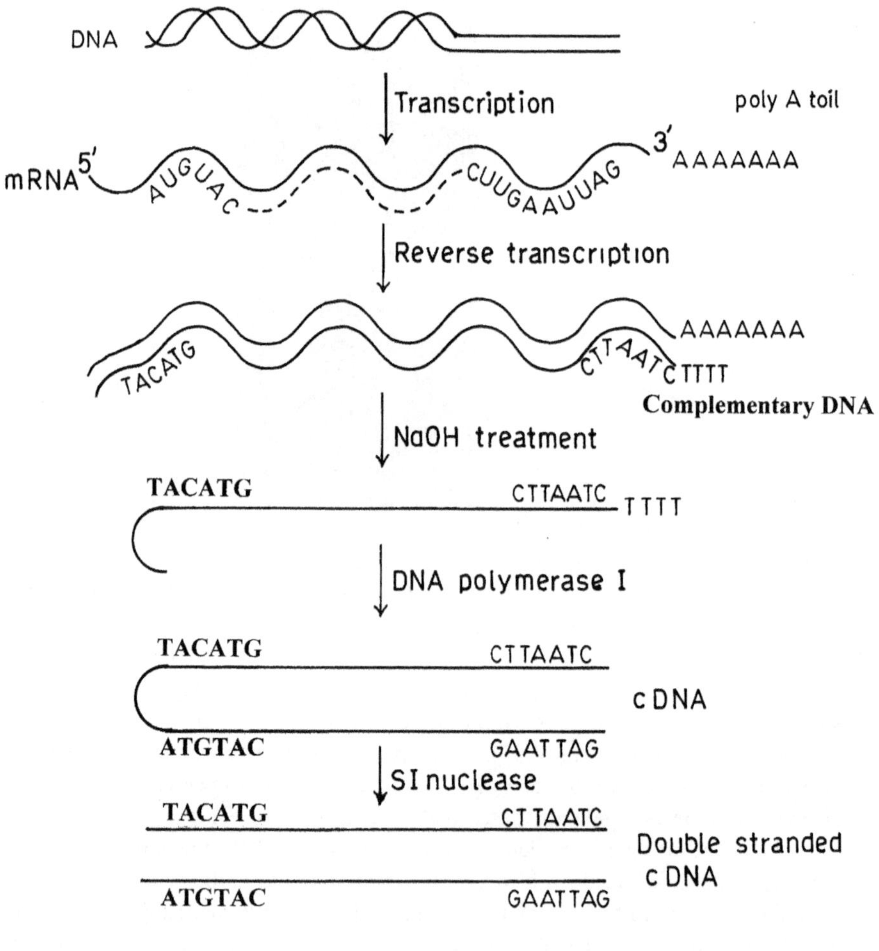

Figure 7.37a. Steps in the synthesis of cDNA. mRNA of an appropriate gene is isolated, poly A tail is hybridized with short oligo-dT chain that serves as a primer for the action of reverse transcriptase, reverse transcriptase synthesizes a complementary DNA (cDNA) strand using RNA as a template. The resulting DNA strand ends in a hairpin loop, mRNA is degraded by alkali treatment, hairpin loop is extended by DNA polymerase I, the double stranded cDNA provides the gene which can be cloned.

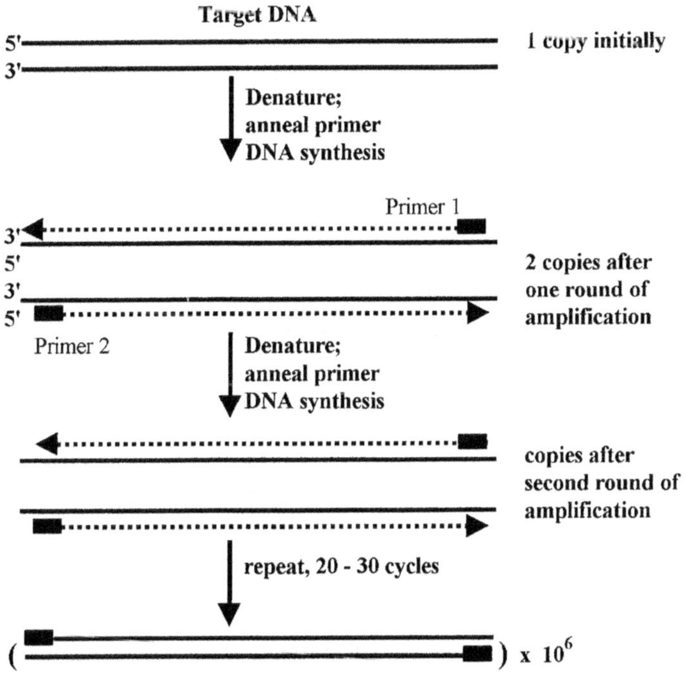

Figure 7.37b. The Polymerase chain reaction (PCR) for amplification of specific DNA sequences. Target DNA is heat denatured for strand separation. Excess of oligonucleotide primers complementary to both the strands is added along with DNA polymerase. Primer annealing is followed by primer extension that yields a copy of the double stranded target DNA. This cycle can be repeated.

allows copies of one of the two strands of plasmid DNA to be produced in sufficient numbers for sequencing or for preparation of radio-labelled single-stranded probes.

Many other vectors carry promoters derived from bacteriophage T3, T7, and/or SP6 adjacent to the polycloning site, allowing the transcription of inserted DNA. In another modification, a different promoter is present at each end of MCS. This enables the preferential transcription of an RNA species depending upon the RNA polymerase used. Such RNA can be used as hybridization probe or can be translated in cell-free protein synthesis systems. Examples of this group are pGEM, and pBluescript (see Fig.7.38). Another set of vectors may carry markers which upon transformation will kill the host cell. Insertion of foreign DNA in this region will, however, permit the direct selection of recombinant clones.

A class of useful vectors have been constructed to contain powerful promoter that generate large amounts of mRNA complementary to cloned sequences of foreign DNA. These vectors called expression vectors express foreign proteins either independently or as fused proteins (Fig. 7.40). Some examples are: pUR and pEX series of plasmids carrying *Plac* and λ PR promoter, respectively.

Often the gene cloned in a cloning vector is transferred to an expression vector to get the expression of the same. Such expression vectors contain the regulatory sequences that permit controlled expression of the gene. One such requirement is to have a strong promoter that ensures the binding of RNA polymerase. Strong *E. coli* promoters used for this purpose are *lacUV5* which controls ß-galactosidase, *Ptrp* controlling tryptophan biosynthesis, and of *ompF* that regulates the production of an outer membrane protein. A special promoter *Ptac*, constructed from part of *lac* and part of *trp* promoters is very useful. It combines the -10 consensus sequence of *lac* and -35 sequence of *trp* (for example pkk vectors).

Another important component is a good ribosome binding site that establishes the appropriate reading frame. To this end, expression vectors have been provided with *lac* promoter, ribosome-binding site, and operator so that *lac* inducer can be used to initiate protein production from the gene/s cloned in an organism.

7.5.2.2. Shuttle Vectors. Many vectors have been constructed to function in different cell types. Such vectors called as shuttle vectors permit the transfer of recombinant DNA from one type of organism to another. For example, the vectors able to multiply both in *Pseudomonas* and *E. coli* or in *E. coli* and yeast, *Saccharomyces cerevisiae,* have been developed (Fig.7.40).

7.5.2.3. Phage Lambda. The use of phage λ as a vector stems from the fact that it has the ability to carry a bacterial gene (specialized transduction). The large size and complexity of the genome, multiple restriction sites for several restriction enzymes and the limitation of the genome size that phage λ can carry, however, does not permit its use directly as a vector.

These problems are overcome by suitably modifying the phage genome. From the genetic studies on specialized transduction, it was known that the central third of the viral genome is not essential for lytic growth and thus can be dispensed with (Fig. 7.41a). Also the cleavage sites for some enzymes were shifted and located in the non-essential region. Recombinants between λ phage and other lambdoid phages (e.g., Ø 80) helped in getting genomes that either contained or lacked particular sites. Together, these methods allowed the construction of two types of vectors: (i) vectors having single site for insertion of foreign DNA or "insertion" vectors, and (ii) vectors having a pair of restriction sites flanking a segment of non-essential regions of λ DNA, that can be replaced by foreign DNA or "replacement" vectors. The region that can be replaced this way is also known as "stuffer" portion.

Some of the popularly used λ based vectors are: the Charon series, EMBL series, λ 2001, λ DASH, and λ FIX as replacement vectors and λ gt series, λ ORF, and λ ZAP as insertion vectors. While the former have been designed to clone large fragments, the latter are more suited for small DNA fragments (Fig. 7.41b).

7.5.2.4. Cosmids. Cosmid vectors are derived from plasmids that carry *cos* sequences from phage λ. This allows the packaging of such a vector into λ particles. They also carry ColEI origin of replication and usually amp^r as a drug resistance marker. Cosmid vectors have been designed to clone large pieces of DNA (~35-45 kb) and are exceptionally useful in cloning and propagating an entire eukaryotic gene or family of genes (see Fig. 7.38).

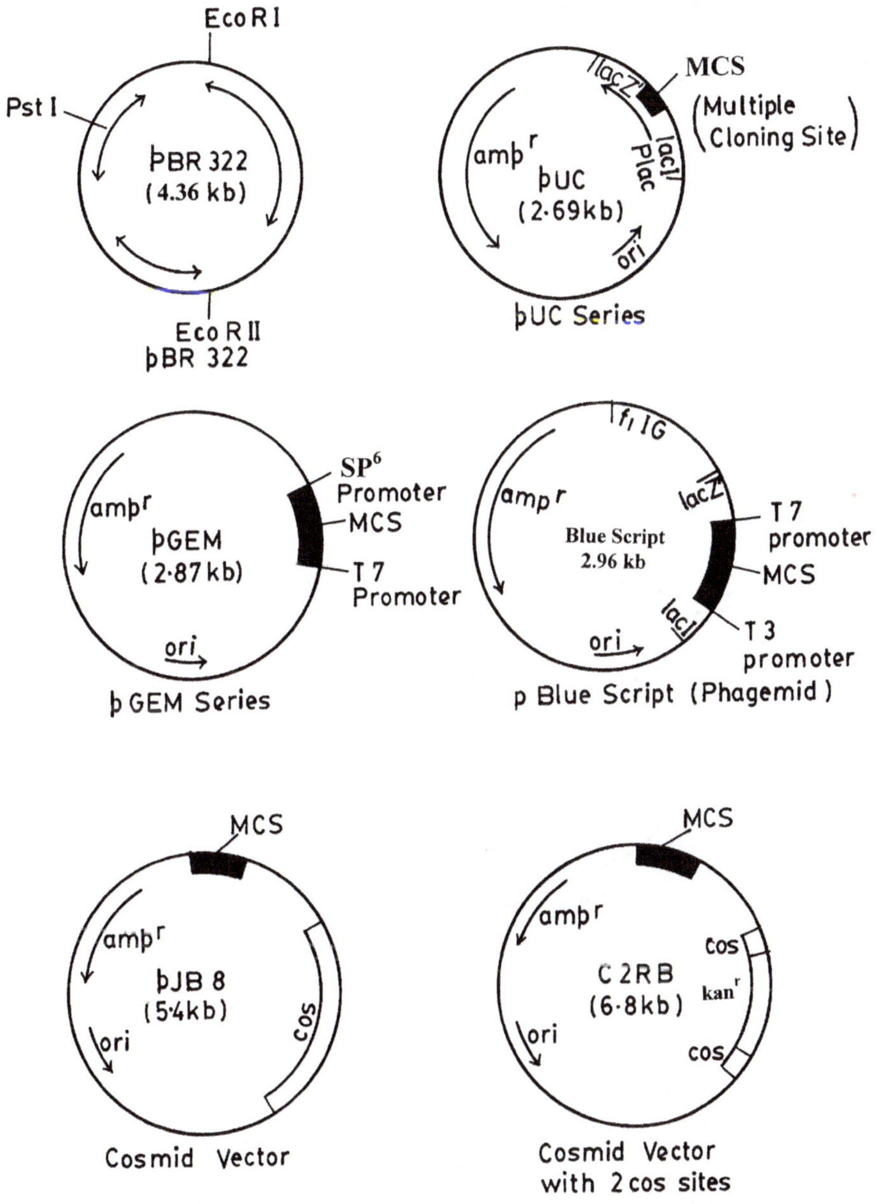

Figure 7.38. Some of the representative vectors, such as, plasmid, cosmid, and phagemid. The ultimate choice of a vector will be determined by a large number of factors (see text).

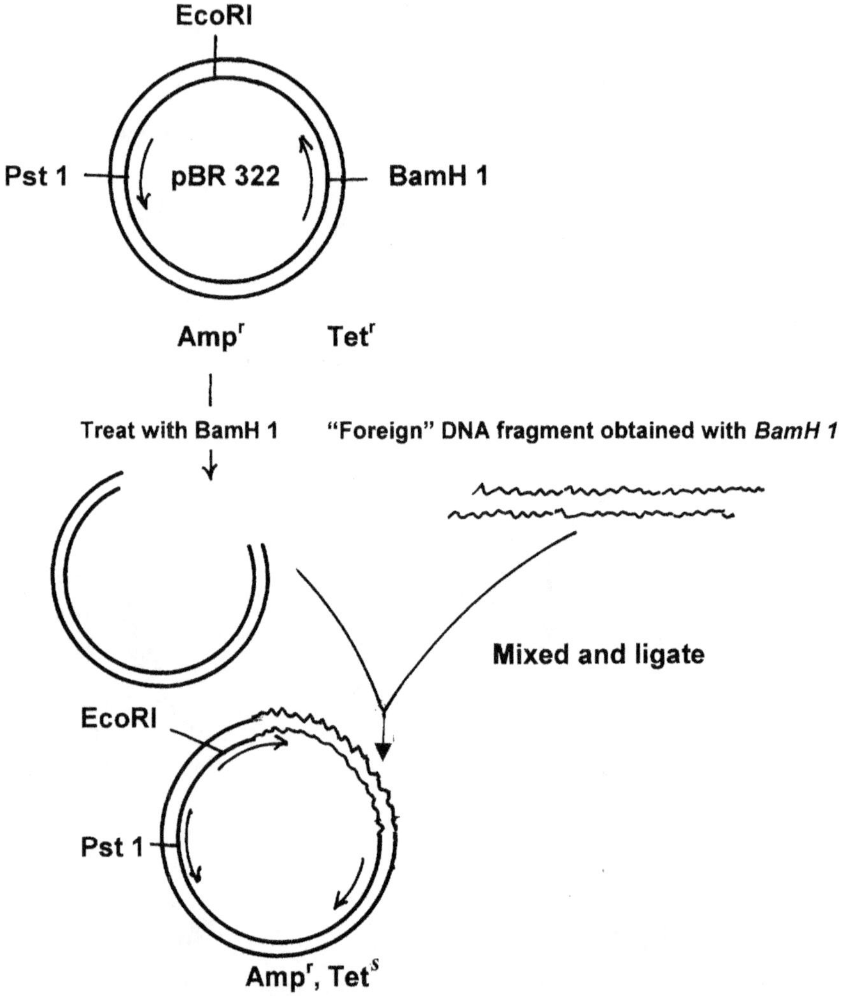

Figure 7.39. One of the ways to choose a restriction enzyme is the one which cuts through one antibiotic resistance gene but leaves the other, also called as insertional inactivation (see also Fig. 7. 45). Insertional inactivation. A piece of foreign DNA generated by BamHI digestion is inserted in BamHI digested pBR322. The insertion will disrupt tet^r gene and the recombinant plasmid will be amp^r but tet^s. A similar effect will be seen if a PstI site is used for cloning except that the recombinant plasmid will show amp^s and tet^r phenotype.

A family of modified cosmid vectors termed "charomids" carry spacer sequences consisting of head-tail repeats of a 2-kb fragment derived from pBR322. Different members carry spacers with different number of repeat units and can accept DNA ranging in size from 2-45 kb.

7.5.2.5. Single Stranded Bacteriophages. Many filamentous coliphages like M13, f1, and fd consist of single-stranded closed circular DNA, ~6.4 kb in length. In M13, all of the ten genes are essential and thus can not be dispensed off in creating a vector. Most of these ten genes are separated by few nucleotides, but two extended intergenic regions exist between genes VIII and III, and II and IV containing *cis* elements that regulate gene expression and DNA replication. Insertion of foreign DNA in the regions II and IV (~500 nt), therefore, severely affects the genome replication. However, it has been observed that such insertions are compensated by accumulation of mutations in gene II or IV restoring the function of these elements. Since single-stranded DNA is a poor substrate for most restriction endonucleases and ligases, double-stranded foreign DNA segment is inserted in vitro into the replicative form of phage DNA and introduced into bacterial cells by usual transformation procedures. One of the commonly used vectors of this group is the mp series of M13 vectors, that carries part of *lacZ* gene of *E. coli*. These vectors are preferred for DNA sequencing, generating probes for DNA hybridization, and site-directed mutagenesis.

7.5.2.6. Phagemid. A group of vectors that combine the desirable features of both plasmids and filamentous phages have been developed and are referred to as phagemid. They are essentially made up of ColE1 origin of replication, a selectable marker in the form of antibiotic resistance and a copy of the major intergenic region of a filamentous phage. Segments of foreign DNA inserted in these vectors can be propagated as plasmids in the conventional way. When infected with a suitable helper phage, the phagemids can be processed and packaged. pBluescript belongs to this series of vectors (see Fig.7.38).

7.5.2.7. Bacterial Artificial Chromosomes (BAC). One of the problems faced in some r-DNA experiments is the cloning and stable maintenance of large (>100kb) inserts in *E.coli*. The most commonly used plasmid vectors can stably carry a size of 10kb or less. A large insert size can pose several problems, such as poor replication and/or segregation of plasmid clone, the poor growth of cell due to production of toxic products by foreign gene expression, and structural changes in the insert DNA such as cruciforms which are readily deleted in *E.coli*.

In 1992, Shizuya and Simon developed vectors that were called bacterial artificial chromosomes or BACs to overcome these problems. These vectors are plasmids that carry *ori* of F factor, a large plasmid that is capable of replicating a molecule larger than 100kb in length. Its low copy number helps in preventing problems of segregation and production of inhibitory products. The F *ori* and the replication fork also provided it a superhelicity different from other high copy number vectors. BAC vectors were further engineered to carry selectable antibiotic resistance genes and restriction enzyme sites to insert or remove foreign DNA. These vectors have been successfully employed in cloning and maintaining mammalian DNA in *E.coli* such as in Human Genome Project.

7.5.3. Combining the "Foreign" and Vector DNA (r-DNA Formation)

In order to create recombinant molecule, the desired foreign DNA and the appropriate vector must be joined. For this joining, the two DNAs will have to be suitably processed by a variety of means.

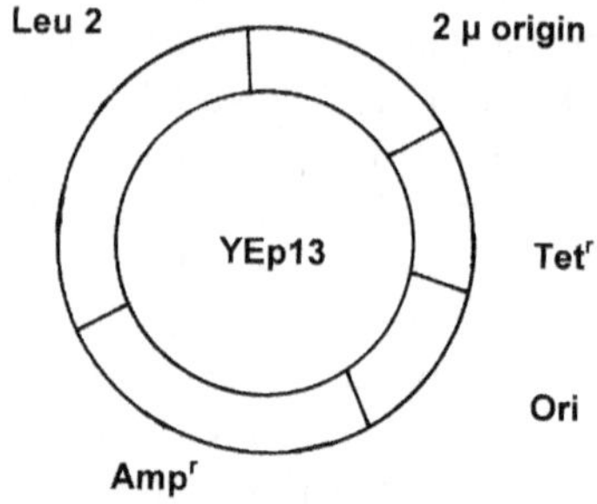

The shuttle vector YEp13. It carries an *ori* (2 μ *ori*) that will function in yeast, *S. cerevisiae* and another in *E. coli*. It contains genes *Leu 2* and amp^r and tet^r of which the first can be selected in yeast and the latter two in *E. coli*

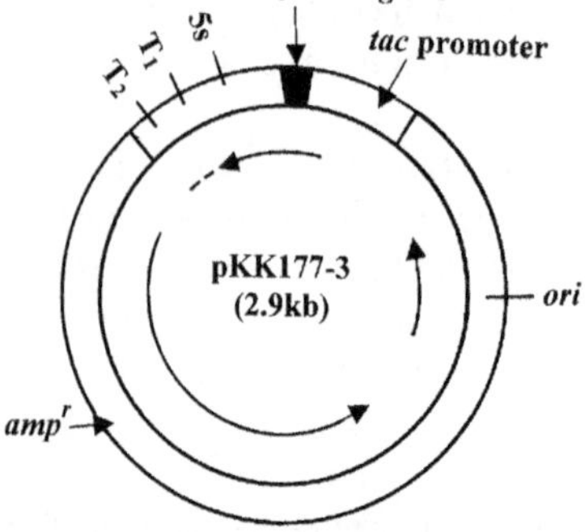

A *tac* (fused *trp-lac* promoter) vector that contains a polycloning site (*Eco*RI, *Sma*I, *Sal*I, *Pst*I and *Hind*III) downstream at which a gene can be cloned. Downstream from these sites is *rrn*B which contains *E. coli* 5S gene and the T1 and T2 terminators. This promoter can be induced by IPTG

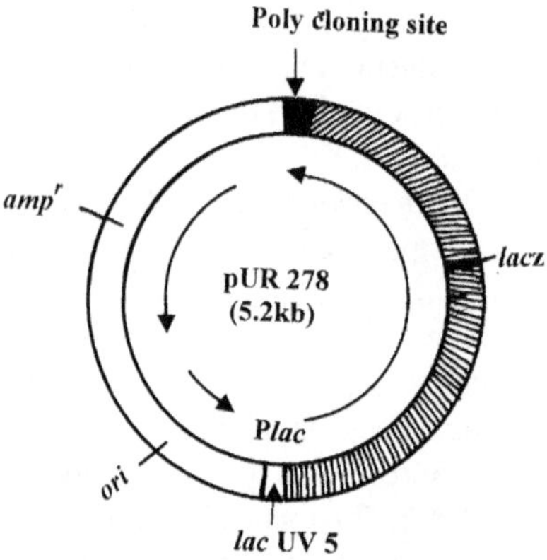

This *lacz* fusion vector has the cloning sites *BamH*I, *Sal*I, *Pst*I, *Xba*I, *Hind*III, and *cla*I in all three reading frames at the 3′ terminus of the *lacz* gene allowing the production of a fused active β-galactosidase and the peptide encoded by cloned gene

Figure 7.40. As per the requirement a large number of vectors have been designed. Those employed for expressing a foreign gene are called expression vectors and those which can be used in different experimental systems are termed shuttle vectors.

As stated earlier, the digestion of a DNA with a particular restriction enzyme generates complementary single stranded cohesive ends. Thus, irrespective of the source of DNA, the ends generated will have complementary sequences. This provides a powerful tool for combining DNA molecules of different organisms by renaturation of the complementary single stranded termini. Once this hybrid (recombinant) molecule is treated with an enzyme, ligase, the paired ends are covalently sealed (ligated) (Fig.7.42).

If the single-stranded cohesive ends are too small or are not produced (blunt ends), an enzyme terminal nucleotidyl transferase can be used. This is an unusual DNA polymerase that adds nucleotides to the 3'-OH end without the requirement of a template. In order to facilitate the functioning of this enzyme, the DNA molecule can be treated with a 5'-specific exonuclease to expose 3'-OH terminus. In the presence of dATP, for example, terminal nucleotidyl transferase will form a poly (dA) tail at both 3'-OH ends. If instead, dTTP is added, a poly (dT) tail will appear at two ends. Combining two such DNAs will become possible as poly (dA) tail will anneal with poly (dT) tail. This method is also known as homopolymer tailing and is completed by any gap filling with DNA polymerase I and sealing the remaining nicks with DNA ligase (Fig. 7.43).

Some restriction enzymes cut DNA with flush ends. Also the blunt ends may be created if the desired sites for a restriction enzyme are not present in the two DNA molecules. In such cases, the two DNAs are cut with different restriction enzymes thus generating non-complementary ends. The single stranded termini are then either filled with DNA polymerase or removed by 3'-5' nuclease. The blunt-ended DNA fragments are then joined by T4 DNA ligase. This ligase produced by phage T4 not only seals the 3'-OH and 5'-P nicks but also DNA with blunt ends. After blunt end ligation, the recombinant site loses the capability of being cleaved by either of the enzymes used originally to generate the fragments (Fig.7.43).

A DNA fragment with cohesive termini can be recombined with a blunt-ended DNA fragment by adding a short DNA segment (a linker) to the latter at both ends. The linker is so designed so as to contain restriction site for an enzyme and the overhangs created by digestion are then used for annealing with the vector (Fig.7.43).

Thus, a recombinant molecule can be generated by combining a fragment of DNA derived from any source into an appropriate vector. Once this recombinant molecule is introduced into a bacterium, the foreign DNA replicates as part of the vector and is thus said to be cloned. Sometimes, a random collection of DNA fragments obtained from a particular source are annealed with a vector, generating a large number of different hybrid molecules. Such a procedure is called "shot gun cloning". When a DNA and the vector is double digested (treated with two restriction enzymes), the annealing of the two will lead to directional cloning as only the appropriate termini will base pair. It has, thus, been possible to create recombinant molecules in which the vector may carry different fragments representing the total cellular DNA. Such collections are called gene banks or gene libraries.

7.5.4. Introduction of Recombinant Plasmid/Phage

The recombinant plasmid/phage is introduced into an appropriate host cell by

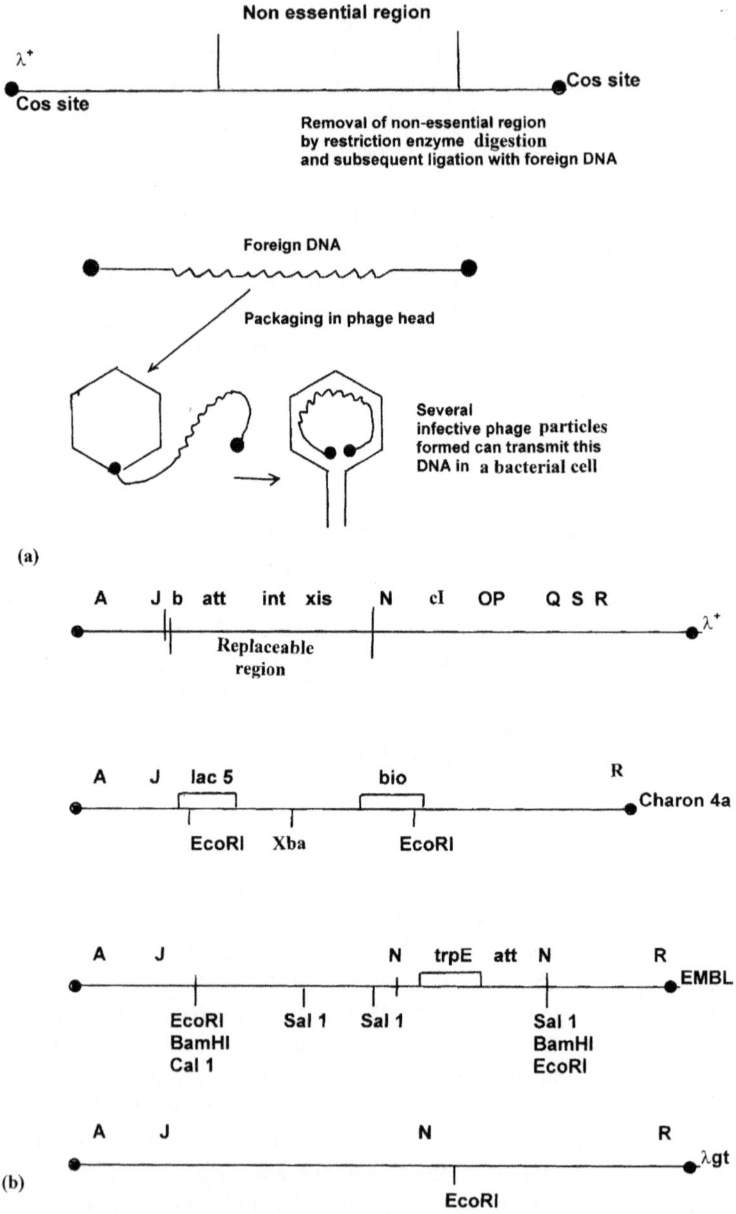

Figure 7.41a. Phage lambda as a cloning vector, b. Some of the λ-derived vectors used for cloning several substitution/deletions are effected in the non-essential (replaceable) region to maintain an approximate size of the genome. Some restriction enzyme sites used for cloning are depicted. • represents cos site.

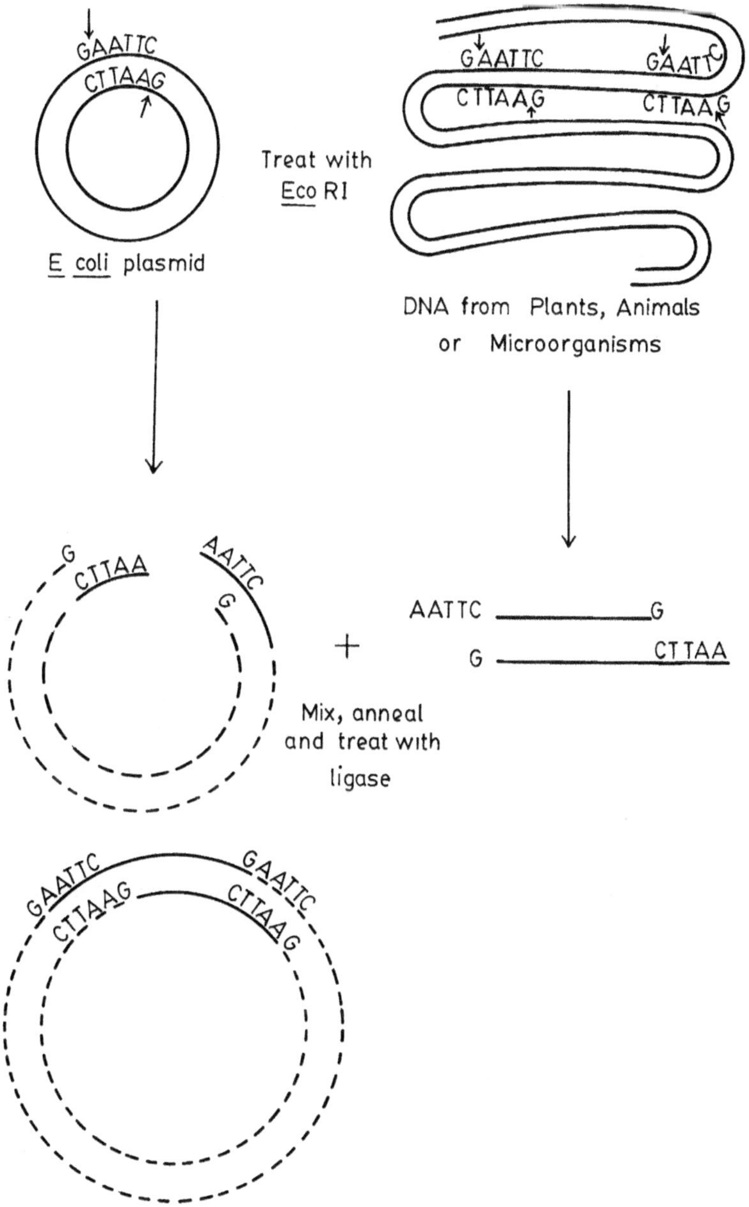

Figure 7.42. Recombinant DNA can be generated by digesting a suitable vector with restriction enzyme and also generating DNA fragments with the same enzyme from theoretically any source. Bringing them together will produce a chimeric molecule (rDNA) that can be ligated finally by the enzyme ligase.

transformation or transfection. We have already discussed these processes earlier. The choice of the host cell depends upon the vector to be used and requires careful planning.

7.5.5. Identification of Recombinant Molecule

The cloning of a DNA fragment in an appropriate molecule, however, requires an appropriate strategy. As the readers may comprehend, the digestion of DNA with restriction enzyme and subsequent ligation can result into three different types of molecule (Fig.7.44): (i) the vector can religate without an inserted fragment, (ii) the vector may ligate with one or more copies of an insert, and (iii) the fragment itself may religate without the vector. While the latter will generate a highly unstable molecule because it is likely to lack the replication machinery, some methods will be required to differentiate between the first two situations.

7.5.5.1. Insertional Inactivation. One commonly used procedure for this purpose is based on insertional inactivation. In this the cloning site is located within one antibiotic resistance marker leaving the cell carrying a recombinant molecule sensitive to that particular antibiotic. Plasmid pBR322 (amp^r, tet^r), for example, has *Bam* H I and *Sal* I sites within the *tet, Pst*I site in *amp* and *Eco*RI site outside both the antibiotic resistance genes. Cloning at first of these two sites will make the cell Amp^r, Tet^s; at *Pst*I site Amp^s, Tet^r; and in the last both the resistances will be retained. Thus, the recombinant plasmid is transformed into an appropriate *E.coli* strain and selected on medium containing the appropriate antibiotic. The resistant colonies are then screened on medium containing other antibiotic. While the recombinants will be sensitive to only one, non-recombinants (vector alone) will exhibit Amp^r Tet^r phenotype (Fig.7.45).

7.5.5.2. Dephosphorylation. In another procedure, the cleaved plasmid is treated with alkaline phosphatase, that removes 5'-P termini, leaving 5'-OH group. Because DNA ligase require 5'-P, a dephosphorylated vector can not self-ligate. However, ligation will be possible with another DNA fragment that has a 5'-P termini. This ensures the formation of recombinant molecule.

7.5.5.3. Reporter Gene. As we have seen earlier, many vectors contain a part of the *lac*Z gene in their polycloning site. Thus, a non-recombinant vector in a suitable host produces blue colonies in the presence of a chromogenic substance (IPTG+X-gal) due to α-complementation. Insertion of foreign DNA into polycloning site disrupts α-complementation leading to white colonies (Fig. 7.46). Such genes, which code for an easily detectable trait are also known as reporter genes. Another useful reporter system involves luciferase operon (*lux*) of *Photobacterium fischeri*. Due to its *lux* genes, this marine bacterium possesses the property of bioluminescence. The presence of a desired DNA sequence can be detected, by linking it to *lux* genes, or in other words by the emission of light.

7.5.5.4. Direct Methods. More direct methods are also available for identifying the clones. For example, a plasmid containing an insert is larger by the size of the insert. The vector and the recombinant molecules can thus be differentiated by agarose gel electrophoresis. Further confirmation can be obtained by restricting the two DNA molecules with the enzymes used for cloning. While the vector will undergo linearization as it preferably contains a single restriction site, the recombinant will release the cloned

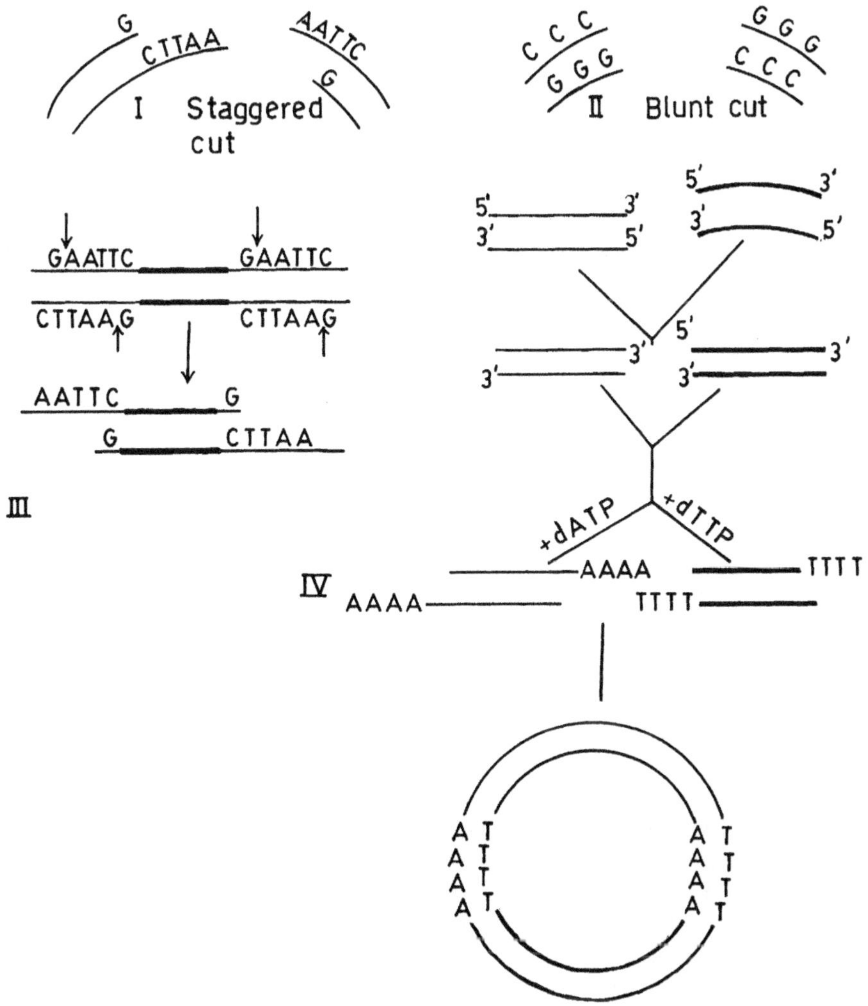

Figure 7.43. Schematic representation of the ways by which recombinant molecules can be formed. I. cohesive ended molecule can be recombined by complementary base pairing (see also Fig. 7. 42), II. two blunt ended molecules can be ligated with the help of ligase, III. a synthetic linker carrying a site for restriction enzyme can be added to the DNA fragment to be cloned, IV. complementary homopolymer tails can be added to the two molecules. The (III) and (IV) are done when either a restriction site is not available or it produces a blunt end molecules, or staggered ends are too small.

fragment. On agarose gel, they will produce one and two bands, respectively (Fig.7.47).

7.5.5.5. Complementation. A cloned gene can also be detected by complementation. For example, a '*trp*' *E. coli* strain can be transformed with a gene from tryptophan biosynthetic pathway cloned in a plasmid vector. The transformants are selected on medium not containing tryptophan, which will arise due to complementation of *trp*$^-$ of *E.coli* by *trp*$^+$ foreign gene. Such a method, however, can not be used for eukaryotic gene as they are either not properly expressed or may not have corresponding homology to the bacterial gene.

7.5.5.6. Hybridization. A number of hybridization procedures are available to identify the desired clone. For this, a labelled oligonucleotide probe is prepared for the desired gene either from a homologous source, or derived from some amino acid sequence of the protein product (Fig.7.48).

In colony or in *situ* hybridization, colonies transformed with recombinant plasmids are replicated onto a nitrocellulose filter. The filter is treated with NaOH to lyse the cells and release the denatured DNA and then dried to fix this DNA to the paper. The filter is flooded with ^{32}P-labelled probe which is complementary to the desired gene. The filter is then incubated for renaturation to occur, washed to remove unbound label, dried and placed on autoradiographic film. The colonies containing the recombinant plasmid appear as a dark spot on the autoradiogram. The corresponding colony can then be picked up from the original plate and maintained as the source of the cloned gene.

A similar assay can be applied to phage vectors where plaques are replicated and therefore the procedure is known as plaque hybridization.

Several other blotting/hybridization techniques are available that not only allow the identification of recombinant DNA but additionally for detecting them in a mixture, mapping of specific sequences in recombinant DNA, and even evolutionary relatedness. All these require the preparation of a sequence specific probe. These probes could either be radiolabelled or even non-radioactive labelling protocols have been developed. Such a technique was originally developed by Southern and is referred to as Southern blotting. In this technique, DNA fragments are separated in agarose gels, denatured by alkali treatment and transferred to nitrocellulose filter on which they can be immobilized by heating. Such nitrocellulose filters, which contain DNA in the same pattern as on the gels, is incubated with ^{32}P-labelled DNA or RNA probes and hybridizing DNA fragment/s are visualized by autoradiography.

Subsequently it became possible to transfer RNA or proteins to nitrocellulose or preferably chemically activated filter and were named as "Northern" and "Western" blotting, respectively. Specific bands on "Northern" filters can be visualized by hybridization with ^{32}P-labelled RNA or DNA probes. Similarly, proteins can be identified by specifically binding ^{125}I-labelled antibodies (Fig. 7.48). Some non-radioactive labelling techniques have also been developed.

For details about these techniques readers can consult any book dealing with Molecular Biology.

7.5.6. Detection of Cloned Gene Product:

Cloning of a gene needs to be followed by the detection of gene expression and the

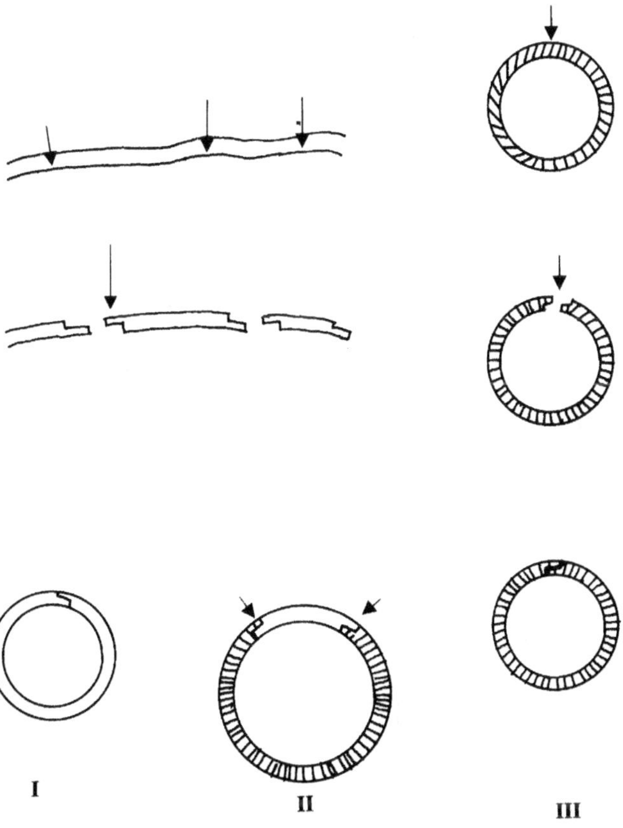

Figure 7.44. Based on the base complementarity of the ends, three possible structures can be formed when restriction digested DNA is mixed with digested vector. The digested fragment from the same source (I) or even vector can religate (III). The former is not likely to survive as it may lack a replication machinery, and that a recombinant molecule (II) is formed.

synthesis of the desired product. Several such techniques are available now.

7.5.6.1. Immunological Techniques. If the host cell is able to synthesize the protein product from the cloned gene, two techniques can be used to identify the "protein-producing" colony. Readers can also consult **Chapter 9** for some details about these techniques.

If the protein is excreted, a radioactive antibody is allowed to react with the colonies and the radioactive signal is perceived through autoradiography. The matching of the radioactive spot with the colony will lead to the identification of the desired clone. Another test called as immunoprecipitation test is carried out by adding the specific antibody directly into the medium. In this test, the antibody will react with excreted protein and form a precipitate called precipitin that exhibits a visible white ring around

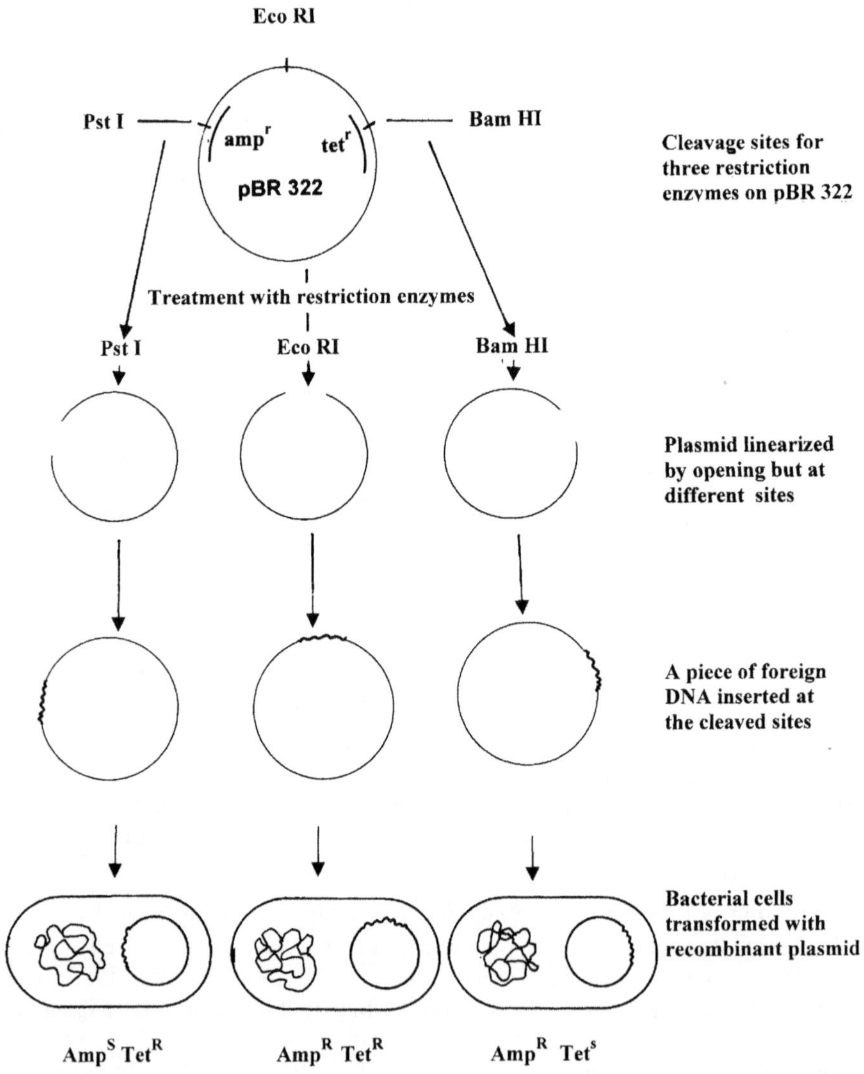

Figure 7.45. Identification of rDNA molecule. Insertional inactivation is one of the ways of identifying a rDNA molecule, as also explained in Fig.7.39. The loss of one or the other antibiotic resistance markers may provide a clue. For other details and methods see text.

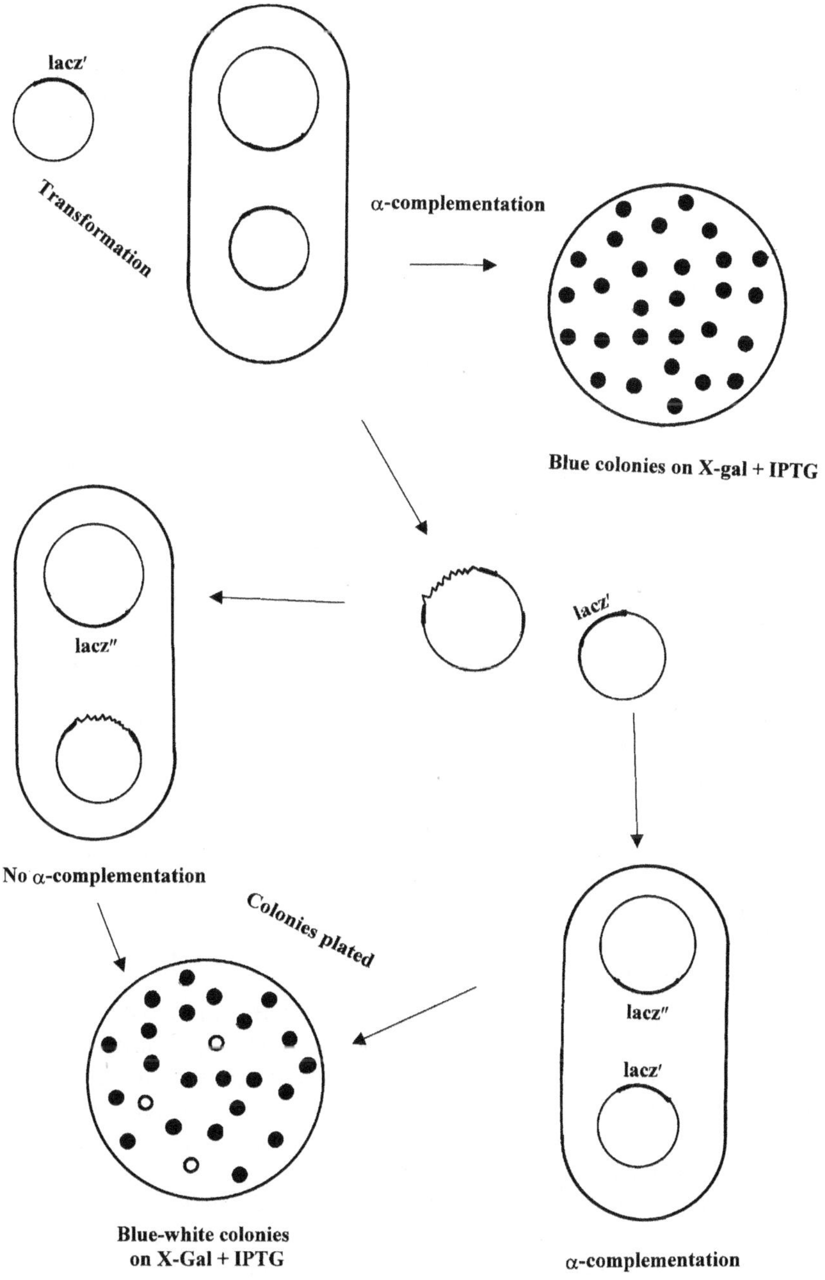

Figure 7.46. Blue-white selection for recombinant clones on chromogenic substrates (IPTG + x - Gal), based on α-complementation.

the colony of interest.

In case the protein product is not excreted, the technique is slightly modified. Cell lysis is brought about either by lysing the cells with lysozyme or using a host cell that is lysogenic to temperature sensitive λ cI857 and inducing the host by raising the temperature. In both cases the foreign protein will be released and react with antibody to give precipitin reaction.

7.5.6.2. Other Techniques. Bacteria have been instrumental in the development of several other molecular biology techniques, such as site-directed mutagenesis, PCR amplification, DNA and genome sequencing, etc., the details of which are outside the scope of this book but are readily available in several text books dealing with molecular biology.

7.5.7. Application of Genetic Engineering

Genetic Engineering or r-DNA technology has provided such a strong tool that it could be applied to innumerable or even unimaginable purposes. Such purposes encompass the field of biotechnology. While the vast multitude of microbial processes of great economic importance can be brought into and refined by biotechnological tools, even the non-microbial functions can be performed by a microbial strain after appropriate manipulation. The most important aspect of these manipulations is that not only the gene/s of economic importance can be handled by this way but also any other critical input from practically any other source can be brought together.

Thus, the application of biotechnology has spanned the fields like production of pharmaceuticals, diagnostics, and of vitamins, amino acids, organic acids, industrial and other enzymes, solvents and fuels. Production of different food products including single cell proteins through genetic manipulations has become a reality. Another interesting area is the production of transgenic plants and animals through which some useful "foreign" genes are directly expressed by the concerned plant or animal. Biotechnology has also been applied to environmental restoration through bioleaching, biodegradation, and bioremediation. We shall discuss these applications in detail in **Chapter 9**.

7.5.8. Synthesis of Cloned Gene Product

The wide-spread applications of rDNA technology can be realized only if the product of the cloned gene can be produced at commercial level. This itself requires the optimization of a number of parameters that defines a process known as fermentation. With the bacteria hosting a number of cloned genes of both "foreign" and microbial origin they have been converted into tiny chemical factories. In the following section, we shall learn about industrial fermentations.

7.5.8.1. Industrial Fermentations. From the foregoing discussion it should become clear that scope for genetic engineering is seemingly unimaginable. While bacteria may play host for cloning and expression of a large number of foreign genes, useful microbial genes can also be cloned and expressed. The latter is attempted when a microbial gene needs either to be shifted to a more convenient background such as that leads to better expression or that lends itself to better process optimization. Readers

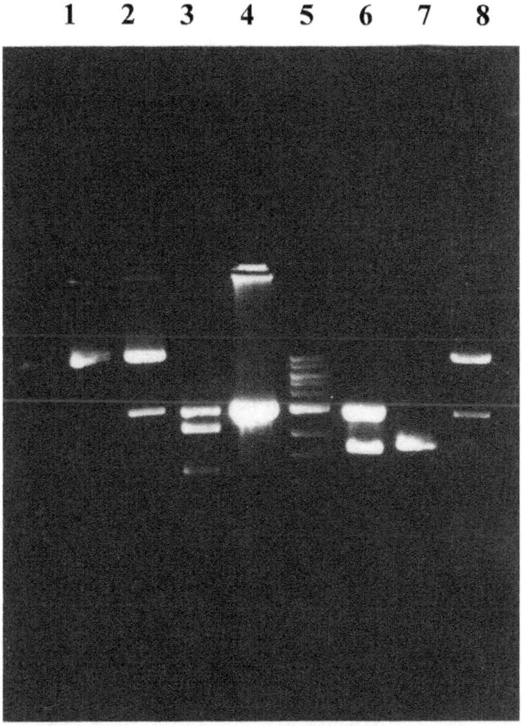

Figure 7.47. Identification of a recombinant molecule by agarose gel electrophoresis of digested DNA. Vector alone will produce one and recombinant two bands. Lane 5- marker (1kb ladder), lanes 1 and 7- undigested recombinant DNA, lanes 2, 3, 6, 8- digested recombinant DNA. Note : In all digested preparations vector is released.

will get a glimpse of the characters of the microbial as well as of the genes of foreign origin that can be exploited by the application of rDNA technology in **Chapter 9**. Any new gene with a commercial potential discovered any time can automatically be brought into the realm of this technology.

When rDNA technology was successfully applied to the production of some useful proteins, it became very clear that the commercial production of a useful product will require an extensive process optimization. Moreover, steps will be required to overcome the inherent regulatory mechanisms that do not allow the overproduction of a desired product.

The commercial production of product or industrial fermentation aims at optimizing the conditions for microbial activity and maintaining a balance between the production cost and the cost of the product.

One of the important considerations in such a process is the selection of the nutrient sources in the fermentation media. For economic considerations, the raw material used are generally the plant materials such as molasses, cellulose, starch, soybean meal, and ammonium or nitrate and phosphate salts. The nutrient source/s should be such that

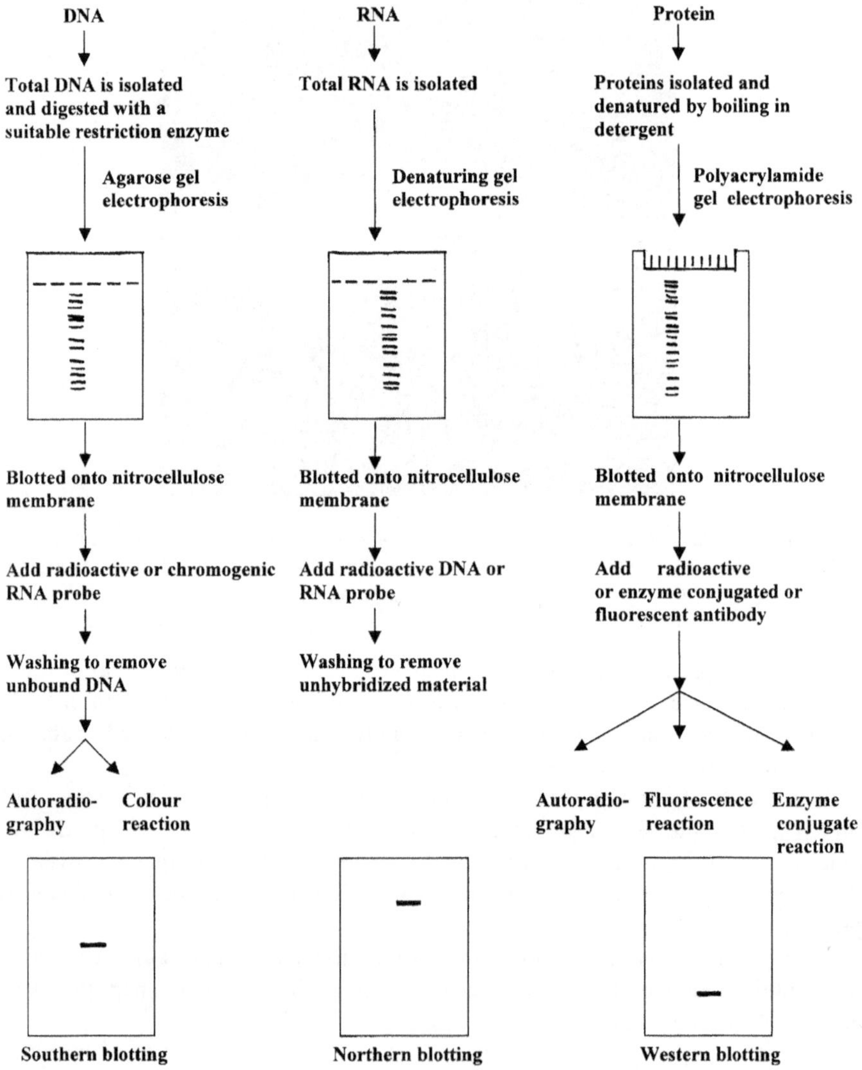

Figure 7.48. Different blotting techniques to identify the recombinant clone and its product.

they satisfy the requirement for all essential macro- and micronutrients. The vessel used for fermenation is a fermenter or bioreactor and the process optimization is gradual. Starting with small lab-scale flasks (batch flasks), to small (less than ten gallons), intermediate (several hundred gallons pilot plant), and to production in large scale fermenters (thousands of gallons) is known as scaling-up. For smooth operation of the process various parameters like pH and O_2 concentration needs regular monitoring.

Many modern day fermenters are provided with computerized monitoring and parametric adjustments. Another important feature is an in-built system of pipes and valves to allow addition or removal of material. Similarly, they are also fitted with system to allow the entry of steam to sterilize the medium ensuring the purity of the desired strain as well as the cooling jackets or coils to maintain the required temperature.

Many of the industrial fermentations are aerobic and therefore maintaining proper oxygen supply is an important parameter. Since oxygen is sparingly soluble in water special attachments are required for the same. On the basis of the system of oxygen supply, four main types of bioreactor designs are available (Fig. 7.49). In a stirred-tank reactor, the medium is agitated with the help of an impeller. A bubble column reactor is provided with a sparger at the bottom through which the air is forced through. In air lift bioreactor, the medium is circulated around two nested columns consisting of an ascending movement or riser and descending movement or downcomer tube. In a packed bed reactor, the column is filled with a solid matrix that contains the microorganisms. In anaerobic fermentation these problems are not encountered as the process does not require oxygen. Instead, heat control is a major parameter.

The fermentation medium is generally buffered so as to overcome the effects of pH changes. The pH change can be brought about either due to microbial growth or by the nature of product (e.g., an acid) both of which will affect the productivity. Temperature changes can also be a major problem during active growth. Cooling and heating coils are often used to achieve desired temperature. The latter is also used for periodic sterilization of the fermenter chamber.

Some bioreactors are designed as batch process that is analogous to a culture flask. In some cases, a flow-through or continuous flow fermenters are preferred which function like a chemostat. Both these systems have their own advantages and disadvantages and the choice between them is often decided by the nature of the product. While the batch process is less prone to contamination, it is more time-taking in starting, incubation, and recovery of the product as well as restarting the process. In a flow through, contamination will have to be thoroughly monitored but they have a higher volumetric efficiency as the product becomes constantly available.

Very often cells or enzymes are immobilized before they are placed in the fermenter. This involves the adsorption or bonding of these materials to a solid support or matrix. Immobilization increases the efficiency and thus the economics of the process. Firstly, the problems of clogging the reactor by the small size of cells can be overcome by immobilization and cost of continuously growing the microorganisms can be substantially reduced. Moreover, the process becomes continuous as substrate can be continuously introduced in the reactor. One set of work is designed towards the use of cheaper matrices for immobilization.

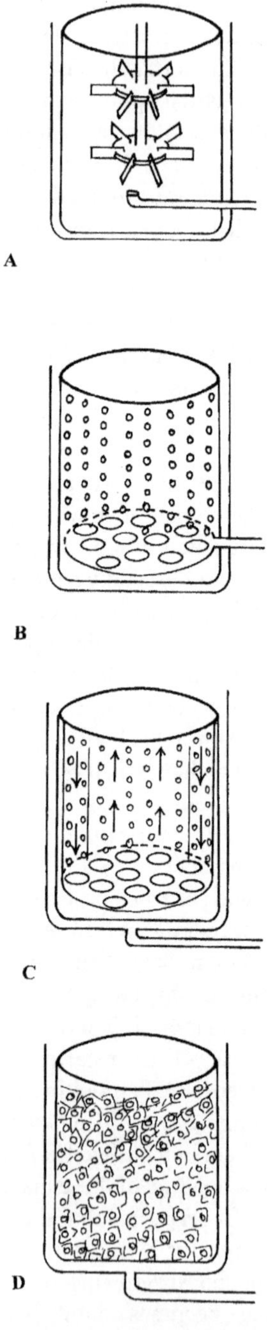

Figure 7.49. Different type of fermenter vessel designs. A. Stirred tank, B. Bubble column, C. Air lift, D. Packed bed.

7.5.8.2. Solid State Fermentation. In another type of set-up, which generally employs fungi, the microorganisms is grown on porous solid substrates, usually in the absence of free water. The water adsorbed on the substrate is sufficient to support the growth of the microorganisms. The growth and secretion of products occur both on the surface of the solid support and within the support matrix. The process, therefore, is known as solid state fermentation. Support substrates used are generally the agricultural residues, such as, wheat bran, wheat straw, brewer's spent grain, rice hulls, bagasse and others. Production of enzymes like proteases, lipases, cellulases, and oxidases can be easily monitored by solid state fermentation. Use of cheap or low grade substrates also makes it a favourable process.

CHAPTER 8
BACTERIA IN NATURE

Bacteria share this earth with all other living organisms. They appeared first about 3.5 billion years ago and monopolized this planet for the next two billion years before other types of single-celled organisms evolved. Multicellular organisms arrived a mere 700 million years ago. Owing to their metabolic versatility, bacteria established themselves not only in the earth's three elements, soil, water, and air but have also occupied almost all possible ecological niches. They are thus ubiquitous in true sense. In order to accomplish successful establishment in a particular environment, all living organisms have to adapt and adjust to the prevailing physico-chemical conditions as well as with the other members of the living populations, and bacteria are no exception. The ecological adaptations may be broadly reflected in two directions: (i) occupying a niche where biological competitions can be avoided. This would be exemplified by extreme environmental conditions such as hot geothermal vents or freezing icy conditions, and (ii) associating with other organisms occupying the same niche.

All living organisms reproduce to form populations or clones and such populations interact with each other to form communities. The location at which a particular community exists is called a habitat, which for microorganisms may be spatially small and often called as microhabitat. The community and habitat together constitute an ecosystem which is functionally a self-supporting system, and provides a mean for energy flow and material cycling. The functions performed by the members of a community are so designed so as to maintain the ecological balance. The microbial populations that occupy a niche represent the autochthonous (indigenous) microorganisms. The best-adapted populations are able to establish themselves most successfully and often replace the less-adapted ones.

8.1. Community Interactions

The members (populations) of a community cannot live in isolation, instead they interact with each other. Broadly speaking, these could occur between different microbial populations, between microbial and plant populations, and between microbial and animal populations. In some cases, these interactions are beneficial or symbiotic, whereas in others, it may be harmful to one or both the populations. The latter is often referred to as competition, predation, or parasitism. Both positive and negative interactions can determine the character of a community. While positive interactions lead to better utilization of available resources and enhance the survival capacity of its members, negative interactions limit population density. Negative interaction is a type of self-regulatory mechanism that also does not allow the instability in a community, such as autochthonous populations are not replaced by allochthonous (foreign) ones. It is very clear that both these types of interactions play a major role in actively growing high density populations. Co-existing populations, therefore, may resort to low population densities often forming resting stage. Under such a situation, populations may completely lack interaction, called as neutralism. Some of these important interactions are depicted in Fig.8.1.

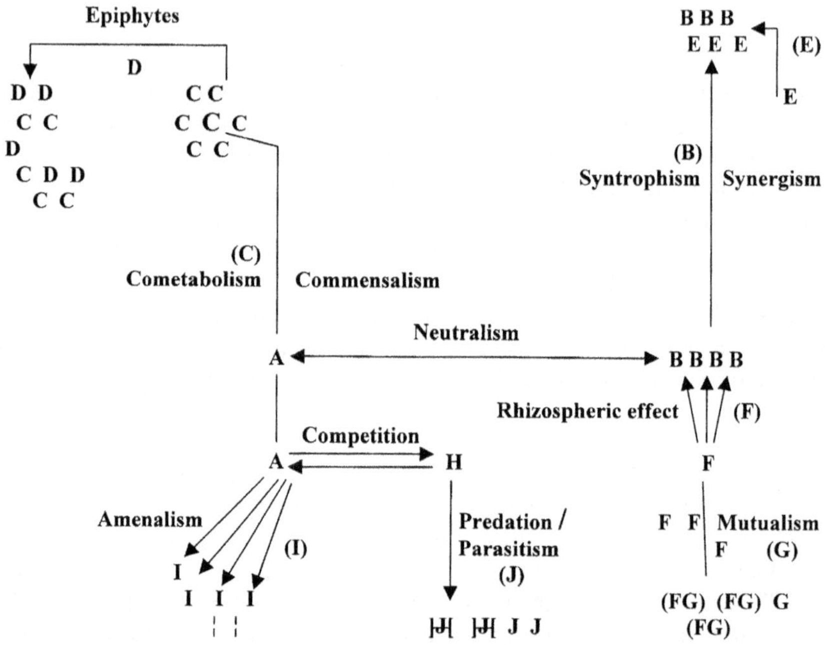

Figure 8.1. Some common interactions between bacterial populations in nature.

8.1.1. Commensalism

In a type of interaction, called commensalism, one population is benefited but the other remains unaffected. Bacterial world is replete with examples of this unidirectional relationship. For example, growth of strict anaerobe, *Bacteroides* species in human gastrointestinal tract is because of facultative anaerobe, *E. coli* which removes oxygen. Secondary colonization by many microorganisms on the bacterial film produced on the hull of a ship, growth of opportunistic pathogens through primary infections caused by a microbial species, growth of a microbe on the waste products of others or the release of growth factors such as amino acid and vitamins by some microbes supporting the growth of others are other examples of commensal interactions. Many of these microorganisms cannot be cultured in synthetic media as their growth requirements are not properly understood.

8.1.2. Cometabolism

Many commensal relationships may be based on cometabolic properties. Cometabolism occurs when an organism growing on a substrate gratuitously transforms another substrate that it cannot but others can assimilate. For example, when *Mycobacterium vaccae* grows on propane it can simultaneously oxidize cyclohexane to cyclohexanol which it does not assimilate further. *Pseudomonas* species, on the other hand, cannot utilize cyclohexane but is able to metabolize cyclohexanol.

8.1.3. Epiphytes

Yet another type of commensal relationship is established when one organism grows on the surface of the other. Bacterial growth on surfaces of algae and plants and on skin surfaces of humans and other animals are examples of this relationship. They derive some of their growth requirements from the surface layers of these organisms.

8.1.4. Synergism

Many a times, two populations benefit from each other's association although both are capable of independent existence. Such an interaction is referred to as synergism or proto-cooperation. Because of this loose relationship sometimes it becomes difficult to distinguish between synergism and commensalism.

8.1.5. Syntrophism

When two populations fulfill the nutritional requirement of each other, they are called syntrophic (=eating together) or cross-feeding. This interaction is based on cooperative metabolism which neither of the members are able to carry out alone.

The conversion of arginine to putrescine by *Enterococcus faecalis* and *E. coli* is an example of syntrophism. *E. faecalis* is able to convert arginine to ornithine which is further converted to putrescine by *E. coli*. Neither of them are able to complete these steps independently. *E. faecalis* requires folic acid and *Lactobacillus arabinosus* phenylalanine. These two organisms can cross feed each other as the former produces phenylalanine and the latter folic acid.

Syntrophism has played an important role in methanogenesis. Many archaea are able to produce methane by involving other bacteria that participate in interspecies hydrogen transfer. For example, many fermentative bacteria produce low molecular weight fatty acids that can be degraded by anaerobic bacteria to generate H_2 and acetate, both can be utilized by methanogenic archaea to produce methane. This chain of reaction, also helps in the removal of end-product inhibitor H_2, so as to facilitate the growth of anaerobic bacteria.

8.1.6. Rhizospheric Effect

The soil region close to the plant root is known as rhizosphere. While plant roots may influence the composition of soil bacteria in rhizosphere, also referred as rhizospheric effect, microbial populations also affect the plant growth. This synergistic action in rhizosphere is based on the release of growth nutrients by plant roots and a number of beneficial processes carried out by microbes. These may be: (i) providing the necessary mineral nutrients by increasing their solubilization and availability, (ii) providing organic nutrients such as vitamins, amino acids, and growth hormones by synthesizing them, and (iii) producing antibiotic or other allelopathic substances that will inhibit the growth of potential plant pathogens.

One action of microorganisms which has made a significant impact on agriculture is their metabolic capability to solubilize several minerals so as to make them available for the plant growth. Increased availability of phosphate, iron, manganese, and calcium and several others are all assigned to microorganisms. Similarly, fixation of atmospheric

N_2 by soil bacteria provide the plants usable forms of nitrogen.

All such interactions are not always beneficial. Immobilization of some micronutrients, such as Zn, or oxidation of Mn by bacteria can lead to some deficiency diseases. Similarly immobilization of N_2 or causing the loss of N_2 by denitrification may prove detrimental to the plants.

8.1.7. Mutualism

Another synergistic interaction may be reflected in mutualism in which both the members are benefited. However, this relationship is obligatory and requires a high degree of specificity between the participating populations. Endosymbiotic association of a microbial population within plant or animal cells is an example of mutualism. Other examples are of lichens (a fungus + an alga), and mycorrhizae (fungus + plant root). Many plant-eating insects actually function in association with microbes that can degrade cellulosic plant materials so as to help the insects to derive nutrition.

One of the best examples of such mutualistic association is observed between ruminate animals and bacteria. Animals like cattles are herbivorous but they do not produce cellulase enzyme for the conversion of cellulosic materials they consume. This is accomplished by the microbial populations maintained in the rumen, the large first chamber within the stomach. The constant temperature and anaerobic environment of rumen suits the growth of microbial populations that can convert cellulose, starch, and other polysaccharides to CO_2, H_2 gas, methane and low molecular weight organic acids. This takes place in two broad steps: firstly the various fermentative bacteria, such as *Ruminococcus,* produce low molecular weight fatty acids, CO_2 and H_2 and subsequently methanogenic bacteria convert them to methane. The overall equation is:

$$\mathbf{57.5\ (C_6\,H_{12}\,O_6) \rightarrow 65\ acetate + 20\ propionate + 15\ butyrate + 60\ CO_2 + 35\ CH_4 + 25\ H_2O}$$

This is the reason why lot of methane is released through cow's belch. The organic acids, on the other hand, are absorbed into the bloodstream of animals, where they enter the aerobic respiration cycle.

Another interesting mutualistic relationship can be seen between luminescent bacteria and marine invertebrates and fishes. Blue-green light emitted by these bacteria, called bioluminescence, involves the action of the enzyme luciferase. This enzyme mediates a reaction involving reduced flavin mononucleotide ($FMNH_2$) with O_2 and an aldehyde to produce FMN in an electronically excited state. When FMN returns to its ground state from excited state, light is emitted

$$\mathbf{FMNH_2 + O_2 + RCHO \xrightarrow{luciferase} FMN + RCOOH + H_2O + light}$$

The light emitted by these bacteria, such as *Photobacterium* and *Vibrio* is used by different fishes differently. While in some they appear to take part in sexual mating rituals, in many nocturnal and deep-sea fishes they may serve the function of flash light helping the fishes to locate food sources and ward off the predators. The luciferase gene has now also been used as a reporter gene in many gene cloning experiments (**see Chapter 7**).

8.1.8. Competition

When microbial populations occupy the same habitat, a competition may occur for the available resources. Although, nutrition may be the most important factor in competition, it may also occur for light and space. As a result of competition, both populations will exist at low densities and may even lead to ecological exclusion of closely related populations.

Competitive exclusion prevents the two populations from occupying the same ecological niche and is overcome by spatial separation. The population which has a higher growth rate under the given set of environmental conditions will outcompete the one which has a lower growth rate.

8.1.9. Amenalism

Sometimes one population may produce substances inhibitory to another population. This interaction is called as amenalism or antagonism. One of the common examples is the antibiotic-producing strains that will gain advantage over the sensitive populations. Lactic acid production by *Streptococcus* and *Lactobacillus* species also eliminate competitors. Use of lactic acid in preserving food products is based on its antimicrobial activity. Several other chemicals, both inorganic and organic, such as O_2, NH_3, mineral acids, H_2S, fatty acids, alcohols also allow a population to antagonize the non-producers.

8.1.10. Predation

Predation involves the eating of a prey species by a predatory population as the former provides the source of nutrition for the latter. Many protozoa species prey on bacteria. This type of relationship can be seen in several lower and higher organisms. Predation is an important process in a food chain within an ecosystem. It not only helps in checking the microbial population but also helps in transferring biomass from the level of microorganisms to that of higher organisms. This type of interaction, therefore, can influence the two types of populations.

8.1.11. Parasitism

In this relationship, one organism (parasite) lives off the other (host). While parasite is benefited, the host is harmed. The parasitic relationship is relatively longer than predation but in the process host may be severely damaged. A large number of bacteria are parasitic to both plants and animals and some of them are obligate parasites (rickettsiae for example). Some bacteria are parasites of other bacteria. For example, *Bdellovibrio* is a parasite of *E. coli*. Parasitism leads to the decline of the host population and may provide a natural mean of controlling population density, thus maintaining ecological stability.

8.2. Non-Culturable Bacteria

Bacteria that we know of, because they could be cultured, constitute a very minute (less than 1%) proportion of the bacteria found in nature (**see also Chapter 2**). A vast majority, therefore, consists of bacteria which cannot be grown on synthetic laboratory media. This also includes several bacteria that live symbiotically or parasitically with plants and animals. As these are viable organisms they are referred to as non-culturable but viable bacteria. In order to study their ecological distributions and biodiversity a number of techniques are employed. These include direct microscopic observations, physiological parameters, and a number of molecular methods such as polymerase chain reaction (PCR) and analyses of 16S rRNA.

Direct microscopic enumerations of microorganisms in samples collected from nature can be facilitated by different staining methods such as those with acridine orange and DAPI. Acridine orange direct counting (AODC) method applied on soil and water samples revealed that actively reproducing cells tend to appear green and those growing more slowly or not reproducing at all at the time of staining appear orange. Both types are, however, viable. DAPI which binds to DNA and produce an intense blue colour has proved particularly useful in visualizing small bacterial cells.

Direct counting procedures were subsequently coupled with reagents that reveal metabolic activities, for example, using INT, in the cells having active dehydrogenases, red dye is deposited, or reduction of 5-cyano-2, 3-ditolyl tetrazolium chloride (CTC) to differentiate respiring cells, or the use of membrane-potential-sensitive fluorochromes to distinguish active, injured (dying), and dead cells. Addition of cell division inhibitor, nalidix acid, to environmental samples will result in actively growing cells appearing elongated. Such studies have discovered a vast world of non-culturable viable bacteria. Moreover, various bacteria such as, *Legionella pneumophila, Salmonella enteritidis, Vibrio cholerae*, and *V. vulnificus* may regularly form viable non-culturable cells.

Molecular techniques have also been applied to study such bacteria. These include development of gene probes for diagnostic sequences to detect selected species, genera, or higher level taxonomic groups or probes for rRNA sequences with an objective to identify closely related organisms. PCR amplification of diagnostic DNA sequences, or targeting of specific mRNA by RT-PCR, has helped in the detection of even small numbers of non-cultural microorganisms in a community.

8.2.1. Most Probable Number

The most probable number (MPN) method is used as an alternative to plate count. In this method, several replicates of different dilution levels of a culture are scored as positive or negative on the basis of growth. These results in conjunction with appropriate statistical analyses are used to give the most probable number of viable organisms.

8.3. Ecological Contributions

The bacterial ecological contributions have some very important implications. It has not only allowed unraveling the important functions that bacteria and other microbes carry out in nature but also to apply these functions for a better use by human beings. We shall consider them under four sections: I) bacterial contribution towards running and maintaining biogeochemical cycles, II) immobilization or fixation of nitrogen, III) environmental restoration, and IV) diseases caused by bacteria in plants and animals. The animal examples, however, will be dealt with separately in **Chapter 9**.

8.3.I- Biogeochemical Cycles

The life on this earth is being sustained due to activities of photosynthetic organisms, both green plants and photosynthetic bacteria. These are, therefore, referred to as producers (also autotrophic) because with the help of solar energy, CO_2 and H_2O (or other chemical compounds) they are able to produce the carbohydrates that provide the material for carbon skeleton of all living organisms. Unlike these autotrophic organisms, heterotrophic (most animals and other microorganisms) organisms derive their materials from the former and are thus called consumers. The continued progression of photosynthetic activities by producers and the transfer of photosynthate to consumers is defined by a food chain in which all types of organisms occupy a definite position. In order to maintain this flow of energy and materials, the ecosystem must be replenished with precursor molecules. Moreover, after the death of the living organisms and the waste products released by the animals, complex organic molecules are returned to the soil. These compounds are degraded by soil microorganisms which furnish the small constituents as well as other intermediates which can be reutilized by plants and microorganisms. This recycling of the matter is known as mineralization (Fig.8.2).

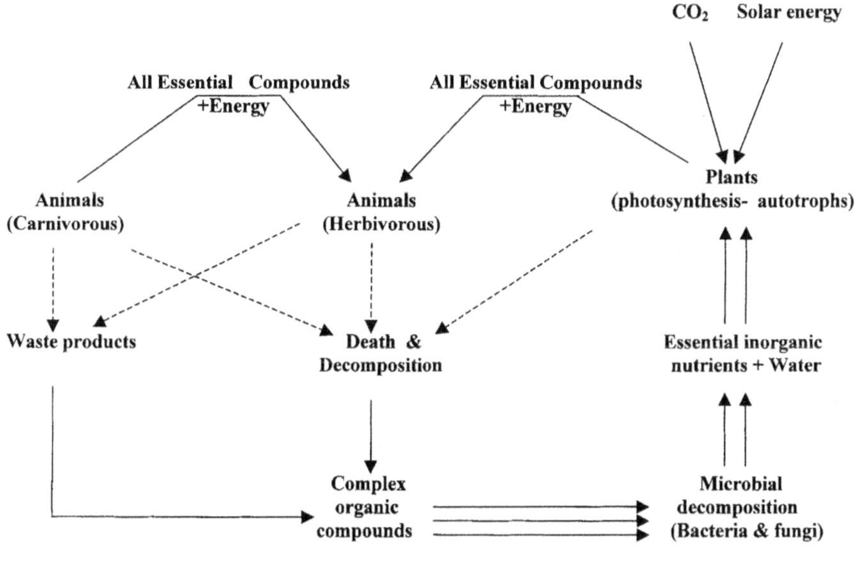

Figure 8.2. Role of microorganisms in mineralization in nature.

Microorganisms, both bacteria and fungi, act as nature's scavengers and through this activity not only keep the earth free of unwanted waste materials but also regenerate the essential precursors which help sustain the life of other members of the biosphere. Their ubiquitous nature, resilience to some very detrimental environmental conditions, high rate of multiplication but most importantly, the diverse metabolic capabilities, have endowed them with the capability to carry out this essential function in nature. As we have seen earlier, soil bacteria do not live in isolation, so biodegradative capacity is a community function. Also, all bacteria do not possess all metabolic pathways. Though the contribution of individual member species in mineralization may be minimal, together they may biodegrade any type of chemical compound. Infact, those which can not be broken down end up as pollutants. A number of microorganisms can be manipulated to handle a whole variety of pollutants, but many may simply turn out to be purely recalcitrant to microbial degradation.

As we discussed earlier, the living cells consist of a large number of elements which remain trapped in the biomass. These elements must go through the "cycle of matter" so that they can return into the environment in order to refind their way into other newly-forming cells. Some of these elements are carbon, oxygen, nitrogen, sulphur, phosphorus and other nutrients. In the following section, we shall discuss the pathways through which they are replenished and recirculated. Since these pathways are cyclic, they are more commonly referred to as cycles.

8.3.I- 1. Carbon-Oxygen Cycle

Recycling of biological matter containing carbon and oxygen is intimately associated with photosynthesis and respiration (Fig.8.3). Through the process of photosynthesis, green plants, autotrophic, and chemolithotrophic bacteria reduce CO_2 to carbohydrates with O_2 being formed by many of them through splitting of H_2O. The source for oxidized carbon for land autotrophs is gaseous CO_2 whereas aquatic organisms use dissolved CO_2 or carbonates. The carbon so fixed not only serves as the source of carbon for primary producers but also for other heterotrophic organisms including human beings directly or indirectly.

The reduced organic compounds are oxidized during respiration to release CO_2, and O_2 is reduced to H_2O. Although this process, undertaken by all life forms continuously release CO_2, a large and significant portion of fixed carbon remains in the body. Upon death and decay of these life forms, scavenging microorganisms (bacteria and fungi) are able to oxidize a variety of organic compounds to release CO_2. In the anaerobic environment, the organic compounds are broken down through fermentation and anaerobic respiration (see **Chapter 6**). This microbial cycle is of extreme importance as it replenishes the atmosphere with CO_2, otherwise the atmospheric level of CO_2 (0.03%) is estimated to have got exhausted within 20 years.

Another source of CO_2 is the combustion of fossil fuels like coal and petroleum. Rapid industrialization as part of the human civilization has led to the high level consumption of these fossil fuels. Not only these precious resources are dwindling but CO_2 along with CH_4, NO, N_2O have contributed significantly towards rising global temperature, an effect known as "green house effect". Other factors which are also responsible for green house effect are soil erosion, water logging, and deforestation.

In aquatic habitats, carbonates combine with calcium to form $CaCO_3$ which either gets deposited as calcareous rocks (lime stone) or in the shells of many aquatic animals, like protozoa, molluscs and corals. Although in this form carbon is wasted, some of it can be retrieved through weathering. Microorganisms are associated with both the processes. Many sulphate-reducing bacteria and denitrifying bacteria create alkaline conditions that help in the precipitation of $CaCO_3$. Others can form acids that leads to solubilization of $CaCO_3$, thereby, releasing trapped carbon.

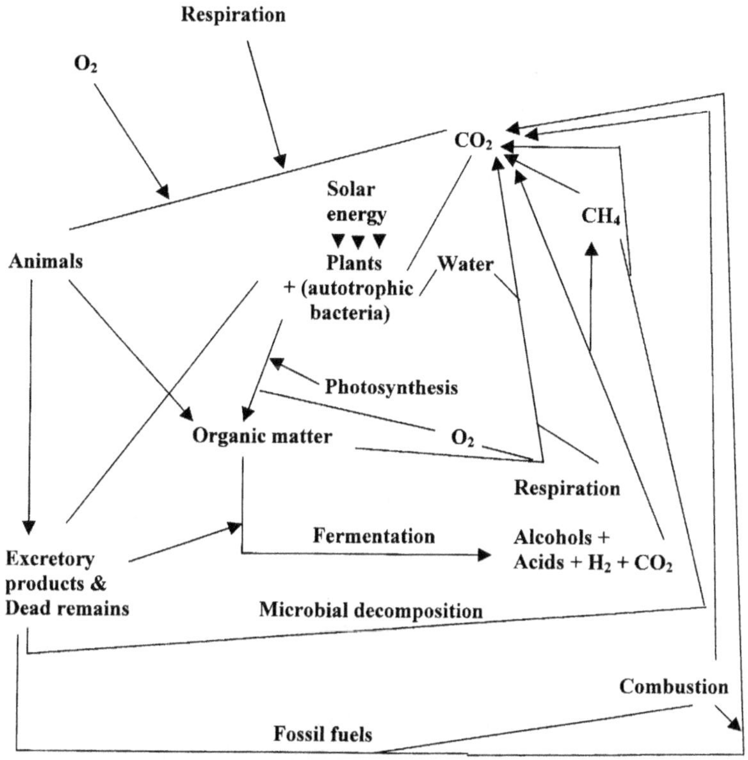

Figure 8.3. Carbon-Oxygen Cycle.

Some natural organic compounds like cellulose, lignin and humic acid are relatively resistant to microbial attack and thus decompose slowly. The organic matter of soil, principally composed of such compounds, is known as humus. The fertility of a soil is dependent upon its humus content. The accumulation of humus coupled with high moisture content, relative anaerobic conditions and abundance of acidic substances, aided by other physical and chemical factors, has resulted in the formation and deposition of peat and coal. Another set of reactions contributed to the formation of natural oils (petroleum and others) and gases. All these constitute major energy sources for mankind, but have taken millions of years to get established. Indiscriminate and extensive use of these resources in the name of industrialization has led to their rapid depletion. Increasing awareness is now being created for judicious use of these resources and find the alternative fuels so that the pace of industrialization could be maintained.

8.3.1- 2. Nitrogen Cycle

Molecular nitrogen is the most abundant substance in earth's atmosphere, yet it is often growth limiting for plants. This is because molecular nitrogen cannot be utilized by most organisms except for some bacteria which can fix nitrogen. Nitrogen fixation is a very important microbial function not only in nature but also of great applied value. We shall, therefore, deal with this process separately in a subsequent section. Most other microorganisms utilize N_2 in the form of NH_4^+, NO_2^- and NO_3^-, as well as organic nitrogenous compounds such as amino acids and proteins. Plants, the primary producers, also prefer to assimilate N_2 in the form of NO_3^-.

Utilization of nitrogenous compounds is based on the changes in its oxidation state that constitutes a nitrogen cycle. The cycle consists of some well-defined processes, interlinked with each other so as to run the nitrogen cycle efficiently (Fig.8.4). These are:

8.3.I- 2.1. Ammonification. Many microorganisms, as well as plants and animals can convert organic amino nitrogen to ammonia by a process called ammonification. This is important because an inorganic form of N_2 is produced from organic molecules. Microbes that can decompose urea release NH_3 to atmosphere. If this reaction takes place in neutral aqueous environment NH_4^+ is formed. Ammonium ions can be assimilated by various organisms.

8.3.I- 2.2. Nitrification. Ammonia or NH_4^+ (oxidation level = -3) is also not the only preferable substrate and thus is further oxidized to first nitrite (oxidation level = + 3) and then to nitrate (oxidation level = + 5). This process is called as nitrification. As we have seen in **Chapter 6**, these two processes are used by chemolithotrophic bacteria to derive energy. The yield of ATP by this method of aerobic respiration is relatively low leading to lower growth rates of nitrifiers.

The two steps in nitrification are carried out by two different microbial populations found in different types of habitat both terrestrial and aquatic- both fresh water and marine. In soil, oxidation of ammonia to nitrite is carried out predominantly by *Nitrosomonas* and of nitrite to nitrate by *Nitrobacter*. Many of them possess extensive internal membrane network to carry out this function. Besides bringing about the oxidation of reduced form of N_2 (NH_4^+) to NO_2^- and NO_3^-, nitrification also converts a cation (NH_4^+) to an anion (NO_2^- or NO_3^-). This has an important bearing on nitrogen status of the soil as the positively charged cations are bound to the negatively charged clay particles whereas negatively charged anions are easily leached out. Thus nitrite and nitrate can easily find their way into ground water supplies and pose a serious health hazard to mankind. Nitrite is highly toxic as it can react with hemoglobin to block the normal gas exchange and forms highly carcinogenic nitrosamines by reacting with amino compounds. Although nitrate is not that toxic, it can be converted into nitrite in the gastrointestinal tract of human infants causing "blue baby syndrome". This reaction does not take place in adults because of the low pH prevailing in the adult gastrointestinal tract.

Application of NH_4-nitrogen fertilizers in combination with nitrapyrin, an inhibitor of nitrification, is recommended to check the leaching problem and improving the soil fertility and thus the crop productivity.

8.3.I- 2.3. Nitrite Ammonification. Some bacteria such as *Clostridium* species can reduce nitrite to ammonium ions by the process known as nitrite ammonification. This ATP generating process uses the electrons from NADH and is responsible for conversion of NO_2^- to NH_3^+ in soil ecosystem.

8.3.I-2.4. Denitrification. Another set of reactions can lead to conversion of fixed forms of nitrogen to molecular nitrogen. Nitrate is often used by some aerobic bacteria as the

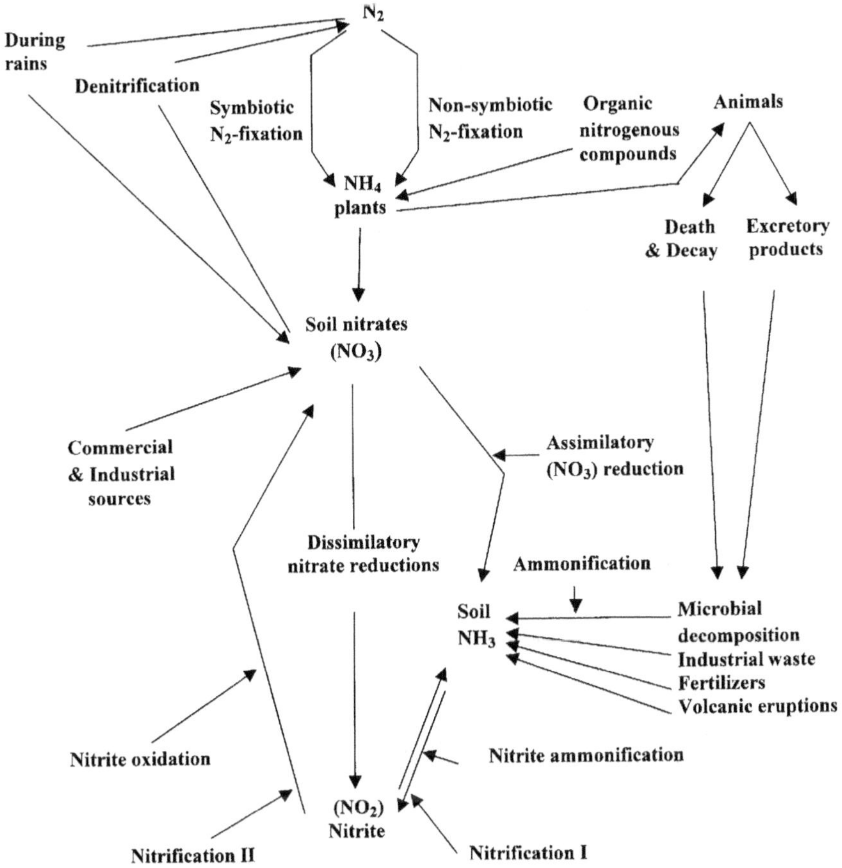

Figure 8.4. The Nitrogen Cycle.

final acceptor in place of O_2, to reduce nitrate by anaerobic respiration. While some bacteria, like *E. coli,* can convert nitrate to nitrite, some others can perform two other steps of anaerobic respiration converting NO_2^- to nitrous oxide gas (N_2O) and subsequently to molecular N_2- Nitrate reductase, the enzyme involved in this process of denitrification that leads to return of N_2 to the atmosphere, is inhibited by O_2.

By interconnecting these processes, a nitrogen cycle is able to maintain a balance of this important component (Fig.8.4). The NH_4^+ ions formed by biological nitrogen fixation, or released from industry, are broken down to NO_2^- and NO_3^- by nitrification. All of them can be absorbed by plant roots to fix them in organic form, and then passed on to animals. Organic nitrogen can also be converted to NH_4^+ by ammonification which re-enters the cycle. Part of the ammonium ions can be transformed to NO_2^- and NO_3^- by nitrification to be reused by plants or microorganisms. Whereas another portion may be converted to N_2 by denitrification.

8.3.I- 3. Sulphur Cycle

Sulphur can exist in different oxidation states within organic and inorganic compounds, and microorganisms facilitate this interconversion to run the sulphur cycle (Fig.8.5). Many microorganisms under aerobic condition remove sulphur from organic compounds (desulfurization) to form sulphate, whereas some others under anaerobic condition produce H_2S. H_2S can also be generated by sulphate-reducing bacteria when they use sulphate as terminal electron acceptor during anaerobic respiration. Hydrogen sulphide though very toxic to most biological systems can be used by photosynthetic sulphur bacteria as an electron donor. Some of them produce elemental sulphur as product whereas others form sulphates. These photosynthetic sulphur bacteria often are found in surface layers of a habitat and more subterranean regions are occupied by sulphate-reducers and anaerobic degraders of organic sulphur-containing compounds. The latter two maintain the supply of H_2S for the former.

Bacteria, like *Beggiatoa* and *Thiothrix,* oxidize H_2S to generate ATP. In the process, they deposit intracellular elemental sulphur granules which can be further oxidized to sulphate. Chemolithotrophic *Thiobacillus* species oxidize sulphur to produce sulphate which is further converted to sulphuric acid. These bacteria are acidophilic growing well at pH 2 to 3 and have the capability of bioleaching many metals from their sulphidic ores.

Coal is often found associated with pyrite (FeS_2). During coal mining, pyrite gets oxidized upon exposure to atmospheric oxygen which supplemented by the activity of sulphur- and iron-oxidizing bacteria produces large amounts of sulphuric acid. The acid drainage from mines not only kills the aquatic life but also renders the water unfit for drinking. Moreover, the low pH associated with acid mine drainage leads to degradation of vegetation and other life forms. The effluent also looks deep brown in colour due to precipitation of Fe $(OH)_3$:

$$\mathbf{2\ Fe\ S_2 + 7.5\ O_2 + 7\ H_2\ O \rightarrow 2\ Fe\ (OH)_3 + 4\ H_2\ SO_4}$$

8.3.I- 4. Phosphorus Cycle

Phosphorus is another essential element found in many biomolecules. It normally occurs as phosphates both in organic and inorganic compounds. As no change in its oxidation state is required, its recycling involve the microbial release of phosphates from organic compounds. In many habitats, phosphate combines with calcium to form calcium phosphate which cannot be utilized directly. Many heterotrophic microorganisms are capable of solubilizing phosphates by producing organic acids. Microorganisms can also remove phosphates rapidly creating phosphate-limiting conditions. When excess concentration of phosphates enter such habitat they result in sudden increase in the growth. This process of nutrient enrichment is known as eutrophication. Eutrophication of rivers and streams permitting excessive growth of photosynthetic organisms is an example of this type of nutrient enrichment.

8.3.I- 5. Other Elements

Various other elements are also recycled by microorganisms to maintain their supply for essential functions.

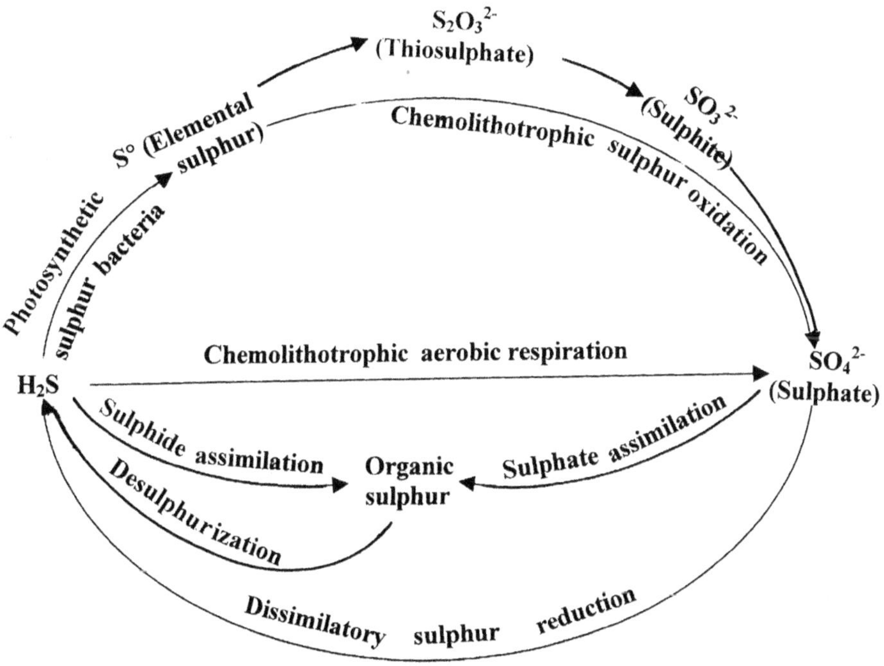

Figure 8.5. The Sulphur Cycle.

Iron, for example is transformed between ferrous (Fe^{2+}) and ferric (Fe^{3+}) oxidation state. Ferric compounds are less soluble than ferrous ones. Iron transformation is important in corrosion and acid mine drainage formation. Bacteria like *Thiobacillus, Galionella,* and *Leptothrix* oxidize iron compounds and some deposit ferric hydroxide so formed in an extracellular sheath. This microbial activity has been responsible for the formation of iron deposits.

Calcium is also cycled between soluble and insoluble forms. Calcium bicarbonate is extremely soluble unlike calcium carbonate. Many bacteria produce acidic compounds to form calcium bicarbonate and calcium carbonates. In the marine ecosystem, cycling of calcium is responsible for coral reef formation, limestone deposits and algal frustule formation. Various algae also participate in precipitating silicon dioxide and forming large deposits.

Manganese exists as a water soluble divalent manganous ion and a relatively insoluble tetravalent manganic ion. Microbial oxidation of manganous forms manganese oxide that accumulates to form manganese nodules. Deep sea manganese nodules are a rich source of manganese.

Many toxic heavy metals like mercury, arsenic and others are managed by microbial cycling. These transformations alter the mobility and thus reduce the toxicity. The reduction of many toxic Hg-compounds to elemental Hg° is important in reducing its toxicity. Mercury that concentrates in many fishes (biomagnification) has led to several disasters. For example, Minamata disease that severely disturbs the nervous system is caused by ingesting the shell-fish containing high levels of methyl mercury.

It should be clear from this discussion, that microorganisms have been solely responsible for maintaining the supply of all essential elements not only for themselves but also for all other life forms. Their importance in any ecosystem, therefore, can not be overlooked.

8.3.II- Nitrogen Fixation

Plants, by photosynthesizing, are the primary producers of food which sustain the life of all other living organisms on this earth. Cultivating plants to serve this function has been a human activity since time immemorial and advancement in this field has been the parameter of human civilization. From the very early times, man has learnt to identify and breed the crops meant for providing food. This demand has grown rapidly over the years to match the multiplying human population. With land becoming increasingly limited, the plant breeders put emphasis on developing high-yielding and disease-resistant varieties. Another line of approach is to improve the nutritional value of the important crops. Quantity and quality of the food go hand-in-hand and is still the biggest challenge facing the mankind.

While the dietary quality of food plants can be improved by a variety of approaches, one important aspect is to satisfy the nutritional requirements of these plants. Being autotrophic, plants and many microorganisms are capable of utilizing gaseous CO_2 to synthesize carbohydrates. Similarly they can use atmospheric oxygen to carry out many metabolic reactions. Several other macro- and micronutrients are supplied by soil. One important component of macronutrients, nitrogen, however, often becomes growth limiting as we have discussed earlier. Although, present in plenty, atmospheric nitrogen cannot be directly utilized by plants. This led to the concept of fertilizer applications that would improve the nitrogen (NH_4^+, NO_3^-) status as well as that of other macro- and micronutrients of soil.

Repeated agriculturing leads to depletion of soil nitrogen making the repeated application of chemical fertilizers imperative. Rising cost of chemical fertilizer production and the consequent influence on the environment has necessitated the need for cheaper and perhaps safer alternatives. In this context, the capability of many bacteria and cyanobacteria to fix atmospheric nitrogen gained immense importance. This importance was realized both by ancient and modern civilization. Many old agricultural practices, based on the observation that legume plants can restore soil fertility, intercropping of leguminous and non-leguminous crops and also rotation of these two types of crops, and mixing of leguminous roots with the soil after harvesting the crop, are still prevalent. By the mid-nineteenth century, however, the plant growth could be assessed in chemical terms. It was found that growth of all plants other than legumes is affected by the nitrogen content of the soil. Legumes, infact, grown on N_2-

poor soil improved the nitrogen content. On closer examination, this property was assigned to root nodules, as the plants failing to form them also failed to express this capability (Fig.8.6).

Microscopic examinations of nodules carried out as early as 1858 by Lachmann and in some detail in 1866 by Woronin showed that they contain structures similar to bacteria in shape and size and were called "bacteroids" (see Fig. 8.7). That bacteroids are capable of nitrogen fixation was demonstrated when surface-sterilized seeds grown in sterile soil failed to produce plants that could fix nitrogen. In 1894, Winogradsky isolated and described a free-living anaerobic, nitrogen-fixing bacterium, *Clostridium pasteurianum*. Since then, two main categories of biological nitrogen fixation (BNF) have been recognized: symbiotic nitrogen fixation and asymbiotic (=free-living) nitrogen fixation. Many other bacteria were subsequently described falling into one or the other of the above categories.

Once recognized, BNF has been hailed as the "greatest discovery of the century". This stimulated a large section of the scientific community to undertake studies on various aspects of BNF. Besides dealing with the basic aspects of this process, physiological and genetic, these studies aimed to reach two major goals: i) to improve the nitrogen status of the soil so as to improve the growth of important plants as well as the quality of dietary proteins, and ii) to transfer the nitrogen-fixing genes (*nif*) to important crop plants so as to make them independent of an external source of nitrogen. This will cut down the cost of fertilizer manufacture and provide relief from environmental pollution which is an integral component of a chemical industry.

Nitrogen, an element that is essential to all life, comprises nearly 80% of the atmosphere. Despite this fact, nitrogen limits the growth and productivity of plants in many ecosystems, as they can utilize only bound form of nitrogen (NH_4^+, NO_3^-). The reduction of dinitrogen (N_2) to NH_3 using the enzyme nitrogenase is a property confined only to prokaryotes. Biological nitrogen fixation carried out by both free-living and symbiotic bacteria is a singular contribution which by allowing the utilization of molecular nitrogen maintains the soil fertility. Biological nitrogen fixation accounts for approximately 2 x 10^8 tons of fixed N per year or about 2/3rd of the total.

Some of the important nitrogen fixing organisms also known as diazotrophs, are listed in Table 8.1.

The interest in nitrogen-fixing bacteria, capable of forming symbiotic relationship with plants has arisen due to the fact that they are likely to contribute the candidate genes for achieving the goals mentioned earlier.

8.3.II- 1. Symbiotic Nitrogen Fixation

Many bacteria are able to form associations with plants and fix nitrogen. Since in this process both the partners are adequately benefited, it is known as symbiotic nitrogen fixation. Three major categories of this association are: *Rhizobium*-Legume, *Anabena-Azolla* and *Frankia*- with some trees and wooden shrubs. Of these, the former is the best understood.

Beijerink (1888) and Prazmowski (1890) succeeded in isolating and culturing bacteria from root nodules of many leguminous plants. In culture, these bacteria showed

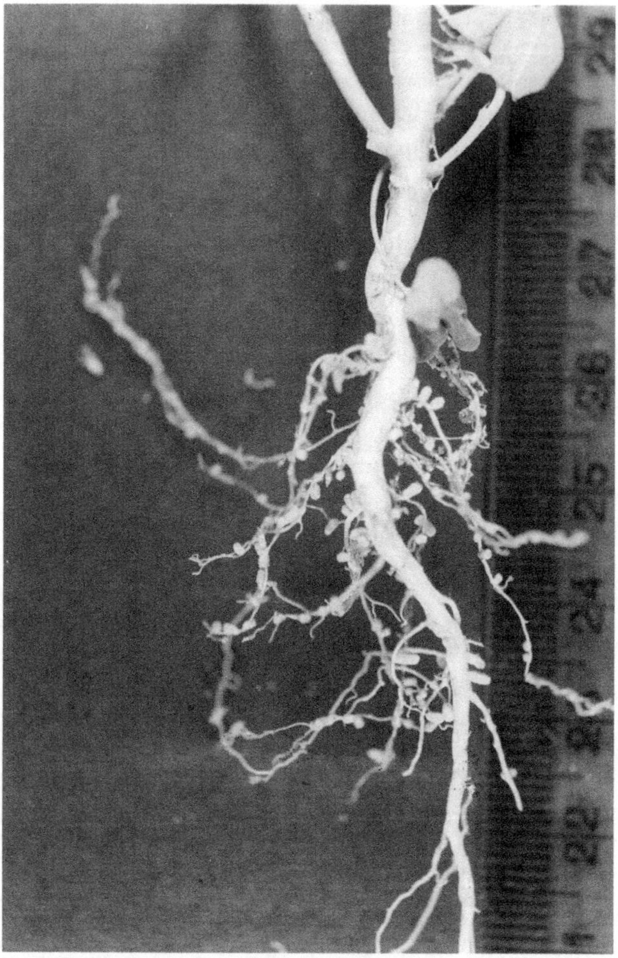

Figure 8.6. Nodules produced on the roots of a leguminous plant. Nitrogen-fixing bacterium, Rhizobium is lodged in these nodules.

a cocco-bacillary form but in the nodules they represented variable shapes, swellings, and branchings. These pleomorphic nodule bacteria of different legumes were initially placed in one genus *Rhizobium* which has now been split into three. While all fast growing species have been placed in the genus *Rhizobium,* the slow-growing ones have been grouped in *Bradyrhizobium,* and the members making stem nodules in place of root nodules come under *Azorhizobium*.

The rhizobia are common inhabitants of soil. Based on the cross inoculation test they have been grouped in 7-8 subgroups, each with a moderate to strict host specificity. Generally speaking, *Rhizobium* species nodulates alfalfa, peas, clover, and numerous

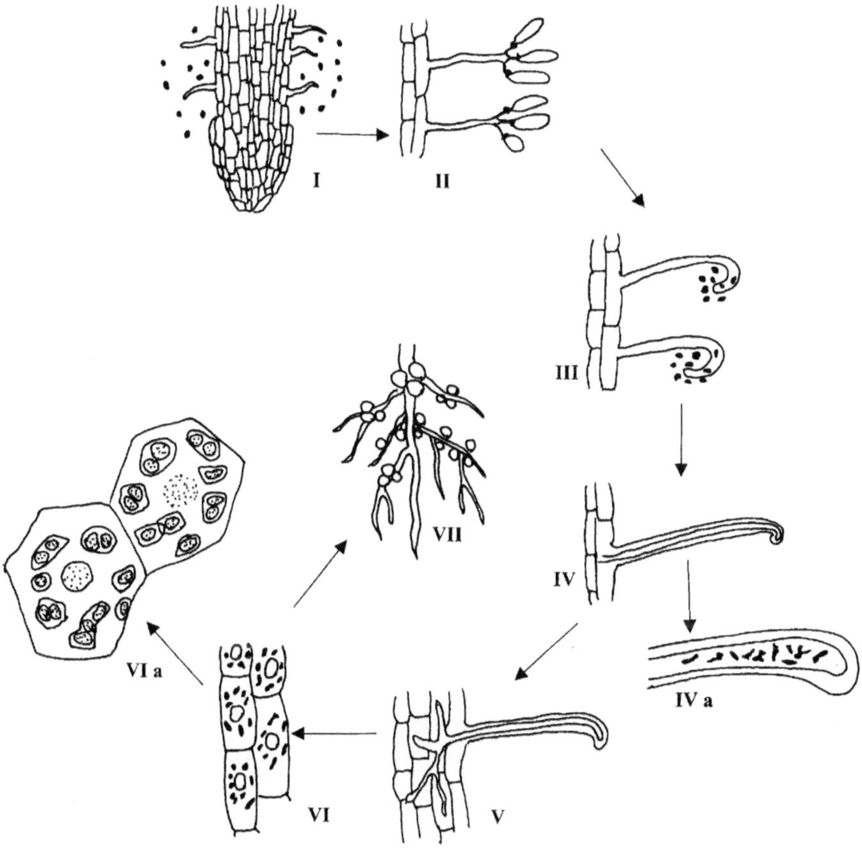

Figure 8.7. Root Nodule Development in Rhizobium-legume Symbiotic Interactions. I. A legume root with rhizobia growing in the proximity of root hairs, II. Root hair producing lectins that help in the recognition and specific association of a rhizobial strain, III. The interaction between the root hair and the rhizobia leads to roots hair curling trapping the rhizobia. This is followed by the dissolution of root hair wall and entry of the rhizobia, IV. The root hair provides an infection thread through which the rhizobia travel inside the root (IV a), V. The infection thread ramifies and penetrates the root cortical cells, VI. The rhizobial cells are liberated individually or in small groups in membrane-bound sacs (VI a), The bacterial cells acquire pleomorphic shapes (bacteroids) and surround the host nucleus, VIII. The host cell nucleus is stimulated to undergo repeated divisions followed by the division of unaffected cells also resulting in nodule formation.

other legumes, whereas *Bradyrhizobium* prefers soybeans, lupins, cowpeas, and various tropical leguminous plants. The nitrogen fixing bacteria are lodged in structures called nodules which in majority of the cases are produced on roots. In this association, bacteria receive nutrition in the form of organic acids or TCA cycle intermediates, amino acids and carbohydrates, whereas the plants, in turn, acquire the fixed form of nitrogen through them.

Table 8.1. Some important nitrogen fixing bacteria

	Name	*Nature*
1.	*Clostridium pasteurianum, C. butyricum*	Free-living, strict anaerobes
2.	*Enterobacter cloaceae*	Free-living, facultative anaerobe
3.	*Klebsiella pneumoniae*	Free-living, facultative anaerobe
4.	*Azospirillum lipoferum, A. brasilense*	Forms association with several grasses, microaerophilic
5.	*Azotobacter vinelandii*	Free-living, obligate aerobe
6.	*Derxia gummosa*	Free-living, obligate aerobe
7.	*Rhodospirillum rubrum*	Photosynthetic, free-living
8.	Many cyanobacteria	Photosynthetic, free-living
9.	*Rhizobium/Bradyrhizobium/Azorhizobium*	Symbiotic with legumes
10	*Anabena*	Symbiotic with *Azolla,* phototroph, aerobe
11.	*Frankia*	Symbiotic with many trees and woody species
12	*Desulfovibrio*	Free-living, strict anaerobe, marine

The genetic and physiological interaction between the plant and rhizobia that leads to nodule formation and nitrogen fixation is a complex process. Several bacterial genes called *nod, exo, nif,* and *fix* are involved in these two processes. The symbiotic association begins with the chemotactic attraction of a specific bacterial strain by the amino acids and dicarboxylic acids released by the host plant. Bacteria then attach to the root surface by recognizing specific binding sites or receptors (lectins). The host plant subsequently secretes flavonoids or isoflavonoids that induce several nodulation or *nod* genes of associated bacteria. The product of the *nod* genes are the enzymes which synthesize the species specific substituted lipo-oligosaccharides or nod factors. These signal compounds elicit the curling and branching of the root hairs and at a later stage the division of the meristematic cells of roots. Bacteria then enter through the root hairs cell wall and progress through the root cells inside a host-derived infection thread. The bacteria are released in the cortical cells which divide and redivide to form the nodules (Fig.8.7).

About two dozen genes called nodulin or *nod* genes have been described which control various steps of nodule development and the physiological functions involved.

Host plant also produces proteins specified by the bacterial *nod* genes which supplement the bacterial functions in the process (Table 8.2). Nodulins not only control the cell division, the metabolism, and morphogenesis of plant root, nutrient exchange and signal transduction but also the specificity exhibited between the two partners in the symbiosis. Based on their temporal requirement they have often been divided into two groups. While the early nodulins ensure the compatibility between the host and the rhizobial partner, the late nodulins specify specific plant proteins, such as leg- hemoglobin, that creates conducive physiological conditions for nitrogen fixation.

Table 8.2. Function of some described nod gene products

Protein	*Function based on sequence homology*
NodA	Unknown
NodB	Deacetylase
NodC	Chitin synthetase
NodD	Transcription activator, LysR family
NodE	β-Ketoacyl synthase
NodF	Acyl carrier protein
NodG	Alcohol dehydrogenase, β-ketoacyl reductase
NodH	Sulphotransferase
NodJ	Capsular polysaccharide secretion proteins
NodK	Unknown
NodL	Acetyltransferase
NodM	D-Glucosamine synthase
NodN	Unknown
NodO	Hemolysin
NodP	ATP-sulphurylase
NodQ	ATP-sulphurylase and APS kinase
NodS	Methyltransferase (Ac)
NodT	Transit sequences
NodU	Unknown
NodV	Sensor (two component regulatory family)
NodW	Regulator (two component regulatory family)
NodX	Acidic exopolysaccharide (encoded by *exoZ*)
NodY	Unknown
NodZ	Unknown

Other late nodulins such as, uricase, glutamine synthetase, and sucrose synthetase play important roles in carbon, and nitrogen assimilation and thus help in maintaining a mutually balanced relationship. Rhizobial genes required for symbiotic nitrogen fixation contribute towards nod factor synthesis, nodule development, synthesis of the

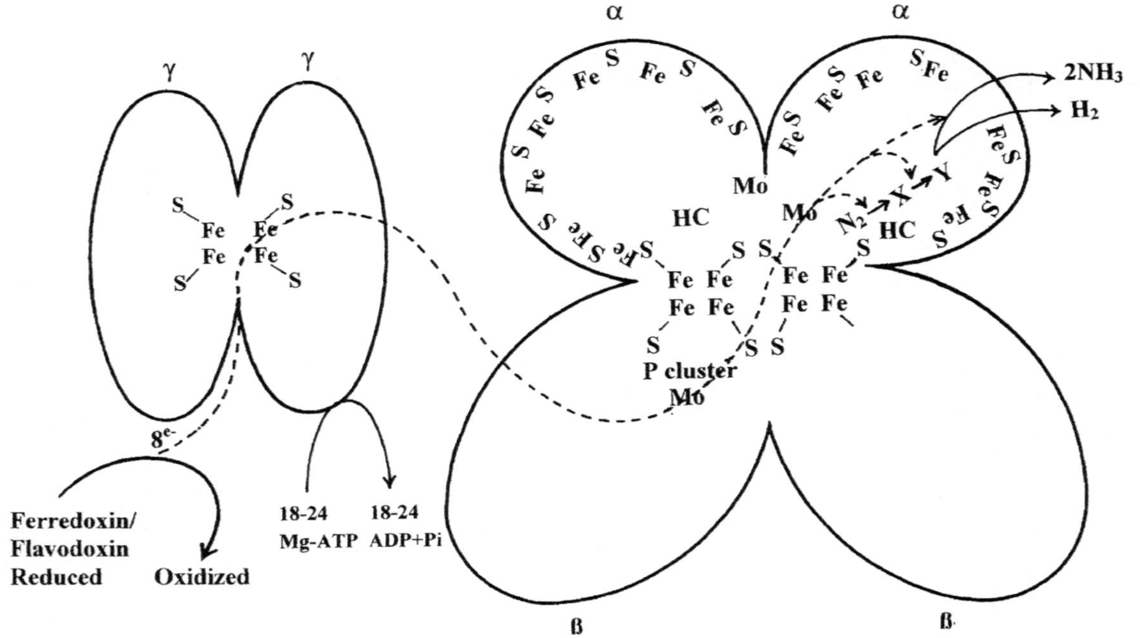

Reaction sequence: $N{=}N \xrightarrow{4e^-} \underset{(X)}{NH{=}NH} \xrightarrow{2e^-} \underset{(Y)}{NH_2{-}NH_2} \xrightarrow{2e^-} 2NH_3$

Figure 8.8. Nitrogen fixation by bacteria through the enzyme nitrogenase. The enzyme nitrogenase consists of two proteins, dinitrogenase reductase (Fe protein) and dinitrogenase (Fe Mo protein). Dinitrogenase reductase is responsible for ATP hydrolysis and for transferring electrons to dinitrogenase. The electrons are first accepted by P-cluster and then subsequently transferred to Fe Mo-co cluster where nitrogen is finally reduced to ammonia. The intermediates represented as X and Y are shown fully in reaction sequence. Fe= iron; S= sulphur; Mo= molybdenum; HC= homocitrate. In symbiotic nitrogen fixation the reductant is obtained from the host.

nitrogen-fixing apparatus, and bacteroid metabolism. Inside the nodule, root nodule bacteria multiply and differentiate into unusually shaped pleomorphic cells, the bacteroids. The bacteroids, once formed lose their capability of independent reproduction. Nodulin genes required by bacteria are classified into two classes.

Genes that specify the biochemical composition of the bacterial cell surface, such as the synthesis of exopolysaccharides (*exo*), lipopolysaccharides (*lps*), capsular polysaccharides or K-antigens, and β-1,2-glucans (*ndv*) are placed in class I. Mutation in these genes disturb the infection process. The *nod* or *nol* genes are grouped in class II, the mutation in which can lead to the absence of nodulation (Nod$^-$) or changes in the host range, among others. Many *nod* genes are common and can be interchanged but others are host-specific *nod* (*hsn*) genes. In most *Rhizobium* species, *nod* along with *nif* and *fix* are carried on large symbiotic plasmids (pSym). In *R. loti*, *Bradyrhizobium* and *Azorhizobium,* these genes are located on the chromosome. In most cases, *nod* genes are not expressed in cultured cells but can be induced by plant flavonoids and transcriptional activator protein NodD. The *nodD* is the only gene expressed constitutively both in symbiotic and free-living rhizobia. NodD protein binds to a conserved DNA sequence upstream of inducible *nod* operons, called *nod* boxes activating all other *nod* genes. The *nodABC* genes have been found in all rhizobia, besides the different *hsn* genes. In addition, all nitrogen-fixing bacteria possess mechanism to protect their nitrogen fixing enzyme, nitrogenase, from oxygen intoxication (inhibition) and *ntr* genes (nitrogen regulatory genes). The *ntr* system is an example of two component regulatory system (**see Chapter 6**) and manipulates nitrogen fixation by responding to the levels of fixed form of nitrogen.

Many different *nif* genes and their functions have been identified in *Rhizobium* and *Bradyrhizobium* which are also shared by free-living N_2-fixers (Table 8.3). The *fix* genes which are not homologous to *Klebsiella* also code for diverse functions required for nitrogen fixation (Table 8.4). This includes bacteroid development and metabolism. Some genes coding for other cellular functions, such as *glyA* (glycine biosynthesis) and *dct*, involved in the transport of dicarboxylic acid, are also involved in symbiosis.

Nitrogen fixing bacterial symbiosis with non-legumes though known, are not as well described as the legume-rhizobial symbiosis. In one such relationship, *Anabena,* a cyanobacterium, forms a symbiotic relationship with water fern, *Azolla*. This filamentous cyanobacterium occupies a cavity located at the base of each dorsal leaf lobe of *Azolla*. *Anabena* forms special cells, called heterocysts, both in symbiotic association and free-living state. Heterocysts are the sites of nitrogen fixation and lack the oxygen-producing photosystem II, thereby protecting the nitrogenase.

Similarly, *Frankia,* an actinomycete, forms symbiotic association with several nonleguminous plants, including various woody shrubs and trees. In the nodules, *Frankia* hyphae differentiates into specialized nitrogen-fixing cells, called vesicles. This actinomycete, however, can form such nitrogen-fixing vesicles in free-living state as well. *Frankia* and its associated nitrogen fixation has played a significant role in forest ecosystem.

8.3.II- 2. Mechanism of Nitrogen Fixation

The biochemistry of nitrogen fixation is best understood in the free-living diazotroph,

Klebsiella pneumoniae. The crucial enzyme involved in this process is the enzyme, nitrogenase.

Table 8.3. nif genes, their function, and distribution

nif Gene	*Molecular weight (kDa) of the product*	*Function*	*Azotobacter*	*Rhizobium*
Q	19.6	Processing of Mo	+	-
B	49.0	Synthesis of Fe Mo-co factor	+	+
A	58.0	Nif specific activation	+	+
L	55.2	Nif specific repression	-	-
F	19.1	Electron transport flavodoxin	+	-
M	30.6	Processing of Fe protein	+	-
Z	16.6	Processing of Fe Mo-co factor into dinitrogenase	+	-
W	10.1	Full activity of dinitrogenase	+	+
V	41.0	Synthesis of Fe Mo-co factor	+	-
S	43.1	Maturation of Fe-Mo protein	+	+
U	29.4	Maturation of Fe-Mo protein	+	-
X	18.2	A possible regulatory function	+	+
N	50.5	Synthesis of Fe Mo-co factor	+	+
E	50.1	Synthesis of Fe Mo-co factor	+	+
Y	24.4	Maturation of Fe-Mo protein	+	-
T	8.3	?	+	-
K	58.4	β-subunit of Fe-Mo protein	+	+
D	54.1	α-subunit of Fe-Mo protein	+	+
H	31.9	Fe protein	+	+
J	128.0	Electron transport: Pyruvate flavodoxin oxidoreductases	-	-

Note: The genes and their products have been identified from *K. pneumoniae*

Nitrogenase is a complex, oxygen labile enzyme that is composed of two protein components, dinitrogenase and dinitrogenase reductase. Dinitrogenase consists of two dissimilar polypeptides with a conformation of $\alpha_2 \beta_2$ and molecular weight of 200-250 kDa. Active dinitrogenase contains two iron-molybdenum (Fe-Mo) cofactors and four iron-sulphur (4Fe-4S) clusters. In addition, dinitrogenase contains 2 Mo and 24-32 non-heme Fe atoms. Dinitrogenase reductase consist of two identical polypeptides. It has a total molecular mass of approximately 60 kDa and contains a single (4Fe-4S) cluster that functions as one e^- donor. Firstly, the electrons are transferred through either ferredoxin or flavodoxin to nitrogenase reductase. The role of dinitrogenase reductase is to transfer these e^- to dinitrogenase which contains an active site where the reduction of nitrogen occurs. The end products of this reaction are NH_3 and H_2 (Fig. 8.8). The generation of H_2 adds on to the ATP requirements of nitrogen fixation.

In the case of *Azotobacter*, a free-living bacterium, three types of nitrogenases have been observed and the synthesis of a particular nitrogenase depends upon the metal content of the growth medium. These consist of: 1) Mo nitrogenase, 2) vanadium nitrogenases, 3) neither vanadium nor Mo nitrogenase.

Each of these nitrogenases is coded by a different set of structural genes. In case of Vanadium (V) and of non-molybdenum nitrogenases, an additional subunit is present, encoded by *vnfG* and *anfG* genes, between the α and β subunits of component 1 of V-and third nitrogenase, respectively. But the component 2 proteins of all nitrogenases are highly conserved.

Table 8.4. Some associated gene functions involved in rhizobial nitrogen fixation

Gene	*Function*
fixABCX	Required for nitrogenase activity. FixX shows similarity to ferredoxins
fixNOQP	Membrane located cytochrome oxidase
fixGHIS	Redox process linked cation pump
fixLJ	Oxygen regulated two component regulatory system involved in positive control of *fixK* and *nifA*
fixK/fixK2	Positive regulator of *fixNOQP, nifA*, $rpoN_1$ and "nitrate respiration"; negative regulator of *nifA* and *fixK*
fixK[1]	Reiterated, functional copy of *fixK* in *Rhizobium meliloti*
fixK[1]	In *Bradyrhizobium japonicum* function unknown
fixR	Function unknown
nrfA	Regulation of *nifA*

The comparative studies of purified Mo- and V-nitrogenases have shown the formation of ethylene in the former system and ethane and ethylene in the latter system as a product of acetylene reduction. When Mo was absent from the growth medium, V-and third nitrogenases were observed suggesting that they are important in organisms living in Mo-deficient soils. V- nitrogenases are more active than Mo-nitrogenases at low temperatures (5-10°C). It can thus be suggested that V-nitrogenases might make a significant contribution to agronomic nitrogen fixation in temperate climates.

As the nitrogenase enzyme system is irreversibly inactivated by O_2, different bacterial systems have evolved different mechanisms to protect their nitrogenase. In root nodule bacteria, the red pigment leghemoglobin which supplies oxygen to bacteria for respiration prevents O_2-denaturation of enzyme. *Azotobacter* has acquired the highest respiration rate which leaves very little free oxygen for inactivation of the enzyme. The heterocystous cyanobacteria carry out nitrogen fixation in heterocysts that lacks oxygen evolving photosystem II and non-heterocystous cyanobacteria function under low oxygen tension. The fixation of nitrogen also requires a high energy input (~30 ATP/N_2 fixed).

The ability of microorganisms to fix nitrogen can be easily detected by acetylene reduction assay. Acetylene, like molecular nitrogen, contains a triple bond and can be readily reduced by nitrogenase to produce ethylene. The latter can be detected by gas chromatography.

8.3.II- 3. Organization of nif Genes

The promoter regions and transcription initiation sites of *ntrC* and *nifA* activated genes are characterized by the sequence CTGGYAYR-N_4-TTGCA in the positions -26 to -10 relative to transcription initiation site (Y=pyrimidine, R=purine), but these sequences are not conserved in all *nif* genes. The *R. meliloti nifA* promoter does not resemble the *nif* consensus promoter nor does the *nifH* and *glnA* promoters of *Anabena* 7120.

Most of our information on biochemistry and genetics of *nif* is derived from the free-living diazotroph, *Klebsiella pneumoniae*. *K. pneumoniae* has 20 genes organized in 8 transcriptional units, of which 6 operons are transcribed in one direction and 2 in opposite direction. The genes are *nif JHDKTYEN XUSVWZMFLABQ* (Fig. 8.9).

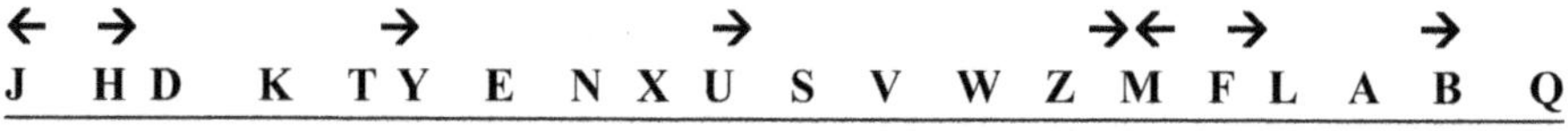

Figure 8.9. Physical organization of nif genes in K. pneumoniae. Arrows indicate the position and orientation of known and proposed transcription initiation sites.

In symbiotic bacteria, the *nif* structural genes have been identified by homology studies with *K. pneumoniae.* The availability of *nif* probes from *K. pneumoniae* and the advent of gene cloning techniques have allowed the analysis of a large number of nitrogen-fixing bacteria, both symbiotic and free-living. Most diazotrophs show striking homology in *nif* components but important differences exist in number, location, and copies of structural and regulatory genes. Symbiotic bacteria acquiring additional requirement to establish and maintain their interaction with the host.

The *nif* and *fix* genes in all the three types of rhizobia are clustered in species-specific pattern. *R. meliloti* carries megaplasmids 1 and 2 known as pSym-a (1,400kb) and pSym-b (1,700kb), respectively. Megaplasmid 1 contains both cluster I (*nifHDKE, nifN, fixABCX, nifA, nifB, frdX*) and cluster II (*fixLJ, fixK fixNOQP, fixGHIS*) with the latter located 200kb downstream of *nifHDKE* and transcribed in opposite direction. A cluster of *nod* genes are lodged in the 30kb region between *nifE* and *nifN*. A duplicated segment covering *fixK* and *fixNOQP* is present upstream of *nifHDKE* (Fig.8.10). Additional genes participating in symbiosis are present on megaplasmid 2 and the chromosome.

Anabena contains *nifH, nifD, nifK,* and *nifS* genes in vegetative cells. The analysis of these genes indicated that *nifH* and *nifD* were adjacent to each other but *nifK* was located about 11 kb to the 3′ end of *nifD* gene. But when the vegetative cells differentiate into nitrogen-fixing heterocysts, the *nifH, nifD* and *nifK* genes get organized together in a single transcriptional unit (Fig. 8.11). This led to the discovery of excision of 11kb fragment separating the *nifD* and *nifK* genes. The *xis* gene responsible for excision event is located 240 bp from the recombination site near the *nifK* gene.

Some methanogenic archaea capable of fixing nitrogen share high degree of homology in their *nif* genes with free-living bacterial nitrogen fixers. *Methanococcus thermolithotrophicus* possesses two dissimilar *nifH* genes.

8.3.II- 4. Regulation of Nitrogenase Activity

Nitrogen fixation is very perfectly regulated in the presence of oxygen, which inactivates nitrogenase proteins, and at high nitrogen status. The mechanism of regulation of *nif* genes has been studied using *Klebsiella* as a model system. There is a two-tier cascade system of regulation, one by *ntr* and *rpoN* genes lying outside the *nif* cluster and another by the product of *nifLA* together with the product of *rpoN*. The essential features of regulation system are similar but due to differences in the diazotrophy conditions, the regulation mechanism of different nitrogen-fixing organisms varies.

In *Klebsiella pneumoniae,* the nitrogen regulation system (*ntr*), sensitive to the level of fixed nitrogen, consists of genes organized in two operons, *gluA-ntrBC* and *ntrA* (=*rpoN*). The two regulatory *nif* genes are *nifA* and *nifL*. While *nifA* gene product is required to activate the operon, the product of *nifL* 'switches off' the *nif* operon. The product of *rpoN* (σ^{54}–factor) is essential for the action of *nifA* product. In addition, NtrC is also necessary for the *nifLA* operon to be read. Thus, both RpoN and NtrC

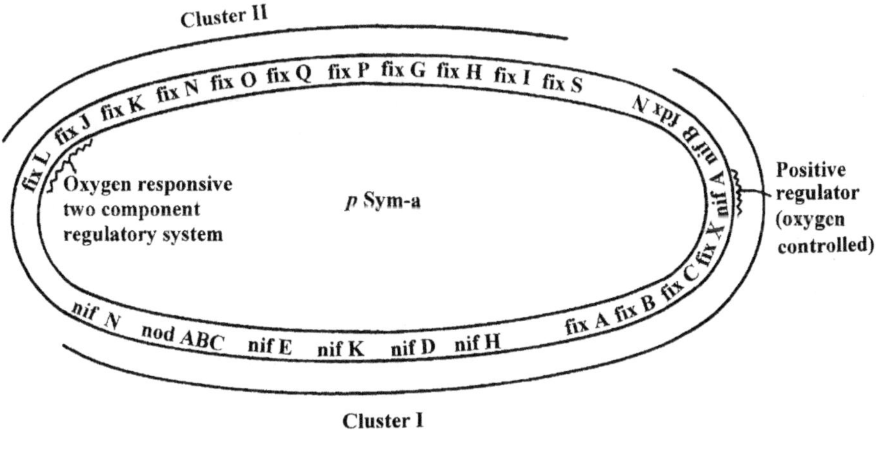

Figure 8.10. Rhizobium meliloti megaplasmid pSym-a *carrying two clusters of nitrogen fixing genes. Additional genes required for symbiosis are located on another megaplasmid (*pSym-b*) and chromosome.*

should be available for *nifLA* to be functional. The formation of *ntrC* product but not of *rpoN* product is regulated by ammonia. When plenty of ammonia is available, NtrC is not formed. On the other hand, *ntrB* product interferes with action of NtrC. It is, thus, clear that *ntr* system provides a coarse control of *nif*, switching it on or off, and the *nifLA* system exerts a finer control. The *ntr* system is involved in a number of other aspects of the nitrogen metabolism in *Klebsiella* (Fig. 8.12).

Oxygen is another regulator of *nif* expression and it does so by bypassing *ntr* system. The product of *nifL* gene again plays a key regulatory role. The NifL is a flavoprotein capable of responding to oxygen levels. When oxygen concentration is high, it gets oxidized and in this form prevents NifA to activate the operon.

Energy status of the cell is also linked to *nif* expression. Energy-starved cells have high ADP which reacts with reduced form of NifL and causes it to block the promoter activation by NifA.

The major regulatory component in symbiotic nitrogen fixation by rhizobia is O_2, as N-compounds are already present in the nodules. The elements which regulate nitrogen fixation genes are *fixL, fixJ, fixK, nifA,* and *rpoN.* Of these, *fixLJ* form two component regulatory system (e.g., *R. meliloti* and *A. caulinodans*) and *nifA* is autoactivated under low oxygen condition (*B. japonicum*).

The FixLJ proteins regulate expression of *nifA* and *fixK* genes. FixL has phosphatase

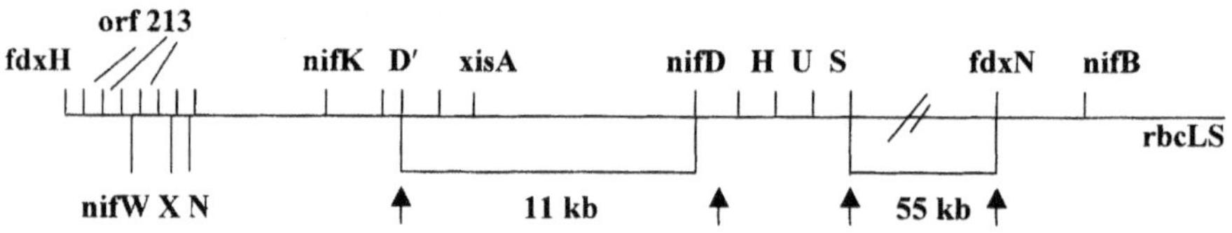

Figure 8.11. Organization of nif genes in Anabena vegetative cells. During heterocyst differentiation two regions → of 11 kb and 55 kb are excised (some 10 kb from nifB is rbcLS operon encoding large and small subunits of Rubisco). Heterocysts are the sites of nitrogen fixation.

and kinase activity. Under low O_2 tension, FixL phosphorylates the FixJ and under high O_2 tension dephosphorylates the same. In some rhizobia, FixK protein exerts negative effect on *nifA* expression and on its own expression (Fig. 8.13).

8.3.II- 5. Future Prospects for Increasing Symbiotic N_2 Fixation

The significance of biological N_2 fixation has led to extensive research to design possible approaches for increasing plant- associated symbiotic N_2 fixation. These are:

- Increasing the number of N_2 fixation genes in the endosymbiont results in the increase in the number of nitrogenase molecules. The *nifH* gene product is the rate limiting step in N_2 fixation. So increasing the copy number of *nifH* gene results in increased turnover rate of nitrogenase.
- With increase in the number of *dct* genes, N_2 fixation also increases.

- Introduction of a more efficient nitrogenase into the endosymbiont can be another approach.
- Creation of plants capable of fixing N_2 independent of symbiotic relationship by introducing nitrogen fixation apparatus into chloroplast in which O_2 generating photosystem II is inactivated.
- Increasing N_2 fixation by any means requires a better understanding of the factors intrinsic to symbiosis.

8.3.II- 6. Improving the Value and Utility of N_2 Fixation by the Legume-Rhizobia Symbiosis

Certain limitations occur which are intrinsic to legume-rhizobia symbiosis and they are unable to fix maximum N_2 under field conditions. Some of the modern approaches to overcome these limitations are:

8.3.II- 6.1. Infection with an "Effective Rhizobial Strain". The successful establishment of legume-rhizobia symbiosis involves an effective association. This can be improved by:

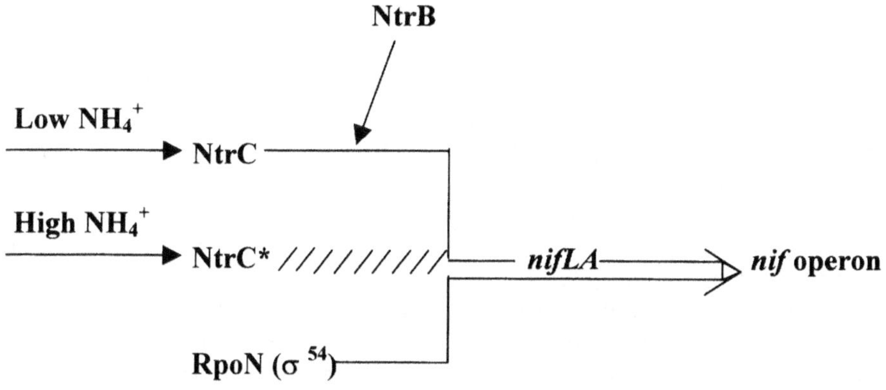

Figure 8.12. Regulation of nif genes by ammonia in K. pneumoniae. NtrC product is not formed under high ammonia (NtrC).*

- selection of locally-adapted strains of rhizobia chosen for their tolerance to extreme soil conditions or unfavourable climates.
- use of novel, genetically engineered strains which contain unidirectional uptake of

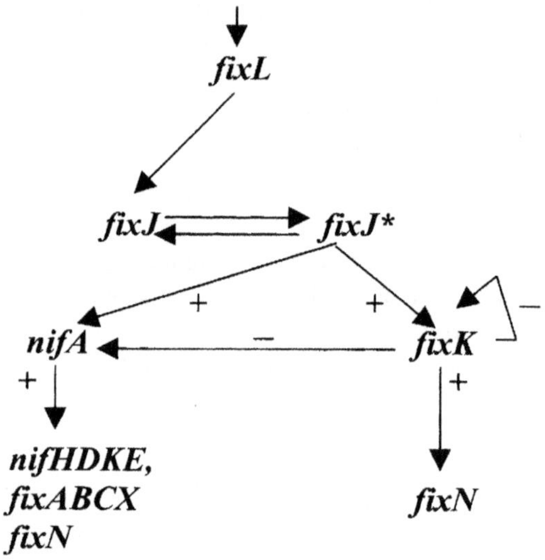

Figure 8.13. Regulation of nif genes in rhizobia.

hydrogenase, express insecticidal proteins to control specific nodule pests or in which host plant specificity is altered, and

- the most important being their ability to survive in soils and to compete successfully against indigenous bacteria for nodule occupancy.

A number of approaches have been proposed to increase the competitive advantage of a rhizobial strain introduced into the soil. Three genetic approaches to enhance competition are:

8.3.II- 6.2. Legume Genotypes with very Restricted Susceptibility to Infection. There are a number of examples where legumes are not infected by indigenous groups of rhizobia in the soil. In tropical soil of West Africa for example, many strains of "cow-pea" rhizobia fail to nodulate the improved US varieties thereby promoting efficient nodulation by *B. japonicum* strains. A number of plant genes have been identified which include a series of *rj* alleles which confer restricted nodulation in soybean as well as traits in alfa-alfa and pea in which selective strain dependence has been demonstrated.

8.3.II- 6.3. Altering the Host Range of the Inoculated Strains. It has become possible as a result of recognition and isolation of a number of bacterial genes involved in nodulation and host range. The nodulation genes in *Rhizobium* are largely located on a high molecular weight symbiotic (sym) plasmid, whereas in *Bradyrhizobium* and *Azorhizobium* species these genes reside on the chromosome. The nodulation genes including common *nod* genes and host specific *nod* genes are regulated by a common *nodD* gene which in turn is directly influenced by plant root exudates which contain phenyl-propanoid molecules (flavones, isoflavones). The *nodD*/flavone interaction is highly species specific (some flavones induce *nodD* while others competitively inhibit *nod* gene activation). A number of chimeric *nodD* genes involving sequences from *R. leguminosarum* and *R. meliloti* have been constructed in which host specificity has been modified. Transfer of *nodD* allele from a broad host range symbiont of the non-legume *Parasponia andersonii* (NGR 234) to *Rhizobium* strain led to nodulation of rice roots. Recent approaches follow the identification of nodulation specific gene products which are necessary in early nodule development. These might be exploited to narrow the host range preferences.

8.3.II- 6.4. Construction of Rhizobial Strains with an Increased Ability to Compete for Nodule Occupancy. Another potential means to enhance competitiveness involve the use of strains which produce toxins or antibiotics inhibitory to infection by other strains. *R. leguminosarum* bv. trifolii strain T24 produces antibiotic trifolitoxin which inhibits growth of other pea strains. The gene responsible for this toxin has been successfully cloned and transferred to other rhizobial strains without affecting their symbiotic association.

Molecular methods of introducing foreign DNA into *Rhizobium* enhances the commercial impact of the inoculants. *Bacillus thuringiensis* subspecies *terebrionis* δ-endotoxin genes (*cry IIA*) have been introduced into *R. leguminosarum* for the control of larvae which feed on legume roots and nodules causing considerable damage. The transgenic *R. leguminosarum* has greater ability to compete for nodule occupancy than the non-transformed parent strain.

8.3.II- 6.5. Nutritional Requirement for Nodule Establishment. The number of bacteroids which occupy the infected cells of functional and mature cowpea nodule ranges from 8-to around $18x10^4$ per cell with 1.5 to more than $13x10^8$ per nodule. This large bacterial population, derived from a small number of infecting bacteria, requires host plant's essential nutrients for their growth and multiplication.

N-hunger, a phenomenon, seen in the seedlings of many grains and pasture legumes during the initial period of nodule formation is due to fixed N being sequestered by bacteroids for their own multiplication resulting in the yellowing of foliage leaves. This can be overcome by adding small amount of starter N (3-5 kg N/ha). In order to find out the partitioning of seed N reserves to bacteroids, the cowpea seeds were ^{15}N labeled and it was found that half of the N in bacteroids come from plant seed reserves.

In addition to C and N, other nutrients required for proper symbiotic development include Mo, Ca, Co, and Cu, P, and Fe.

Important Functions of Various Nutrients:

Ca: is involved in both nodulation and nodule functioning. The adsorption of rhizobial cells to roots of alfalfa and to the root hairs of pea have been shown to be Ca-dependent. The deficiency of Ca in soils is associated with high levels of aluminium and acidity.

Cu: is required for nodule functioning rather than nodule initiation and development. The Cu-deficient nodule contains less cytochrome C oxidase activity.

Co: is a constituent of cobalamine coenzymes required for a number of enzyme catalysed reactions. One of these, ribonucleotide reductase is required for the incorporation of ribonucleotides into the deoxy-derivatives in DNA. Thus in Co-deficient nodules, multiplication is likely to be limited.

Mo: is a functional component of the Mo-Fe protein of nitrogenase and in ureide-forming nodules and xanthine dehydrogenase.

Fe: is an important component of Fe and Fe-Mo proteins of nitrogenase, ferredoxins, and heme moiety of leghemoglobin. Rhizobia also possess transport system for Fe (siderophores). *R. meliloti* mutants defective in transport system for Fe indicate that Fe translocation from host to bacteria is essential for full symbiotic development. Therefore, a potent limitation to symbiotic development occurs under the soil nutrient-deficiency conditions.

8.3.II- 6.6. Role of O_2 in Regulating N_2-Fixation in Nodules. As described earlier, almost all diazotrophs are extremely sensitive to O_2 and are irreversibly inactivated at high O_2 concentration. The intracellular concentration of free O_2 is kept at low level by a high rate of O_2 consumption and at a high flux by the reversible oxygenation of leghemoglobin.

A step-wise increase in external O_2 pressure results in an increase in gaseous diffusive resistance of nodules. These changes in gaseous permeability is reversible and is associated with number of treatments (like drought stress, application of fertilizer NO_3^-, inhibition of phloem transport etc.) which inhibit nitrogenase.

An O_2 diffusional control exists in the inner cortex region of nodule. There have been proposals that osmotic changes in cells of the cortex constitute the basis for a "variable barrier". A number of sensors within legume nodules have been proposed which can alter diffusion barrier resistance and include sensing O_2 concentration, phloem translocation and N assimilation. These sensors in turn change the water-solute relation of the inner cortex and alter gas movement in nodules.

8.3.II- 6.7. Effect of Combined N on N_2 Fixation. N_2 fixation is completely inhibited in N-rich soils or following application of nitrogenous fertilizers. The inhibition may be direct; due to action of nitrate on root infection, nodule initiation, nitrogenase activity, or on redox state of leghemoglobin or indirect effect resulting from changes in hormone balance and partitioning of assimilates to nodules.

Mutants of pea, common bean and soybean have been isolated which form nodules at inhibitory concentration of NO_3^-. These mutants were termed as supernodulators. The trait of nitrate tolerant symbiosis (nits) in soybean and some pea mutants is controlled through basipetal translocation of specific growth regulating substances.

The strains which have limited uptake and/or assimilation of NO_3^- provides another means for effective symbiosis in presence of combined N, but progress in this area has not been rapid.

8.3.II- 6.8. Premature Senescence of Nodules. Grain legumes form protein-rich seeds leading to "self destruction" of their photosynthetic potential during reproductive growth. Leaf N provides a significant proportion of the seed N. As a consequence legumes suffer premature leaf senescence because of reduced assimilate supply, leading to nodule senescence. The process of leaf senescence is triggered by a "senescence signal" translocated from the developing seeds to the rest of the plant.

8.3.II- 6.9. Transfer of Symbiotic Traits to Non-legumes. Two distinct approaches in transferring symbiotic traits to non-legumes include (i) non-genetic means for induction of nodulation, and (2) genetic manipulations to transfer traits for nodulation.

8.3.II- 6.9.1. Non-genetic Manipulation Approaches. Hydrolytic enzymes were used to remove the physical barriers to infection of non-legume roots by rhizobia. Nodular structures were reported in associations between *Rhizobium* and *Bradyrhizobium* spp., and range of non-legume crops (rice, wheat) but the levels of nitrogenase activity were extremely low; typically 0.1% or less than that found in legume nodules.

Paranodular structures on wheat roots were observed particularly when roots were treated with hormonal herbicide, 2,4-D, but no nitrogenase activity was observed. Nodulation of barley roots with *R.astragali* was observed under the influence of permanent magnetic field as it might act to remove physical barriers between the heterologous partners. Once again, no nitrogenase activity was detected in such pseudonodules.

8.3.II- 6.9.2. Genetic Manipulation Approach. This involves direct manipulation of the host plant genome by transferring traits from legumes to non-legumes. A pea lectin gene transferred into white clover, facilitated the latter to be infected and nodulated by *R. trifolii*. Mutations in the strains of *R. sesbania* isolated from *Sesbania cannabina* were able to nodulate varieties of rice.

Transconjugants between *R. phaseoli* and *R. leucaena* crossed with *R. trifolii* are able to nodulate large varieties of wheat in presence of plant exudates.

One of the basic limitations in the operation of symbiotic N_2 fixation is the establishment of an internal microaerobic environment in nodules. So a better understanding of the plant-microbe symbiosis in the plant host is necessary.

With the availability of genetic engineering methods, introduction of the capacity to fix nitrogen by plants like wheat, rice, and corn was considered to be the most important goal. The information available on genetics and biochemistry of rhizobial infection and nitrogen fixation has been utilized by many groups.

In early experiments, *nif* genes were conjugationally transferred from *K. pneumoniae* to *E.coli* converting the latter to fix nitrogen. Some of these *E.coli* hybrids showed integration of *Klebsiella nif* genes into the *E.coli* chromosome but some formed plasmids carrying *nif* genes. A *nif* plasmid (pRD1) construct in *E.coli* has been employed in many subsequent studies. With nitrogen-fixing *E.coli*, it became possible to transfer these genes to *Salmonella typhimurium, Serratia marcescens*, *Erwinia herbicola* and *K.aerogenes* obtaining effective nitrogen fixers. However, a similar exercise in many other bacteria has not provided the desired results.

In another experiment, a chimeric plasmid was created between an *E. coli* plasmid and of yeast. This plasmid could integrate in yeast chromosome. The *nif* genes (coding for 17 proteins) were isolated from *Klebsiella* chromosome recombined with a plasmid vector and reintroduced into the yeast cell. The latter contained the *E.coli* plasmid sequence which could be recognized by the hybrid plasmid and the same is integrated. However, nitrogen fixing genes were not expressed in yeast. In order to achieve this goal, a thorough understanding of the molecular biology of *nif* gene expression and its regulation, physiological and environmental condition requirements, will have to be reached. Until then, it remains one of the most cherished dreams of plant genetic engineers.

8.3.III- Environmental Restoration

As we have seen earlier, the biomass of all living forms returns to nature through death, decay and decomposition. Microorganisms by supporting a specific biogeochemical cycle are able to decompose the organic substances into simpler form for their reuse, so as to maintain a balance between their availability and consumption.

The rapid progress of human civilization has disturbed this balance in two major ways: i) extensive urbanization has led to the huge piles of wastes, and ii) development and use of some inert materials like glass, plastics, and metals and some recalcitrant materials (xenobiotics) resistant to microbial decomposition, has led to large scale environmental deterioration. This has created the problems of environmental pollution that threatens the existence of all life forms including humans. It is still gratifying that many microorganisms are capable of and thus are employed for decomposition of wastes to help restore the health of the environment.

Disposal of Solid Waste: A large proportion of urban waste consists of degradable organic garbage emanating both from industry as well as household. In order to enhance the efficiency of decomposition they need to be separated from inert material. The sewage sludge produced during treatment of liquid wastes, and animal waste obtained from farms generate lot of solid waste. At a small scale these wastes are reutilized as fertilizers but in large urban centres and agricultural and animal farms, the amount of waste generated need proper disposal methods.

8.3.III- 1. Land Fills

As the dumping of waste in water bodies or on land can cause severe pollution and health problems, a simple and cheap process of landfills has been devised. In this, both organic and inorganic wastes are deposited in low-lying land that is not to be used for any purpose. Many times such landfills covered with a layer of soil, so as to detract insects and rodents, and to avoid a fire hazard, are called as sanitary landfills. In a span of over 30 to 50 years, the organic waste is slowly degraded anaerobically producing CO_2, H_2O, CH_4, and various alcohols and acids. Huge amounts of CH_4 so produced can be used as a potential source of natural gas. Once the decomposition slows down the site can be used for recreational and real estate purposes. One biggest drawback of this method is the seepage of the products of decomposition, heavy metals and other recalcitrant hazardous materials into underground potable water supply. The shrinking land availability also makes it a less popular solution.

8.3.III- 2. Composting

The degradation of heterogenous solid organic matter by combined action of aerobic, mesophilic, and thermophilic microorganisms is known as composting. The compost is a stable, sanitary, humus-like product which can be mixed with soil to improve its nutritional status.

Composting is a sequential process which may be initiated by mesophilic heterotrophs but is soon taken over by thermophilic microbes as the temperature rises. Temperature, humidity, aeration, and carbon-to-nitrogen ratio all play an important role in composting. Compost is sold as a good soil conditioner and a source of plant nutrients. High concentration of potential toxic heavy metals restricts its use in agriculture.

8.3.III- 3. Treatment of Liquid Waste

Industries, agriculture and other anthropogenic activities including domestic sewage generate huge quantities of liquid waste. Their disposal in natural water bodies can cause serious health problems to all living organisms. Self purification is an inherent property of the natural waters, in which interpopulation interactions between indigenous and nonindigenous members help maintain the quality of water. The latter reflects a number of physical and chemical characteristics. A heterogenous microbial community grows in water receiving sewage effluent but one popular member is *Sphaerotilus natans*, a filamentous aerobic bacterium also called "sewage fungus". The organic load of a contaminated water sample is measured by biological oxygen demand or BOD. A high BOD reflects higher organic content and, therefore, leads to higher oxygen consumption. The latter produces a detrimental effect on other life forms. Combinations of physical, microbiological, and chemical methods can be used to reduce the BOD. These are known as primary, secondary, and tertiary sewage treatments, respectively (Fig. 8.14). The primary sewage treatment removes the suspended solids in settling tank or basins which is removed and treated separately as solid waste. Heterotrophic bacteria are employed to degrade sewage organic matter as secondary treatment in oxidation ponds or stabilization ponds and lagoons. The bacterial activity produces conditions conducive for algal growth which in turn replenishes oxygen for continued decomposition. The effluents containing oxidized products are periodically removed so that fresh sewage can be received in the pond.

Various methods, such as trickling filter and biodisc system have been designed to facilitate aerobic sewage treatment. In these, the sewage is distributed by a rotating sprinkler or a biodisc system. This allows the formation of a microbial film. The microbial community in the film adsorbs and mineralizes the dissolved organic matter in the sewage. Another system known as activated sludge treatment system employs an aeration tank which receives the influent from primary treatment tank. Extensive aeration and agitation favours complete degradation of organic matter by microorganisms. The biomass so produced settles in a secondary settling tank. A portion of this activated sludge is refed into the aeration tank for a new batch of sludge. From the secondary settling tank, relatively clean water can be obtained. The combined effects of competition, adsorption, predation, and settling during this process reduces not only the BOD but also the load of pathogenic microbes. Similarly, anaerobic digestion can be carried out in big digestors for processing the sewage sludge further. Through this the volume of residue is further reduced and gases like CO_2 and CH_4 produced can be used as natural gas.

After these two treatment steps, tertiary treatment is carried out to remove non-biodegradable organic pollutants and mineral nutrients especially nitrogen and phosphorus salts. Activated carbon filters are normally used to remove toxic organic pollutants such as chlorophenols and polychlorinated biphenyls. Another process of reverse osmosis to eliminate organics and inorganics is often limited by the problems of microbial fouling. Removal of phosphates, by precipitation with calcium, aluminum, or iron and of nitrogen by volatilization as NH_3 can check eutrophication. Ammonia is removed by addition of hypochlorous acid, a process known as breakpoint chlorination.

The final step in the sewage treatment process is disinfection employed to kill enteropathogenic bacteria and viruses. This is commonly done by chlorination using chlorine gas (Cl_2) or hypochlorites of sodium or calcium, in which, the former eventually produces hypochloric acid. Hypochlorite is a strong oxidant that explains its anti-bacterial action. The disadvantage of this method is the production of toxic partially chlorinated forms of resistant lipids and hydrocarbons. Also trace amounts of trihalomethane (THM) are generated which are suspected carcinogens. Alternative methods of ozonation which is expensive, and treatment with chloramines, relatively cheaper, are now available.

Fecal contamination of the treated water is another safety parameter. This is estimated by coliform counts and has to be maintained under permissible limits. The limit may vary as to the uses of this treated water. The coliforms are detected by a three stage biochemical test consisting of presumptive, confirmatory, and completed tests. Initially, the water samples are inoculated in lactose broth tubes, and the positive ones (gas production) are followed on EMB-agar. On this medium, the coliforms acquire a characteristic greenish metallic sheen which are once again incubated into lactose broth. Gas production, if shown, completes the testing for such bacteria. Several alternative quick procedures are now available. One test uses a defined substrate to detect characteristic enzymatic activities of fecal *E. coli*. A medium containing isopropyl β-D-thiogalactopyranoside (IPTG) to detect β-galactosidase activity is characteristic of total coliform bacteria. The medium containing O-nitrophenol-β-D-galactopyranoside-4-methyl umbelliferyl-β-D-glucuronide (MUG) detects the enzyme β-glucuronidase, diagnostic of *E. coli*. However, the chances of an outbreak of the enteroviruses-mediated infections, from the water containing permissible coli counts, can not be ruled out, as viruses can not be detected by the conventional tests.

8.3.III- 4. Biodegradation of Pollutants

As we have seen earlier, microorganisms, both bacteria and fungi, have unique capability of degrading almost all types of complex biomolecules. In recent years, however, this capability has been put to a strict test as a large number of man-made novel chemicals have entered the environment. Although microorganisms have shown immense versatility, they have to adopt an altogether different strategy for non-biodegradable compounds. The microbial decomposition is a community function and therefore is a slow process. When the concentration of a compound is too high to be degraded quickly or if it is non-degradable, it becomes a pollutant. Many of these compounds are toxic to living organisms and their presence in aquatic and terrestrial habitats often have serious ecological consequences. With an increasing world-wide concern for the environment, it is no longer acceptable to allow the accumulation of a wide variety of wastes lying untreated. Conventional techniques of digging of the waste followed by incineration or chemical treatment has not only become expensive but tends to generate

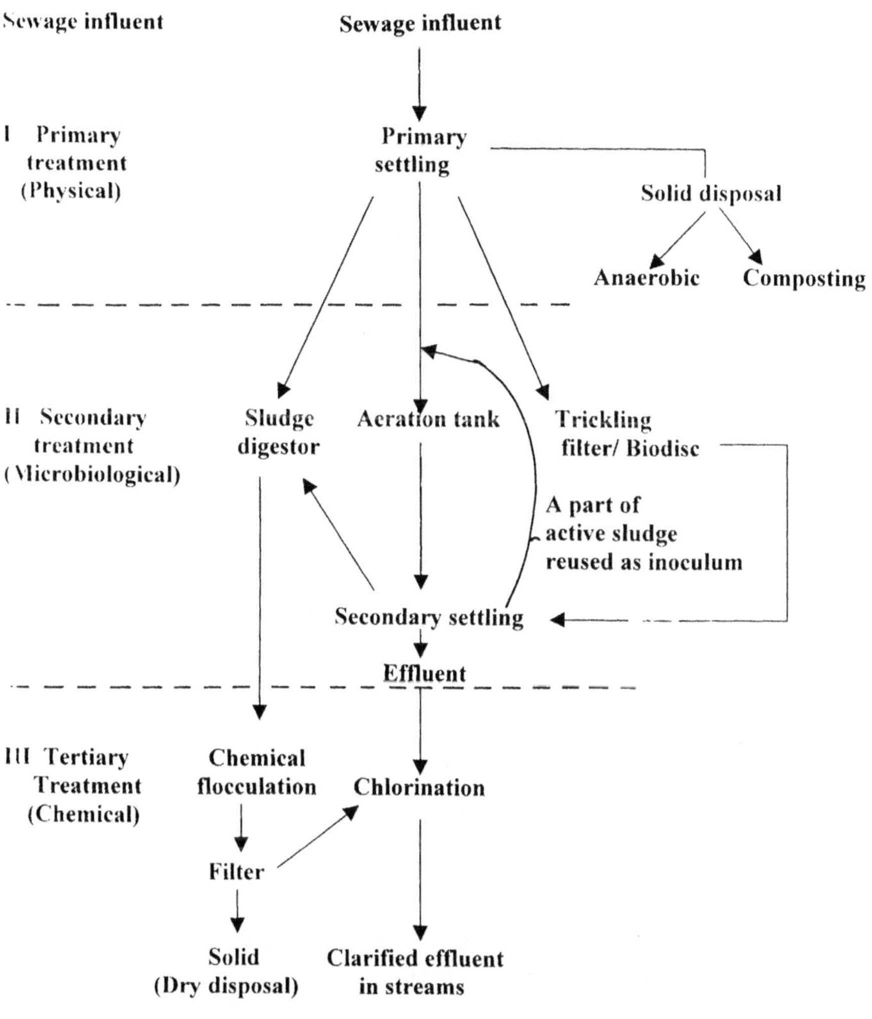

Figure 8.14. Treatment of wastes.

fresh classes of pollutants. Microbial decomposition, therefore, is not only an inexpensive alternative but is also environmentally benign.

8.3.III- 4.1. Organic Pollutants. A large number of such compounds like anionic detergent, alkylbenzyl sulphonates (ABS), a number of pesticides especially the chlorinated hydrocarbons, and oil pollutants can severely damage an ecosystem. The non-linear branched ABS is used as a laundry detergent and its release into water bodies cause extensive foaming. This molecule is resistant to microbial decomposition and is thus hazardous. By converting this synthetic molecule into a linear ABS, the detergent can be more easily biodegraded.

By understanding the role of microbial decomposition of wastes in maintaining environmental quality, it has been argued that synthetic polymers be replaced by biodegradable ones. Various plastics, including polyethylene, polystyrene, and polyvinylchloride which are resistant to microbial attack make a point. As they are not degraded they are piling up in environment posing serious health hazard. The use of biodegradable plastics is already being advocated for this reason.

Many chlorinated hydrocarbons, such as DDT, have been successfully used as pesticides. However, their persistence in the soil due to their resistance to microbial degradation allows them to get accumulated in microbial cells as well as plants. This accumulation which may be one to three orders of magnitude higher than those in the surrounding habitat is known as biomagnification. When these microbes are ingested by the members of the next higher trophic level in the food web, the pesticide gets transferred and further concentrated. Thus, while travelling through the members of different trophic levels the pesticide molecule gets biomagnified to reach a body burden exceeding 10^4 to 10^6 times the environmental concentration. At this concentration, it may turn very hostile. DDT and other chlorinated hydrocarbons have been implicated in the death or reproductive failures of several birds and other wild-life. Since human beings are also associated with the food web, the chances of pesticide accumulation and toxicity are also very high in them. DDT has, therefore, been banned by many developed countries but in developing countries, it is unfortunately still in use. The formulation of a new synthetic biodegradable pesticide or a microbial degradation test for those in use is very important so that the presumptive environmental damages can be assessed and controlled.

The biodegradability is often affected by structural constraint of a molecule and modulating the same may break its recalcitrancy. For example, the herbicide 2,4-D is easily biodegraded in comparison to 2,4, 5-T which has an additional chlorine at the meta-position. In some cases, one portion of the pesticide molecule is susceptible to degradation and another one is recalcitrant. When acylanilide herbicides are attacked by microorganisms the aliphatic moiety is converted to CO_2 and H_2O through initial amidase reaction but its aromatic moiety resists mineralization.

Marine, other aquatic as well as the soil environment can be seriously threatened by the accidental spillage of oil and several aliphatic, alicyclic, and aromatic hydrocarbons. Although microbial biodegradation of petroleum pollutants is a viable solution to tackle this problem, the metabolic capabilities of several microorganisms (consortium) will have to be utilized. The susceptibility of these hydrocarbons depend upon several factors. Firstly, the alkane chain length plays an important role; the longer the chain more resistant it is to microbial degradation. Moreover, the lack of suitable growth temperature and of available supplies of fixed forms of nitrogen, phosphate, and molecular oxygen in the oceans, often limit the breakdown of oil pollutants. Besides killing birds, fishes and other marine organisms, even low concentration (ppb) of petroleum can disrupt the chemoreception of some marine organisms that seriously hampers their feeding and mating responses.

Many different microorganisms, especially *Pseudomonas* and *Acinetobacter* species can degrade oil components in contaminated ecosystems. Although superbugs combining the abilities of different species have been created, several parameters still need to be standardized before their large scale applications.

8.3.III.5. Management of Metal Ions

Besides the organic molecules, another important class of pollutants which is recalcitrant to microbial degradation consists of different metals. Metals have existed on this earth from the very beginning and have played significant role in various life processes such as growth, differentiation, and reproduction. They are generally classified into essential and non-essential elements with the former further divided into macro-and microelements. Both types of metals generally occur at low natural abundance but extensive industrialization, modern agricultural practices and faulty waste disposal methods have led to a spurt in their levels which is further compounded by their concentration build-up due to bionondegradability.

Microorganisms play a predominant role in the solubilization, transport, and deposition of metals and minerals in the environment. Two microbial functions, viz., metal leaching and sorption of metals from low grade ores, ore tailings, and industrial and mine effluent has led to the proposal of innovative microbe-based technologies as economic and safe alternatives for today's mining, and mineral and waste water treatment industries, especially at a time when high grade mineral resources are becoming scarce, energy costs are rising and adverse environmental effects are becoming too apparent.

8.3.III- 5.1. Metal Leaching. In order to satisfy the ever-increasing demand of important minerals and metals, man has exploited the ore deposits. As a result, the high grade ores are gradually getting depleted. It has thus become imperative that low-grade resources are tapped and other sources are recycled. The normal metallurgical routes of extraction of metals from low-grade sources are highly uneconomical. This has necessitated the employment of microbial metabolic capabilities to recover metals from such sources, a process known as bioleaching.

Metals can be extracted economically from low-grade sulphide ores with the help of thiobacilli, specially *Thiobacillus ferrooxidans*. This process is currently commercially exploited for the extraction of copper and uranium and has also shown immense promise in recovery of nickel, zinc, cobalt, tin, cadmium, molybdenum, lead, antimony and aresenic from their low-grade sulphidic ores. The action of thiobacilli is based on the oxidation of insoluble sulphide ores into water soluble sulphate ones, or by oxidizing ferrous component of the ore to ferric, which in turn chemically oxidizes the metal in a soluble form. The metal thus gets collected in a leachate. *T. ferrooxidans* grows at 30°-50°C, at pH 2.3 to 2.5 and at an iron concentration of 2-4g/l of leach liquor. These bacteria can produce high temperatures during leaching which is detrimental for continuation of this activity. *Sulfolobus* species which are acid-tolerant thermophiles and are also capable of oxidizing sulphur and ferrous can be employed under such conditions.

Copper bioleaching at commercial scale is being carried out by both direct and indirect oxidation of sulphide, such as pyrite ($CuFeS_2$), to form $CuSO_4$. Copper is then recovered by solvent extraction or by using scrap iron:

$$\mathbf{CuSO_4 + Fe^{\circ} \rightarrow Cu^{\circ} + FeSO_4}$$

Similarly, insoluble tetravalent uranium oxide (UO_2) found in low grade ores can be converted to leachable hexavalent form (UO_2SO_4) by indirect oxidation of ferrous to ferric by *T. ferrooxidans*.

8.3.III- 5.2. Metal Removal. Microorganisms possess unique capability of accumulating metals from low grade sources and wastes. The underlying mechanisms range from purely physico-chemical interactions to dedicated metabolic pathways. The most successful biotechnological process, to date, utilize biosorption and bioprecipitation, but binding to specific macromolecules may have future potential. Microorganisms and microbial products are avid bioaccumulators of soluble and particulate form of metals (especially from dilute external concentrations). This covers several different processes such as adsorption, surface binding or complexation, precipitation and/or mineral nucleation to internal sequestration (Fig. 8.15). Such novel capabilities of microorganisms are of immense industrial interest because the removal of potential hazardous heavy metals and radionuclides from industrial effluents and waste waters can serve the dual purpose of bioremediation and thus environmental restoration and recovery of important metals including the precious ones.

8.3.IV- Bacterial Diseases of Plants

Bacteria due to their small size and invisibility have always been shrouded with myths. Their one action which gets highlighted and thus has brought them the notoriety and familiarity even among the lay population is their disease causing nature. They are often associated with morbidity, disease and overall misery. It is a fact that many bacteria along with other microbes are capable of inflicting diseases in plants and animals. In this section, we shall consider the plant diseases, the animal diseases have been dealt

Table 8.5. Some plant diseases caused by bacteria

Disease	*Host*	*Causative Organism*
Crown gall	Several plant species	*Agrobacterium tumefaciens*
Hairy root	Apple	*A. rhizogenes*
Plant wilt	Tomato	*Corynebacterium michiganese*
Ring rot	Potato	*C. sepedonicum*
Fire blight	Pears and apples	*Erwinia amylovora*
Stalk rot	Maize	*E. carotovora*
Yellowing of plants	Aster, Peach	*Mycoplasma* species
Leaf spot	Tobacco	*Pseudomonas angulata*
Plant blights	Pea	*P. pisi*
	Lilac	*P. syringae*
Slippery skin	Onion	*P. marginalis*
Brown rot	Potato	*P. solanacearum*
Scabs	Potato	*Streptomyces scabies*
Blights	Beans, Walnut	*Xanthomonas phaseoli, X. juglandis*
Cankers	Citrus	*X. campestris*

with in the following chapter (**Chapter 9**).

A large number of plant species are affected by pathogenic bacteria (Table 8.5). The symptoms produced by them are often used to characterize the disease and the phytopathogen. Depending upon the pathogen, the host-parasite relationship, and the environmental conditions, the symptoms may fall under the following categories: (i) leaf spots, (ii) blights, (iii) soft rots, (iv) cankers, (v) vascular diseases and necrosis, and (vi) tumors and galls (Fig. 8.16).

The entry of a phytopathogenic bacterium results in extensive cell damage or necrosis that leads to spots, blights or scabs. In the latter, the lesions may be slightly raised or sunken. In some cases, the warty outgrowths or cankers are formed which are filled with the dead host tissues. Bacteria causing rots produce an enzyme that decomposes pectic substances leading to the disintegration of host cells. The wilts were earlier thought to be due to the blockage of the xylem vessels but it has been subsequently shown that in several cases, the pathogen releases a wilt toxin to carry out this effect. The bacterium, *Agrobacterium tumefaciens* induces rapid cell division in the affected parts that leads to tumorous growth or gall formation. The tumor inducing and maintenance functions have been assigned to a fragment (T-DNA) of a large tumor-inducing plasmid (Ti). T-DNA is not only transferred to the plant cells but also integrates into the chromosome at several sites. This DNA also carries the oncogenes, the expression of which leads to crown gall formation. T-DNA is a case of prokaryotic DNA that has crossed the domain barrier and integrates in eukaryotic DNA (plant chromosome) **(see also Chapter 9)**.

Psuedomonas syringae and *Erwinia herbicola* produce a protein that can initiate ice crystal formation. These bacteria are called conditional pathogens as the frost damage induced by them can take place at a temperature that can initiate freezing. Respite from this damage has been provided by mutants that lack this protein and do not form ice crystals until the temperature goes down to -7° to -9°C.

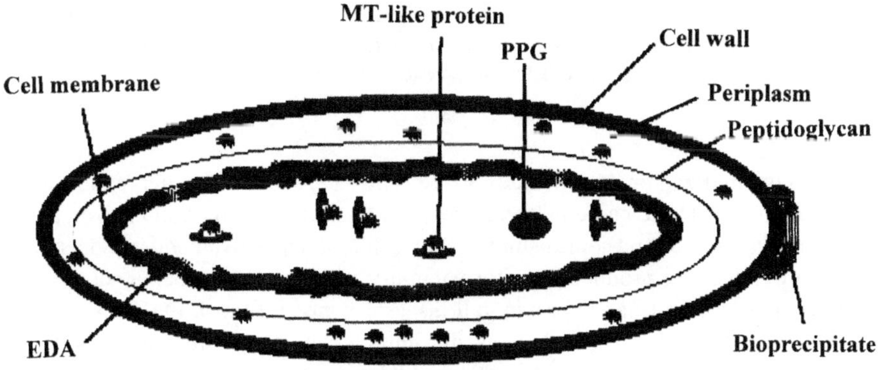

Figure 8.15. Schematic representation of metal accumulating sites in bacteria. PPG: Poly phosphate granules, EDA: Electron dense aggregates, MT-like protein: Metallothionein like protein.

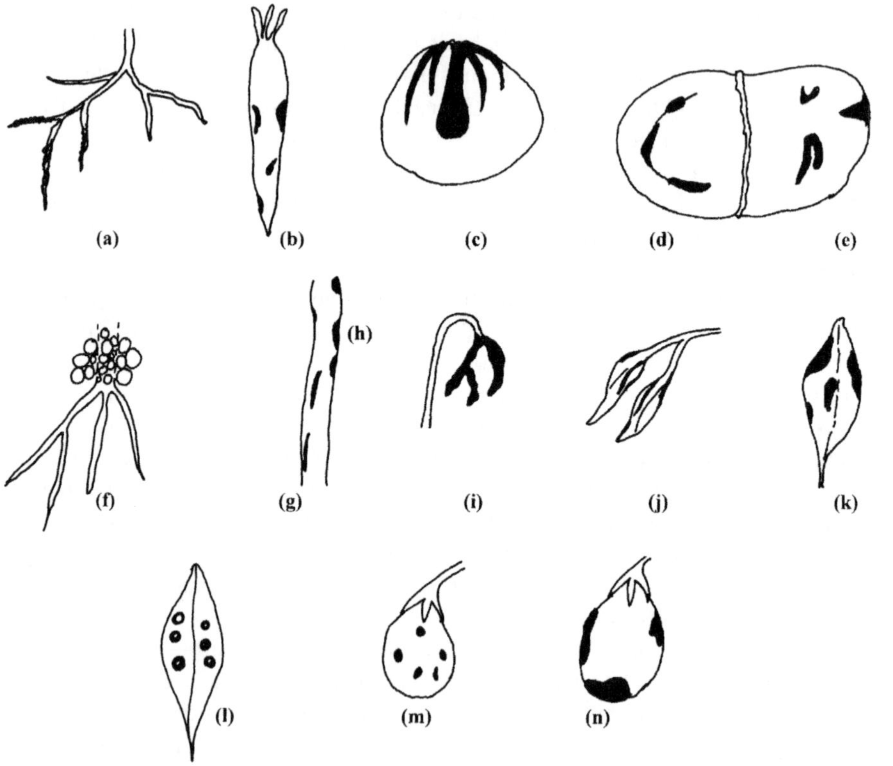

Figure 8.16. Symptoms of some common bacterial diseases of plants. a. root rot, b. soft rot of root, c. bulb rot, d. ring rot of potato, e. potato scab, f. crown gall, g. vascular wilt, h. stem canker, i. shoot blight, j. leaf wilt, k. leaf blight, l. leaf spot, m. fruit spot, n. fruit rot.

The cuticularized walls and thick cork cells provide a natural barrier to the entry of phytopathogen. These bacteria, therefore, preferentially use non-cutinized areas like root hair, stigma, etc., and some natural openings like stomata, hydathodes and lenticels. Wounds or punctures produced by physical injury or by insects and nematodes provide another route for entry as in crown gall disease. Once inflicted, the disease can be disseminated through seed, soil, air, water, insects, and small animals. Besides the surface barriers, some plants may produce antibacterial substances as part of the plant's defense mechanism.

The control of bacterial diseases as other microbial diseases require appropriate management procedures. Some of these are:

1) Quarantine or restricted movement will check the spread of pathogens.
2) Sanitary practices can be employed to prevent infection of plants with contaminated soils and tools.
3) One important way of controlling the plant diseases is to develop and grow disease-resistant varieties.
4) To limit the contact between the host plant and the pathogen, the crop rotation practices can be employed.
5) To control the population of pathogens, pesticides can be used. In this case, the effects of pesticides on the plant and the environment should be assessed before application.

CHAPTER 9
BACTERIA AND HUMANS

Bacteria have served the mankind since time immemorial. In almost all walks of human life, one or the other bacterial (microbial) process has contributed significantly from the time even when the basis was not clearly understood (refer also **Chapter 8**). When wide spectral metabolism of bacteria was revealed man learnt to exploit these processes for the production of a variety of useful chemical compounds. But once again, it is their role as dreadful disease-causing organisms that has caught the attention of humanity more than anything else. In this Chapter we shall discuss various aspects of human life in which bacteria are involved directly or indirectly. These aspects have been classified under eight different sections and include both beneficial and harmful activities of bacteria. Readers may recall that we have already dealt with some of their activities in nature in **Chapter 8**, which also have an important bearing on human life. And, with bacteria playing host to a large number of recombinant DNA experiments, innumerable possibilities have opened up for their exploitation. Thus, even if one is not dealing with a microbial function directly, the channel may be created through the microbial route.

9.I- Bacterial Pathogenicity

Like plants, humans and other animals can also be inflicted with bacterial diseases. A very large number of bacteria have been reported to cause diseases, literally affecting every part of the human body. The study of the incidence and spread, and prevention and control of infectious and non-infectious diseases has been so crucial to the human health that a science of Epidemiology has been carved out.

Pathogenic microorganisms are able to gain access to the body through some defined routes known as portals of entry. These are generally the mucosal surface of the respiratory, gastrointestinal and genitourinary tracts, and skin and wounds. These portals of entry may be so specific that if a wrong route is used, the microorganism may not only fail to establish itself but also fails to cause the disease.

While the onset of a particular disease depends upon several factors, the outbreak can be traced to a single source of exposure either with rapid onset of disease or more slowly when it is transmitted from one individual to another. In the latter case, it may be directly transmitted from one individual to another or indirectly through another living agent, called a vector. The vector may consist of several insects, fleas, rodents, cattles, other wild animals and even pet animals like, cats, dogs, horses, and birds. Another important source of disease transmission could be inanimate objects like food and water or even air (Table 9.1). When a microorganism enters and multiplies in a body it causes infection. The infections may be localized or may spread to different parts of body causing generalized or systemic infection. All invading microorganisms do not cause a disease and many constitute the normal microbiota of the body playing an important role in host defense mechanism. Under certain conditions, however, some of them may turn pathogenic and are referred to as opportunistic pathogens. Some microbes are pathogens as they can cause disease. A disease is an infection that brings about a change in the normal physiology of the body. The relative pathogenicity or virulence varies with the extent of damage that a pathogen can bring about in its host.

Accordingly, they can be classified into highly virulent, moderately virulent or even avirulent strains.

A disease can be often recognized by characteristic signs and symptoms. While the former represents the objective changes (e.g., rashes or fever) heralding the onset of disease, the latter represents the subjective changes in the body functions (Table 9.1). A characteristic group of signs and symptoms constitute a disease syndrome. The progression of a disease is a sequential process and occurs in stages. For example, after the pathogen enters it requires an incubation period, varying in different diseases, to establish itself. Following this is the prodromal stage, when it starts giving the associated discomfort followed by a stage when it multiplies enough to produce signs and symptoms of a particular disease. Then comes the period of illness (acute stage) when the disease is most severe. This may be short in acute diseases but prolonged in the chronic ones. Recovery from the acute stage depends on whether the host defense (immune) system and/or medical treatments are adequate. If the disease is not fatal or chronic, signs and symptoms begin to disappear during the period of decline leading to convalescence or recovery.

Disease-causing microorganisms (pathogens) possess properties, called virulence factors or genes. The expression of virulence genes may allow the pathogen to establish an infection (invasiveness) or result in the production of toxins that will disrupt the body functions. Regulation of many of these genes is based on two component signal systems.

9.I- 1. Invasiveness

The invasion of host cells and tissues by a pathogen is essential to establish infection. This can be brought about by a number of ways.

Many pathogens adhere to host's mucus membranes to initiate infection. They may produce specific factors or adhesins to facilitate this process. Bacterial cell surface layers such as capsules, slime layers, and pili may all contribute towards adherence. For example, enteropathogenic and uropathogenic strains of *E. coli* and cholera-causing *Vibrio cholerae* possess pili and associated adhesins that play a key role in their attachment to host cell surface. *V. cholerae* also produces a hemagglutinin and certain membrane proteins that help in adhesion. Filamentous hemagglutinin and pertussis toxin produced by *Bordetella pertussis* helps it to cling on to host's ciliated cells and macrophages.

Streptococcus species causing dental caries and streptococci and lactobacilli associated with dental plaques produce dextran sucrase that forms extracellular glucans from dietary sucrose. The glucans help in the attachment of these bacteria to tooth surface and the acid produced by them leads to the dissolution of tooth matrix thus forming caries.

Some pathogenic bacteria produce surface proteins called invasins following adhesion. Invasins function by stimulating phagocytosis by human cells so that the pathogen is internalized. Thus, invasins help the bacterial cells to bypass immune defenses (**see later section**) and to escape getting transported to other tissues. The expression of invasins is controlled by prevailing environmental conditions. Many bacteria infecting the gastrointestinal tract, such as *Shigella* species use invasins to

Table 9.1. Some common human diseases caused by bacteria, the causative organisms and the disease-associated chararcteristic symptoms

Disease (Transmission/Acquisition)	*Causative organism*	*Some characteristic symptoms*
I. Contaminated Food and Water		
(1). Enteric fever gastroenteritis, Typhoid, Bacterial dysentry, Cholera	*Salmonella typhimurium, S. schottmulleri, S.typhi, Shigella dysenteriae, Vibrio cholerae*	Gastroenteritis, fever, weakness, abdominal pain inflammation of intestine, severe diarrhoea, dehydration may lead to shock and death
(2). Food poisoning, Botulism	*Staphylococcus aureus, Clostridium botulinum*	Nausea, vomiting, abdominal pain, toxemia and in infants sudden death
II. Air borne-diseases (droplet inhalation)		
(3). Tuberculosis	*Mycobacterium tuberculosis,*	Formation of tubercles in lungs and inflammation
(4). Pneumococcal pneumonia	*Streptococcus pneumoniae,*	Inflammation of bronchia, bronchioles, and alveoli of lungs
(5). Diphtheria	*Corynebacterium diphtheriae,*	Inflammation of pharynx, fever, cough, sore throat, difficulty in swallowing
(6). Other respiratory infections, Whooping cough	*Haemophilus pneumoniae, Klebsiella pneumoniae, Bordetella pertussis*	Inflammation of respiratory tract , symptomatic cough sometimes associated with vomiting, general symptoms of cold
III. Disease transmitted by animal bite, insects, ticks etc.		
(7). Plague	*Yersinia pestis*	Inflammation of lymph nodes, progressive infections of other organs and tissues
(8). Lyme disease	*Borrelia burgdorferi*	Skin lesions, later arthritis, cardiac and neurological abnormalities
(9). Typhus fever	*Rickettsia prowazekii*	Fever, headache, rashes, lesions in heart and kidney
IV. Disease transmitted through direct contact/sexual contact		
(10). Leprosy (skin contact)	*Mycobacterium leprae*	Skin lesions, damage to nervous system, some physical deformities
(11). Gonorrhoea (sexual contact with infected person, from birth through infected mother)	*Neisseria gonorrhoeae*	Infection of genitourinary tract with symptoms on other parts as well

attack the target proteins, called integrins, to gain entry into the cells.

Many pathogens produce enzymes whose activities are employed for invasiveness. Several *Clostridium* species, including *C. perfringens* that causes gas gangrene, produce a number of enzymes that help in rapid spread of infections. These include tissue necrosins and haemolysin, and lecithinase that hydrolyzes lecithin of the eukaryotic membrane. Disruption of cytoplasmic membrane, lysis of blood cells, release of iron from lysed blood cells, tissue necrosis are all associated with this fatal wound infection, if not treated rapidly. Some pathogenic *Staphylococcus* and *Streptococcus* species produce fibrinolysins (staphylokinase and streptokinase) that cause the dissolution of fibrin clots. Thus by inhibiting the fibrin in the host, the pathogen can freely spread to surrounding areas. Some *Staphylococcus* species such as *S. aureus* produces coagulase which converts fibrinogen to fibrin. It appears that deposition of fibrin around the cell protects them from host's circulatory defense mechanism. These pathogens also produce the enzyme hyaluronidase (spreading factor) that breaks hyaluronic acid, the substance that holds together the cells of connective tissues. Some clostridia also produce collagenase that breaks the proteins of collagen tissue. All these enzymes facilitate the spread of the pathogens and provide them protection from the host defense.

9.I- 2. Avoidance of Host Immune Defenses

Pathogenic microorganisms have evolved several complex mechanisms to overcome inhibition by host defense mechanisms. Formation of a capsule enhances infectivity and serves as virulence factors. Formation of different types of proteins that coats the cells or those which disrupt the structure and function of phagocytic cells protects them from phagocytosis as in staphylococci, streptococci, *Yersinia pestis* and *Neisseria meningitidis*. Moreover, capsules consisting of sialic acid as in *N. meningitidis* or hyaluronic acid as in *Streptococcus pyogenes* do not act as antigens and thus do not elicit antibody-mediated defense. *N. gonorrhoeae* is able to thrive within phagocytic cells as its outer membrane protein (P_1) prevents phagolysosome formation. Typhoid causing *Salmonella* species are also able to survive within phagocytes as they possess a complex genetic network to ensure survival. This includes a two component regulatory system (PhoP/PhoQ) which controls the expression of genes contributing towards cellular responses to reactive forms of O_2, low pH, carbon availability, and defensins. LPS of *Salmonella* species also provide protection against host-mediated defenses.

Another way by which capsule enhances the virulence of a pathogen is by its ability to interfere with phagocytosis. Infection of lung tissue by *Legionella pneumophila* is based on its macrophage potentiator gene that allows it to grow within macrophage without being destroyed by the antibodies. *M. tuberculosis* survives and grows within alveolar macrophage as the mycolic acid of its cell wall protects it from killing by toxic forms of oxygen and other enzymes. Capsules of *E. coli* K1 and *N. meningitidis* B mimic the host's neural cell adhesion molecules and evade recognition by the immune system. Encapsulated cells of *P. aeruginosa* and *S. aureus* grow in the form of biofilms that not only provide them a mechanism to escape the immune system but also a capability to infect medical devices such as catheters, artificial heart valves or artificial hips etc. The group of pneumonia-causing organisms, viz., *Streptococcus pneumoniae,*

Haemophilus influenzae, Klebsiella pneumoniae and even wall-less *Mycoplasma pneumoniae,* all possess capsules helping them to survive and reproduce in lung tissues and spread rapidly especially when the host defense mechanisms are compromised. The pili of *Neisseria* retard phagocytosis providing the longer persistence of the pathogen. This is supplemented by the secretion of a protease that degrades and destroys IgA, a secretory antibody that protects the mucus membrane surfaces. *Staphylococcus aureus* produces a toxic protein, leukocidin, that destroys phagocytes. The pus formed due to infection by this bacterium consists of dead phagocytes. The growth of many pathogens in blood and other body tissues is retarded by the non-availability of free iron. Some pathogens produce low molecular weight siderophores, which can remove iron from transferrin or other iron-binding compounds in blood. The same function in *Neisseria* and *Mycobacterium* is carried out by an outer membrane protein.

9.I- 3. Dysfunctional Immune Reactions

The physiological damages and the disease caused by a pathogen can often be traced to immune reactions triggered by these infections. One such response known as inflammation is the cause of damage to lung tissue by tuberculosis-, and pneumonia-causing bacteria. Various other microorganisms including fungi work by triggering allergic (hypersensitivity) responses. In many cases, the processing of bacterial cell antigens by antigen presenting cells (APCs) (**see section on B-cell activation**) creates surface antigens on APC that resemble a host cell antigen. This leads to T-cell activation against host cells, a process that mimics autoimmunity. The damage to heart in people suffering from rheumatic fever is caused by such a response. Similarly, highly conserved heat-shock proteins (HSP), which are also produced when a bacterial cell is attacked by phagocytic cells, can trigger an "autoimmune response". We shall discuss the process of dysfunctional immunity in some detail in a later section.

9.I- 4. Bacterial Toxins

Many pathogenic bacteria express their virulence by producing toxin that destroys the normal functions of a cell. In some cases, as in the case of *Clostridium botulinum*, the bacteria do not get associated with the host but the toxin produced by them enters human cells through infected food with lethal effects. Similarly, *C. tetani*, causing tetanus, invades the tissue only slightly but its toxin causes widespread effects at sites far removed from the initial site of infection. In past, they have been classified as endotoxin, which constitute the heat-stable part of a cell, or exotoxin, the heat-labile protein that is secreted. These terms are now used to define the biochemical nature. While LPS constitutes endotoxins, the exotoxins consist of proteins. They can be further classified based on their site and mode of actions. These are neurotoxins (*C. botulinum, C. tetani*), enterotoxins (*Vibrio cholerae, Shigella dysenteriae, Staphylococcus aureus*), cytotoxins (*Corynebacterium diphtheriae*) and haemolysins (*Streptococcus* species and *C. perfringens*).

The LPS component of the Gram-negative eubacterial cell wall, especially the lipid portion or lipid A, can act as a toxin. LPS toxin of *Salmonella* and *Shigella* may supplement the exotoxins produced by them in causing gastroenteritis. Release of these LPS toxins in some cases may have far fetched effects like septic shock. This is a complex response consisting of blood clotting that restricts the normal flow of blood, damage to endothelial cells and leakage of fluids from the blood vessels, resulting in acute respiratory distress and multiple organ failure such as of lungs, kidney, and brain.

The proteins or exotoxins are more specific. These are secreted and are readily inactivated by heat. They consist of a receptor protein component (B region) that binds to host cell targets and a toxic component (A region) that enters and destroys the normal cell activity. While the A region provides an enzymatic function, some exotoxins may act through membrane disruption or by acting as superantigens. Such toxins cause diseases like cholera, diphtheria, whooping cough, bacterial dysentery, tetanus, botulism, gas gangrene, and staphylococcal food poisoning and toxic shock syndromes and are often referred to by the name of the disease.

9.1- 5. Biological Warfare

The disease-causing nature of bacteria, viruses, and other microbes has been unfortunately exploited in developing frightful biological weapons.

In biological warfare or bioterrorism, the disease-causing microorganisms are dispersed deliberately in an area in order to incapacitate human beings or to reduce their powers of resistance. Apart from the intentions to harm humans, they can also be used against domestic animals and crops which in trun can have serious implications on the livelihood and economy of a country.

World has only recently faced the scare of such a warfare in the form of human anthrax. Because anthrax spores can be spread by aerosols, pulmonary anthrax has been cited as a possible biological weapon. The incidence of human anthrax is rare but occurrence in grazing animals is quite common. People handling animal hides, wool, and other animal products are at higher risk of contracting this disease. The bacterium gains the entry through the cuts and abrasions in the skin and causes pustule formation. If these pustules are not contained by the defense mechanisms of the body, development of septicemia is possible. If bacterial endospores are inhaled a dangerous form of pneumonia called Woolsorter's disease or pulmonary anthrax can develop. The symptoms of this disease are high fever, difficulty in breathing and chest pain, eventually resulting in septicemia and high rate of mortality. Penicillin is the drug of choice only if septicemia is not well advanced. Though the bacterium acts more slowly than most chemical agents and does not spread from person to person, the persistence and proliferation of such agents is definitely a cause for worry.

The successful use of biological warfare dates back to the time when Conquistadors deceitfully disseminated the white man's disease among native Indians by distributing infected garments. However, today's world faces a greater challenge of germ warfare as many countries are known to develop and possess pathogens which can wipe-off the enemies. Hard evidences exist that some developed countries acquired and used biological warfare technology during the Second World War. And there is no reason to disbelieve that many others may have joined this programme since then. Unlike the conventional weapons, biological weapons will have the aggressive property of self-proliferation after initial application. The effects are going to be far-reaching as these weapons will not only debilitate the humans but also his animals and plants. It is left to the good sense of humans not to deploy such biological weapons as it will have an unimaginable consequence on its own race.

9.I- 6. Host Defense Mechanisms

From the above discussion, it becomes very clear that a human body can be targetted and attacked by a number of pathogenic bacteria. Similarly, the diseases can also be caused to other animals. However, humans have evolved a complex and interactive defense network that wards off the pathogens. This consists of two lines of defenses, the non-specific and specific.

The non-specific barrier is composed of physical components such as skin, and cellular such as, resident microbial flora, and phagocytic cells, and chemical components which are the various antimicrobial chemicals that repel the pathogen. The specific defense constitutes the immune system consisting of dedicated responses that help the body to recognize the foreign substances, such as a pathogen, and to destroy them. The host cells possess genetic information to produce specific antibodies and maintaining a memory system for recognizing and responding to foreign antigens in future as well. Breakdown of one or both the barriers renders one susceptible to an infectious disease.

9.I- 6.1. Non-specific Defense

Humans are endowed with several modes of non-specific defense against pathogenic microorganisms. It consists of physical, chemical and cellular components with each having a significant role in overall defense. In a majority of cases, they are innate and start operating right from the child birth. Some mechanisms prevent the invasion and, therefore, the establishment of a pathogen. General health, nutrition, and stress levels play important role in determining body's ability to defend against infection. Age also has an effect and so also the genetic make up of an individual.

9.I- 6.1.1. Physical Barriers. These constitute the first line of defense against microorganisms and work through physically blocking the entry of pathogens. Intact skin is impenetrable by most microorganisms. The outer surface of skin layer is composed of keratin which is not degraded readily by microbial enzymes. Moreover, the outermost layer may consist of dead cells that are continuously being shed. Dead cells prevent infection by many microbes which require live cells for their growth as with some viruses. Many pathogens, therefore, enter through wounds where protective barrier of skin is broken allowing microorganisms to penetrate other body tissues or/ and enter the circulatory system.

Several organs such as respiratory and digestive tracts are lined by mucus membrane. These cells secrete mucus that traps the microorganisms or prevent their attachment. Similarly some body tissues, like oral cavity, eye, and urinary tract are flushed with body fluids. Thus, secretions like saliva, tears, and urine help prevent the adherence of pathogenic microorganisms to the respective body tissues.

9.I- 6.1.2. Chemical Defenses. Several body fluids including blood contain antimicrobial substances. These substances act either by inhibiting the growth or killing the pathogen.

Saliva, mucus, tears and colostrum (milk secreted a few days before and after child birth) owe their antimicrobial protection due to presence of lysozyme that degrades the cell wall of bacteria. Ova, which are surrounded by mucus are also protected by the presence of lysozyme. Similarly, swallowing, sneezing and coughing can expose bacteria to such body fluids that contain lysozyme. Various body tissues are protected by the low pH environment created due to acid production. The resident microflora often produces acidic conditions to ward off the pathogens. The production of fatty acids by indigenous microflora of skin, formation of lactic acid by *Lactobacillus* and *Streptococcus* species in vagina and production of lactic and acetic acid by resident microbes in lower intestinal tract all contribute towards low pH and protection against pathogens. The hydrochloric acid present in stomach also prevents microbial invasion. Some cells of small intestine produce lysozyme as well as a small peptide called cryptins. Cryptins protect stem cells which are needed to refurbish mucosal cell lining of intestine as it is regularly sloughed off.

The formation of iron-binding lactoferrin and transferrin creates iron-limiting conditions and limit the growth of microorganisms. Lactoferrin is found in tears, semen, breast milk, bile, and nasopharyngeal, bronchial, cervical, and intestinal mucosal secretions and transferrin is present in serum, cerebrospinal fluid, sweat, and intercellular spaces of many tissues and organs. Interferons (IFNs) protect the body against viral infections such as common cold, as they can block viral replication. IFNs are a family of inducible glycoproteins produced by eukaryotic cells and besides protecting against viral infections regulate the complex defense network that provides protection against other kinds of infections and also the development of malignant cells. Looking at its multifarious applications, the human genes coding for IFN have been cloned in *E. coli*.

Human blood also contains a family of glycoproteins known as complement, which are important in defense mechanism. More than 20 such complement glycoproteins, designated as C_1,C_2, and so on, are known which act in an autocatalytic cascade. They complement or augment other defense pathways. The cascade may be triggered by LPS toxin, teichoic acid of Gram-positive cell wall, fungal cell wall carbohydrates, viral envelope (alternative pathway) or by specific antigen-antibody complex (classical pathway). While in the former, the complement C_3 is the central activator which initiates a chain of splitting and activation of other complements, the latter is initiated by C_1,C_2 and C_4. This enables the system to provide protection either by forming a membrane-attack-complex (MAC) that creates pores in the membrane of the pathogen or by coating the bacterial cell to facilitate phagocytosis, also known as opsonization (Fig. 9.1).

9.1- 6.1.3. Indigenous Microbiota. Normal human microbiota form an important component of defense mechanism. It is said that an adult human being is composed of 10^{13} cells to which are associated 10^{14} cells of both prokaryotic and eukaryotic microbes (Table 9.2). They are usually non-pathogenic and provide protection against pathogens by producing allelopathic substances. These are substances produced by an organism that kill or inhibit other organisms. The use of drugs like antibiotics can disturb this natural flora leading to other infections. For example, use of antibiotics may destroy the microbial community of the gastrointestinal tract thus permitting the growth of *Clostridium difficile* resulting in severe and sometimes fatal infection of gastrointestinal tract.

Table 9.2. Resident microflora of human body

Site	*Indigenous microbial population*
Skin	*Staphylococcus* and *Micrococcus, Corynebacterium, Brevibacterium, Propionibacterium*, few Gram-negative bacteria and some fungi.
Oral cavity	*Streptococcus* species, *Lactobacillus* species, *Velionella, Bacteroides.*
Nasal cavity, Nasopharynx	*Streptococcus, Staphylococcus, Neisseria, Haemophilus, Bacteroides.*
Gastrointestinal tract	*Bacteroides, Fusobacterium, Escherichia, Lactobacillus, Enterococcus, Clostridium, Proteus, Klebsiella, Enterobacter, Salmonella.*
Urethra (distal end)	*Streptococcus, Lactobacillus, Bacteroides, Clostridium.*
Vaginal tract	coliforms, spirochaetes, some yeasts, like *Candida.*

Note: Lower respiratory tract, upper urinary tract (kidney and bladder), some body fluids like urine and blood, as long as they are not exposed, are considered sterile

The resident microflora is also responsible for providing several nutritional supplements that maintain the general well being of the host. For example, they synthesize and supply to the body biotin, riboflavin, pantothenate, pyridoxine, and vitamin K.

9.I- 6.1.4. Phagocytosis. Another mechanism for handling the deadly microbes is by their engulfment and ingestion followed by destruction. This process known as phagocytosis is of utmost importance once a pathogen invades the circulatory system. Different types of leukocytes (white blood cells) participate in this response. Some leukocytes containing cytoplasmic granules are called granulocytes and can be further classified on the basis of staining behaviour, into basophils, eosinophils, and neutrophils (also known as polymorphonuclear neutrophils, polymorphs or PMNs). The other leukocyte is agranulocyte which is further differentiated into monocytes (macrophages) and lymphocytes. While the former are important in non-specific immune response, the latter plays the significant role in specific immune response. Monocytes, macrophages and PMNs are the most efficient phagocytic cells. As we have seen earlier, many pathogenic bacteria have acquired structures (capsule) or mechanisms that help them resist phagocytosis.

Phagocytic cells are rich in lysosomes that contain a number of hydrolyzing enzymes. During phagocytosis, the microorganism is engulfed by the pseudopodes of the phagocytic cells and is transported inside a vacuole called phagosome. Phagosome fuses with lysosome to create phagolysosome within which the destruction of the pathogen is carried out by the lysosomal enzymes (Fig. 9.2). These killings may involve

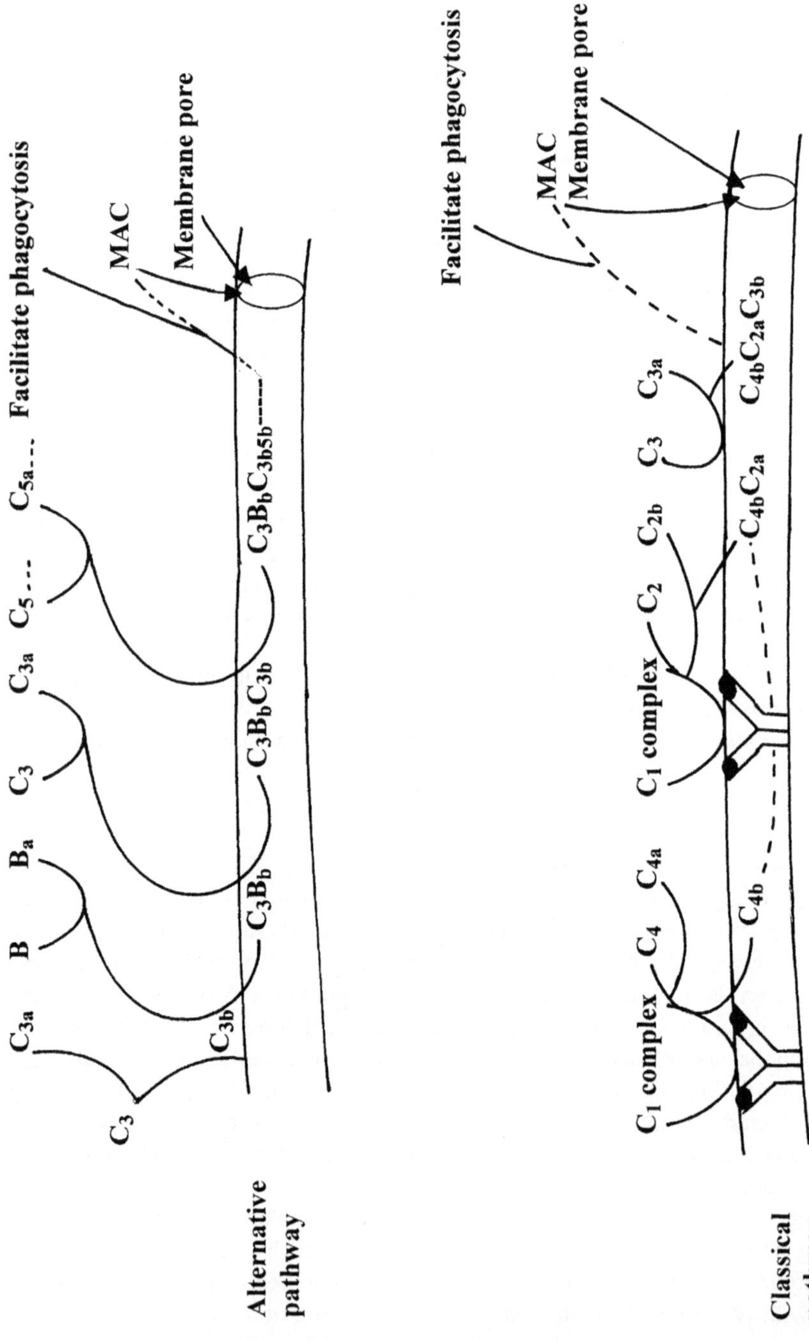

Figure 9.1. Role of complement cascade in nonspecific defense.

oxygen-dependent, oxygen-independent, and nitrogen-dependent mechanisms. Oxygen-dependent pathways deal with the generation of superoxide, hydrogen peroxide, singlet oxygen and hydroxyl radicals, all of which can metabolically poison the invader. Oxygen-independent processes involve enzymes like phospholipases, proteases, RNAase that can destroy the structure of a cell. Moreover, a shift from respiratory to fermentative mode leads to low pH which is again toxic to the microbes. PMNs also produce peptides called defensins that make the bacterial and fungal cell membranes leaky leading to lysis. Some phagocytic cells produce reactive nitrogen intermediates (RNIs), such as nitric oxide (NO), nitrite (NO_2^-) and nitrate (NO_3^-). All are antimicrobial but NO is most effective killing agent.

The dead cellular material of the microorganism is either absorbed within the phagocytic cells or transported outside by exocytosis.

9.I- 6.1.5. Inflammatory Response. Microbial infections and tissue damages are often reflected as inflammatory response. This serves a variety of functions, such as removal of cellular debris, restriction of invading microorganism and arrest of spread of infection. This is characterized by reddening of the affected area, swelling, pain, and high temperature. The dialation of the capillary allows more blood to flow to the damaged tissue leading to reddening and associated high metabolic activities in neutrophils and macrophages cause elevation of temperature. Moreover, the dilation is accompanied by increased capillary permeability of the blood vessels, as a result, fluids accumulate in the spaces surrounding the tissue. Lysis of blood cells triggers the appropriate nervous system response that results in pain.

9.I- 6.1.6. Fever. Fever is another feature of microbial infections. This is due to an abnormal increase in body temperature for homeothermic animals (those maintaining a constant range of body temperature) like humans. Pathogenic microorganisms release fever causing chemicals, pyrogens, that directly or indirectly stimulate hypothalamus that results in fever.

Bacterial endotoxins and cell wall peptidoglycans when ingested by phagocytic cells cause them to release interleukin-1 (IL-1), also called as endogenous pyrogen. Similarly, bacterial exotoxin when enters the blood stream stimulates macrophages and monocytes to release IL-1. IL-1 stimulates hypothalamus to secrete prostaglandins that regulate the body temperature in a way that elevated levels of fever is caused. This is accompanied by constriction of peripheral blood vessels, high metabolic rates, and muscular contractions that is reflected as shivering. The depletion of IL-1 from the blood is responded by dilation of blood vessels, and sweating to lose the heat and return to the normal temperature. Many staphylococci produce specific pyrogenic exotoxin, such as toxic shock syndrome, toxin-1 and streptococci produce erythrogenic toxin causing fever.

Fever constitutes the body's natural defense mechanism as it stimulates phagocytosis, increases the rate of enzymatic reactions that may degrade the pathogen or stimulate tissue repair, intensifies the action of interferons, and causes a reduction in blood iron levels that may cause growth inhibition of many bacteria.

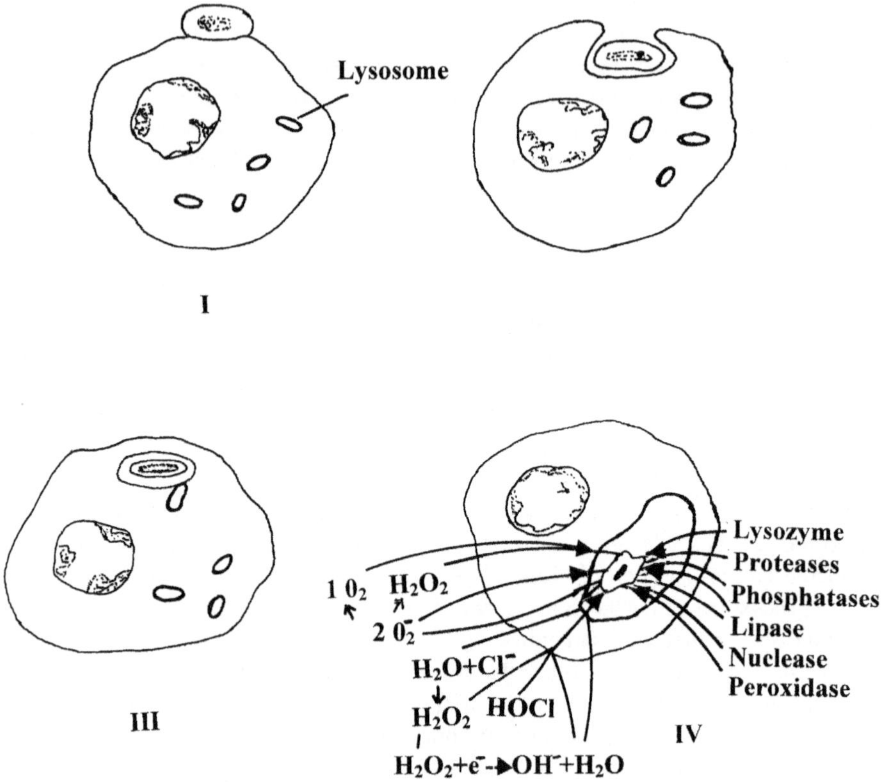

Figure 9.2. Steps in phagocytosis. I. A pathogen gets attached to a phagocyte, II. Pathogen is engulfed by the invagination of the cytoplasmic membrane, III. The entire complex pinches off and subsequently fuses with lysosome, IV. In the structure, now called phagolysosome, the pathogen is attacked by lysosomal enzymes and toxic oxygen species and destroyed. Note: The organelles are not drawn to the scale.

9.1- 6.2. Specific Immune Response

Unlike the general non-specific defense, a host may exhibit a very specific response against an invading pathogen. This, known as immune response, is based on the recognition of a foreign substance (a pathogen or its toxin) or an antigen by the body and provides a defense by producing specific antibody to counter the antigen.

Discovery of this type of mechanism dates back to an old observation that people recovering from and surviving an epidemic showed remarkable resistance to a second outbreak of the same disease. The acquired immunity was traced to an earlier infection. In the early part of the 16th century, several attempts of deliberate immunization were made. Many people allowed themselves to be inoculated with pustular material from mild cowpox to protect themselves from more severe small pox. It was only in 1796, when Jenner repeated these experiments more scientifically that he established the acquisition of immunity through vaccination (latin, Vacca=Cow). Jenner's vaccination got a sound footing when Pasteur, and Koch demonstrated the microbial basis of many diseases, and Pasteur devised methods of strain attenuation (loss of virulence). Attenuation through which the pathogen loses its virulence can be brought about by a number of methods such as prolonged incubation or heating. This creates a safe strain which can be inoculated into a person to stimulate the immune response without causing any discomfort.

Various views were put forward to explain the fundamental nature of immunity. According to humoral theory proposed by Büchner, Pfeiffer and others, the immunity is attributed to the soluble substances in the blood serum. This was based on the observation that many bacteria are killed and destroyed when injected into an immune animal. On the other hand, Metchnikoff's cellular theory emphasized upon phagocytosis to provide immunity and recovery from a disease. In 1903, Wright and Douglas brought these two theories together to propose their Opsonic theory. According to them, phagocytosis is greatly stimulated in the presence of a component, supplied by the serum, they called as opsonin. Opsonin is the other name applied to antibody.

Antibodies were discovered in 1890 by E. von Behring and S. Kitasato, when they demonstrated that the blood serum of an immune animal contains antibodies which when transferred to another animal provide a similar immunity. An animal receiving this immune serum or antiserum acquires passive immunity in contrast to the animal developing it in response to a pathogen or the toxin, referred to as active immunity. The mechanism of immune response is based on the recognition of a pathogen or its toxin as a foreign substance or antigen and respond by producing antitoxin or antibodies. The antibodies are highly specific and each type of antigen or closely related antigens may elicit a specific type of antibody to combat them. An underlying memory system also appears to exist as once acquired, the immunity may be recalled during re-exposure. Vaccination infact stimulates this memory on which acquired immunity is based. The immune response appears to be of recent origin in evolutionary times as invertebrates and some lower vertebrates do not possess this defense mechanism.

9.I- 6.2.1. Structure of Antibody. Antibodies are globular glycoproteins, also called as immunoglobulins, present in blood serum, tissue fluids and some other secretions. They are synthesized by plasma cells derived from activated B lymphocytes in response to an antigen and subsequently react with the same antigen.

There are five major classes of immunoglobulin, viz., IgG, IgM, IgA, IgD, and IgE. IgG is the commonest of the immunoglobulins comprising ~80% of the body's immunoglobulins. It is predominantly circulating and can pass through the walls of small vessels into extracellular body spaces. It can cross the placenta and provide immunity to the foetus up to first few months after birth. It is also involved in reaction with surface antigens on bacteria by activating the complement system and crowding of the additional neutrophils at the site of infection. IgG can neutralize or inactivate the toxins, prevent the systemic spread of infection through the body, and facilitate recovery from many infectious diseases. IgA is found in mucus and in secretions such as saliva, tears and sweat. It plays the important role of preventing the adherence of the infecting microorganisms to respiratory, gastrointestinal, and genitourinary tracts. IgA is found in human breast milk and thus protects the nursing newborns. It is the major immunoglobulin providing immune response to protect the external body surfaces.

IgM is a high molecular weight immunoglobulin, existing as a pentamer. Because of the high number of antigen-binding sites, it is important in the initial response to a bacterial infection. It primarily occurs in the blood serum and along with IgG effectively tackles the circulation of infectious microorganisms. IgD along with monomeric IgM are present on the surface of some lymphocyte cells. It appears to play a role as an antigen receptor in lymphocyte activation and suppression. It is short-lived because of its susceptibility to proteolytic degradation. IgE is normally present in very low proportion. The ratio of IgG to IgE may be 50,000:1. The main role of IgE is in the protection of external mucosal surfaces by attracting the phagocytic cells and the initiation of inflammatory response. IgE is also involved in hypersensitivity reactions such as hay fever and other allergic reactions.

All the five classes of immunoglobulins have the same basic structure consisting of four polypeptides, two each of identical heavy, and identical light chains joined by disulphide bridges (Fig. 9.3). The term heavy and light reflects the molecular weight differences. The enzyme papain cleaves the immunoglobulins into two identical Fab (fragment for antigen binding) fragments which contain the antigen binding site and an Fc (fragment crystallizable) fragment which can be crystallized. The amino acid sequence of Fc portion suggests that it may be used to anchor the immunoglobulins to the cytoplasmic membrane facilitating its action by binding to complement molecules or phagocytic cells.

There can be variants in each of the major classes of immunoglobulins. IgG for example, can be grouped into four subclasses, IgG 1, IgG 2, IgG 3, and IgG 4 differing in their heavy chains. IgA has two subclasses. Immunoglobulins differing in their heavy chains but belonging to the same class are referred to as isotypes. Allotypes are derived from genetically controlled allelic forms of the immunoglobulins which can be differentiated on the basis of variations in the constant region of the heavy chains. Idiotypes represent the individual specific immunoglobulin which differ in certain regions of Fab fragment. These variations are important in establishing the diversity required for an effective immune response.

Two types of light chains of immunoglobulins are kappa (k) and lambda (λ). Each consists of roughly 110 amino acids and comprises two domains. In both, the amino-terminal end having a high degree of variability in the amino acid sequence is called variable region (V). Three hypervariable domains can be identified in the V region having even higher variability than the other regions of V, and differences in them result in idiotypes. This variability also helps in determining the variety of the immunoglobulins as this region forms the antigen recognition and combining site. The carboxy-terminal end comprises the constant region (C) which is highly conserved between the two light chains (Fig. 9.4).

Heavy chains have been classified into five types: alpha (α), gamma (ϒ), delta (δ), epsilon (ε) and mu (μ). Differences in the heavy chains lead to isotype differences. Heavy chains also consist of domains, each containing ~110 amino acids. The ϒ, δ, α

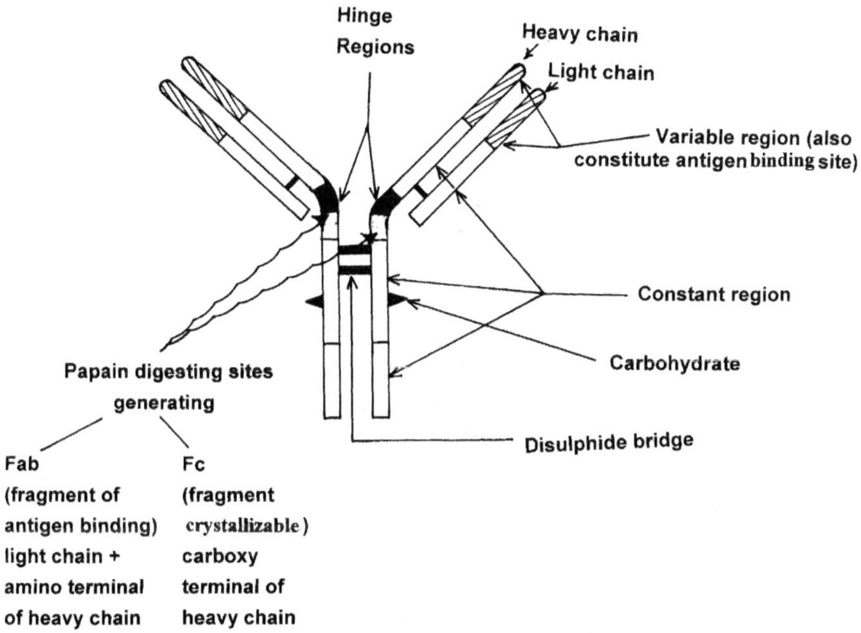

Figure 9.3. The basic structure of Immunoglobulin G (IgG).

form three domains and ε and μ form four domains. Like the light chains, N terminal end is the variable end with three hypervariable regions. The remaining domains constitute the constant region, C, but conserved nature of this is restricted within a particular class (Fig. 9.4). Between constant domains 1 and 2, a more flexible region called hinge region is located that may help the antibody molecule to stretch in order to cross link more than one antigen. The μ, α and δ have additional amino acids at the carboxy terminus, constituting the tail piece. The tail piece allows the binding of an additional polypeptide chain, the joining chain (J) which helps in the polymerization of the immunoglobulins.

9.I- 6.2.2. Immunoglobulin Diversity. The human system has to have virtually unlimited diversity in their antibody types to recognize millions of potential antigens. The genetic basis of immunoglobulin diversity has been a subject matter of intense research that led to various hypotheses. The earlier concept of each cell containing a separate gene for each antibody type has been rejected on the ground that human cell does not contain those number of genes required for such variability. Similarly the high mutability of fewer number of immunoglobulin genes to generate the required variability has also not found experimental support.

The currently accepted theory is that there are separate group of genes for variable and constant regions of the antibodies. There are three cluster of genes, one for heavy chain and two for light chain synthesis. Within each cluster, there are gene fragments containing the information for complete immunoglobulin molecules. Recombination permits shuffling and joining of the components so that billions of different combinations can be generated.

The active gene for light chain is assembled and expressed by somatic recombination and RNA splicing. The genetic region for light chain is divided into different variable regions (V), the joining sequence (J) and the constant region (C). A large number of

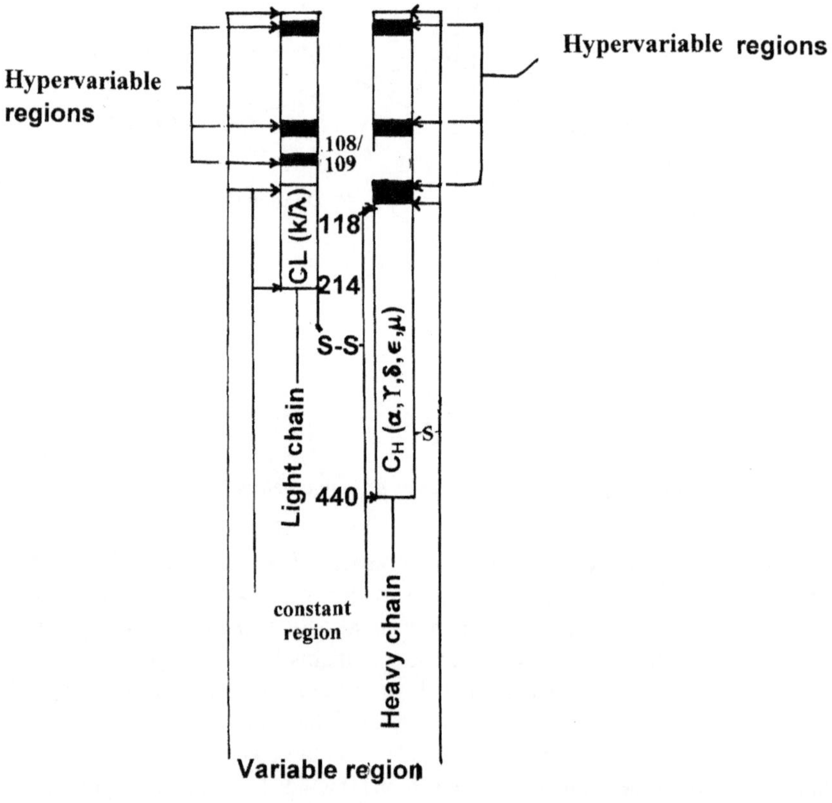

Figure 9.4. Detailed structure of heavy and light chains.

variable genes (~150) with closely related nucleotide sequences, each separated from a leader sequence (L) by a short intervening sequence have been identified in embryonic DNA. The L/V segments are separated from five J sequences by a long non-coding sequence. J consisting of short sequences that is repeated several times and showing slight but significant variation, codes for rest of the variable region, and is separated from a C gene by another intervening sequence. During lymphocyte development one V gene with its L sequence, recombines with one of the J sequence along with the single C gene to create one active light chain gene. The variability that is possible is implicit in the multiplicity of both V and J and imprecision in V/J recombination site. The active gene is transcribed into hnRNA which after the removal of intervening sequences gives rise to mRNA. The leader is cleaved away during the maturation of light chain precursor (Fig. 9.5).

The diversity in the variable region of heavy chains is brought about by mechanisms similar to light chain. Additional diversity comes from a sequence of embryonic DNA, the diversity gene (D), which accounts for a significant portion of the hyper-variable regions of heavy chains. An active heavy-chain gene is assembled by recombination of one of the L/V sequences, a D gene, and a J gene to code for variable region and a C gene to code for constant region. In heavy chain, there are eight different C sequences (Fig. 9.6).

Assuming multiplicity of each one of these regions and combining it with the variabilities created in the light chains as described above, the readers should get the feel of the immense or rather unlimited variability that can be created in the antibodies. Human beings are thus endowed with the required variability in antibodies to tackle myriads of antigens. With the genetic basis established, one should not forget the mutations that will contribute additional variability in the immunoglobulins.

9.I- 6.2.3. Lymphocytes. The immune defense system involves the participation of lymphocytes. These have been classified as B lymphocytes (B cells), T lymphocytes (T cells) and natural killer cells (NK cells). Both B and T lymphocytes originate from bone marrow stem cells and get differentiated during maturation at different locations of the body (Fig. 9.7). B cells develop in the bone marrow, and undergo secondary activation within lymphatic tissues, including spleen, tonsils, and lymph nodes. Within these tissues, B cells give rise to antibody-secreting plasma cells in response to antigenic stimulation and thus are mainly responsible for antibody-mediated immunity (humoral immunity). The precursors of T cells after a passage through liver and spleen are processed in the regions of thymus gland independent of that of B cells. The differentiation of T cells occurs during childhood and by puberty a whole complement of T cells are present in the secondary lymphoid organs of the body. The T cells generally circulate throughout the body and are responsible for cell mediated immunity (cellular immunity).

The T lymphocytes have been classified on the basis of their functions in immune response and carry different types of surface receptors, different from immunoglobulins or antibodies found on the surface of B cells. As the surface marker identifies a specific cell line or a stage of cell differentiation by interacting with a group or cluster of individual antibodies, it is called a CD (cluster of differentiation) antigen.

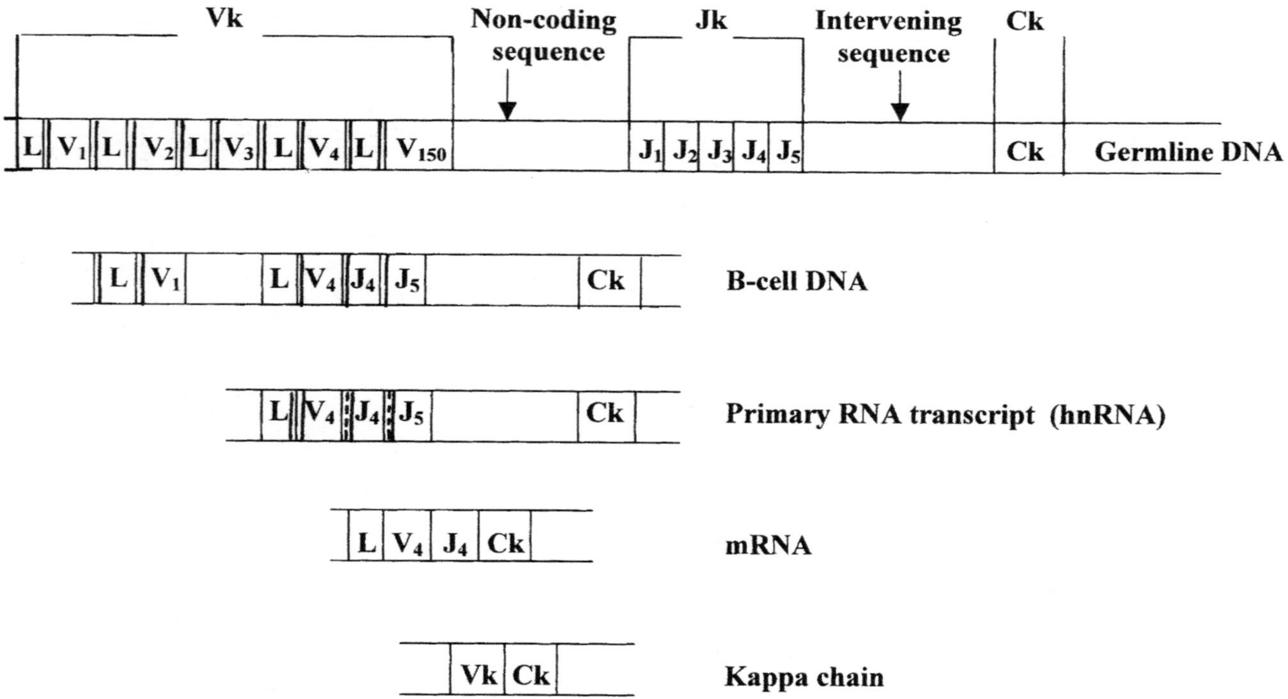

Figure 9.5. Arrangement, rearrangement and expression of kappa light chain immunoglobulin genes. The light chain genetic region present in germline DNA consists of variable (~150 alternative V sequences), and joining sequences that makes the variable region and the constant region encoded by C gene. Each V sequence is preceded by a leader (L) sequence separated by a short intervening sequence. The L/V segments are separated from J region by a long non-coding sequence, and similarly, an intervening sequence is present between J and C genes. In B-cell a rearrangement leads to joining of one V segment with J. This realigned sequence is then transcribed to give hnRNA which on RNA splicing produces the active mRNA and subsequently kappa chain. The other light chain gene goes through similar rearrangement.

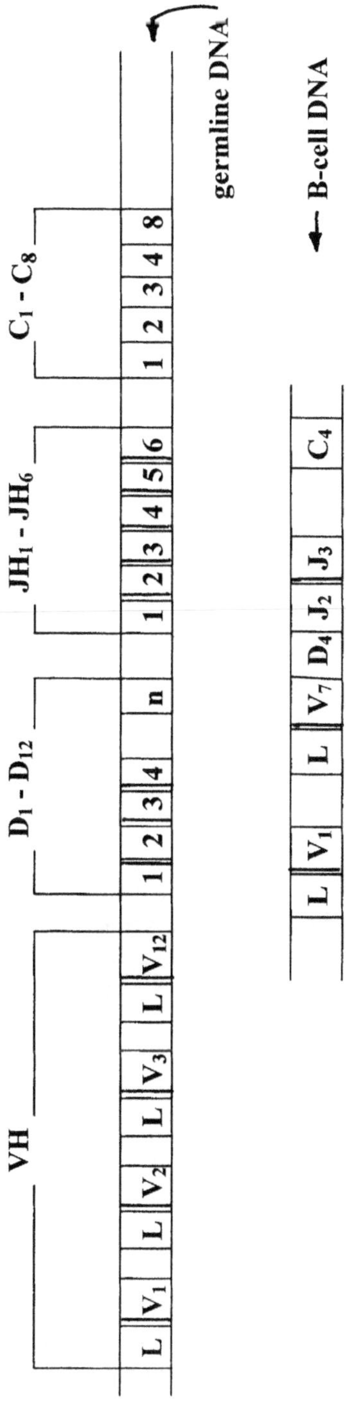

Figure 9.6. *Generation of heavy chain variability. The arrangement of variable segments (V), the diversity gene (D), the joining segment (J) and gene for constant region (C) are shown. Rearrangements in B-cell results in jointing one V, with D, J and C genes through recombination.*

Some T lymphocytes, and the T helper cells (T_H) interact with B cells and activate the latter to produce antibody and thus initiate the immune response. These cells contain CD3 and CD4 antigens. The cytotoxic T cells (T_C) recognize the body cells that may have been invaded by an intracellular parasite, e.g., viruses and some microorganisms. They also have the capability to recognize tumor cells and in the process can lyse and destroy them, thus forming a part of the defense mechanism against viral infection and cancer. Suppressor T lymphocytes (T_S) are involved in shutting down the immune response. T_C and T_S carry CD3 and CD8 antigens. Another subset of lymphocytes called natural killer cells (NK cells) or null cells are responsible for lysis of tumor cells. They neither form antibodies nor do they carry CD receptors. They also have an important role in tissue grafts.

9.I- 6.2.4. B Cell Activation. Each individual B lymphocyte cell contains the genetic information for initiating an immune response to a single specific antigen. The antigen receptor is located within the cytoplasmic membrane of the differentiated B lymphocyte that helps in the selection of a specific pre-existing B cell by the antigen. The process also called as B cell activation leads to the formation of plasma cells that secrete antibody and to a clone of genetically identical B cells called B memory cells. The latter also known as clonal expansion is brought about by the cell division of a specific B lymphocytes that responds to the same antigen that initiated its activation (Fig. 9.8).

During activation, the antigen cross links with membrane immunoglobulins that leads to internalization of this complex within the B cells, where subsequent processing of the antigen takes place. A proteinaceous antigen, for example, is proteolytically broken down and the resulting polypeptides become associated with the major histocompatibility (MHC) class II proteins of B cells. Such B lymphocytes known as antigen presenting cells (APC) are responsible for presenting the antigen to T_H cells. The T_H cells by binding to APC get activated and release substances called cytokines or lymphokines such as interferon and interleukins, that specifically stimulate the cell growth and proliferation of B lymphocyte. The events leading to antibody production and clone of B memory cells are called B cell maturation. In the case of B cell maturation, IgD and IgM disappear from the cell surface and IgM, IgG, IgE, or IgA is secreted. During different immune reactions, the same variable region appears in association with different heavy chain constant regions by a process called heavy chain class switching.

Initially, the precursor of antibody-producing cells, the pre-B lymphocytes, make μ heavy chain constant region linked to a specific variable region (a product of V/D/J recombination). Once the light chain and δ heavy chain synthesis takes place they all combine to form complete IgM and IgD molecules which finally show up on the cell surface. Once stimulated by an antigen these mature B cells undergo an isotype switching to form an activated B cell with IgG receptors. Such B cells give rise to plasma cells that secrete large amounts of antigen-specific antibody, and a population of memory cells (Fig. 9.8).

At the time of first exposure to an antigen, the primary immune response is characteristically slow, producing relatively low yields of antibody. This is because relatively few B cells with the appropriate receptors are present at this time. Upon

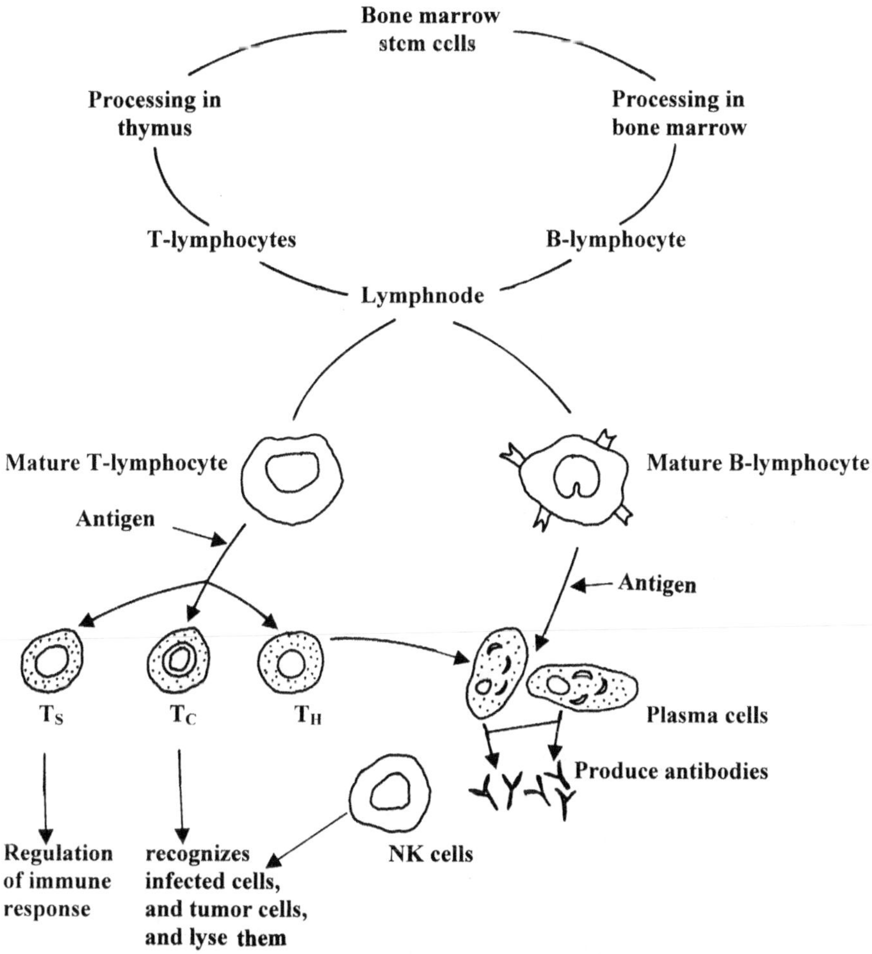

Figure 9.7. Formation, maturation, and role of B- and T-lymphocytes in immune response.

subsequent exposure to the same antigen, even years later, the memory cells, produced during first exposure, are activated to divide rapidly producing a burst population of plasma cells that produce the characteristic antibody. This is known as secondary immune response and since it involves exposure to similar antigen it is also called memory or amnestic response.

9.1- 6.2.5. Antigen. Any macromolecule that not only elicits the formation of an antibody but can also subsequently react with the same is known as an antigen. Almost all macromolecules, like all proteins, most polysaccharides, nucleoproteins, lipoproteins, and various small molecules attached to proteins or polypeptides can act as antigens. Two important properties of an antigen is its immunogenicity, i.e., ability to stimulate antibody formation, and reaction with the same. The portion of the antigen that reacts

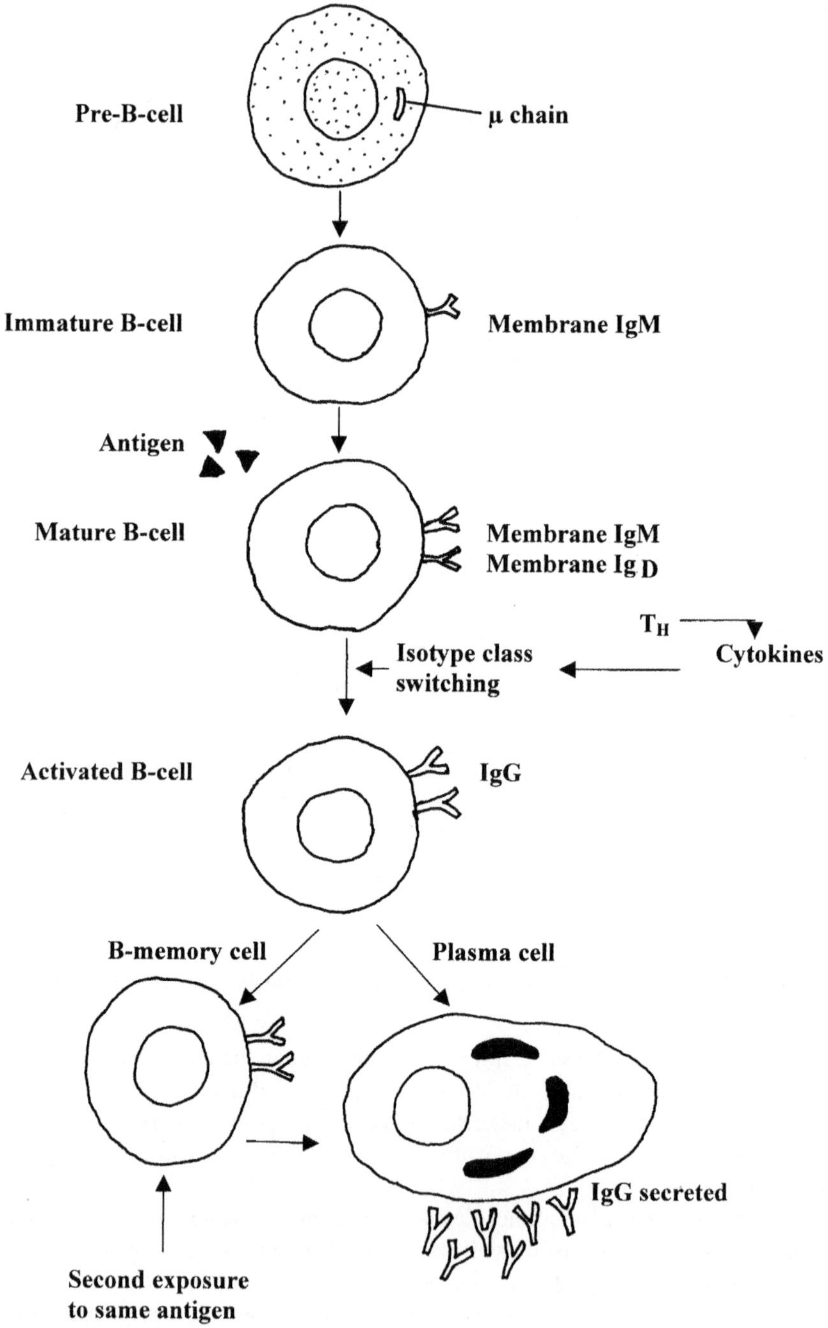

Figure 9.8. Steps in B-cell activation and synthesis of antigen-specific antibodies.

with antibodies is known as the antigenic determinant or epitope. An antigen may be multivalent, having multiple epitopes, or monovalent, having only one epitope. The former generally elicit stronger immune response.

In many cases, antigens are associated with cell surface and thus called surface antigens. Human blood types, for example, are determined by the presence or absence of antigens A and B on the surface of red blood cells.

Living organisms of all kinds are composed of a whole variety of antigenic materials. Bacteria and viruses represent a mosaic of antigens, so much so that even the whole organism is able to elicit immune response. Characteristic antigens are found in the flagella (H antigen of *Salmonella*), capsule, cell wall lipopolysaccharide (somatic or O antigen), and cell interior as well as some are secreted outside. The toxins released by many pathogens are strongly antigenic. Many denatured proteins retain their antigenic properties but fail to elicit the disease. Such proteins known as toxoids have been exploited for developing vaccines such as for diphtheria, tetanus, etc.

Some molecules called haptens are immunogenic but are too small to elicit the formation of antibodies. Once complexed with a larger molecule, or a carrier, it can be converted into a full-fledged antigen. Penicillin is an example of hapten that sometimes complexes with proteins and forms an antigen.

9.I- 6.2.6. Antigen-Antibody Reactions. Antibody-mediated immunity is important in several microbial infections. IgA molecules for example can prevent the adsorption of many potential pathogens on the surface tissues, thus, checking the establishment of infections. IgG antibody, which acts as anti-toxins, neutralizes toxin molecules, thus preventing the onset of disease. Poisonous cobra venom can also be neutralized by appropriate antibody molecules. The antibody can also interact with the surface antigens of bacterial cells making them susceptible to phagocytosis. This process called as opsonization, is important in the destruction of pathogenic bacteria.

The antigen that induces a specific antibody is known as homologous antigen. In many cases, a multivalent antigen, also known as heterophile, heterologous or Frossman antigen can react with antibodies produced in response to different antigens. This leads to an interaction called cross-reaction. While the antigens carry antigenic determinant sites, the antibodies offer the active sites. These two sites are involved in binding in which ionic, hydrogen or hydrophobic bonds are used but covalent bonding is absent. The preponderance of hydrophobic bonds in antigen-antibody reactions suggest that the amino acids occupying the active site may be the one which carry nonpolar side chains.

When both antigens and antibodies are multivalent their interaction may result in a complex or lattice. In a lattice, the antibodies may be heterogenous because of the different antigenic determinants carried by the antigen. Thus, when antibodies combine with soluble multivalent antigen, precipitation reaction can take place. Such antibodies are called precipitating antibodies or precipitin. This forms the basis of an useful immunochemical assay to detect the antigen-antibody interaction, known as Ouchterlony double diffusion technique (Fig. 9.9). When surface bound antigens or other insoluble particles react with antibodies, molecular aggregation or clumping may take place. This forms the basis of blood typing in human beings such as ABO, MN etc. Aggregation or agglutination plays an important role in host defense. Agglutinated microbial cells are more readily trapped in the alveoli of lungs or in the sinuses of the lymph nodes slowing their spread in the body.

Besides precipitation and agglutination, two techniques used for the identification of antigen-antibody complexes are Enzyme Linked Immuno Sorbent Assay (ELISA) and Radioimmuno Assay (RIA). While the first one uses an enzyme as a ligand to the secondary antibody, whose reaction produces a coloured measurable product, in the second, a radioactive ligand, such as ^{125}I is used and the reaction is assessed by radioactivity measurement (Fig. 9.9). Several variations of these techniques are being used. Besides these techniques, immunoelectrophoretic techniques are also available. These have been developed by combining the separation by electrophoresis and identification by double immunodiffusion.

An antigen mixture is separated by charge through electrophoresis. Slots are cut into the agar gel parallel to the direction of the electric field and these are filled with antiserum. The agar gel is then incubated in a humid chamber to allow the diffusion of both, antigen and antibody. Formation of a precipitin band identifies the specific antigen component.

A slightly refined technique that can be applied for quantification of the antigen is known as rocket electrophoresis. In this, the negatively charged antigen is electrophoresed on a gel containing antibody. The precipitate formed between antigen and antibody takes the shape of a rocket, the height of which is proportional to the concentration of the antigen.

In another modified method known as two-dimensional immunoelectrophoresis, the antigen mixture is first separated by electrophoresis. The gel is then laid over by another gel containing antiserum and electrophoresis is repeated at right angles to the first one. Precipitin peaks similar to rocket electrophoresis allows quantitation of a number of proteins in a complex mixture.

9.I- 6.2.7. Dysfunctional Immunity. As discussed above, specific immune response plays a critical role in protecting humans and other higher vertebrates from numerous infections and diseases. Any failure of this response can have serious consequences as the ability of humans to defend themselves against infections will be compromised. Moreover, such failures may also lead to auto-immune diseases in which, ability to differentiate properly between self and non-self is hampered resulting in reactions with self antigens and killing of one's own cells.

9.I- 6.2.8. Immunodeficiencies. Several types of immune deficiencies lead to non-recognition and improper response to the antigens offered by pathogenic microorganisms. Such individuals are highly prone to infection as compared to normal ones. These deficiencies normally affect the cells that are involved in immune response systems. They include complement, myeloid cell, B cell, T cell, and stem cell marrow.

The most severe type of immunodeficiency is known as severe combined immunodeficiency that results from a failure of stem cells to differentiate properly. In its most severe form, the affected individual neither has B nor T lymphocytes and thus fails to invoke any immunological response. Under such a condition any exposure to pathogen can be fatal as microbial growth will go on unhampered. Such individuals can only be kept alive in sterile environment where they are protected from any exposure

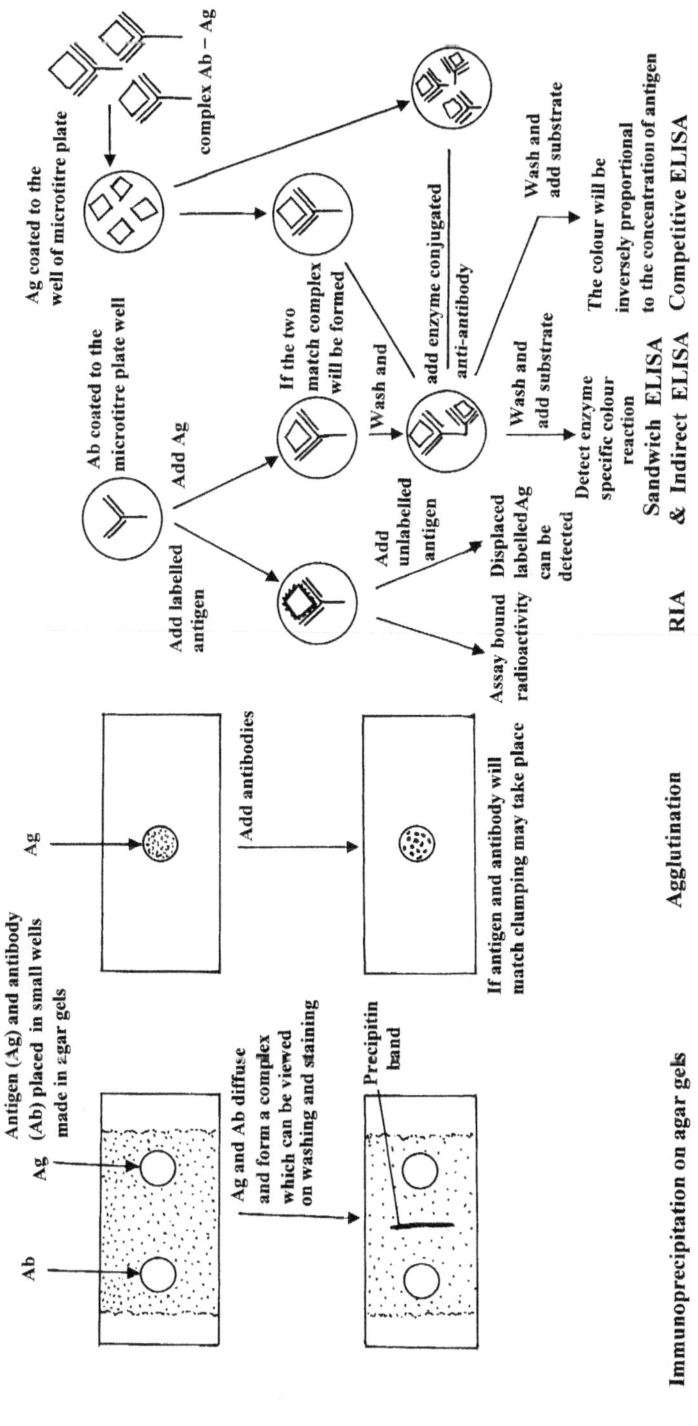

Figure 9.9. Different methods of detecting antigen-antibody reactions.

to microorganisms. Bone marrow grafts can also be employed to establish normal immune functions but these marrows have to be derived from histocompatible siblings.

Lesser severe forms are those in which either T cell function (Di George Syndrome) or B cell function (Bruton's congenital agammaglobulinemia) is lacking. In the former, cell-mediated immunity is missing and the affected individuals are prone to viral and other intracellular infections. Also because T helper cells are involved in enhancing antibody production by B cells, the antigen-mediated humoral response is also depressed. In the second type, where B cells fail to differentiate and thus produce antibodies, cell-mediated response is normal. Bruton's globulinemia is sex-limited and affects only males. Repeated administration of IgG to maintain adequate levels of antibody is the only line of effective treatment.

More common form of immunodeficiencies arise from deficiency of circulating B cells and/or B cell with IgG surface receptors (late-onset hypogammaglobulinemia) or failure of synthesis of specific classes of antibodies (such as IgA). Similarly, failure of complement system, deficient functioning of monocytes, neutrophils, and macrophages, and lysosomal function deficiency in phagocytic cells can all lead to various types of immunodeficiencies.

The most dreaded form of immune deficiency that surfaced in 1979 is the acquired immune deficiency syndrome or AIDS. Caused by the retrovirus, human immunodeficiency virus (HIV), the disease is transmitted through sexual contact, infected blood and blood products and from the infected mother to the foetus. The virus multiplies in T cells thus the affected individuals have more T suppressor than T helper cells leading to immunosuppression. These individuals are prone to a wide-variety of infections and a form of cancer, with the disease turning fatal ultimately.

9.I- 6.2.9. Hypersensitivity. Antibody formation against a specific antigen is one of the strongest host's defense measures. However, in some cases the host may be damaged by the association of antigen-antibody. Such a response is known as hypersensitivity. Various types of allergies are the most familiar examples of this response. It is common knowledge that some people are unusually sensitive (allergic) to a number of substances, like foodstuff, pollen, and even those of microbial origin (known as allergy of infection).

For all cases of hypersensitivity, the individual must be sensitized by exposure to an antigen (allergen). In response to this exposure, the plasma cells produce specific IgE antibodies. It is during the second exposure that different types of hypersensitivity responses can be recognized. If tissue damage takes place immediately after the second exposure it is known as immediate hypersensitivity, and if the effect is manifested only after 24 to 72 hours the same is called delayed hypersensitivity.

The most severe form of immediate hypersensitivity, type 1 or anaphylactic hypersensitivity is anaphylaxis, when the second exposure to antigen is followed by violent symptoms and often death. The release of histamine, heparin, and serotonin by the concerned tissue is the cause of anaphylaxis. Moreover, they in combination with antigeñ-antibody complex, bring about the inflammatory changes in blood vessels making them more permeable, and/or muscle contraction.

Antibody-dependent cytotoxic hypersensitivity (type 2 hypersensitivity) occurs by the reaction of an antigen located on the surface of the cell with an antibody that stimulates phagocytic attack or complement pathway that results in cell lysis. This type of response is commonly seen after transfusion with incompatible blood types.

Formation of a complex between antigen-antibody, and complement may initiate an inflammatory response known as immune complex-mediated hypersensitivity or type 3 hypersensitivity. This is more prevalent in the presence of large excesses of the antigen. In such a situation antigen-antibody complement complex may circulate and get deposited in various tissues and start inflammatory response. They can cause physiological damage to kidney, joints and skin. Administration of large doses of antisera to protect against tetanus and diphtheria or for immunosuppression to protect against rejection of transplanted tissues may initiate this type of response. Several fungal spores can also induce this type of reaction.

Cell-mediated delayed hypersensitivity or type 4 hypersensitivity is characterized by the accumulation of inflammatory cells at the site of antigen injection or localized infection. Inflammation is followed by tissue necrosis in many cases. Two common types of delayed hypersensitivity are allergy of infection and some contact allergies. These reactions involve activated T lymphocytes.

Tuberculosis is the disease caused primarily due to delayed hypersensitivity. An infection with tubercle bacilli hypersensitises the host against bacterial antigen, "tuberculin". Subsequent exposure to tuberculin or release of tuberculin by the death of the bacterial cells causes severe tissue necrosis. In delayed hypersensitivity, the antibodies are often associated with the cells (they are not circulating) and binding of antigen to the same leads to local inflammation and cell death. As we have seen earlier, the inflammation produced during delayed hypersensitivity also contributes towards host's defense mechanisms.

9.1- 6.3. Hybridomas and Monoclonal Antibodies

A typical immune response in an animal results in a mixture of antibodies consisting of different classes, affinities, and specificities. Such a mixture of antibodies is referred to as polyclonal antibodies. Experimental production of one type of antibody (monoclonal) for the study of its structure and the antibody-antigen interactions as well as therapeutic applications has always been hampered by the fact that each specific antibody is synthesized by a different cell line of B lymphocytes and their derived plasma cells. The cell culture techniques could also not provide the answer as most normal cells tend to die off after a few rounds of culturing. Myeloma cells, however, are an exception as they can be cultured indefinitely in tissue cultures. In 1975, Milstein and Kohler fused B lymphocytes producing a known antibody with myeloma cells. This cell fusion created a population of hybrid cells, that were immortalized. These were called hybridomas with the capacity to produce large amounts of monoclonal antibody of a specific type (Fig. 9.10). A monoclonal antibody shall react with a specific type of antigen.

For the formation of hybridomas, myeloma cells lacking thymidine kinase and hypoxanthine-guanine-phosphoribosyl transferase activities ($TK^-/HGPRT^-$), and thus unable to synthesize nucleotides/DNA were used. These myeloma cells also could not grow on a medium containing aminopterin which is known to block nucleotide synthesis. The B lymphocytes were prepared by injecting a specific antigen into mouse, and a population of B lymphocyte producing antibodies against this antigen were obtained from the spleen. The fusion between these two cell populations was achieved by using Sendai virus (today polyethylene glycol is the preferred fusogen) and cells were placed on a selection medium called HAT (containing hypoxanthine, aminopterin, and thymidine). While the normal B cells will not survive for more than one or two weeks, the myeloma cells will die due to mutagenic lesions. However, fused hybridoma cells grew as they had the TK and HGPRT genes from B cells and the ability to grow in tissue culture from the myeloma cells. The clones of these hybridoma cells will expectedly produce large amounts of a specific monoclonal antibody (Fig. 9.10).

The monoclonal antibodies are already in therapeutic uses such as diagnosis of allergies and infections, diseases like hepatitis, rabies, and some veneral diseases. They may also be used in detecting some cancers at early stages and if used alone or chemically attached to drugs could also be helpful in locating and destroying a cancer cell or pathogenic microorganisms specifically.

9.II- Bacteria and Food

One area in which microorganisms have contributed immensely is in the food industry especially for food production but bacteria along with other microbes may also cause food spoilage. We shall cover these two aspects in two different sections. The latter is discussed under "Biodeterioration and Spoilage". So important is the impact of microbes on food that it has led to the creation of a separate branch called "Food Microbiology".

Many traditional foods and beverages have been based on microbial metabolic activities. Generally it is the fermentative metabolism of microorganisms which is exploited in the production of these food items. This activity either produces the final product or imparts characteristic flavour or other beneficial properties. In order to harness this microbial function, not only the microbes but also the proper substrate and environmental conditions need optimization. Besides the microbially-produced food products, microbial biomass is also considered as a rich source of protein that can be exploited to meet the food needs of growing world population. Although, such products have not been accepted for direct human consumption, their use as the animal feed supplement is gaining ground.

9.II- 1. Fermented Products

One of the biggest class of such products are the dairy products obtained through the fermentation of milk. These include buttermilk, sour cream, yoghurt and cheese. The basis of this fermentation is the production of lactic acid from milk sugar lactose by lactic acid bacteria. The lactic acid acts as a natural preservative and aroma and flavour comes from other fermentation products (Table 9.3).

The important bacteria taking part in these fermentations consist of *Lactobacillus* species such as *L. bulgaricus* and *L. acidophilus*, *Leuconostoc* species and *Streptococcus* species like *S. thermophilus*, and *S. lactis*. Different *Leuconostoc* species provide the characteristic flavour compounds. For example, characteristic flavour and aroma in butter and yoghurt is because of diacetyl formation from citric acid, and lactic acid, and

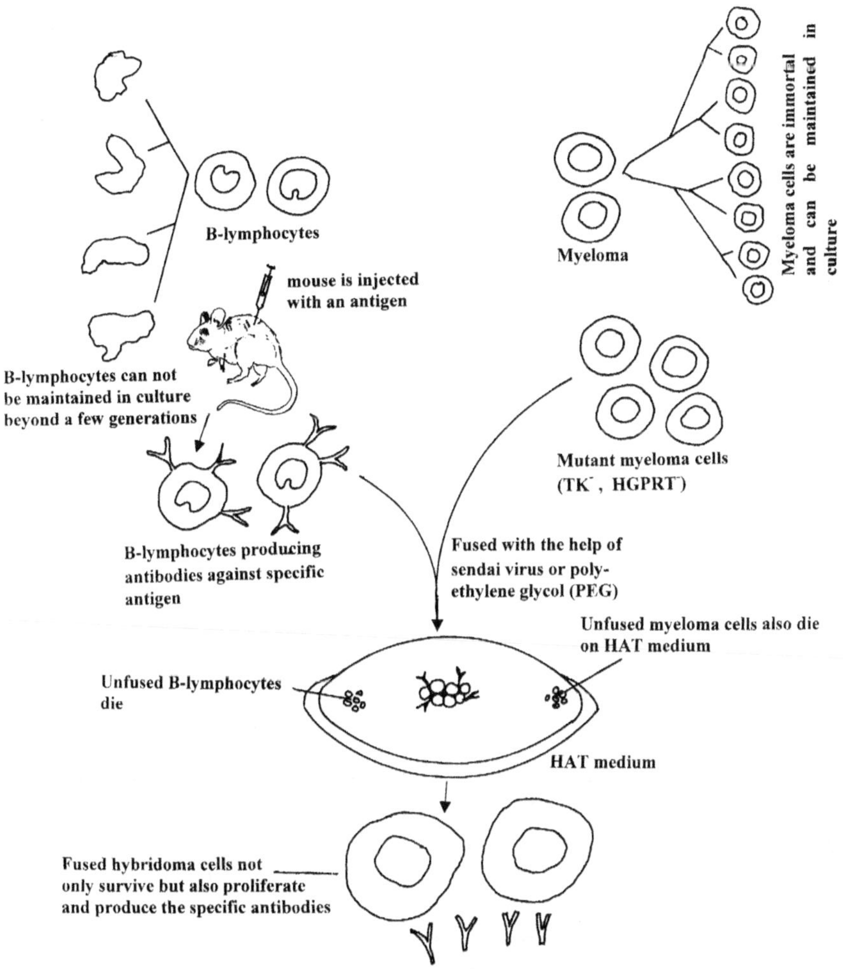

Figure 9.10. Formation of hybridomas and production of monoclonal antibodies.

acetaldehyde production, respectively. Many yeasts also contribute their metabolic power in these processes. In cheese production, the initial curdling step is contributed by lactic acid bacteria but during subsequent ripening, the involvement of other bacteria and fungi produces different types of cheese. Propionic acid fermentation by *Propionibacterium* species and fungi like *Penicillium* species produce the popular swiss cheese, and Roquefort and Camembert cheese varieties. Commercial production of cheese employs addition of chymosin or rennin obtained from calves for milk curdling. The gene for this enzyme has been cloned in *E. coli* and yeast, but problem of yield and full activity still persists. An enzyme with similar activity is found in fungi, such as *Mucor* and *Endothia* sps. This with some desired modification could replace rennin.

Conversion of whey, a waste byproduct of cheese manufacturing and a rich source of sugars and minerals, through lactic acid fermentation, into a beverage is also being attempted.

*Table 9.3. Some lactic acid bacteria and *fungi used in food industry*

Species	*Uses*
Lactobacillus lactis	Buttermilk, sourcream, initial steps of cheese production, sourdough, fermented vegetables, silage production
L. diacetilactis	Sourcream, butter
L. bulgaricus	Bulgarian buttermilk, sourdough, kefir
Leuconostoc species	Cultured buttermilk, fermented vegetables, pickles, silage production
Streptococcus species	Sourcream, cultured buttermilk, silage production
S. thermophilus	Kefir, cheese ripening, sourdough
Pediococcus cerevisiae	Fermented meat, pickles, maturation of soysauce
Propionibacterium species	Ripening of cheese (swiss cheese)
Acetobacter/Gluconobacter	Vinegar production
**Penicillium* species	Ripening of cheese (Roquefort & Camembert)
*Yeasts (*S. cerevisiae*)	Alcoholic beverages, leavening of bread

Fermented meat products like sausages and salamis are produced by allowing the meat to undergo heterolactic acid fermentation, a process known as curing. Once again lactic acid bacteria are predominantly present but *Pediococcus cerevisiae* is exceptionally useful in controlling the levels of this fermentation.

In bread making, the major role is played by the yeast, *Saccharomyces cerevisiae*. However, the requirement of CO_2 production for bread leavening can also be fulfilled by bacteria like coliform and *Clostridium* species. In many bread varieties, sour dough is used in which souring is caused by several lactic acid bacteria.

9.II- 2. Alcoholic beverages

Beverages like wines, beer and ale and distilled liquor are produced by alcoholic fermentation in which the central role is played by the yeast that ensures the conversion of sugars to alcohol and CO_2. Moreover, some higher alcohols, acids and esters produced during the process impart characteristic flavour. The finished product comes after a slow aging process.

9.II- 3. Vinegar

Vinegar is produced in two steps. Initially carbohydrate is converted to alcohol by

S. cerevisiae and then alcohol is oxidatively transformed to acetic acid by *Acetobacter* and *Gluconobacter*.

9.II- 4. Fermented Vegetables

Production of fermented vegetables like pickles, olives, sauerkraut, and soysauce and several others are once again based on lactic acid fermentation. In this process, salt plays an important role. Salting may itself provide a preservative effect but also creates conditions that favours the growth of lactic acid bacteria and prevent other bacteria from growing on the starter materials. The sourness of these products reflect the amount of lactic acid that accumulates during fermentation.

Silage- Silage or preserved plant materials as animal fodder is also derived by lactic acid fermentation. Materials which are good for silage are rich in carbohydrates (corn, moist cereals, sunflower, turnip leaves, etc.) but low in protein and water content. These materials are stacked in a low structure or a pit called silo so as to create anaerobic conditions. Initially, the coliform and other aerobic bacteria grow and consume oxygen that makes conditions conducive for homofermentative lactobacilli, *Streptococcus, Lactococcus,* and *Leuconostoc* species, which produce lactic acid.

9.II- 5. Single Cell Protein

In recent times, microorganisms have been advocated as a source of single cell protein (SCP). Since they can be grown rapidly in large numbers they can yield a high-protein crop. Various bacteria, algae, and fungi are potential sources of SCPs and their proteins contain all essential amino acids. Algae like *Scenedesmus* and the cyanobacterium, *Spirulina* have been grown in warm ponds as a food source. Being autotrophic, their culturing greatly reduces the cost.

Imperial Chemical Industries (ICI) in England produces Pruteen, an SCP derived from *Methylophilus methylotrophus* that can grow on methanol. However, the increasing price of oil has made the petroleum hydrocarbon very expensive and therefore search for cheaper substrates is very much required. In this context, yeasts have been recommended as excellent candidates for SCP production. Firstly, they can be grown on cheaper or even waste substrates and secondly, that they have a high vitamin content.

SCPs have not been used presently for direct human consumption. This is because they have a high nucleic acid content which may lead to increased serum levels of uric acid causing kidney stone formation, gout, and possible allergic and gastrointestinal reactions such as diarrhoea and vomiting. This is one area in which lot of research interests have been directed. Their use as animal feed supplement, however, is significant.

9.II- 6. Food Supplements

Microorganisms are a rich source of various food and feed supplements such as vitamins, amino acids, and organic acids.

Vitamin B_{12} can be produced as a byproduct of *Streptomyces* antibiotic fermentation. Commercial production, however, employs direct fermentation by *Propionibacterium shermanii* or *Paracoccus denitrificans*. Riboflavin is a byproduct of acetone-butanol fermentation by various *Clostridium* species, but commercial production employs several fungi.

Many microbial strains have been employed for the production of amino acids, although the strains may have to be suitably modified to overcome the natural regulatory controls, or enhance the efficiency of production. For example, the essential amino acid lysine is not present in sufficient quantities in grains to meet the animal's nutritional requirements. Microbially produced lysine can be used as feed supplement or additive. Commercial production involves carbohydrate fermentation using a homoserine-requiring auxotroph or an analogue (S-aminoethylcysteine) resistant mutant of *Brevibacterium flavum*. *Corynebacterium glutamicum* produces glutamic acid. During glutamic acid fermentation, this bacterium produces tetramethylpyrazine which itself is a powerful aroma compound. Glutamic acid is principally used as monosodium glutamate (MSG), that is widely used as a flavour enhancer. L-glutamic acid and MSG can be produced by direct fermentation using *Brevibacterium, Arthrobacter,* and *Corynebacterium*. Microbially produced amino acids such as cysteine, lysine, glutamate, methionine, proline, threonine, phenylalanine, and tryptophan have been employed as food enrichments and flavorants. Similarly, alginate, cellulose, curdlan, levan, dextran and xanthan are used as thickening agents in processed food. Aspartame, a low calorie sweetner, is synthesized from L-phenylalanine and L-aspartic acid and is a common ingredient of sugar-free products. Many amino acids like cysteine, tryptophan and histidine also act as antioxidant and are used in many products.

Organic acids like acetic, gluconic, citric, itaconic, gibberellic, and lactic acid can be produced by microbial fermentations. Many of them are used in food industry as preservative and acidulants. Two principal bacterial fermentations involve acetic acid production by *Acetobacter* and lactic acid production by *Lactobacillus delbrückii*. Similarly, many lactic acid bacteria produce secondary metabolites like bacteriocins which have the potential to inhibit a wide range of undesirable microorganisms in food. These bacteriocins can be exploited as biopreservatives to maintain the safety of food. Whole microbial cells such as that of yeasts, lactobacilli, and streptococci are used in brewing, baking and wine and cheese making.

9.III- Bacteria and Industry

If one considers the metabolic versatility of the microorganisms they can be equated with one room chemical industry. Bacteria along with fungi are capable of synthesizing a wide range of chemicals and enzymes. A better understanding of these metabolic pathways has been exploited in the commercial production of several of these useful compounds. Compared to chemical industries the microbial processes are not only efficient but are also cost effective and environment friendly (no pollution generating).

9.III- 1. Pharmaceuticals

One of the biggest industry, pharmaceutical industry deals with the production of drugs to control various diseases. These include antibiotics, steroids, vaccines, vitamins etc.

Antibiotics are one of the major microbial products which are being manufactured at commercial scale. Antibiotics are the compounds produced by some microorganisms, both bacteria and fungi, that can inhibit or kill other microorganisms. Of more than 6000 antibiotics known, only some 150 are being produced commercially (Table 9.4).

The medium used for the production of antibiotic depends upon the microorganism selected as also the growth conditions. Most antibiotics are secondary metabolites, and their production takes place at the end of an active growth phase, called idiophase. The active growth phase is the trophophase. The genetic manipulations of the producer strains and appropriate manipulations of the culture conditions are continuously sought to increase the production. At the completion of fermentation, the desired antibiotic is recovered from the broth and purified.

Many antibiotics are derived from the basic structure by appropriate chemical modifications or by microbial enzymes. This leads to antibiotic derivatives with wider and superior biological activity. They not only vary in their chemical form but also in their action spectrum. At the same time, the search for new and more effective antibiotics continue.

Some of the brightly coloured green- and purple-sulphur bacteria living symbiotically with sponges and sea squirts produce antiviral compounds. A new compound isolated from deep sea bacteria living in sediments more than 300m deep inhibits the replication of HIV. Many of them are also tipped as potential sources of new antibiotics to combat the drug-resistant pathogens.

9.III- 2. Steroids

Transformation of steroids to give different commercially valuable steroid preparations is another important microbial function exploited by pharmaceutical industry. Steroids have a number of therapeutic uses, such as in relieving pain associated with rheumatoid arthritis, and alleviating several allergic and inflammatory responses. Moreover, many steroids regulate the sexuality and reproductive cycles in humans. Some of them are thus used in oral contraceptive formulations.

In many of these transformations, the microbial involvement not only makes the process simpler and thus cost effective but in combination with appropriate chemical steps can lead to a whole variety of derivatives for different medicinal uses.

9.III- 3. Vaccines

Vaccines, extremely useful in preventing the onset of various serious diseases constitute an important component of pharmaceutical industry. The production involves the culturing of microorganisms possessing the antigenic properties that can elicit a primary immune response. The microorganisms used are the mutants or attenuated or inactivated virulent pathogens whose antigenic properties remain intact. The application of genetic engineering has led to the production of entirely new kinds of vaccines in terms of efficiency, fewer side effects, and newer and multiple targets. Strict quality control and appropriate safety test procedures are mandatory before releasing a vaccine.

9.III- 4. Human Proteins

One area in which the application of recombinant DNA technology has proved its worth is in the production of several human proteins by microorganisms. For this, the gene/s responsible for the said proteins have been isolated from an appropriate source or chemically synthesized and inserted in the bacterial genome by rDNA technology (**see Chapter 7**). The necessary products can then be recovered by growing these manipulated bacteria in commercial fermenters. The production of such human proteins for use as prophylactic drugs (=to prevent disease) and diagnostic kits has reduced the production time and cost considerably.

Many of these proteins which are commercially produced are listed in Table 9.5. These include hormones, growth factors, tumor necrosis factor, blood clotting factors, vaccines, antiviral compounds and many others.

It is also very clear that production of important human proteins through rDNA technology is a rapidly growing field and more and more products are likely to find place in Table 9.5 in near future.

9.III- 5. Industrial Chemicals

As we have seen earlier, many bacteria and fungi produce number of organic chemicals, such as organic acids, solvents and enzymes with a wide-variety of applications. These act as chemical feed stocks for various chemical processes including pharmaceuticals, detergents, textile, resins, leather, plastics, and agrochemical industry.

9.III- 6. Microbial Enzymes

The metabolic versatility of a microorganism depends upon a whole panoply of enzymes. Many of them have found useful application in different industries ranging from analytical to food and beverages, detergent, leather, paper, pharmaceutical and clinical, photographic, and textile etc. (Table 9.6). As enzymes are often products of single gene, both selection and rDNA technology is employed to increase the production, modify their specificities, and the process efficiency of such enzymes. In order to reduce the cost of producing and using enzymes, focus is on microbial strain improvement,

Table 9.4. Some of the common antibiotics produced by microorganisms

	Antibiotic	*Producing Microorganism*
1.	Amphotericin B	*Streptomyces nodosus*
2.	Bacitracin	*Bacillus licheniformis*
3.	Chloramphenicol	*Streptomyces venezuelae*
4.	Erythromycin	*Streptomyces erythreus*
5.	Kanamycin	*Streptomyces kanamyceticus*
6.	Neomycin	*Streptomyces fradiae*
7.	Polymixin B	*Bacillus polymyxa*
8.	Streptomycin	*Streptomyces griseus*
9.	Tetracycline & its chloro-derivative	*Streptomyces auerofaciens*
10.	Penicillin & its derivatives like ampicillin*	*Penicillium chrysogenum*
11.	Cephalosporin* (*derived from fungi)	*Cephalosporium acremonium*

cell and enzyme immobilization and increased stability. Moreover, the controlled protein engineering and construction of artificial enzyme complexes are being attempted to get the desired results. Although this field is dominated by several fungi, bacteria also provide some essential enzymes.

One such enzyme is bacterial alkaline protease derived from *Bacillus licheniformis* and *B. subtilis*. Its major use is in detergent industry as a stain remover but also finds application in baking, digestive aids, meat tenderization, chill proofing beer, cheese, and flavour production, leather softening and textile sizing. Salt resistant, protein digesting enzyme isolated from marine bacteria is a potential ingredient of detergent formulations. Recombinant DNA technology has been employed to develop a bacterial strain that produces kerazyme. This enzyme breaks down hair protein, keratin and thus cleans the hair-clogged drains.

Various types of amylase, such as α, β, and glucamylase produced by bacteria and fungi is also an enzyme with multiple applications. All three enzymes are used in the production of syrup and dextrose from starch. Amylases are used in the preparation of sizing agents, and in textile and paper industry. Additional applications are in bread making, manufacture of corn and chocolate syrup and removal of food spots. Glucose isomerase obtained from *Bacillus coagulans, Streptomyces* species and *Nocardia* species converts glucose into fructose. Fructose is sweeter than sucrose and glucose and is used in many foods and beverages.

Various other microbial enzymes such as lipases, pectinases, invertases and others have been used in different industries. Production of rennin (chymosin), the milk curdling enzyme used in manufacture of cheese, by two species of *Mucor, M. pussilus* and *M. meihei* can effectively cut down the cost as presently the enzyme is obtained from calf's stomach. Marine bacteria possess unique enzymes that could combine biological molecules with halogens (chlorine, iodine). In Japan, they are being used in medical, cosmetic, and food industry. Many enzymes obtained from extremophiles will undoubtedly find more applications in industry.

Many microbial enzymes are expected to replace the conventional chemical methods as these processes would be economical and environmentally clean.

Thermally stable DNA polymerase (*Taq* polymerase) from *Thermus aquaticus* or genetically engineered *E. coli* cells is applied in polymerase chain reaction (PCR). Numerous restriction endonucleases (**refer Chapter 7**) and other enzymes required for genetic engineering are also derived from different bacterial sources. Marine organisms have turned out to be rich sources of enzymes. A number of thermostable enzymes have been harvested from bacteria living near thermal vents on the ocean floor. These are marketed as vent and deep vent polymerases with potential use in molecular biology and PCR. Once again many modified enzymes are being created that have superior properties to those of natural enzymes. Such protein engineering is not only aimed at enhanced production of enzymes with better properties but also with respect to stability and performance in specific industrial applications, and environmental considerations.

9.IV- Solvent and Fuels

Several organic solvents are also produced as part of the microbial fermentation. These

Table 9.5. Some human proteins commercially produced by recombinant microbial cells and their functions

1.	Insulin	Regulates blood sugar levels. Used in treatment of diabetes.
2.	Human growth Hormone (HGH)	Controls body growth, used in treatment of dwarfism.
3.	Bone growth factor	Stimulates bone cell growth. Used in the treatment of Osteoporosis.
4.	Interferon α, β, and ϒ	Cytokines of immune system. Used in treatment of cancer and viral diseases.
5.	Tumor necrosis factor	Cytokines of immune system. Causes death of malignant cells. Used in treatment of cancer.
6.	Interleukin-2	Cytokines of immune system. Used in treatment of immune deficiencies and cancer.
7.	Tissue plasminogen activator	Dissolves blood clots. Used in treatment of heart diseases and during heart surgery.
8.	Blood clotting factor III	Stimulates blood clotting. Used in treatment of hemophilia.
9.	Erythropoietin	Stimulates red blood cell formation. Used in treatment of anemia in dialysis patients.

include ethanol, acetone, butanol, glycerol etc. Although many of these are produced as products of the petrochemical industry, the rising cost of petroleum, their rapid depletion, and the environmental degradation due to industrial releases may force the microbial processes to become more attractive.

Production of ethanol by microbial fermentation is an age old practice. Several bacterial strains with higher efficiency than yeasts are now known. The microbial production of acetone and butanol can be carried out by *Clostridium* species, *C. acetobutylicum* and *C. saccharoacetobutylicum.* The remarkable point is that they can be grown on cheaper and easily available substrates (starch, molasses). Similarly, butanol, and glycerol can be produced by using yeast, *S. cerevisiae* and bacterium, *Bacillus subtilis.* Glycerol is widely used in food, pharmaceuticals and cosmetics. Most of these processes can be suitably modified to make them more competitive than chemical synthesis.

Rising cost of petroleum and rapid depletion of this and other natural non- renewable resources has created an energy crisis. This has led to extensive research in developing alternative or synthetic fuels. Microorganisms have provided a solution in this area whereby waste materials or cheaper substrates can be converted into a wide range of fuel substitutes.

Ethanol from plant residues or biofuels by microbes is currently under production. Although still an expensive process, various innovations are being carried out to bring down the cost. The finding that the bacterium *Zymomonas mobilis* is two times quicker than yeast in fermenting carbohydrates to ethanol or *Thermoanaerobacter ethanolicus* which carry out the same process at high temperature may go a long way in enhancing the process efficiency as well as product recovery. The major part of the carbohydrate content in plants is not in the form of starch or sugar but in cellulose or lignocellulose. A 2-step fermentation of cellulose to ethanol is also being suggested to make it economical. Lignocellulose is hard to break down and cellulose break down alone releases a mixture of sugars such as glucose, xylose, mannose, galactose and arabinose. No single microorganism has been found in nature that can rapidly and efficiently metabolize all these sugars. The development of a microbial strain that can ferment lignocellulose remains a major research goal. A solution from rDNA technology is being expected. Marine amoebas that feed on cellulose of seaweeds might be the potential source of an effective enzyme system. A mutant amoeba strain has been developed that can degrade other stable molecules such as PVC.

Whey as a waste product of cheese manufacturing is also advocated as a source of complex polysaccharides, lipids and proteins. Once hydrolysed to simpler components, various bacteria and yeasts can be employed to produce butanol, acetone, ethanol, methane and other useful compounds.

Methane produced from anaerobic decomposition of waste material by methanogenic archaea offers another solution. Methane can be used to generate electrical and heat energy. This is a community function by a number of microbes and can take care of waste disposal as well (**see next Section**).

Hydrogen produced by many photosynthetic microorganisms is another potential energy source. In this, the abundant solar energy can be converted in an energy form which can be stored and used for generating electrical energy. This microbial function is a subject matter of intensive research. Since 1970s, when first of such alarms about energy crisis were raised, a lot of effort has been devoted towards the development of alternative fuels. At that time hydrogen was touted as the "fuel of the future". Although interest in this area has fluctuated with the prices of oil, the interest was rekindled by the fact that fossil fuels may not only be a major cause of air pollution and global climate change, they are not going to last very long.

Hydrogen can be produced chemically, electrochemically, as a by-product of oil/coal processing or by the use of certain microorganisms. The microbial production of hydrogen is significant for several reasons more so because it will be produced from renewable resources and shall be environmentally benign. One could also visualize virtually unlimited production of hydrogen from the earth's most plentiful of available resources-water, solar energy, and carbohydrates. Currently this process may involve various types of photosynthetic and non-photosynthetic bacteria as well as algae (Table 9.7).

The current research efforts are targeted towards photolysis of water into H_2 and O_2 by photosynthetic systems. The electrons so generated are passed on to protons to make hydrogen through hydrogenase. However, O_2, the other product, is a strong inhibitor of hydrogenase.

In order to circumvent this problem, heterocystous cyanobacteria can be used. They compartmentalize O_2 generation and CO_2 fixation in vegetative cells, and nitrogenase containing heterocysts where H_2 evolution can take place.

Table 9.6. Industrial applications of some microbial enzymes

Industry	*Enzyme*	*Applications*
Food, Dairy, Beverages	Amylase	Baking, syrup and glucose manufacture
	Glucose oxidase	Glucose and oxygen removal, preservation of flavour and colour in eggs, food, and fruit juices
	Glucose isomerase	High fructose corn syrup in soft drinks
	Lipase	Dairy products
	Pullulanase	Production of beer, maltose and glucose
	Protease	Baking, meat tenderizing
	Invertase	Candy manufacture
	Rennin	Cheese manufacture
	Pectinase	Pressing, clarification of wine, fruit juices
Paper, Textile, Detergent	Amylase	Starch coating on paper, textile desizing, cold-swelling and laundry starch
	Protease	Spot removal in dry cleaning, textile desizing, house-hold detergent
	Lipase	Laundry
	Cellulase	Laundry detergent, fabric softening and brightening
Pharmaceutical	Amylase	Digestive aid, dental hygiene
	Glucose oxidase	Test paper for diabetes
	Protease	Wound cleansing
	Asparaginase	Anti-tumor agent
Research biochemicals	All the above and enzymes used for different reactions	Biochemical, and molecular biology work

Alternatively, non-heterocystous nitrogen-fixing bacteria can also be employed where these two reactions are temporally or spatially separated. One problem with nitrogen-fixing bacteria is the high metabolic energy demand, that lowers the potential solar energy conversion efficiency.

Another interesting alternative can be to use photosynthetic bacteria which can convert organic substrates, including many wastes, quantitatively in H_2 and CO_2 in the presence of light. The involvement of nitrogenase with high ATP requirement and low light intensities at which these bacteria operate makes it less favourable.

The capability of some photosynthetic bacteria to convert carbon monoxide to hydrogen (microbial "shift reaction") can be an exciting alternative. Although large scale application of this process is still fraught with some problems it could be most effectively employed during biomass gasification.

Another potential approach can be the dark fermentative production of hydrogen. This could either come as a byproduct of a high value product, such as gluconic acid or by conversion of low cost substrates. The latter has already been employed for methane production. It has been estimated that hydrogen production cost by this method may be 2-3 folds lower than methane production.

One viable option could be the production of a mixture of methane and hydrogen through biomass gasification (**see next Section**). Such a mixture has a lower pollutant value than pure methane when used as fuel in internal combustion engines. Although the large scale application of this process is still unclear, answers can be sought through the use of thermophilic bacteria, nutrient limitations and most importantly genetic and metabolic engineering to manipulate the metabolic pathways towards higher hydrogen production.

Similarly, many microorganisms can synthesize complex hydrocarbons. The formation of petroleum deposits is a microbial process in conjunction with some physico-chemical processes. Better understanding of these processes may provide a means for hastening the process or genetically engineer a microorganism to produce synthetic sources of petroleum and other hydrocarbons.

9.V- Bacteria and Resource Recycling and Management

The world's fast growing human population has put tremendous pressure on natural resources. While on one hand there has been an increasing demand for more and more food, the other resources used up by the human civilization are getting rapidly depleted. We have already discussed the energy crunch that the humanity is facing and the role of microorganisms in providing solutions to some of these problems. Also we have seen, the microbial role in recycling of complex organic molecules to regenerate the simple molecules for utilization by all living organisms. In present times, resource recycling and utilization of low grade resources has become a reality.

Two main problems encountered in the new century are the availability of resources for chemicals and energy, and environmental pollution. The provision of resources has largely centered around non-renewable materials especially oil. Diminishing reserves of these, together with uncertainties of supply and cost have generated great interest in renewable resources. These are largely lignocellulosic materials (e.g., wood and straw) which are available through natural biomass turnover, farming, and forestry and as waste generated from industrial processes (kraft lignin).

Environmental pollution is no longer accepted as inevitable for a technological society. Over the past decade, there has been a tremendous increase in awareness of the adverse and long lasting effects of pollution and public pressure has mounted on both industry and government. To be realistic, however, it is not possible to replace all processes generating polluting waste with clean alternatives. Instead, treatment of pollution, both at source and after an incident are currently acceptable and in many instances a great deal of emphasis is lately being placed on these.

9.V- 1. Biomass Recycling

While considering the technologies currently available to utilize renewable resources and remove pollutants from the environment both physical and chemical methods are often unsatisfactory and uneconomical. There has, therefore, been a movement towards biological conversion and remediation procedures which can be more specific and convenient though in many cases it is yet to be economical.

Earth is replete with biomass which can be defined as the living matter or the residues derived from the same. It has been estimated that world-wide photosynthetic activity can store 17 times as much energy in the biomass as is annually consumed by all the nations of the world. Biomass is, therefore, a renewable and perpetual resource which could be profitably exploited for a variety of uses including energy.

This biomass can be broadly classified into three categories: (i) wood biomass derived from forests and agro-forestry that is used as source of timber, raw materials for pulp, paper, plywood, and as cooking fuel, (ii) agricultural and agro-industry residues such as saw dust, rice husk, groundnut shells, bagasse, bamboo dust, coir, pith, coffee husks, etc. These may also be used as cooking fuel but frequently cause disposal problems, and (iii) animal waste and urban waste.

Conversion of biomass into useful forms of energy can be brought about by two methods, thermochemical and biochemical. The thermochemical methods demand high temperature treatment and with the help of oxygen, pyrolysis, combustion, and gasification occur. Because of the costs involved, this is an expensive process.

Biochemical route relies on the action of bacteria in degrading complex molecules of biomass into simpler ones, and producing biogas in the process. Biogas is a mixture of methane and carbon-dioxide and is normally produced by anaerobic digestion of animal dung. Other organic wastes can also be treated to get biogas production. As methane is inflammable, it can be used as a source of energy. The process of methane generation is known as biomethanation.

Due to highly complex and varied nature of the initial substrate, diverse groups of microorganisms will have to be employed for biomethanation. They comprise hydrolytic fermentative microorganisms, acetogenic fermentative bacteria, and methanogenic bacteria.

The hydrolytic fermentative microorganisms hydrolyze polymeric substances like proteins, polysaccharides and lipids into amino acids, oligo-and monosaccharides, and free fatty acids, respectively. Some of the dominant members of this group are: *Fibrobacter succinogens, Ruminococcus albus, R. flavefaciens, Butyvivibrio fibrisolvens, Prerotella ruminicola* and *Selenomonas ruminatium*. The smaller molecules so generated are utilized by fast-growing fermentative bacteria which produce short-chain volatile fatty acids (VFA), organic acids like acetic, propionic, butyric and small quantities of formic, valeric, and caproic acids. Ammonia, sulphate, and some

Table 9.7. Possible pathways of biological hydrogen production process

I. **Direct Biophotolysis:**

$$H_2O \xrightarrow{\uparrow O_2} \text{photosystems} \longrightarrow \text{ferredoxin} \longrightarrow \text{hydrogenase} \longrightarrow H_2$$

II **Heterocystous Cyanobacteria:**

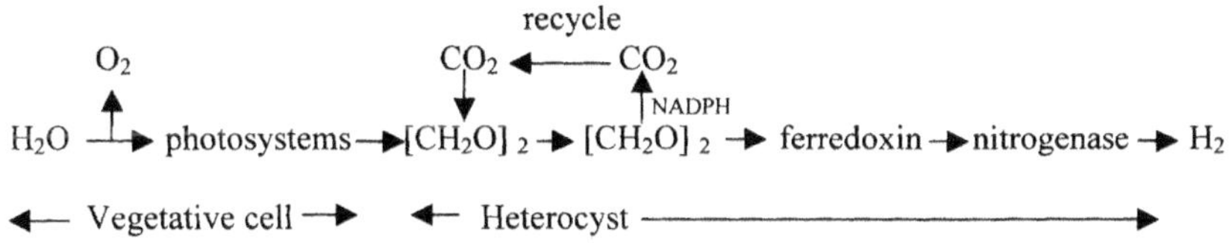

III **Indirect Biophotolysis-Non-heterocystous Cyanobacteria:**

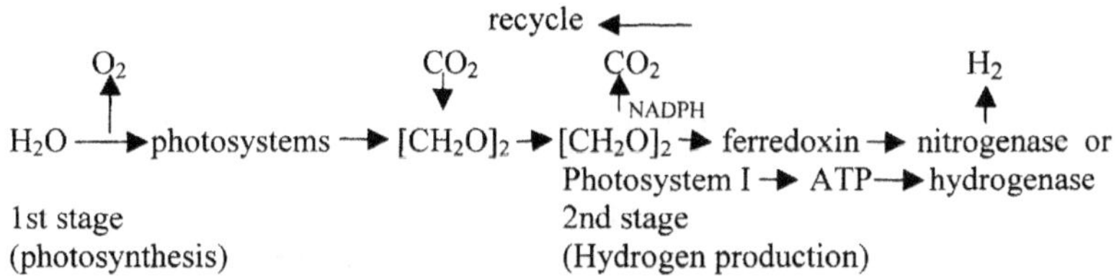

IV **Photofermentation: Photosynthetic bacteria:**

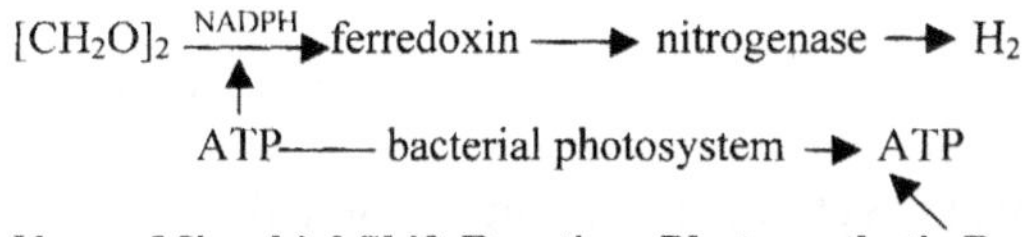

V **Microbial Shift Reaction: Photosynthetic Bacteria:**

$$CO + H_2O \longrightarrow H_2 + CO_2$$

VI. **Dark fermentation:**

$$[CH_2O]_2 \longrightarrow \text{ferredoxin} \longrightarrow \text{hydrogenase} \longrightarrow H_2$$

aromatic compounds may also be formed. As the acid production is the major step, this process is also known as acidogenesis. These bacteria, in the presence of autotrophic methanogens and sulphate-reducing bacteria, may also produce hydrogen.

Acetogenic bacteria thrive on energy derived from the oxidation of organic acids, alcohols and VFAs with each member contributing a specific step. The products are acetate, formate, bicarbonate, and hydrogen. The commonly found members are *Acetobacterium woodii, Clostridium aceticum, C. formicoaceticum, Eubacterium limosum, Acetoanaerobium noterae, Butyribacterium methylotrophicum* and others. Some acetogens can ferment xenobiotic compounds such as polyethylene glycol and phenolic compounds, some can produce acetate from trimethoxybenzoate, trimethoxy cinnamate and some methoxylated aromatics, and still others can reductively dechlorinate tetrachloromethane. This unique capability of acetogens give them a competitive advantage in the conditions where normal substrates may be in association with such toxic compounds, and the same can also be exploited through biotechnological means. Acetogens are extremely sensitive to hydrogen and are thus found in association with methanogens which swiftly remove hydrogen and thus help acetogens to grow.

The methanogenic bacteria or methanogens are the terminal members of this sequence as they can utilize acetate, formate, methanol, methylamine, and hydrogen and carbon-dioxide as carbon and energy source. Some are autotrophic and others reduce CO_2 using hydrogen. They produce the most reduced form of carbon, methane. Some aceticlastic, chemo-organotrophic archaebacteria split acetate into methane and carbon dioxide, a predominant reaction during anaerobic digestion. Major methanogenic bacteria belong to *Methanosaeta, Methanothrix, Methanosarcina* species.

Biomethanation has gained high importance as an industrial process as it provides an economic possibility for energy generation as well as waste recycling. The efficiency of this process largely depends upon required microbial population, anaerobic conditions, temperature, pH, substrate composition, carbon: nitrogen (C:N) ratio, micronutrients, presence of toxic materials in the substrate, and accumulation of inhibitory by-products and intermediates.

In the earlier section, we have already discussed the production of ethanol from plant residues (biofuels). This success story has been amply demonstrated in Brazil, where fuel alcohol has been produced for years from fermented sugar cane juice. The cane debris left behind is used as boiler fuel, supplying steam for stills, sterilization of equipment and for local energy generation. Corn is another easy to grow crop with high starch content. Corn starch is first converted to sugar to initiate fermentation. Presently, energy requirement for distillation is as much as they get from the ethanol produced. It is, therefore, important to select potential fuel crops and engineer them for faster growth and high content of easily fermentable tissues. Microbes can also be manipulated for more efficient conversions and to broaden the range of materials that they can use. Similarly the utilization of waste biomass for hydrogen generation will go a long way in providing long term solutions to energy crisis and resource management (**see earlier Section**).

9.V- 2. Minerals and Metals

Another such resource is mineral and metal deposits. Metals are used in all walks of human life. Several of them are also essential nutrients contributing towards growth, metabolism, differentiation and reproduction of all life forms. Microorganisms have also been involved in formation and deposition of the minerals in the earth's crust since geologically ancient times. Thus, most metals are found locked in the form of ores and are extracted from high grade ores by appropriate pyrometallurgical processes. However, low grade ores and the left-over processed ores (ore tailings) are potential sources of metals on which the application of conventional extraction techniques would be highly uneconomical. These and other metal bearing industrial effluents and wastes are major sources of metal pollution. Metals at high concentration are so toxic that it can lead to large scale environmental degradation. Two types of microbial processes have been envisaged in dealing with the metal-bearing wastes. These are metal leaching and metal recovery.

Many bacteria have helped the mining operations as they could leach the metals from insoluble ores. The earliest record of application of bioleaching dates back to 1000 B.C. The ancient romans, the Welsch in 16th century, and Spanish in 18th century certainly used the microbial leaching for recovering metals. Biological leaching is under successful operation in many countries to recover metals like copper and uranium, although cobalt, nickel, zinc, lead, silver, and gold can also be obtained. According to an estimate made by Gorham International Inc., the annual production of the world wide microbial metal industry stands at ~90 billion $ by the year 2000.

The bacterium responsible for leaching of metal sulphide ores is the acidophilic *Thiobacillus ferrooxidans*. Other species, such as *T. thiooxidans, and T. acidophilus* have often been used together with *T. ferrooxidans* in industrial leaching processes. Non-acidophilic *T. novelus* and facultative organotroph *T. cuprinus,* on the other hand, may provide necessary attributes for genetic improvement of Thiobacilli. Some thermophilic archaea, such as *Leptospirillum ferrooxidans, Sulfolobus* and *Acidianus brierley* as well as some sulphate-reducing bacteria have shown great potentials in bioleaching. Scarce genetic information on these bacteria has led to a fresh search for more amenable strains and species. *Alcaligenes, Pseudomonas, Bacillus, Salmonella, Staphylococcus, Sphaerotilus* and *Leptothrix* have been identified as promising strains.

The ability of the microorganisms to grow and survive under metal-polluted conditions is attributed to certain molecular mechanisms. Of the several mechanisms, some bestow upon the microorganism the capability to remove, immobilize or detoxify heavy metals and radionuclides from the natural environment. Todate, the most successful biotechnological processes are based on biosorption, bioprecipitation, and bioaccumulation, but others such as binding by specific macromolecules may have future potential **(see also Chapter 8)**

The mechanism of removal of heavy metals and other toxicants may range from purely physico-chemical interactions to those which are metabolism-dependent. Microbial cell surfaces provide a number of anionic sites which can be engaged in passive adsorption of metal cations. Many a times this binding may facilitate subsequent binding of more metals to effect metal nucleation. Formation of paramanganese nodules in deep oceans has involved this microbial characteristic. Bioaccumulation is carried out by certain metal binding proteins or metallothionein-like proteins. Bioprecipitation is brought about by release of certain counter ions like sulphides, phosphates, carbonates, and hydroxides, and some organic acids like citric and oxalic acid. Removal of metals and radionuclides from industrial effluents, waste waters, and low grade sources can serve the dual purpose of environmental restoration and recovery of precious important metals. The former is also referred to as bioremediation that will be dealt with separately.

9.V- 3. Recovery of Oil

The highly industrialized society of today largely depends upon oil as the major energy source. These sources are fast dwindling because of over-exploitation. Besides looking for alternative sources, the process manipulation for the enhanced recovery of oil is the need of the day. Large amounts of carbonates and pyrites harden the shales making the recovery of oils difficult. Removal of these minerals will increase the porosity and thus the ease with which the oil can flow out. Bacterium like *Thiobacillus* species can grow on the sulphur and iron present in the pyrite thus providing the desired effect.

Other biological and chemical means can also be employed for tertiary recovery of petroleum. This involves the use of solvents, surfactants, and other polymers which can dislodge oil from geological formations. Xanthan gums produced by *Xanthomonas campestris* are promising compounds for such applications. These polymers with high viscosity and flow characteristics can be combined with water flooding operations. Thus, when water is pumped into oil reservoirs, xanthan gums can pass through the small pores so that oil can be pushed out from the rock deposits.

9.VI- Bacteria and Environment

We have already discussed the role of bacteria along with other microorganisms in maintaining the health of the environment in **Chapter 8** and to some extent in the earlier **Sections** of this **Chapter**. Here we shall deal with two aspects bioremediation and synthesis of environment-friendly compounds. Environmental problems are linked to resource misuse or overuse. We humans are unique as we demand too much, too fast and produce waste at each step. In nature, there is no such thing as waste. There is a regular flow of molecules from rocks and soils to plants and animals to air and water and back again, largely through the efforts of the microbes. Unseen and unsung these original recyclers have been ignored until recently.

9.VI- 1. Bioremediation

One of the biggest banes of extensive industrialization is the generation of a class of toxicants known as pollutants. These are the compounds which are either released in such high concentrations that they can not be managed by natural cycle of biodegradation or are non-biodegradable such as plastics, glass and metals. Microbial degradation of wastes and its role in pollution abatement has been amply recognized.

Microbes deal with poisonous chemicals with as much ease as we and other animals process the food. There is practically nothing that cannot be viewed as food by one or the other microbe. There are microbes that can feed on toxic chemicals such as methylene chloride, detergents, creosote, pentachlorophenol, sulphur, PCBs, toluene etc. Although the capability of each strain may be limited, together (consortia) they can lead to complete detoxification.

Microorganisms can be employed to remove pollutants from the environment by a process known as bioremediation. Bioremediation is based either on hastening the natural process of biodegradation or some type of bioaccumulation and biotransformation. The remarkable metabolic capability of diverse microorganisms can be encashed to biodegrade toxic pesticide or petroleum hydrocarbons into carbon dioxide and water or even into useful intermediates or transform obnoxious inorganic substances such as metals into less toxic form. This area, therefore, has lot of scope for genetic engineering to create efficient strains both for waste treatment and pollution abatement.

Toxic petroleum hydrocarbons can easily enter the environment through storage tank leakages and major oil spills. They contain benzene, toluene, and xylenes (BTX), all of which are highly hazardous to living organisms. Indigenous microflora can degrade many of these hydrocarbons but too slowly to overcome the degradative effect on the ecosystem. Still fresh in our memory are major oil spills such as the one from Exon that occurred in Alaska in 1987 and at the coast of Kuwait during the war with Iraq, all of which have caused major havoc to the marine flora and fauna. A genetically engineered pseudomonad, the first organism to be patented, can degrade petroleum hydrocarbon but the world-wide stringent environment regulation has not allowed the wide-spread use of engineered microorganisms. Currently, either the mixed cultures or indigenous microorganism are being employed for oil spill remediation. In order to stimulate the growth of indigenous hydrocarbon-degrading microbes, nitrogen and phosphorus is supplied in the form of fertilizer formulations often in an oleophilic (oil-loving) form. The requirement for molecular oxygen by most of these microorganisms is fulfilled by tilling, forced aeration, and addition of peroxides in soil and water.

Another important class of pollutants comprises chlorinated compounds such as polychlorinated biphenyls (PCBs) and trichlorethylene (TCE) and a number of pesticides. They are common pollutants of soil and water and are highly toxic as they can be accumulated or biomagnified posing a serious threat not only to the environment but also to human health. PCBs occur as mixtures of different isomers carrying varying numbers of chlorine. Those with higher numbers of chlorine substituents (5 or more) can be anaerobically dehalogenated. The resulting PCB with fewer chlorides (3 or less) are aerobically degraded.

TCE, on the other hand, can be degraded by bacteria using methane monooxygenase or toluene monooxygenase. The product TCE epoxide is further metabolized into various compounds such as formic acid and glyoxylic acid, and to hydrochloric acid, water and carbon dioxide. In some remediation strategies, methane and toluene are supplied along with phosphate to stimulate the growth of bacteria.

In **Chapter 8,** we have already discussed the breakdown of certain chlorinated pesticides. One practical difficulty encountered in employing microorganisms to carry out such jobs is of tracking their function outdoors, that is, in nature. One method of monitoring, widely tested, involves linking the genes that empower bacteria to degrade contaminants with genes for bioluminescence. For example, naphthalene degradation has been coupled with *lux* genes. Alternatively, coupling of oxidative reactions such as of CCl_4, toluene, and xylene with a dye will permit the change of colour that can be monitored.

9.VI- 2. Biodegradable Polymers

From the above discussion, it should become clear that bioremediation will demand an evergoing microbial function. Moreover, as more and more man-made chemicals become available their biodegradation will require appropriate metabolic activities of a microorganism. And many of them may turn out to be completely recalcitrant.

In recent times, a new concept of "green products" has emerged. These products being biodegradable will not cause any harm to the environment. One important product on this list is plastic. Plastic is of multiple uses in the modern society and this synthetic polymer is nonbiodegradable. Bacterium *Alcaligenes eutrophus* produces poly β-hydroxybutyrate or PHB which has properties similar to plastics and is completely biodegradable. This gene from *A. eutrophus* has been introduced on a multicopy plasmid and transferred to *E. coli*. The recombinant *E. coli* contains 90% of its dry weight in the form of PHB granules. In order to facilitate the release of the polymer, phage T7 lysozyme gene was also introduced in this *E. coli* strain. Treatment with EDTA allows the lysozyme to lyse the cells releasing the polymer. A Japanese strain of *A. eutrophus* produces a copolyester of polyalkanoates consisting of varying levels of 3-hydroxybutyrate and 4-hydroxybutyrate with properties ranging from hard plastic to elastic rubber.

Similarly, xanthan gums with a wide variety of applications in industry are being produced from bacterial source, *Xanthomonas campestris*. It is used as stabilizer, thickner, gelling or suspending agent in various processed foods and fermented products. Its other uses as a suspending agent include paints, printing, dyeing, glazing, and insecticide spray, and as a gelling agent in explosives, detergents and deodorants. We have already discussed the application of xanthan gum in oil drilling. Other polysaccharides of industrial use derived from bacteria are dextran, alginate, curdlan and gellan etc. Of these dextran has been used as stabilizer in some food products and extender of blood plasma. Dextran derivative under the name of Sephadex is employed as a molecular sieve for separation of macromolecules and molecular weight determination.

9.VI- 3. Biopesticides

Manufacture of pesticides is an important industry but the deleterious effects of these pesticides have led to the identification of natural pesticides obtained from biological sources. *Bacillus thuringiensis* and *B. sphaericus* make specific proteins with potent insecticidal activities. These toxins are also called as Bt toxins and are stored in the spores (Fig. 9.11) in the form of protoxin. The protoxin is converted into a full toxin by a protease in the insect midgut destroying the gut epithelial cells. The bacterium thrives then on dead insects. The bacterial spores are commercially produced to be used as bioinsecticide. These are either sprayed on plants to kill the agricultural pests or added to water to kill mosquitoes. Different strains of *B. thuringiensis* produce toxins with different specificities. Many types of toxic crystals are known. Bipyramidal crystal coded by *cryI* genes are toxic to Lepidoptera (butterflies & moths) and cuboidal crystals coded by *cryII* works against Lepidoptera and Diptera (flies, mosquitoes).The *cryIII* product is Coleoptera specific and *cry IV* crystal protein is effective against Diptera. Different variants of each of these four major classes of insecticidal proteins have also been described. They are non-toxic to humans and other animals and are readily biodegradable. The *Bt* genes (*cry* for crystal protein) have now been engineered in the plants directly as described later.

9.VII- Bacteria and Agriculture

Soil bacteria have played a pivotal role in agriculture. This role has been multifarious leading to the carving of a branch known as Agricultural Microbiology. Once again it should be emphasized that through biogeochemical cycle they replenish the soil with all important nutrients which are finally utilized by the plants. In this connection, the importance of biological nitrogen fixation has been realized for a long time since this process is instrumental in providing fixed forms of nitrogen in the soil essential for plants. Other essential macro- and micronutrients also become available through microbial activities. Many other microbial functions may also provide other benefits like providing tolerance to pH, salinity, and other environmental stresses, as well as protection against pathogens. They are collectively known as plant growth promoting rhizobacteria (PGPR).

9.VII- 1. Plant Biotechnology

The basic information available on plant-microbe interactions led to the studies related to plant genetic manipulations for transferring useful traits to plants. Although one of the basic goals of plant biotechnology in terms of transferring nitrogen fixation (*nif*) genes to plants directly has remained unrealized, the benefits have been accrued in terms of making efficient associations. For example, bioinoculants consisting of correct associations between a legume and a *Rhizobium* species, or biofertilizers in terms of nitrogen-fixing bacteria like *Azospirillum* and cyanobacteria, and phosphate solubilizing bacteria may improve crop productivity significantly. One percent improvement in the efficiency of nitrogen use could mean a substantial saving on the amount spent on fertilizers. Bacterization of certain seeds with such beneficial microorganisms along with some biocontrol microbes can improve the soil fertility and crop productivity. A mention here about the growth promoting characteristics provided by some mycorrhizal fungi will not be out of place.

With the advent of genetic engineering techniques, plant genetic engineering for some useful traits became a field of extensive research. The objectives of these studies were to provide plants with useful traits such as inherent tolerance to various abiotic and biotic stresses that may affect the plant productivity. The source for most of these genes were attributed to bacteria because their metabolic versatility and genetic basis were better known.

Plant biotechnology and its role in agriculture got the most important boost when a natural and novel DNA transfer mechanism between bacteria and plants was discovered. Two bacterial species, *Agrobacterium tumefaciens* and *A. rhizogenes* causing crown gall tumors and hairy root disease on some solanaceous plants contain plasmids Ti, and Ri, respectively. Of the two, Ti plasmid is better known and serves as a vehicle for introducing genes in plants (Fig. 9.12). A segment of Ti, the T-DNA, is transferred and inserted into plant chromosomes, that is flanked by two 25 bp sequence referred to as left and right borders. This natural transfer system is unique as it has crossed the barrier between prokaryotes (bacteria) and eukaryotes (plants).

Agrobacterial infections initiate at wound sites on the host. Infact, the phenolic compounds released by the wounded plant cells act as signal for *vir* (virulence) genes induction. Formation of crown gall due to integration of T-DNA results from expression of genes for the plant growth hormones, auxin and cytokinin, located on T-DNA (called *onc* genes). This causes rapid proliferation of plant cells giving rise to tumorous growth. T-DNA also carries genes that code for synthesis of small compounds called opines (octopine, nopaline). Genes coding for enzymes that degrade a particular opine are located outside the T-region. The bacterium uses the opines as carbon, nitrogen, and energy source, and for stimulating conjugation between bacteria. Thus, Ti plasmid carries genes which express only in bacteria and those which express only in plants. The *vir* (virulence) genes (~24 of them), carried by Ti plasmid are analogous to *tra* genes of transmissible plasmids as they are responsible for transfer and infection. The transfer involves a two-component signal system in which *virA* (histidine kinase) acts as sensor and *virG* as response regulator. Activation of VirG initiates transcription of *vir* genes, of which *virD* gene products (VirD1, and VirD2) recognize the border sequence and produce a single-stranded (SS) endonucleolytic nicking of T-DNA. This is followed by unwinding and injection of single strand (T strand) of the DNA to plant cell, which is assisted by *virE* product VirE2. VirE2 is an inducible SS nucleic acid binding protein which coats the T strand all through the transfer process. In fact VirD2 protein binds to 5'-P end of T-DNA and leads it into the plant cell. The T-strand along with VirD2 and VirE2 is termed the T-complex. The transport and injection is also facilitated by product of *virB* gene consisting of at least eleven ORFs. The product has features of membrane associated export apparatus. Once into the plant cell, single stranded T-DNA is converted into double strand and gets integrated into chromosome perhaps through an illegitimate recombination step.

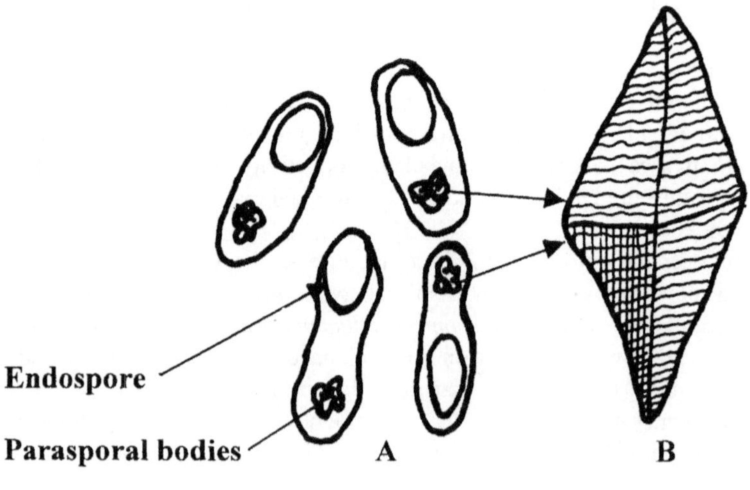

Figure 9.11. Bacillus thuringiensis. A. Cells showing the formation of endospores and parasporal bodies, B. A purified Btδ-endotoxin crystal with a strong insecticidal activity.

The insertion of T-DNA is facilitated by two repeat sequences flanking T-DNA. The right border is not only the end which enters first but is also the absolute requirement for insertion. Another 24 bp region known as transfer enhancer promotes the T-DNA transfer. Any sequence placed between these two border sequences can be transferred to a plant cell by Ti delivery system.

The availability of T-DNA transfer system has stimulated the research to generate transgenic plants. These are the plants containing and expressing new genetic information of foreign origin. Towards this objective, Ti delivery system has been suitably modified to accommodate foreign DNA. These consist : (i) disarming the vector by eliminating all the tumor inducing genes, (ii) including a suitable selection marker, such as antibiotic or herbicide resistance to select the transgenic plant cells, (iii) having a multiple cloning site at which foreign gene can be introduced, and (iv) having specific promoters for efficient expression of foreign genes.

A foreign DNA can be directly cloned in the T-DNA or a binary vector system can be used in which one plasmid carry the *vir* genes and the other has the desired gene, or a cointegrate plasmid can be created by combining the two (Fig. 9.13). With this kind of vector, literally any gene of interest can be transferred and integrated into a plant creating a transgenic plant.

Although Ti transfer system has worked very well with several plants but limitation still persists in universal application of this technique. Direct DNA transfer techniques, therefore, have been explored that consists of:

1) Transformation of plant protoplast by foreign DNA employing polyethylene glycol
2) Use of electric current or electroporation
3) Direct bombardment with DNA-coated microprojectiles

9.VII- 1.1. Transgenic Plants

In several cases, transgenic plants have been generated carrying useful traits. Some examples are cited here.

9.VII- 1.1.1. Herbicide Resistance. Genes for herbicide glyphosate tolerance have been transferred to some crop plants using Ti-transfer system. Such tolerant plants are not affected by the application of herbicide for killing the weeds. In fact, glyphosate resistant soybean, cotton, corn cultivars and others have been developed by expressing glyphosate tolerant 5-enolpyruvylshikimate-5-phosphate (EPSP) synthase of *Agrobacterium* species and glyphosate oxidizing enzyme of *Achromobacter* in the same plant. Similarly, a synthetic bacterial gene coding for acetolactate synthase protein confers resistance to sulfonylurea, triazolopyrimidine, sulphonamide and imidazolinone herbicides in many plants, when transferred to them.

9.VII- 1.1.2. Resistance to Pathogens. Many plants produce phytoalexins that protect them against various fungal infections. These genes have been cloned into important crop plants to overcome the losses due to disease. Some other genes coding for enzymes like chitinase, glucanase, peroxidase or pathogen related proteins that confer pathogen resistance, have been cloned. For example, the chitinase gene from *Serratia marcescens* has been introduced into tobacco, potato, lettuce, and sugar beet. As the chitinase can break down the chitin in the cell wall of fungi, these engineered plants are resistant to

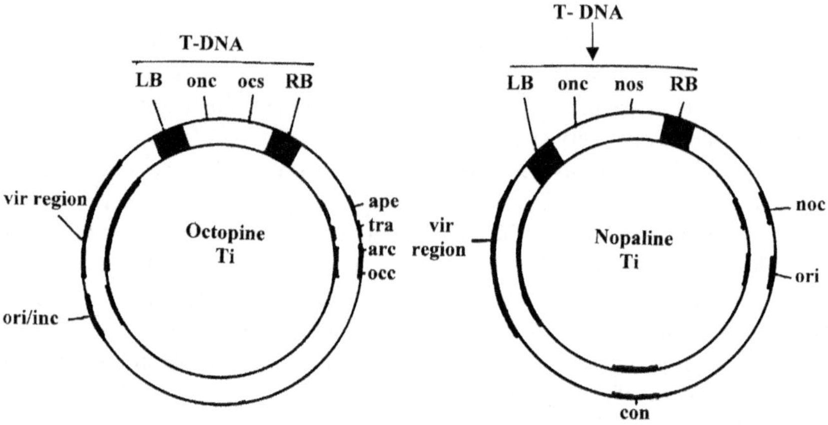

Figure 9.12. Two types of Ti plasmid of wild type Agrobacterium tumefaciens. T-DNA segment of Ti plasmid transferred to plant cells; LB & RB= left and right border flanking sequences, respectively; onc= gene for oncogenesis that leads to crown gall formation; ocs= octopine synthetase gene; nos= nopaline synthetase gene; arc, occ, noc= genes coding for catabolism of arginine, octopine, and nopaline, respectively; ori/inc= origin of replication/ incompatibility; vir region= region carrying virulence genes; tra= plasmid transfer; con= plasmid conjugation; ape= phage exclusion gene.

soil-borne phytopathogens *Rhizoctonia solani,* powdery mildews, and *Botrytis*. In some cases, antisense mRNA gene has been introduced which checks the expression of pathogenicity related genes. Transgenic plants resistant to many viral, bacterial, and fungal diseases are available now.

9.VII- 1.1.3. Resistance to Insects. As described earlier, *Bacillus thuringiensis* produces a crystalline toxin during sporulation. This is a δ-endotoxin which is highly toxic to several insects. Insect-resistant tobacco, cotton, tomato and other plants have been generated by transferring the *Bt* genes through Ti plasmid. Different strains of *B. thuringiensis* produce toxins with different spectra of activities and creation of a toxin of a broad range spectrum is also being carried out to get the desired results. Some of these genes have been introduced into rhizospheric bacteria like other strains of *B. thuringiensis* and fluorescent pseudomonads. Such recombinant strains are grown, killed, and applied to crops. The killed cells encapsulating the deadly protein kill the feeding insects, thereby protecting the crops. Successful examples are available from cabbage, broccoli, lettuce, and other vegetable crops, as well as many others.

Pht protein from *Photorhabdus luminescens* is another strong candidate as bioinsecticide.

9.VII- 1.1.4. Frost Resistance. Frost damage to plants is a major cause of crop loss in many regions of the world. Formation of ice crystal is often initiated by bacteria that grow on the plant surfaces, as they provide a nucleation site. Bacteria that can form ice crystals belong to *Xanthomonas campestris, Erwinia herbicola, Pseudomonas fluorescens* and *P. syringae*, often referred to as Ice^{+}. These bacteria produce a surface protein that facilitates ice crystallization at temperatures of -5° to 0°C, causing excessive damage to the plant cells, and facilitating the growth of bacteria on plant organic matter. From Ice^+ bacteria, Ice^- variant has been isolated which allow the plants to face temperatures several degree lower than that with Ice^+ strains. Spraying of plants with Ice^- strains provides them frost resistance as the Ice^+ bacteria are outcompeted in colonization. Genetically engineered strains of *P. syringae* are able to protect strawberries from frost damage.

The gene coding for an anti-freeze protein similar to that of winter flounder fish has been synthesized and expressed in vegetable plants and yeast. Such engineered plants withstand freeze-thaw cycles better and without the loss of flavour or texture. This anti-freeze protein can also be added to ice creams and ice milk to prevent ice cystal formation.

9.VII- 1.1.5. Gene Conferring Tolerance to Environmental Stresses. Deleterious environmental conditions can seriously hamper plant growth and productivity. Since microorganisms can withstand harsh environmental conditions they may provide rich source for such genes that could be transferred and expressed in a desired crop. The citrate synthase gene of *Pseudomonas aeruginosa* has been introduced into tobacco and papaya. Such transgenic plants secrete citrate from their roots which chelates aluminium. These plants thus could withstand high levels of aluminium without affecting the yield. Many other traits of agronomic importance are also being studied.

9.VII- 1.1.6. Biodegradable Polymers. Plant based commodities are no longer looked upon as merely foodstuffs. Great potential exists to convert this renewable raw material into industrial products, such as chemicals and polymeric substances. A boom can be created in the agricultural economy by engineering the plants to produce high-value, environmentally safe industrial products. This will enhance the profitability and sustainability of agriculture.

Many peptides and proteins of pharamaceutical value, like interleukin, enkephalins, human serum albumin, antibodies, and vaccines have been produced in transgenic plants. Similarly, the genes for biosynthesis of a biodegradable polyester PHB have been successfully engineered into *Arabidopsis*. Yet another example is of protein based polymers (PBPs) which have remarkable elastic properties. They offer a whole range of materials similar to that of oil based polymers such as hydrogels, elastomers, and plastics. PBPs have a range of medical appplications including the post-surgical adhesions, tissue reconstructions and programmed drug delivery, as well as non-medical applications such as biodegradable plastics, transducers, molecular machines, super-absorbent agents and controlled release of agrochemicals like pesticides, growth factors and fertilizers. Their ability to break down in saline conditions provide an easy disposal.

Transgenic plant researches have in recent times thrown up some major concerns especially those related to escape of foreign genes through pollen dispersal from transgenic crop plants to their wild relatives. This may lead to the creation of super weeds, or causing gene pollution among other crops, or toxicity of transgenic pollen to non-target and sometimes beneficial insect species. Maternal inheritance of chloroplast genes has offered chloroplast genetic engineering instead of nuclear ones, as the viable alternative.

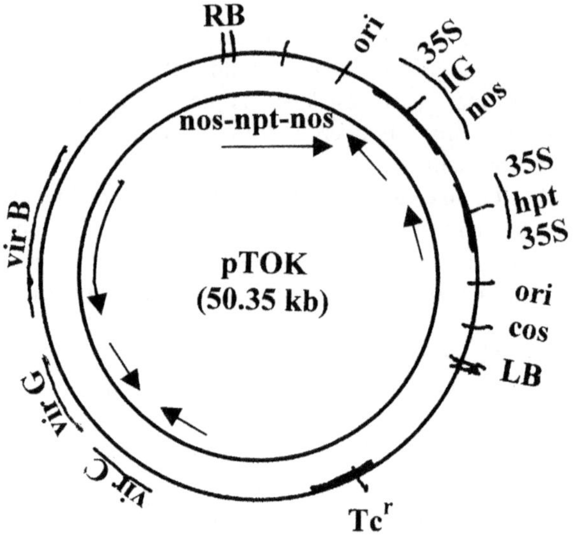

Figure 9.13. A Ti- plasmid-based vector construct used for plant genetic engineering experiments.

9.VIII- Microbial Biodeterioration and Spoilage

In the earlier sections, the metabolic versatility of microorganisms has been documented in detail. The voluptuous range of microbial metabolic degradation is also instrumental in bestowing upon them the capability of biodeterioration. Biodeterioration is a process that leads to spoilage of a large number of materials and can be defined as biodegradation in the wrong place. The spoilage leads to the changes in the nature of a particular material affecting its usage. These may include food, paper, textile, wood, paint, metals, fuels and others. Both fungi and bacteria are highly efficient deteriogens, i.e., they are agents that can cause large scale deterioration of a large number of materials.

9.VIII- 1. Food

The microbial spoilage of food renders the same undesirable and unsafe for human consumption. Based on their susceptibility to microbial attack, food can be classified as perishable (most fruits and vegetables, meat, poultry, fish, and milk), semiperishable (some fruits and vegetables) and non-perishable (dry products like flour, sugar, spices). However, non-perishable foods can suffer spoilage if not stored under proper conditions such as under high humidity. The composition of the food and some environmental conditions may contribute towards spoilage. For example, the carbohydrates, proteins, and lipids present in food items are the compounds that can be easily metabolized by several bacteria and fungi. The environmental factors which are important for microbial growth, such as relative humidity, temperature, pH, and oxygen concentration also influence spoilage. All these may also be provided along with biochemical composition as inherent properties of the food suitable for microbial growth and spoilage. Spoilage of food not only leads to the lowering of nutritional status but may also cause undesirable taste, odour, colour and texture. Production of toxins or growth of pathogenic microorganisms make the food unsafe.

Cereal grains, fruits and vegetables, and products derived there from are rich sources of carbohydrates and support the growth of large number of fungi and bacteria. Some microbes produce pectin esterases and polygalacturonases that can hydrolyze pectins and result in soft spots in fruits and vegetables. Many microorganisms produce low molecular weight acids and alcohols, which leads to souring. Souring of milk and milk products which is a composite food consisting of carbohydrates, proteins and lipids, is a sign of spoilage and occurs due to accumulation of lactic acid. Meat and meat products, fish and other sea foods and poultry and eggs are rich source of protein and lipids and attract a number of bacteria and fungi. The breakdown of proteins in these foods through proteinases by anaerobic bacteria is known as putrefaction. This is characterized by the emission of noxious odour emanated by hydrogen sulphide, ammonia, amines and mercaptans derived from breakdown of amino acids. In canned food, such a spoilage is referred to as sulphide stinkers. Under aerobic condition, the spoilage results from accumulation of surface slime contributed by several bacteria. Microbial degradation of fats and oils produces rancidity, caused by oxidation or hydrolysis of fats. Rancidity of butter is due to hydrolysis of fat that produces free fatty acids and glycerol. Spoilage of fish is also due to a similar process. Some microorganisms may alter the colour of food. Bright red bread, blue milk, black, and white spots on meat, green patches on bread and cured meat are all caused by different deteriogens (Table 9.8).

9.VIII- 2. Food Poisoning

Severe changes in food can be brought about by a number of bacteria that produce toxins. Such foods are unsafe for consumption and many produce a variety of effects. In the most serious types of food poisoning, Clostridia, Staphylococci, and Salmonellas are involved. Strains of *C. botulinum* produces different types of neurotoxins which are responsible for botulism. Contaminated fruits, vegetables and fishes are the main sources. With some types of toxin, that leads to paralysis of respiratory muscles the mortality rate may be very high. In adults, *C. botulinum* is not able to establish itself but in infants sudden infant death syndrome is attributed to this bacterium. A less severe form of food poisoning is caused by *C. perfringens* that occur after ingesting contaminated meat. The effect of toxin appears as abdominal pain and diarrhoea.

Staphylococcal food poisoning is caused by *S. aureus* which may be found in bakery and dairy products, processed meat, salad and canned foods. The enterotoxin leads to nausea, vomitting, abdominal pain and diarrhoea.

Infected poultry and eggs lead to salmonellosis caused by several species of associated *Salmonella*. The bacterium leads to an inflammation of the intestinal lining or gastroenteritis. The symptoms are abdominal pain, fever, and diarrhoea. Similar inflammation and symptoms are also produced by species of *Shigella* (shigellosis),

Table 9.8. Microbial deterioration of food*

S.No.	*Process of spoilage*	*Targeted food*	*Some deteriogens*
1.	Low nutrition content	Cereals and products, fruits & vegetables, milk, meat and poultry etc.	*Aspergillus, Fusarium, Penicillium, Bacillus, Acetobacter, Pseudomonas, Achromobacter, Salmonella*
2.	Souring	Milk and milk products, canned food	*Streptococcus, Lactobacillus, Penicillium*
3.	Putrefaction	Meat and meat products, poultry, fish, eggs	*Micrococcus, Pseudomonas, Cladosporium, Bacillus, Leuconostoc*
4.	Rancidity	Fats and oils	Many bacteria capable of hydrolysing fats
5.	Soft rot	Fresh fruits and vegetables	*Flavobacterium, Lactobacillus, Saccharomyces, Bacillus*
6.	Colouration or spots	Bread, milk, meat	*Serratia, Pseudomonas,* heterolactic fermenters, *Penicillium, Cladosporium*

**Includes both, bacteria and fungi*

Campylobacter fetus, Bacillus cereus, Yersinia enterocolitica, and some invasive strains of *E. coli*. Contaminated food, sea food, milk preparations and water are the major sources of these infections.

Preventing the contamination of food products with pathogenic and spoilage microorganisms is an important aspect of food preservation. We have already discussed various methods of controlling microbial growth in **Chapter 5**. The choice of method will depend upon the nature of the food to be preserved.

9.VIII- 3. Paper

Paper is one product that is consumed in large amounts by the civilized societies. This is also one article which is readily degraded by microorganisms. Microorganisms attack paper at all stages of production: they can degrade wood that is used as raw material or tarnish the quality of the final product, leading to huge losses. The reduction in the quality of pulp wood by microbes especially wood rotting fungi, may lead to poor quality of cellulose fibres, low strength in the paper and several blemishes like spots (discolouration) and imperfection. Many antimicrobial agents like copper sulphate is added to minimize such losses.

High levels of cellulose also allow the growth of microorganisms on paper mill equipment surfaces and solutions used in making paper, contributing to the losses